# Numerical Weather and Climate Prediction

This textbook provides a comprehensive, yet accessible, treatment of weather and climate prediction, for graduate students, researchers, and professionals. It teaches the strengths, weaknesses, and best practices for the use of atmospheric models, and is ideal for the many scientists who use such models across a wide variety of applications. The book describes different numerical methods, data assimilation, ensemble methods, predictability, land-surface modeling, climate modeling and downscaling, computational fluid-dynamics models, experimental designs in model-based research, verification methods, operational prediction, and special applications such as air-quality modeling and flood prediction. The book is based on a course that the author has taught for over 30 years at the Pennsylvania State University and the University of Colorado, Boulder, and also benefits from his wide practical modeling experience at the US National Center for Atmospheric Research.

This volume will satisfy everyone who needs to know about atmospheric modeling for use in research or operations. It is ideal both as a textbook for a course on weather and climate prediction and as a reference text for researchers and professionals from a range of backgrounds: atmospheric science, meteorology, climatology, environmental science, geography, and geophysical fluid mechanics/dynamics.

**Tom Warner** was a Professor in the Department of Meteorology at the Pennsylvania State University before accepting his current joint appointment with the National Center for Atmospheric Research and the University of Colorado at Boulder. His career has involved teaching and research in numerical weather prediction and mesoscale meteorological processes. He has published on these and other subjects in numerous professional journals. His recent research and teaching has focussed on atmospheric processes, operational weather prediction, and arid-land meteorology. He is the author of *Desert Meteorology* (2004), also published by Cambridge University Press.

# Numerical Weather and Climate Prediction

## THOMAS TOMKINS WARNER

**National Center for Atmospheric Research, Boulder, Colorado**

**and**

**University of Colorado, Boulder**

## CAMBRIDGE
### UNIVERSITY PRESS

University Printing House, Cambridge CB2 8BS, United Kingdom

Published in the United States of America by Cambridge University Press, New York

Cambridge University Press is part of the University of Cambridge.

It furthers the University's mission by disseminating knowledge in the pursuit of education, learning and research at the highest international levels of excellence.

www.cambridge.org
Information on this title: www.cambridge.org/9780521513890

First published 2011

*A catalogue record for this publication is available from the British Library*

*Library of Congress Cataloguing in Publication data*
Warner, Thomas T.
Numerical weather and climate prediction / Thomas T. Warner.
p.  cm.
Includes bibliographical references and index.
ISBN 978-0-521-51389-0 (hardback)
1. Weather forecasting – Mathematical models.   2. Climatology – Mathematical models.   I. Title.
QC995.W27 2011
551.63′4 – dc22      2010035492

ISBN  978-0-521-51389-0  Hardback

Lewis Fry Richardson is arguably the father of numerical weather prediction.
In addition to his great interest in methods for modeling the atmosphere,
he was equally passionate about developing mathematical equations
that could predict wars, with the hope that they could thus be avoided.
Let us all, in small or large ways, follow LFR's passions.

With gratitude
to
John Hovermale,
who wanted to write this book

# Contents

# Preface

This textbook provides a general introduction to atmospheric modeling for those using models for either operational forecasting or research. It is motivated by the fact that all those who use such models should be aware of their strengths and limitations. Unlike the many other books that specialize in particular aspects of atmospheric modeling, the aim here is to offer a general treatment of the subject that can be used for self study or in conjunction with a course on the subject. Even though there is considerable space devoted here to numerical methods, this is not intended to be the major focus. As the reader will see, there are many other subjects associated with the modeling process that must be understood well in order for models to be used effectively for research or operations. For those who need more information on particular topics, each chapter includes references to specialized resources. It is assumed that the reader has a Bachelors Degree in atmospheric sciences, with mathematics through differential equations.

Abbreviations or acronyms, as well as symbols, will be defined in the text the first time that they appear, and for future reference they are also defined in the lists that appear before Chapter 1. Even though the student should focus on concepts rather than jargon, a technical vocabulary is still necessary in order to discuss these subjects. Thus, commonly used, important terms will appear in italics the first time, in order to identify them as worth remembering.

There has been no attempt to provide an exhaustive list of references for any particular topic. The reader should refer to the more-recent references, or one of the review papers recommended at the end of the chapters, for a thorough list of historical references. Because World Wide Web addresses tend to change frequently, none are provided here. Instead, the reader should use an available search engine to access current information about model specifications or data sources.

Many colleagues provided tangible and moral support during the production of this book. Cindy Halley-Gotway skillfully and patiently produced the graphic art for the figures and for the cover. Gregory Roux ran model experiments that served as the basis for plots of shallow-fluid-model solutions, and also generated graphical displays of some of the functions in Chapter 3. Many individuals shared their time by engaging in very helpful technical discussions, where special thanks go to George Bryan, Gregory Byrd, Janice Coen, Joshua Hacker, Yubao Liu, Rebecca Morss, Daran Rife, Dorita Rostkier-Edelstein, Robert Sharman, Piotr Smolarkiewicz, Wei Wang, and Andrzej Wyszogrodzki. Those who donated their time and skills by reading and editing chapters include Fei Chen, Luca Della Monache, Joshua Hacker, Andrea Hahmann, Thomas Hopson, Jason Knievel, Yubao Liu, Yuwei Liu, Linlin Pan, Daran Rife, Robert Sharman, David Stensrud, Wei Wang, Jeffrey Weil, and Yongxin Zhang. Christina Brown efficiently managed the process of obtaining copyright permissions, and technical assistance with manuscript preparation was provided by Carol Makowski. Leslie Forehand and Judy Litsey of the library of the National Center for Atmospheric Research assisted with reference material. And, John Cahir offered useful comments on the organization of the chapters, which led to a more logical presentation. Lastly, valuable assistance in many forms was provided by Matt Lloyd, Editor; Laura Clark, Assistant Editor; and Abigail Jones, Production Editor, of Cambridge University Press.

# Acronyms and abbreviations

| | |
|---|---|
| 3DVAR | Three-Dimensional VARiational data assimilation |
| 4DVAR | Four-Dimensional VARiational data assimilation |
| AC | Anomaly Correlation |
| AGCM | Atmospheric General Circulation Model |
| AGL | Above Ground Level |
| ALADIN | European NWP project |
| AOGCM | Atmosphere-Ocean General Circulation Model |
| AR4 | Assessment Report number 4 |
| ARPEGE | Action de Recherche Petite Echelle Grande Echelle (Research Project on Small and Large Scales) |
| ARPS | Advanced Regional Prediction System |
| ARW | Advanced Research WRF model |
| ASL | Above Sea Level |
| BB-LB | Big-Brother–Little-Brother experiment |
| BS | Brier Score |
| BSS | Brier Skill Score |
| CAM | Community Atmospheric Model, of NCAR |
| CAPE | Convective Available Potential Energy |
| CCA | Canonical Correlation Analysis |
| CCM | Community Climate Model, of NCAR |
| CCN | Cloud Condensation Nucleus |
| CCSM | Community Climate System Model |
| CFD | Computational Fluid Dynamics |
| CFL | Courant–Friedrichs–Lewy numerical stability criterion, which requires that $U\Delta t / \Delta x \leq 1$ |
| CFS | Climate Forecast System of the US NCEP |
| CIN | Convective INhibition |
| CMAP | CPC Merged Analysis of Precipitation |
| CMC | Canadian Meteorological Centre |
| CMIP | Climate Model Intercomparison Project |
| COAMPS | Coupled Ocean–Atmosphere Mesoscale Prediction System, of the US Navy |
| COLA | Center for Ocean–Land–Atmosphere studies, USA |

| | |
|---|---|
| CPC | Climate Prediction Center |
| CRMSE | Centered Root-Mean-Square Error |
| CSI | Critical Success Index |
| CSIRO | Commonwealth Scientific and Industrial Research Organisation, Australia |
| DCISL | Departure Cell-Integrated Semi-Lagrangian finite-volume method |
| DEMETER | Development of a European Multimodel Ensemble system for seasonal to inTERannual prediction |
| DMO | Direct Model Output |
| DNS | Direct Numerical Simulation |
| DSS | Decision Support System |
| ECHAM | Global climate model developed by the Max Planck Institute for meteorology |
| ECMWF | European Centre for Medium-range Weather Forecasts |
| ECPC | Experimental Climate Prediction Center, US Scripps Institution of Oceanography |
| EKF | Extended Kalman Filter |
| EL | Equilibrium Level |
| EML | Elevated Mixed Layer |
| EnKF | Ensemble Kalman Filter |
| ENSO | El Niño - Southern Oscillation |
| EOF | Empirical Orthogonal Function |
| ERA | ECMWF global reanalysis |
| EROS | Earth Resources Observing System, of the US Geological Survey |
| ESA | European Space Agency |
| ETKF | Ensemble Transform Kalman Filter |
| ETS | Equitable Threat Score |
| F | False-alarm rate |
| FAR | False-Alarm Ratio |
| FASTEX | Fronts and Atlantic Storm Tracks EXperiment |
| FDDA | Four-Dimensional Data Assimilation |
| FFSL | Flux-Form Semi-Lagrangian finite-volume method |
| FIM | Flow-following finite-volume Icosahedral Model, of the US NOAA |
| GABLS | Global Energy and Water-cycle EXperiment (GEWEX) Atmospheric Boundary-Layer Study |
| GCM | General Circulation Model |
| GEM | Global Environmental Multiscale model of the Meteorological Service of Canada |
| GEOS | Goddard Earth Observing System, of NASA |
| GFS | Global Forecasting System, of the US NCEP |
| GLDAS | Global Land Data Assimilation System, of the US NOAA and NASA |
| GME | Global model of the German Weather Service |

| | |
|---|---|
| GOES | Geostationary Operational Environmental Satellite |
| GPI | GOES Precipitation Index |
| GPS | Global Positioning System |
| GSS | Gilbert Skill Score |
| H | Hit rate |
| HIRLAM | HIgh-Resolution Limited Area Model |
| HRLDAS | High-Resolution Land Data Assimilation System, part of the WRF system |
| HSS | Heidke Skill Score |
| IC | Initial Conditions |
| IN | Ice Nucleus |
| IPCC | Intergovernmental Panel on Climate Change |
| IRI | International Research Institute for Climate and Society |
| KE | Kinetic Energy |
| KF | Kalman Filter |
| LAM | Limited-Area Model |
| LBC | Lateral-Boundary Condition |
| LCL | Lifting Condensation Level |
| LDAS | Land Data-Assimilation System |
| LES | Large-Eddy Simulation |
| LFC | Level of Free Convection |
| LM | Lokal Modell, of the German Weather Service |
| LSM | Land-Surface Model |
| MADS | Model-Assimilated Data Set |
| MAE | Mean Absolute Error |
| ME | Mean Error |
| MERRA | Modern Era Retrospective-analysis for Research and Applications, of NASA |
| MET | Model Evaluation Toolkit |
| MICE | Modeling the Impact of Climate Extremes |
| MM4 | Penn State University–NCAR Mesoscale Model Version 4 |
| MODIS | MODerate-resolution Imaging Spectroradiometer |
| MOS | Model Output Statistics |
| MRF | Medium-Range Forecast model, of the US NWS |
| MSC | Meteorological Service of Canada |
| MSE | Mean-Square Error |
| NAM | North American mesoscale Model, of the US NCEP |
| NAO | North Atlantic Oscillation |
| NARR | North American Regional Reanalysis |
| NASA | National Aeronautics and Space Administration, of the USA |
| NCAR | National Center for Atmospheric Research, of the USA |
| NCDC | National Climatic Data Center, of NOAA |

| | |
|---|---|
| NCEP | National Centers for Environmental Prediction, of NOAA |
| NESDIS | National Environmental Satellite, Data, and Information Service, of NOAA |
| NetCDF | Network Common Data Format |
| NMC | National Meteorological Center, predecessor of NCEP |
| NNMI | Nonlinear Normal-Mode Initialization |
| NNRP | NCEP-NCAR Reanalysis Project |
| NOAA | National Oceanic and Atmospheric Administration, of the USA |
| NOGAPS | Navy Operational Global Atmospheric Prediction System, of the USA |
| NSIP | NASA Seasonal-Interannual Prediction Project |
| NWP | Numerical Weather Prediction |
| NWS | National Weather Service, of the USA |
| OI | Optimal Interpolation |
| OLAM | Ocean–Land–Atmosphere Model |
| OLR | Outgoing Longwave Radiation |
| OMEGA | Operational Multiscale Environment Model with Grid Adaptivity |
| OSE | Observing-System Experiment, Observation Sensitivity Experiment |
| OSSE | Observing-System Simulation Experiment |
| PC | Proportion Correct |
| PCA | Principal Component Analysis |
| PCMDI | Program for Climate Model Diagnosis and Intercomparison |
| PDF | Probability Distribution (or Density) Function |
| PILPS | Project for Intercomparison of Land-surface Parameterization Schemes |
| POD | Probability Of Detection |
| PP | Perfect-Prognosis |
| PRUDENCE | Prediction of Regional scenarios and Uncertainties for Defining EuropeaN Climate change risks and Effects |
| PV | Potential Vorticity |
| QA | Quality Assurance |
| QC | Quality Control |
| QPF | Quantitative Precipitation Forecast |
| RAMS | Regional Atmospheric Modeling System, of Colorado State University |
| RANS | Reynolds-Averaged Navier–Stokes equations |
| RASS | Radio Acoustic Sounding System |
| RCM | Regional Climate Model |
| RFE | Regional Finite Element model, of Canada |
| RH | Relative Humidity |
| RMS | Root-Mean-Square, error or difference |
| RMSE | Root-Mean-Square Error |
| ROC | Relative Operating Characteristic |
| RPS | Rank Probability Score |

| | |
|---|---|
| RPSS | Rank Probability Skill Score |
| RSM | Regional Spectral Model, of NCEP |
| RTG | Real-Time Global analysis, of the Marine Modeling and Analysis Branch of NCEP |
| RUC | Rapid Update Cycle model, of the US NCEP |
| RUC-2 | RUC, version 2 |
| SC | Successive Correction |
| SCIPUFF | Second-order Closure Integrated PUFF model |
| SEVIRI | Spinning Enhanced Visible and InfraRed Imager |
| SFS | SubFilter Scale |
| SGMIP | Stretched-Grid Model Intercomparison Project |
| SL | Starting Level |
| SLP | Sea-Level Pressure |
| SNOTEL | SNOw TELemetry |
| SOM | Self-Organizing Map |
| SREF | Short-Range Ensemble Forecasting |
| SS | Skill Score |
| SSM/I | Special Sensor Microwave Imager |
| SST | Sea-Surface Temperature |
| STARDEX | STAtistical and Regional dynamical Downscaling of EXtremes |
| STATSGO | State Soil Geographic data base |
| SVD | Singular Value Decomposition |
| TKE | Turbulent Kinetic Energy |
| TOMS | Total Ozone Mapping Spectrometer |
| TRMM | Tropical Rainfall Measurement Mission satellite |
| TS | Threat Score |
| UCM | Urban Canopy Model |
| UKMO | United Kingdom Meteorological Office |
| UMOS | Updatable MOS |
| WRF | Weather Research and Forecasting model |
| WSR-88D | Weather Service Radar, 1988, Doppler |

# Principal symbols

*Roman capital letters*

| | |
|---|---|
| $\mathbf{A}$ | covariance matrix of the analysis errors |
| $B$ | Planck's function |
| $\mathbf{B}$ | background covariance matrix |
| $C$ | phase speed |
| | cloud fraction |
| | thermal capacity, or heat capacity |
| | economic cost of protecting against a weather event |
| $C_G$ | group speed |
| $C_P$ | phase speed |
| $C_R$ | real part of a phase speed |
| $D$ | rate of water loss through drainage within the substrate |
| $D_\Theta$ | soil-water diffusivity |
| $E$ | evaporation rate |
| $ET$ | evapotranspiration rate |
| $F$ | all terms on the right side of a prognostic equation |
| | flux |
| $Fr_x$ | frictional acceleration in the $x$ direction |
| $G$ | sensible heat flux between the surface and subsurface |
| $H$ | rate of gain or loss of heat |
| | sensible heat flux between the surface and the atmosphere |
| | mean depth of a fluid |
| | scale height |
| $\mathbf{H}$ | forward operator, observation operator |
| $H_S$ | heat flux within the substrate |
| $I$ | longwave radiation intensity |
| $I{\downarrow}$ | downward-directed longwave radiation intensity |
| $I{\uparrow}$ | upward-directed longwave radiation intensity |
| $J$ | cost function |
| $K$ | highest permitted wavenumber |
| | transfer coefficient |

| | |
|---|---|
| **K** | Kalman gain matrix |
| | Weight matrix of analysis |
| $K_\Theta$ | hydraulic conductivity |
| $K_{Hs}$ | thermal diffusivity of a substrate |
| $L$ | domain length |
| | latent heat of evaporation |
| | horizontal length scale |
| | economic loss from a weather event |
| $L_R$ | length scale of the Rossby radius of deformation |
| **M** | model dynamic operator |
| $P$ | wave period |
| | rate of water input through precipitation |
| **P** | error covariance matrix |
| $Q$ | direct-solar radiation intensity |
| $Q_v$ | rates of gain or loss of water vapor through phase changes |
| **Q** | covariance matrix of the model forecast errors |
| $R$ | rhomboidal truncation |
| | gas constant for air |
| | Rossby radius of deformation |
| | net-radiation intensity |
| | rate of water loss through surface runoff |
| | radius of influence |
| **R** | covariance matrix of the observation errors |
| $RH$ | relative humidity |
| $S$ | source or sink of water substance |
| $T$ | temperature |
| | turbulent, eddy, or Reynolds' stress |
| | triangular truncation |
| $T_a$ | atmospheric temperature a short distance above the surface |
| $T_g$ | temperature of the ground surface |
| $T_s$ | temperature within the substrate |
| $U$ | mean wind speed |
| V | value, economic value |
| $\vec{V}$ | velocity vector |
| $V_T$ | terminal velocity |
| **X** | vector of atmospheric state variables |

*Roman small letters*

| | |
|---|---|
| $a$ | radius of Earth |
| $c$ | specific heat |

| | |
|---|---|
| $c_p$ | specific heat at constant pressure |
| $e$ | Coriolis parameter |
| | base of natural logarithms |
| $f$ | Coriolis parameter |
| | generic dependent variable |
| $g$ | acceleration of gravity |
| $h$ | depth of a fluid |
| $i$ | $\sqrt{-1}$ |
| $k$ | wavenumber |
| | kinetic energy |
| | von Karman constant |
| | weighting coefficient in statistical analysis |
| $k_s$ | soil thermal conductivity |
| $l$ | length scale of energy-containing turbulence |
| $m$ | map-scale factor |
| | integer wavenumber |
| $n$ | integer wavenumber |
| $o$ | observation |
| $p$ | pressure |
| | probability |
| $p_s$ | pressure at the land or water surface |
| $p_t$ | pressure at the top of a model |
| $q$ | specific humidity |
| | diffuse solar radiation |
| $q_s$ | saturation specific humidity |
| $r$ | radius of Earth |
| | radial distance |
| $t$ | time |
| $u$ | east–west component of wind |
| $u_*$ | friction velocity |
| $v$ | north–south component of wind |
| $w$ | vertical component of wind |
| $x$ | east–west space coordinate |
| | general space coordinate |
| $\mathbf{x}$ | state vector |
| $y$ | north–south space coordinate |
| $\mathbf{y}$ | observation vector |
| $z$ | vertical space coordinate – distance above or below surface of substrate |
| $z_o$ | roughness length |

*Greek capital letters*

| | |
|---|---|
| $\Delta$ | change or difference in some quantity, operator |
| | spatial filter length scale |
| $\Delta x$ | grid increment |
| $\Theta$ | volumetric soil-moisture content |
| $\Omega$ | rotational frequency of Earth |

Greek small letters

| | |
|---|---|
| $\alpha$ | albedo |
| | generic dependent variable |
| $\gamma$ | vertical lapse rate of temperature |
| $\gamma_d$ | dry adiabatic lapse rate of temperature |
| $\delta$ | Kronecker delta |
| $\varepsilon$ | alternating unit tensor |
| | emissivity |
| | error |
| $\theta$ | potential temperature |
| $\lambda$ | longitude |
| | amplification factor |
| | wavelength |
| $\mu$ | dynamic viscosity coefficient |
| | thermal admittance |
| $\pi$ | pi |
| $\rho$ | density |
| $\sigma$ | Stefan–Boltzmann constant |
| | terrain-following vertical coordinate |
| | standard deviation |
| $\tau$ | momentum stress, or shearing stress |
| | relaxation coefficient |
| $\phi$ | latitude |
| $\omega$ | frequency of a wave |

*Common subscripts and superscripts*

| | |
|---|---|
| $E$ | applies on Earth's surface |
| $G$ | applies on a grid |
| $I$ | imaginary part of a number |

| | |
|---|---|
| $R$ | real part of a variable |
| $T$ | transpose |
| $a$ | analysis |
| | atmosphere |
| $b$ | background |
| $g$ | ground or substrate surface |
| $i$ | grid-point index in $x$ direction |
| $j$ | grid-point index in $y$ direction |
| $k$ | grid-point index in $z$ direction |
| $m$ | wavenumber |
| $o$ | observation |
| $p$ | wavenumber |
| | applies at constant pressure |
| $s$ | saturation |
| | surface |
| | substrate or soil |
| $\tau$ | point on the discrete time axis |

# 1 Introduction

When Phillip Thompson began to write the first widely read textbook[1] on numerical weather prediction[2] (NWP), the subject was in its infancy, even though an earlier book, *Weather Prediction by Numerical Process* by L. F. Richardson (1922), presaged what was to come later in the century after the advent of electronic computers. The availability of computers increased greatly in the 1960s, and universities began to offer courses in atmospheric modeling, but most modelers had to also be model developers because the untested codes had many errors, the numerical schemes for solving the equations and the physical-process representations were not well tested and understood, lateral-boundary conditions for limited-area models produced noisy solutions, and codes for defining the initial conditions needed to be further developed. These early practitioners learned the basics of atmospheric modeling from each other, through journal articles, in seminars and conferences, and from early courses on the subject. During the last 30 years of the twentieth century, graduate-level courses in atmospheric modeling flourished at many universities. And because computer modeling of the atmosphere was increasingly becoming an important tool in research and operational weather prediction, these courses were typically filled. Nevertheless, atmospheric modeling was still somewhat of a specialty, and models were not very accessible beyond national centers and a few research universities. Smagorinsky (1983), Thompson (1983), Shuman (1989), Persson (2005), Lynch (2007), and Harper (2008) should be consulted for additional history on atmospheric modeling.

In contrast, most of today's modelers are model users only, not developers, and have available, at no cost, well-tested community, global and limited-area models with complete documentation, regular tutorials, and help desks. Some models are being touted as "turn-key" systems that can be run on desk-top computers, and they are accessible to anyone in the meteorological and nonmeteorological communities having little experience in atmospheric modeling and knowledge of the model limitations. There are, of course, still the developers working on the next-generation in modeling capabilities, but they are distinct from the much-more-numerous model users who simply want to employ the model as

---

[1] Thompson (1961)

[2] Historically the expression "numerical weather prediction" has been used to describe all activities involving the numerical simulation of atmospheric processes, whether or not the models were being used for research or operational forecasting. But, some reserve the use of this reference only for model applications to forecasting. In this book we will use the term "numerical weather prediction" to refer to all types of model uses.

a tool to address practical questions related to physical processes, policy, or operational prediction.

The range of time and space scales simulated by contemporary models is great. Regarding time scales, in some cases models are used as the basis of data-assimilation schemes where the objective is to simply define the current state of the atmosphere in a way that is consistent with the data and the model dynamics. Model-based "nowcasts" have time horizons of 1–2 hours. Deterministic predictions of weather (i.e., specific meteorological events) extend to weeks, while interseasonal predictions of weather trends are produced with coupled ocean–atmosphere models. On the longest end of the time spectrum, climate models are integrated for hundreds of years of simulated time. Resolved spatial scales are shrinking as well. Some models that span the globe have sufficient horizontal resolution to simulate mesoscale processes. Other models can simulate winds in urban street canyons and in the wakes of buildings, in some cases quickly enough to be useful for operational applications.

With the growing skill of atmospheric models, and the availability of cheap computing power, a variety of new applications has emerged for specialized and standard versions of the models. When coupled with air-quality models, they are applied to regional airsheds to help government and business develop strategies for managing regional air quality. They are used by governments and private industry for operational prediction of weather to which agriculture is sensitive, for purposes of estimating crop-disease spread, timing planting and harvesting operations, and scheduling irrigation. Militaries employ models for producing specialized forecasts of weather that affects the conduct of their operations on the land and sea, and in the air. Models are used for planning the emergency response to accidental or intentional releases of hazardous chemical, biological, or radiological material into the atmosphere. And they predict quantities such as wind-shear, turbulence, cloud ceiling, visibility, and aircraft icing that affect the safety and efficiency of commercial and private aviation. Atmospheric models are coupled with river-discharge models for prediction of floods. Wind-energy companies use models to "prospect" for the best places to locate farms of wind turbines. Energy companies use atmospheric models to predict cloud cover, temperature, and other quantities that influence the near-future demand for electricity for heating and cooling. And, there are dozens of other sectors of industry and government that have found that model-based weather forecasts improve the profitability and safety of their operations. In general, it has been found that better weather predictions lead to better decisions.

Global atmospheric models have been at the center of the climate-change challenge and controversy for decades, and our increasing confidence in their skill is mirrored in the worldwide call to reduce emissions of carbon dioxide and other greenhouse gases. Even though climate-change processes are of global proportions, there is evidence that the specific manifestations (precipitation and temperature changes) will vary greatly from region to region. Thus, high-resolution regional models are being embedded within the global models in order to provide specific guidance to local decision makers. The models can also be used to better understand and anticipate climate change that is unrelated to greenhouse-gas concentrations. For example, worldwide land-use degradation and modification, such as from deforestation and urbanization, are known to have significant effects on atmospheric processes. Thus, "what if" experiments are performed in which different scenarios are assumed for the landscape change, and the model is run for short or long periods to define the effects of the change on precipitation, for example. The results can be used as motivation for reversing those trends that have negative consequences.

A traditional use of global and regional models has been for basic research on atmospheric processes. Special field programs are very expensive to perform, and they only sample a small area of the atmosphere for a short period of time. Thus, it has been common practice in the research community to augment observations with model simulations. If the model reproduces the atmospheric conditions reasonably well at the observation locations and times, it is assumed that the model is also skillful elsewhere. Thus, the gridded four-dimensional (three space, and time) model data set is used as a surrogate for the real atmosphere, where the advantage, in addition to low cost, is that the availability of data on a regularly spaced grid, at high temporal frequency, makes it much easier to diagnose atmospheric structures and physical processes. However, it will be noted strongly in Chapter 10, about experimental designs in model-based research, that we should first thoroughly analyze all available observational data, and learn everything we can in that process, before running a model. Figure 1.1 emphasizes that observations and theory are as important as models, as research tools that we have at our disposal. And we should avoid the tendency to start running the model before we have learned all that we can from theory and observations. Indeed, it is the author's experience that using the model early in the process only prolongs the amount of time required to complete a research project, or a thesis.

Even though the historical trend has been to use specialized models for different scales and forecast durations, the cost of maintaining multiple modeling systems has

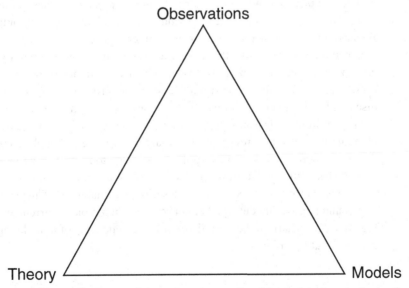

Fig. 1.1 Illustration of the equal importance of observations, theory, and models as tools in atmospheric research.

led to a trend toward a "unified" modeling approach by national meteorological services and other organizations. For example, instead of developing different models for mesoscale and global-scale applications, a single flexible system can be used for both. Similarly, weather-prediction and climate-simulation models used to be distinct, but there are efforts to merge the models used for these two purposes. Lastly, operational models have often not been used by the research community, which has meant that there has not been a straightforward path for operational implementation of improved numerical methods, physical-process parameterizations, initialization schemes, etc. But, there are now a number of examples where operational and research activities use the same models.

This book begins with a review of the governing equations that serve as the basis for atmospheric models (Chapter 2). It is assumed that the reader already has a good understanding of atmospheric dynamics, and the meaning of the various terms in these equations. One goal of the book is to educate the model user about the various components of the modeling process, and how the errors in those components affect the solution. Thus, the well-known sources of error will be described: the numerical approximations in the dynamical core (Chapter 3), the physical-process parameterizations (Chapters 4 and 5), the lateral-boundary conditions (Chapter 3), and the initial conditions (Chapter 6). The discussion of ensemble methods in Chapter 7 responds to the fact that most models, the operational ones at least, use this approach in order to provide valuable information to the model user about uncertainty in the forecast. The inherent predictability of the atmosphere has profound implications regarding the skill that we can expect from models, so this is discussed in Chapter 8. This is followed in Chapter 9 by the related topic of how we can best verify the skill of models. This is important for comparing different models, and for determining whether changes that we make in a single model have a positive or negative effect on the quality, and therefore the utility, of the output. Chapters 10 and 11 summarize common practices in designing research experiments with models, and the techniques for analyzing model output, respectively. Because models used for operational weather prediction often have different requirements and constraints than those used for research, some common differences are discussed in Chapter 12. The post processing of operational-model output to correct for biases and to make the forecast fields easier to interpret and support decision making is discussed in Chapter 13. As noted above, atmospheric models are sometimes coupled with other models that provide information about specialized processes, and these coupled applications are reviewed in Chapter 14. Even though computational fluid-dynamics models are normally applied on scales too small to be called weather, they nevertheless still simulate atmospheric processes, and are becoming more routinely used for a variety of purposes, so they are described in Chapter 15. Chapter 16 discusses how global and regional models are being used for simulation of current and future climates. Figure 1.2 summarizes the overall structural components of a modeling system, and the chapters that describe them.

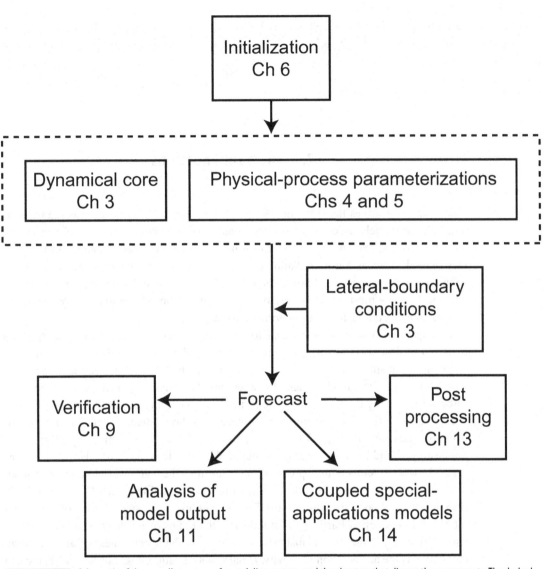

Schematic of the overall structure of a modeling system, and the chapters that discuss the components. The dashed line encloses the two major components of the model code.

# The governing systems of equations

## 2.1 The basic equations

This chapter describes the governing systems of equations that can serve as the basis for atmospheric models used for both operational and research applications. Even though most models employ similar sets of equations, the exact formulation can affect the accuracy of model forecasts and simulations,[1] and can even preclude the existence in the model solution of certain types of atmospheric waves. Because these equations cannot be solved analytically, they must be converted to a form that can be. The numerical methods typically used to accomplish this are described in Chapter 3.

The equations that serve as the basis for most numerical weather and climate prediction models are described in all first-year atmospheric-dynamics courses. The momentum equations for a spherical Earth (Eqs. 2.1–2.3) represent Newton's second law of motion, which states that the rate of change of momentum of a body is proportional to the resultant force acting on the body, and is in the same direction as the force. The thermodynamic energy equation (Eq. 2.4) accounts for various effects, both adiabatic and diabatic, on temperature. The continuity equation for total mass (Eq. 2.5) states that mass is neither gained nor destroyed, and Eq. 2.6 is analogous, but applies only to water vapor. The ideal gas law (Eq. 2.7) relates temperature, pressure, and density. The variables have their standard meteorological meaning. The independent variables $u$, $v$, and $w$ are the Cartesian velocity components, $p$ is pressure, $\rho$ is density, $T$ is temperature, $q_v$ is specific humidity, $\Omega$ is the rotational frequency of Earth, $\phi$ is latitude, $a$ is the radius of Earth, $\gamma$ is the lapse rate of temperature, $\gamma_d$ is the dry adiabatic lapse rate, $c_p$ is the specific heat of air at constant pressure, $g$ is the acceleration of gravity, $H$ represents a gain or loss of heat, $Q_v$ is the gain or loss of water vapor through phase changes, and $Fr$ is a generic friction term in each coordinate direction.

$$\frac{\partial u}{\partial t} = -u\frac{\partial u}{\partial x} - v\frac{\partial u}{\partial y} - w\frac{\partial u}{\partial z} + \frac{uv\tan\phi}{a} - \frac{uw}{a} - \frac{1}{\rho}\frac{\partial p}{\partial x} - 2\Omega(w\cos\phi - v\sin\phi) + Fr_x \quad (2.1)$$

$$\frac{\partial v}{\partial t} = -u\frac{\partial v}{\partial x} - v\frac{\partial v}{\partial y} - w\frac{\partial v}{\partial z} - \frac{u^2\tan\phi}{a} - \frac{uw}{a} - \frac{1}{\rho}\frac{\partial p}{\partial y} - 2\Omega u\sin\phi + Fr_y \quad (2.2)$$

---

[1] In this text, the noun *simulation* refers to a model solution that is obtained for any purpose other than estimating the future state of the atmosphere (for example, for research). An estimate of the future state of the atmosphere is referred to as a *forecast*.

$$\frac{\partial w}{\partial t} = -u\frac{\partial w}{\partial x} - v\frac{\partial w}{\partial y} - w\frac{\partial w}{\partial z} - \frac{u^2 + v^2}{a} - \frac{1}{\rho}\frac{\partial p}{\partial z} + 2\Omega u\cos\phi - g + Fr_z \tag{2.3}$$

$$\frac{\partial T}{\partial t} = -u\frac{\partial T}{\partial x} - v\frac{\partial T}{\partial y} + (\gamma - \gamma_d)w + \frac{1}{c_p}\frac{dH}{dt} \tag{2.4}$$

$$\frac{\partial \rho}{\partial t} = -u\frac{\partial \rho}{\partial x} - v\frac{\partial \rho}{\partial y} - w\frac{\partial \rho}{\partial z} - \rho\left(\frac{\partial u}{\partial x} + \frac{\partial v}{\partial y} + \frac{\partial w}{\partial z}\right) \tag{2.5}$$

$$\frac{\partial q_v}{\partial t} = -u\frac{\partial q_v}{\partial x} - v\frac{\partial q_v}{\partial y} - w\frac{\partial q_v}{\partial z} + Q_v \tag{2.6}$$

$$P = \rho RT \tag{2.7}$$

A complete model will also have continuity equations for cloud water, cloud ice, and the different types of precipitation (see Chapter 4). See Dutton (1976) and Holton (2004) for discussions of this set of prognostic,[2] coupled, nonlinear, nonhomogeneous partial differential equations. The equations are called the *primitive equations*, and models that are based on these equations are called *primitive-equation models*. This terminology is used to distinguish these models from ones that are based on differentiated versions of the equations, such as the vorticity equation. Virtually all contemporary research and operational models are based on some version of these primitive equations. Note that the terms in the equations related to diabatic effects (*H*), friction (*Fr*), and gains and losses of water through phase changes ($Q_v$) must be defined within the model. This particular example of the primitive equations has pressure as the vertical coordinate, but other options will be discussed in the next chapter.

## 2.2  Reynolds' equations: separating unresolved turbulence effects

The above equations apply to all scales of motion, even waves and turbulence that are too small to be represented by models designed for weather processes. Because this turbulence cannot be resolved explicitly in such models, the equations must be revised so that they apply only to larger nonturbulent motions. This can be accomplished by splitting all the dependent variables into mean and turbulent parts, or, analogously, spatially resolved and unresolved components, respectively. The mean is defined as an average over a grid cell, as described by Pielke (2002a). For example:

$$u = \bar{u} + u',$$

$$T = \bar{T} + T', \text{ and}$$

$$p = \bar{p} + p'.$$

---

[2]  The word *prognostic* implies that an equation is predictive, in contrast to a *diagnostic* equation, which has no time derivative and simply relates the state of variables at the same time. For example, the ideal gas law is diagnostic.

These expressions are substituted into Eqs. 2.1–2.7, producing expansions such as the following one for the first term on the right side of Eq. 2.1:

$$u\frac{\partial u}{\partial x} = (\bar{u} + u')\frac{\partial}{\partial x}(\bar{u} + u') = \bar{u}\frac{\partial \bar{u}}{\partial x} + \bar{u}\frac{\partial u'}{\partial x} + u'\frac{\partial \bar{u}}{\partial x} + u'\frac{\partial u'}{\partial x}. \tag{2.8}$$

Because we want the equations to pertain to the mean motion, that is, the nonturbulent weather scales, we apply an averaging operator to all the terms. For the above term, we have

$$\overline{u\frac{\partial u}{\partial x}} = \overline{\bar{u}\frac{\partial \bar{u}}{\partial x}} + \overline{\bar{u}\frac{\partial u'}{\partial x}} + \overline{u'\frac{\partial \bar{u}}{\partial x}} + \overline{u'\frac{\partial u'}{\partial x}}. \tag{2.9}$$

Note that the last term on the right is a *covariance term*. Its value depends on whether the first quantity in the product covaries with the second. For example, if positive values of the first part tend to be paired with negative values of the second, the covariance, and the term, would be negative. If the two parts of the product are not physically correlated, the mean has a value of zero. We then simplify the equations using Reynolds' postulates (Reynolds 1895, Bernstein 1966). For variables $a$ and $b$,

$$\overline{a'} = 0,$$

$$\overline{\bar{a}} = \bar{a} \text{ and } \overline{\bar{a}\bar{b}} = \overline{\bar{a}\bar{b}} = \bar{a}\bar{b}, \text{ and}$$

$$\overline{\bar{a}b'} = \overline{\bar{a}\bar{b}'} = \bar{a}\overline{b'} = 0.$$

Given these postulates, the terms in Eq. 2.9 become

$$\overline{u\frac{\partial u}{\partial x}} = \bar{u}\frac{\partial \bar{u}}{\partial x} + \bar{u}\frac{\overset{0}{\partial u'}}{\partial x} + \overline{u'\frac{\overset{0}{\partial \bar{u}}}{\partial x}} + \overline{u'\frac{\partial u'}{\partial x}} = \bar{u}\frac{\partial \bar{u}}{\partial x} + \overline{u'\frac{\partial u'}{\partial x}}. \tag{2.10}$$

Before we show how to apply these methods to all the terms in Eqs. 2.1–2.7, let us rewrite Eq. 2.1 with a typical representation for the friction terms, $Fr_x$, without the Earth-curvature terms, and with only the dominant Coriolis term. In these equations, which explicitly represent turbulent motion, subgrid friction results only from viscous forces, which are a consequence of molecular motion.

$$\frac{\partial u}{\partial t} = -u\frac{\partial u}{\partial x} - v\frac{\partial u}{\partial y} - w\frac{\partial u}{\partial z} - \frac{1}{\rho}\frac{\partial p}{\partial x} + fv + \frac{1}{\rho}\left(\frac{\partial \tau_{xx}}{\partial x} + \frac{\partial \tau_{yx}}{\partial y} + \frac{\partial \tau_{zx}}{\partial z}\right). \tag{2.11}$$

Here, $\tau_{zx}$ is the force per unit area, or the momentum or shearing stress, exerted in the $x$ direction by the fluid on one side of a constant-$z$ plane with the fluid on the other side of the $z$ plane, and $\tau_{xx}$ and $\tau_{yx}$ are the forces in the $x$ direction across the other two coordinate planes. In hypothetical, inviscid fluids, there would be no "communication" between the flow on either side of a plane. But, in real fluids, the molecular motion, or molecular diffusion, across each of the coordinate surfaces will allow for the exchange of properties. A typical representation for the stress is

$$\tau_{zx} = \mu \frac{\partial u}{\partial z},$$

where $\mu$ is dynamic viscosity coefficient. This is called Newtonian friction, or Newton's law for the stress. Referring to the two (infinitesimally shallow) layers of fluid on either side of the $z$ plane, if there is no shear in the fluid, viscosity produces no stress, or force per unit area, of one layer on the other. Substituting these expressions for the Newtonian friction into the terms for $Fr_x$ in Eq. 2.11, we have

$$\frac{\partial u}{\partial t} \propto \frac{1}{\rho}\left(\mu \frac{\partial^2 u}{\partial x^2} + \mu \frac{\partial^2 u}{\partial y^2} + \mu \frac{\partial^2 u}{\partial z^2}\right) = \frac{\mu}{\rho}\nabla^2 u. \tag{2.12}$$

Now apply the averaging process to all the terms in Eq. 2.11. In particular, we represent each dependent variable by the sum of a resolved mean and an unresolved turbulent component, and then apply the averaging operator. Using Reynolds' postulates, and the assumption that $\rho' \ll \bar{\rho}$, we obtain

$$\frac{\partial \bar{u}}{\partial t} = -\bar{u}\frac{\partial \bar{u}}{\partial x} - \bar{v}\frac{\partial \bar{u}}{\partial y} - \bar{w}\frac{\partial \bar{u}}{\partial z} - \frac{1}{\bar{\rho}}\frac{\partial \bar{p}}{\partial x} + f\bar{v} - \overline{u'\frac{\partial u'}{\partial x}} - \overline{v'\frac{\partial u'}{\partial y}} - \overline{w'\frac{\partial u'}{\partial z}} + \frac{1}{\bar{\rho}}\left(\frac{\partial \bar{\tau}_{xx}}{\partial x} + \frac{\partial \bar{\tau}_{yx}}{\partial y} + \frac{\partial \bar{\tau}_{zx}}{\partial z}\right).$$
$$\tag{2.13}$$

Stull (1988) uses a scale analysis to show that, for turbulence scales of motion, the following continuity equation applies:

$$\frac{\partial u'}{\partial x} + \frac{\partial v'}{\partial y} + \frac{\partial w'}{\partial z} = 0. \tag{2.14}$$

Multiply this by $u'$, average it, and add it to Eq. 2.13 to put the turbulent advection terms into *flux form*:

$$\frac{\partial \bar{u}}{\partial t} = -\bar{u}\frac{\partial \bar{u}}{\partial x} - \bar{v}\frac{\partial \bar{u}}{\partial y} - \bar{w}\frac{\partial \bar{u}}{\partial z} - \frac{1}{\bar{\rho}}\frac{\partial \bar{p}}{\partial x} + f\bar{v} - \frac{\overline{\partial u'u'}}{\partial x} - \frac{\overline{\partial u'v'}}{\partial y} - \frac{\overline{\partial u'w'}}{\partial z} + \frac{1}{\bar{\rho}}\left(\frac{\partial \bar{\tau}_{xx}}{\partial x} + \frac{\partial \bar{\tau}_{yx}}{\partial y} + \frac{\partial \bar{\tau}_{zx}}{\partial z}\right).$$
$$\tag{2.15}$$

By analogy with the molecular viscosity-related stresses, we define turbulent stresses (also, eddy stresses or Reynolds' stresses) as follows:

$$T_{xx} = -\bar{\rho}\overline{u'u'},$$
$$T_{yx} = -\bar{\rho}\overline{u'v'},$$
$$T_{zx} = -\bar{\rho}\,\overline{u'w'}.$$

Substituting these expressions into Eq. 2.15, and assuming that the spatial derivatives of the density are much smaller than those of the covariances, we have

$$\frac{\partial \bar{u}}{\partial t} = -\bar{u}\frac{\partial \bar{u}}{\partial x} - \bar{v}\frac{\partial \bar{u}}{\partial y} - \bar{w}\frac{\partial \bar{u}}{\partial z} - \frac{1}{\bar{\rho}}\frac{\partial \bar{p}}{\partial x} + f\bar{v}$$

$$+ \frac{1}{\bar{\rho}}\left(\frac{\partial}{\partial x}(\tau_{xx} + T_{xx}) + \frac{\partial}{\partial y}(\tau_{yx} + T_{yx}) + \frac{\partial}{\partial z}(\tau_{zx} + T_{zx})\right). \tag{2.16}$$

This equation is the same as Eq. 2.11, except for the turbulent-stress terms and the mean-value symbols. The mean-value symbols are rarely used with the primitive equations, but it is still understood that the dependent variables represent only nonturbulent motions. And, the turbulent stresses are much larger than the viscous stresses, so the latter terms are usually not included. The turbulent-stress terms are sometimes represented symbolically as "$F$", referring to friction. The representation of the turbulent stresses in terms of variables predicted by the model is the subject of turbulence parameterizations for the boundary layer, or for above the boundary layer, described in Chapter 4.

## 2.3  Approximations to the equations

There are a few reasons why we might desire to use approximate sets of equations as the basis for a model.

- Some approximate sets are more efficient to solve numerically than the complete equations. For example, the hydrostatic, *Boussinesq*, and *anelastic approximations* described below do not permit sound waves in the solutions, which, for reasons that will be explained in the next chapter, means that less computing resources are required to produce a simulation or forecast of a given length.
- The complete equations describe a physical system that is so complex that it is challenging to use them in a model for research, to better understand cause and effect relationships in the atmosphere. Thus, sometimes specific terms and equations (and the associated processes) are removed from the set of equations. For example, removing equations for water in all its phases, and the thermodynamic effect of phase changes, allows the study of processes in a simpler setting.
- Very simple forms of the equations are more amenable for pedagogical applications and for initial testing of new numerical algorithms. For example, the shallow-fluid equations, described below, are used as the basis for "toy models" in NWP classes (and in this text). But, they contain enough of the dynamics of the full set of equations that they can be profitably used to test new differencing schemes, which can later be evaluated in complete models.

The approximations described in the following subsections are commonly used in research and operational models.

### 2.3.1 Hydrostatic approximation

The existence of relatively fast-propagating sound waves in a model solution means, as will be explained in the next chapter, that short time steps are required in order for the model's numerical solution to remain stable. The consequence of the short time step is that many more will be required in a model integration of a specific duration, and more computing resources will be required. Because sound waves are generally of no meteorological importance, it is desirable to use a form of the equations that does not admit them. One approach is to employ the hydrostatic approximation, wherein the complete third equation of motion (Eq. 2.3) is replaced by one containing only the gravity and vertical-pressure-gradient terms. That is

$$\frac{\partial p}{\partial z} = -\rho g\,.$$

This implies that the density is tied to the vertical pressure gradient. Because the propagation of sound waves requires that the density adjust to the longitudinal compression and expansion within the waves, sound waves are not possible in a hydrostatic atmosphere. For the hydrostatic assumption to be valid, the sum of all the terms eliminated in the complete equation must be, say, at least an order-of-magnitude smaller than the terms retained. Stated another way

$$\left|\frac{dw}{dt}\right| \ll g\,.$$

A scale analysis of the third equation of motion (e.g., Dutton 1976, Holton 2004) shows that the hydrostatic assumption is valid for synoptic-scale motions, but becomes less so for length scales of less than about 10 km on the mesoscale and convective-scale. Thus, coarser-resolution global models will tend to be based on the hydrostatic equations, while models of mesoscale processes will not. It will be shown in the next chapter that there are other approaches for dealing with the computational effects of fast waves on the model grid.

### 2.3.2 Boussinesq and anelastic approximations

As with the hydrostatic assumption, the Boussinesq and anelastic approximations are part of a family of approximations that directly filter sound waves from the equations by decoupling the pressure and density perturbations. However, their use is not limited to modeling larger horizontal length scales, as is the case with the hydrostatic approximation. Indeed, these approximations are widely used in models of mesoscale or cloud-scale processes. The Boussinesq approximation (Boussinesq 1903) is obtained by substituting the following for Eq. 2.5, the complete continuity equation:

$$\frac{\partial u}{\partial x} + \frac{\partial v}{\partial y} + \frac{\partial w}{\partial z} = 0\,.$$

This amounts to substituting volume conservation for mass conservation. For the anelastic approximation (Ogura and Phillips 1962, Lipps and Hemler 1982),

$$\frac{\partial}{\partial x}\bar{\rho}u + \frac{\partial}{\partial y}\bar{\rho}v + \frac{\partial}{\partial z}\bar{\rho}w = 0$$

is substituted for the complete continuity equation, where $\bar{\rho} = \bar{\rho}(z)$ is a steady reference-state density. In addition, both approximations involve simplifications in the momentum equations (see Durran 1999, pp. 20–26). Another type of approximation in this class is the pseudo-incompressible approximation described by Durran (1989).

### 2.3.3 Shallow-fluid equations

The *shallow-fluid equations*, sometimes called the shallow-water equations, can serve as the basis for a simple model that can be used to illustrate and evaluate the properties of numerical schemes. Inertia–gravity, advective, and Rossby waves can be represented. Not only is such a model useful for gaining experience with numerical methods, the fact that the equations represent much of the horizontal dynamics of full baroclinic models makes it a useful tool for testing numerical methods in a simple framework. For example, Williamson *et al.* (1992) used a shallow-fluid model applied to the sphere to test numerical methods that were proposed for climate modeling.

The name "shallow fluid" refers to the fact that the wavelengths simulated must be long relative to the depth of the fluid. There are various forms of this set of equations (Nadiga *et al.* 1996), but here the fluid is assumed to be autobarotropic (barotropic by definition, not by virtue of the prevailing atmospheric conditions), homogeneous, incompressible, hydrostatic, and inviscid. The homogeneity condition means that the density does not vary in space, and incompressibility means that density does not change in time following a parcel. The equations from which we begin the derivation are

$$\frac{\partial u}{\partial t} + u\frac{\partial u}{\partial x} + v\frac{\partial u}{\partial y} + w\frac{\partial u}{\partial z} - fv + \frac{1}{\rho}\frac{\partial p}{\partial x} = 0, \tag{2.17}$$

$$\frac{\partial v}{\partial t} + u\frac{\partial v}{\partial x} + v\frac{\partial v}{\partial y} + w\frac{\partial v}{\partial z} + fu + \frac{1}{\rho}\frac{\partial p}{\partial y} = 0, \tag{2.18}$$

$$\frac{\partial p}{\partial z} = -\rho g, \text{ and} \tag{2.19}$$

$$\frac{d\rho}{dt} + \rho\left(\frac{\partial u}{\partial x} + \frac{\partial v}{\partial y} + \frac{\partial w}{\partial z}\right) = 0. \tag{2.20}$$

Now, incompressibility and homogeneity imply

$$\frac{d\rho}{dt} = 0, \tag{2.21}$$

$$\rho = \rho_0, \text{ for } \rho_0 \text{ a constant, and} \tag{2.22}$$

$$\frac{\partial u}{\partial x} + \frac{\partial v}{\partial y} + \frac{\partial w}{\partial z} = 0. \tag{2.23}$$

The hydrostatic equation can thus be written

$$\frac{\partial p}{\partial z} = -\rho_0 g. \tag{2.24}$$

Differentiating Eq. 2.24 with respect to $x$, and using the fact that the right side is a constant, yields

$$\frac{\partial}{\partial x}\left(\frac{\partial p}{\partial z}\right) = \frac{\partial}{\partial z}\left(\frac{\partial p}{\partial x}\right) = 0, \tag{2.25}$$

which means that there is no horizontal variation of the vertical pressure gradient or vertical variation of the horizontal pressure gradient (the definition of barotropy). Because the pressure-gradient force generates the wind, and the resulting Coriolis force, all forces are invariant with height. Integrating Eq. 2.24 over the depth of the fluid,

$$\int_{z(P_S)}^{z(P_T)} \frac{\partial p}{\partial z}dz = -\rho_0 g \int_{z(P_S)}^{z(P_T)} dz, \tag{2.26}$$

where $P_T$ and $P_S$ represent the pressure at the top and bottom boundaries of the fluid, respectively, yields

$$P_S - P_T = g\rho_0 h, \tag{2.27}$$

for $h$ equal to the depth of the fluid. If $P_T = 0$, or $P_T \ll P_S$,

$$\frac{P_S}{\rho_0} = gh \text{ and} \tag{2.28}$$

$$\frac{1}{\rho_0}\frac{\partial P_S}{\partial x} = g\frac{\partial h}{\partial x}. \tag{2.29}$$

This statement that the horizontal pressure gradient at the bottom of the fluid is proportional to the gradient in the depth of the fluid provides a new form of the pressure-gradient term in Eqs. 2.17 and 2.18. The incompressible continuity equation (Eq. 2.23) can also be rewritten by integrating it with respect to $z$:

$$\int_0^z \frac{\partial w}{\partial z}dz = w_z - w_s = -\int_0^z \left(\frac{\partial u}{\partial x} + \frac{\partial v}{\partial y}\right)dz. \tag{2.30}$$

If $u$ and $v$ are initially not a function of $z$, they will remain so because the pressure gradient is not a function of $z$. And, because $u$ and $v$ are not a function of $z$, neither are their derivatives, so that

$$w_h - w_S = -\left(\frac{\partial u}{\partial x} + \frac{\partial v}{\partial y}\right)h \qquad (2.31)$$

for $z = h$. For a horizontal lower boundary, the kinematic boundary condition $w_S = 0$ prevails. Recognizing that

$$w_h = \frac{dh}{dt} \qquad (2.32)$$

leads to a new continuity equation. There are now three equations in three variables, $u$, $v$, and $h$.

$$\frac{\partial u}{\partial t} + u\frac{\partial u}{\partial x} + v\frac{\partial u}{\partial y} - fv + g\frac{\partial h}{\partial x} = 0, \qquad (2.33)$$

$$\frac{\partial v}{\partial t} + u\frac{\partial v}{\partial x} + v\frac{\partial v}{\partial y} + fu + g\frac{\partial h}{\partial y} = 0, \qquad (2.34)$$

$$\frac{\partial h}{\partial t} + u\frac{\partial h}{\partial x} + v\frac{\partial h}{\partial y} + h\left(\frac{\partial u}{\partial x} + \frac{\partial v}{\partial y}\right) = 0. \qquad (2.35)$$

For simplicity, a one-dimensional version of this system of equations is frequently used. In order to permit a mean $u$ component on which perturbations occur, a constant pressure gradient of the desired magnitude is specified in the $y$ direction. The one-dimensional equations are

$$\frac{\partial u}{\partial t} + u\frac{\partial u}{\partial x} - fv + g\frac{\partial h}{\partial x} = 0, \qquad (2.36)$$

$$\frac{\partial v}{\partial t} + u\frac{\partial v}{\partial x} + fu + g\frac{\partial H}{\partial y} = 0, \qquad (2.37)$$

$$\frac{\partial h}{\partial t} + u\frac{\partial h}{\partial x} + v\frac{\partial H}{\partial y} + h\frac{\partial u}{\partial x} = 0, \text{ where} \qquad (2.38)$$

$$\frac{\partial H}{\partial y} = -\frac{f}{g}\overline{U} \qquad (2.39)$$

and $\overline{U}$ is the specified, constant mean geostrophic speed on which the $u$ perturbation is superimposed. Obviously there are limitations to the degree to which this system of equations can represent the real atmosphere, but one step toward more realism is to define the fluid depth to be consistent with the layer being represented, such as the boundary layer or the troposphere. The depth of the total atmosphere can be represented by the scale height

$$H = \frac{RT_0}{g},$$ (2.40)

where $T_0$ is the surface temperature and $H$ is about 8 km. If the model atmosphere is to represent the troposphere, it can be assumed that the active fluid layer of depth $h$ is surmounted by an inert layer (Fig. 2.1) that represents the stratosphere. This exerts a buoyant

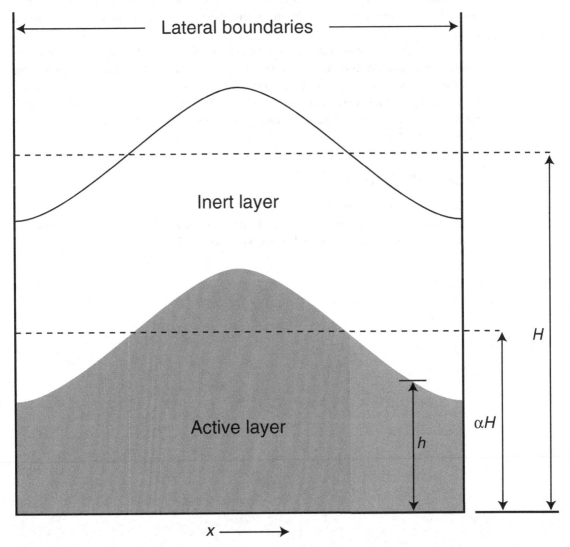

**Fig. 2.1** Schematic showing the vertical structure of a shallow fluid model, for a situation where a wave ridge is centered in the computational domain. The lower shaded layer represents the active fluid for which the depth ($h$) and wind components are simulated. The depth, $H$, is the scale height of the atmosphere, and $\alpha$ is the factor by which the depth is reduced to account for the buoyancy of an inert layer above.

force on the lower layer that can be represented in the model by a reduced gravity. But this would impact the geostrophic relationship, so a better approach is to proportionately reduce the depth of the active layer. This is justified by the fact that, in the linear solution for the phase speed of external gravity waves, the acceleration of gravity is multiplied by the mean depth of the fluid. Application of either method would have the same effect of decreasing the phase speed of external gravity waves to one that is more characteristic of the internal waves at the layer interface. It can be shown that the gravity or layer depth should be reduced by a factor $\alpha = (\theta_T - \theta_B)/\theta_B$, which is based on the mean potential temperatures of the top and bottom layers. For the example where the lower layer represents the troposphere, this ratio is ~0.25 and the layer mean depth would be defined as 2 km.

When the above nonlinear shallow-fluid equations are used as the basis for a model, an explicit numerical diffusion term will need to be added to each equation to suppress the short wavelengths that will grow through the aliasing process, which will be described in the next chapter. Additional information on the shallow-fluid equations, and their numerical solution, may be found in Kinnmark (1985), Pedlosky (1987), Durran (1999), and McWilliams (2006).

## PROBLEMS AND EXERCISES

1. Derive Reynolds' equations for Eqs. 2.2–2.7.
2. Reproduce the development of Reynolds' equations using tensor notation, and note the relative simplicity compared to the process in Section 2.2.

# 3

# Numerical solutions to the equations

The current chapter summarizes various topics related to the numerical solution of the model equations, for resolvable scales of motion. This part of an atmospheric model that treats the resolvable scales is called the *dynamical core*, and is distinct from the representations of the subgrid-scale, parameterized physical processes. An especially important topic is how the numerical approximations that are used to solve the equations can affect the model solution. These nonphysical effects should be thoroughly understood by all model users. Even though basic concepts are described here, and examples provided, this presentation of numerical methods is far from exhaustive. A comprehensive text on this subject, such as Durran (1999), should be consulted if more depth is needed. Step-by-step derivations are frequently left to the reader.

Numerical methods used for solving the equations have naturally evolved over the last few decades, partly because of the results of research and partly because of changes in the available computational resources. Various factors are involved in the decision about the numerical methods to use for a particular modeling application, including computational efficiency (speed), accuracy, memory requirements, and code-structure simplicity. The last factor is especially important if the model is going to be used for research, especially by students. Simple methods that are not typically used in current operational models are sometimes described here for pedagogical purposes.

## 3.1 Overview of basic concepts

The following brief overview of concepts will help the reader to better understand the specialized material in later sections.

### 3.1.1 Grid-point and spectral methods for representing spatial variations of the atmosphere

The model equations are often solved at points defined by a quasi-regular, three-dimensional spatial grid. Section 3.2.1 reviews the different options for the structure of these grids. The term "quasi-regular" is used here as an acknowledgment that the points are typically not exactly equally spaced, when the grid is defined on a map projection where the Earth-distance between grid points varies from place to place. Sometimes the points are very nonuniformly spaced, for example when using latitude–longitude coordinates or with adaptive grids where the resolution is increased in areas of strong gradients.

The time axis is also defined by discrete, evenly spaced points. The time and space deriva-tives in the equations can be approximated using finite-difference methods (Section 3.3), which introduce nonphysical properties to the model solution, and have stability criteria that limit the time step (Section 3.4). As an example of solving an equation on a grid, the first equation of motion (Eq. 2.1) will be represented using a simple three-point centered-difference approximation in time and space, such as

$$\frac{\partial}{\partial y} f(x, y, z, t) = \frac{f_{i,j+1,k}^{\tau} - f_{i,j-1,k}^{\tau}}{y(j+1) - y(j-1)} = \frac{f_{j+1}^{\tau} - f_{j-1}^{\tau}}{2\Delta y},$$

where $f$ is any dependent variable; $\tau$ defines a discrete point on the time axis; $i, j, k$ define coordinates on the $x, y, z$ space axes, respectively; and $\Delta y$ is the distance between two adjacent points on the $y$ axis. The first equation of motion, Eq. 2.16,

$$\frac{\partial u}{\partial t} = -u \frac{\partial u}{\partial x} - v \frac{\partial u}{\partial y} - w \frac{\partial u}{\partial z} - \frac{1}{\rho} \frac{\partial p}{\partial x} + fv + Fr_x,$$

is a nonlinear, nonhomogeneous, partial differential equation that cannot be solved analyt-ically. With the above, three-point, finite-difference approximations, it becomes the fol-lowing solvable arithmetic equation

$$\frac{u_{i,j,k}^{\tau+1} - u_{i,j,k}^{\tau-1}}{2\Delta t} = -u_{i,j,k}^{\tau} \frac{u_{i+1,j,k}^{\tau} - u_{i-1,j,k}^{\tau}}{2\Delta x} - v_{i,j,k}^{\tau} \frac{u_{i,j+1,k}^{\tau} - u_{i,j-1,k}^{\tau}}{2\Delta y}$$

$$-w_{i,j,k}^{\tau} \frac{u_{i,j,k+1}^{\tau} - u_{i,j,k-1}^{\tau}}{2\Delta z} - \frac{1}{\rho_{i,j,k}^{\tau}} \frac{p_{i+1,j,k}^{\tau} - p_{i-1,j,k}^{\tau}}{2\Delta x} + fv_{i,j,k}^{\tau} + Fr_{x(i,j,k)}^{\tau-1}, \quad (3.1)$$

where $\Delta x$ and $\Delta y$ are often assumed to be the same, and $Fr$ is a frictional-dissipation term. This equation is solved for $u_{i,j,k}^{\tau+1}$ on the left side, and, for each grid point, the right side is evaluated based on values of the dependent variables from the two previous time levels $\tau$ and $\tau-1$. The other equations are similarly solved for the $\tau+1$ values of the dependent variable.

The value of $\Delta x$ is chosen so that there is a sufficient number of grid points to ade-quately represent the smallest meteorological feature of interest, for the particular appli-cation of the model. Section 3.4.1 on the concept of *truncation error* quantifies the accuracy associated with representing continuous functions with a finite number of points. A rule of thumb is that 10 grid points are needed to reasonably resolve a wave. So, depending on whether the purpose of the model is to simulate synoptic-scale Rossby waves or mesoscale convective complexes, the grid increment must be chosen accord-ingly. An alternative approach is to represent the spatial variation of the dependent varia-bles using global or local functions, and calculate the derivatives analytically. Such approaches include the spectral and finite-element methods described in Sections 3.2.2 and 3.2.3, respectively.

There are two general types of models in terms of the spatial extent of the computational volume. If the model calculations span the sphere, the model is referred to as a *global model*. If the model applies only to a particular regional subvolume of the atmosphere, it is called a *limited-area model*.

### 3.1.2 The time integration

Section 3.3.1 reviews different methods for integrating the equations in time. The approach shown in Eq. 3.1, for all terms except the friction term, is known as leap-frog, or three-point centered, time differencing because the value of the derivative is calculated at a time ($\tau$, right side of equation) that is centered between the initial ($\tau - 1$) and final ($\tau + 1$) times of the extrapolation. Figure 3.1 illustrates this time-differencing method. Note that a forward time step is required at the beginning of the integration, before the leap-frog process can be used. For the friction term, forward differencing is used, where the derivative is calculated at the point from which the extrapolation originates. This is the only method that is stable. For many differencing schemes, the time step is constrained by a limiting value of the *Courant number*, defined as $U\Delta t/\Delta x$, where $U$ is the horizontal speed of the fastest wave on the grid, and $\Delta x$ was chosen, as described earlier, to allow resolution of the relevant meteorological processes. If the time step is too long, a stability criterion is violated, nonmeteorological features grow exponentially in the solution, and floating-point overflows in the computer will cause the integration to stop. The stability requirement for the advection term represented in an Eulerian framework (the equations are solved at grid points), for the combination of space and time differencing methods used in Eq. 3.1, is called the Courant–Friedrichs–Lewy (CFL) criterion, which requires that $U\Delta t/\Delta x \leq 1$. The concept of numerical stability is further developed in Section 3.4.2.

In order to understand what controls the computational requirements for running a model, assume that a forecast is needed over a certain limited geographic area. Also, for simplicity, assume that the grid points are regularly spaced in the horizontal. The horizontal grid increment is chosen such that the features of meteorological interest are well represented by a sufficient number of points over the length of a wave. The chosen vertical distribution of points will similarly depend on the vertical structures that need to be resolved. Given the types of atmospheric waves admitted by the model equations, it is possible to estimate the fastest wave on the grid. For example, if sound waves and external gravity waves are not part of the model solution, the fastest advective wave (e.g., in a jet

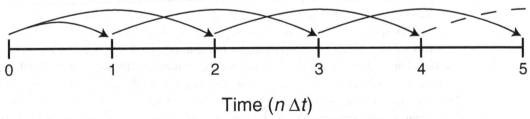

## Time ($n \Delta t$)

**Fig. 3.1** Centered-in-space time-differencing schematic, except for the initial forward-in-time step to extrapolate from $n = 0$ to $n = 1$.

stream) can be estimated based on knowledge of the local atmosphere. Thus, the only remaining parameter in the expression for the Courant number is the time step. This must be chosen such that the numerical solution will remain stable. The cost of running the model on this grid will be based on the number of times the full set of algebraic equations, of which Eq. 3.1 is one, must be solved during the forecast period. This is linearly proportional to the number of calculation points, sometimes called nodes, in the three-dimensional grid, because the equations need to be solved at each point. The cost is also related to the fact that the equations need to be solved for each time step, at each grid point, so the number of time steps in the forecast period will also control the computational requirements. Thus, for a given area, the smaller the grid increment in the vertical and the horizontal, and the smaller the time step, the more computationally demanding the model will be to run. For a given area, a fine-mesh mesoscale model will have many more grid points than will a model designed for synoptic-scale processes, and thus it will be much more costly to run. To reveal the nonlinearity of the dependence of computational cost on resolution, assume that we want to double the horizontal resolution of a grid over a given area. This will require four times as many grid points, and because the stability criterion (e.g., based on the Courant number) needs to be satisfied, the time step will probably need to be halved. Thus, increasing the horizontal resolution by a factor of two will cause an increase in the computational expense by a factor of eight. It is for this reason that NWP research has often focussed on the development of more efficient numerical schemes for solving the equations of motion.

### 3.1.3 Boundary conditions

Solving the model equations represents both a boundary-value (lateral, upper, lower) problem and an initial-value problem. For a global model, there are no lateral boundaries because the computational area is naturally periodic. For limited-area models (i.e., not global), the equations cannot be solved for points on the edge of the grid because there are no points beyond the boundary to use for evaluating the derivative perpendicular to it (see Section 3.5 for a discussion of lateral-boundary conditions). The values of dependent variables at these boundary points need to be externally specified. For operational forecasting with limited-area models, the lateral-boundary values must be defined by interpolation from grid points of a previously run global forecast model. For research applications, archived, gridded regional or global analyses of observations may be used.

In addition to these lateral-boundary conditions, there are also upper and lower boundary values that must be specified with both global and limited-area models. Because the model atmosphere cannot extend to infinity as does the real atmosphere, and because we sometimes want to limit our computations to the troposphere, it is necessary to define an artificial upper-boundary condition for models. Approaches for doing that, which minimize downward reflections, are discussed in Section 3.6. Another major challenge is defining the fluxes of heat, moisture, and momentum at the land and ocean surface. Because the midlatitude planetary circulation and monsoons are driven by gradients in sensible heating at the surface, it should be obvious that models must treat this process reasonably well. In addition, mesoscale boundary-layer circulations result from horizontally differential heating at coastlines and at other landscape boundaries, so small-scale variations in sensible-heat fluxes need to be modeled accurately as well. And, the sensible- and latent-heat fluxes compete for the solar-energy input at the surface, so the latent-heat fluxes can greatly influence boundary-layer winds and thermal properties. The modeling of land-surface processes and surface fluxes is discussed in Chapter 5 on land-surface modeling and in Section 4.4 on boundary-layer parameterizations, respectively.

### 3.1.4 Initial conditions

Because atmospheric modeling is an initial-value problem, the state of the dependent variables at the beginning of the integration of the equations must be specified (the left-most point in Fig. 3.1). This process is called *initializing* the model, and is discussed in Chapter 6. How well this is accomplished has important consequences regarding the accuracy of the forecast. First, except for locally forced processes (e.g., forced by orography, coastlines), it is reasonable to assume that forecast quality can generally be no better than that of the initial conditions. Second, if the mass and momentum fields are far out of balance (e.g., geostrophic) relative to what should prevail for the physical processes as rendered by the model equations, inertia–gravity waves are created by the model fields adjusting after the initialization. These waves, which have no counterpart in the atmosphere, can sometimes have sufficient amplitude to obscure real features in the model solution, at least until they have been damped or have propagated away from the source region. Lastly, if the initial conditions do not contain realistic vertical motions associated with orography or convection, or realistic mesoscale coastal or mountain-valley circulations, the model has to *spin up* these features during the forecast. The above adjustment issues led to the historical situation where forecasters did not use the forecast for the first 12–24 h after initialization.

There are two general types of initializations: *static initializations* and *dynamic initializations*. In the former, observations applicable at the initial time are objectively analyzed to the model grid, perhaps some balance constraint is applied, and the model forecast is begun. These static initializations that do not provide spun-up vertical motions and ageostrophic circulations are referred to as *cold starts*. For the dynamic initialization, the name implies that there is a dynamic constraint, typically from a model, that is aimed at ensuring that the model solution is spun up, or nearly so, at the initial time of the forecast. One approach would be to perform a static initialization, 12–24 h before the desired start time of the forecast, and run the model to allow the solution to spin up during the preforecast period. A variant of this is for the model to assimilate observations during the preforecast period of integration. Or, a common technique is to use an existing spun-up model forecast, which is valid at the initialization time, as the *first guess* for an objective analysis that incorporates observations made within some time window (e.g., perhaps ±1 h) of the initial time. That is, the observations are used to adjust a model forecast that is valid at the initialization time. For example, Fig. 3.2 illustrates a series of forecasts of 24-h duration, where the forecasts are initialized at a 6-h interval. In this example, it would be said that the model runs with a 6-h forecast cycle. For initialization of the fourth forecast in the cycle, the 12-h forecast

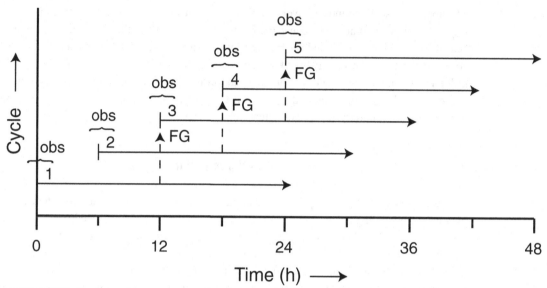

Illustration of forecast cycling, where the model is initialized and a new forecast is launched every 6 h. The vertical dashed lines show that output from a previously initialized forecast is used as a first guess (FG) for an objective analysis that employs observations (obs) that are available within a time window around the initialization time. The cycle number is shown at the initial time of each forecast.

from the second forecast in the series (which should be spun up by that time) could be used as a first guess in an objective analysis of observations. This process by which observations are merged with an analysis at a sequence of times is called *intermittent* or *sequential data assimilation*. That is, data are incorporated into the model intermittently at the initialization times only. Alternatively, there are techniques where observations are ingested by a continuously running data assimilation system, as they become available. This is called *continuous data assimilation*. Initializations using conditions where small-scale circulations are spun up to varying degrees are called *warm starts* or *hot starts*.

Because radiosonde soundings are still the only generally available, and somewhat spatially and temporally regular, sources of three-dimensional atmospheric data over land, they are a primary source of information. Unfortunately, these profiles are many hundreds of kilometers apart, so a typical situation is one in which synoptic-scale processes may be represented reasonably, but those on the mesoscale are not. Even though there are currently many other sources of observational data from satellites, radars, commercial aircraft, etc., the reliance on radiosondes explains why operational forecast models still use 0000 UTC and 1200 UTC as two of the times when forecasts are initiated daily (most operational models employ four cycles per day, initialized at 0000 UTC, 0600 UTC, 1200 UTC, and 1800 UTC).

### 3.1.5 Physical-process parameterizations

The terms in Eqs 2.1–2.7 that represent the effects of turbulent mixing of heat, water vapor, and momentum; moist convection; cloud-microphysical processes; and solar and atmospheric radiation are very complex to include in a model, and often require considerably more arithmetic operations than the rest of the terms in the equations combined. The parameterization of these physical processes is treated in Chapter 4. Parameterization involves the representation of a process in terms of its known relationship to dependent variables resolved on the model grid. For example, we cannot resolve individual turbulent eddies, but we can develop relationships between turbulence intensity and model-resolved wind shear and static stability. There are typically three reasons why we parameterize a process: we do not understand the process well enough to represent it directly through physical relationships, the process is of sufficiently fine scale that we cannot resolve it on the model grid, or the physical relationships are so complex that they would require a prohibitive amount of computing resources to treat explicitly.

## 3.2  Numerical frameworks

There are four different modeling frameworks described here for dealing with the space dependence in the nonlinear partial differential equations of atmospheric dynamics and thermodynamics:

- finite difference, or grid point;
- spectral;
- finite element; and
- finite volume.

This section does not focus on the details of the methods used to approximate the space derivatives in the equations, but rather on the overall approaches.

### 3.2.1  Spatial finite-difference/grid-point methods

Over the past half century, atmospheric scientists and oceanographers have developed numerous approaches for applying grid-point methods to the solution of the equations of fluid flow over part or all of the sphere. These methods include the use of map projections, latitude–longitude grids, and spherical geodesic grids. In each case, a procedure is defined for organizing grid points in a systematic way over the area of the sphere for which the atmosphere is to be modeled. The choice of which method to adopt in a particular modeling application depends on a variety of factors including whether the model has a limited area or global computational area, and the degree to which the code needs to be easy to modify for research purposes.

Computational grids may be classified as structured or unstructured. Traditional grids are structured in that they consist of an array of cells that are arranged in a regular pattern in two or three dimensions. In contrast, unstructured grids are defined by collections of elements, such as triangles, in an irregular pattern. This provides for greater flexibility in discretizing complex domains, and allows for the convenient use of adaptive-meshing techniques where cells can be added or subtracted. Unlike structured grids, unstructured grids require a list of mesh connectivities.

## Map projections

Map projections are geometric, and therefore mathematical, relationships that transform atmospheric properties defined on a quasi-spherical surface, such as Earth's surface or a 500-hPa surface, to a flat surface, such as a geographic map, a weather map, or a model grid. Figure 3.3 shows the geometric relationships between a spherical surface and a surface on which properties defined on the sphere are projected (the image surface), for the three types of projections commonly used in atmospheric modeling. In each case, imagine a set of rays drawn from a common origin, so that the rays connect points on the sphere

### Mercator

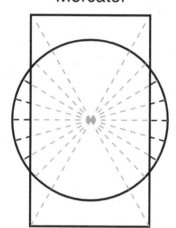

### Lambert conformal

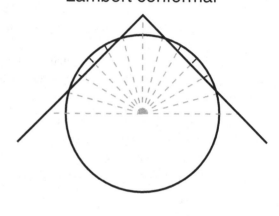

### Polar stereographic

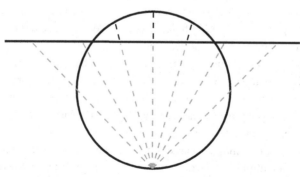

**Fig. 3.3**      Three map projections commonly used in atmospheric modeling. The cylinder (Mercator), right-circular cone (Lambert conformal), and plane (polar stereographic) are the surfaces on which the information on the sphere is projected. The radial lines connect points on the sphere and points on the projection surface. The axes of the cylinder and the cone, and the perpendicular to the plane, are parallel to Earth's axis of rotation. In these images, we are thus viewing Earth from over the Equator.

and points on the surface of the projection. For example, the Mercator projection is defined using a cylinder whose central axis passes through the center of the sphere. The conditions on the sphere are mapped to the cylinder, and the cylinder can be cut, opened, and made flat. Similarly, a flat surface and a cone define the projection surfaces for polar-stereographic and Lambert-conformal projections, respectively. The plane and the cylinder may be viewed as special cases of the cone, with vertex angles of 180° and 0°, respectively. In atmospheric-modeling applications, the axes of the cylinder and cone, and the perpendicular to the plane, are virtually always coincident with the axis of rotation of Earth. For each type of projection, the projection surface may intersect the sphere, as in the figure, or it may be tangent to the sphere. In the former case it is called a secant projection, and in the latter a tangent projection.

It is not possible to preserve all geometric properties on the sphere (e.g., area, shape, angles) during a projection. For example, Fig. 3.3 shows that distances and areas of high-latitude features are exaggerated with the Mercator projection, as are low-latitude features with the polar-stereographic projection. In fact, only at the lines or points of intersection of the sphere and the projection surface (the standard parallels) are all properties preserved. But, the three projections described above are all *conformal* in that they everywhere preserve angles between two curves, and the distance distortion is the same in all directions at a point.[1] For meteorological applications where preserving the angles of atmospheric features is important (e.g., the angle between isobars and wind vectors), conformal projections are desirable. The lack of distance and area preservation with conformal projections is dealt with by applying them only for latitudes where the distortion is small.

Map projections are needed for virtually all atmospheric models. For a global model that uses spherical coordinates, visualizing the output on paper or on a computer screen requires that the atmospheric conditions, and associated georeference information, such as political boundaries and natural features, be defined on one of the map projections. Mathematical transformations for each of the projections convert from spherical coordinates (latitude and longitude) to Cartesian coordinates on the projection surface. This is the process by which geographers transfer properties defined on Earth's surface to a map.

For limited-area models, which employ Cartesian coordinate systems and solve the equations on planar surfaces, the transformation between the sphere and projection surface becomes an intimate part of the modeling process and the equations themselves. In particular, observations whose locations are defined in latitude–longitude coordinates need to be applied at the appropriate coordinates of the Cartesian model grid that is defined on the projection surface. And, because of the distance distortion, the grid increment used in the finite-difference equations needs to reflect the true horizontal distance between points. Figure 3.4 shows how the grid increment is affected by the distance transformation between the sphere and a projection surface on which a computational grid is defined. The points on the computational grid defined on the projection surface are equidistant, but the

---

[1]  If $x$ is a horizontal displacement from a point, $\delta x_E / \delta x_G$ is the same regardless of the direction of the displacement, where $E$ refers to Earth and $G$ to the grid.

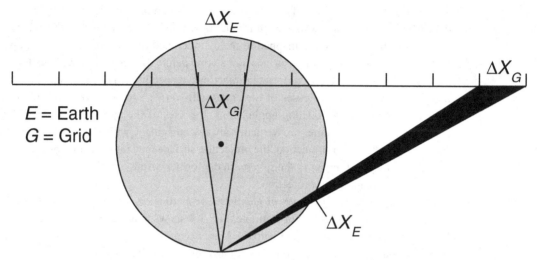

$\Delta X_E$

$\Delta X_G$

$E$ = Earth
$G$ = Grid

$\Delta X_G$

$\Delta X_E$

Fig. 3.4 The relationship between the grid increment on Earth and on the computational grid, for a polar-stereographic, secant projection.

physical distance represented by each increment is generally different and dependent on the location on the grid. A measure of this distortion is the map-scale factor, defined in Eq. 3.2, using the notation in Fig. 3.4, as the ratio of distance on the projection to distance on the sphere:

$$m = \frac{\Delta x_G}{\Delta x_E}. \tag{3.2}$$

The map-scale factor is the same along each latitude circle. Figure 3.5 shows this scale factor as a function of latitude for the three projections. The direction of the departure of $m$ from unity depends on whether the image surface is above or below Earth's surface. For each point on a model grid, the map-scale factor is precalculated, and used in the equations to account for the varying distance between grid points. Equation 3.3 shows the first equation of motion, with Earth-curvature terms, for a popular limited-area model (Dudhia and Bresch 2002).

$$\frac{du}{dt} = -\frac{m}{\rho}\frac{\partial \rho'}{\partial x} + \left(f + u\frac{\partial m}{\partial y} - v\frac{\partial m}{\partial x}\right)v - ew\cos\alpha - \frac{uw}{r} + D_u. \tag{3.3}$$

Here, $m$ is the map-scale factor given by, for example,

$$m = \frac{1 + \sin 60°}{1 + \sin\phi}$$

for a polar-stereographic projection that is true at $60°$; $e$ and $f$ represent the full Coriolis force, where $f = 2\Omega\sin\phi$ and $e = 2\Omega\cos\phi$; $D$ is a diffusion, or frictional-dissipation,

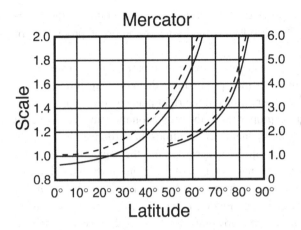

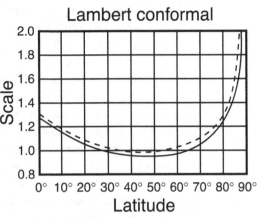

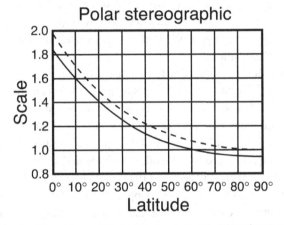

**Fig. 3.5**    Map-scale factors for different tangent (dashed lines) and secant (solid lines) projections as a function of latitude. For the secant projections, the conical surface intersects the sphere at 30° and 60° (north or south) latitude, the plane intersects at 60° (north or south) latitude, and the cylinder intersects at 20° (north and south) latitudes. From Saucier (1955).

term; $r$ is the radius of Earth; $\rho$ is density; and $\alpha(x, y)$ is the angle between the local meridian and the $y$ axis. Figure 3.5 shows that the differential-map-scale terms in Eq. 3.3 are smallest for this projection near the poles and largest in equatorial areas.

Because of the varying effective distance between grid points, computations are really being performed on a "stretched" grid, and this leads to spatial contrasts in the numerical properties (i.e., the errors) of the solution to the equations. As will be seen later in this chapter, this means that the same wave on the grid will have different phase and group speeds depending on latitude. And there will be latitudinal differences in the conditions needed to maintain stability of the numerical solution to the equations. Thus, objectives in the decision about the best choice of map projection to use for a particular model application are to minimize (1) the departure of the map-scale factor from unity over the grid and (2) the latitudinal derivative of the map-scale factor. In general, these conditions can be best satisfied by using the Mercator projection for grids in tropical latitudes, the polar-stereographic projection for high-latitude grids, and the Lambert-conformal projection for midlatitude grids (see Fig. 3.5).

Even when a reasonable map-projection choice is made for a particular application of a model, and the transformations are properly incorporated into the model equations and the initialization process (i.e., getting the observations in the right place on the computational grid), there are a few ways in which the properties of the projection may impact the model user. One is that the $u$ and $v$ velocity components in the atmosphere (defined in terms of the east–west and north–south directions on the sphere) are not the same as the $u$ and $v$ components on the computational grid (defined in terms of grid-point rows and columns). This issue must be dealt with when initializing the model. Another is that the time step that is chosen by the user, or automatically by the model, is based on the grid increment in order to maintain a stable solution to the equations (see Section 3.4.2). Because the true horizontal grid increment varies spatially, some areas of the grid may have sufficiently large values that stability criteria are locally violated. Evidence of this would be unrealistic-appearing (e.g., small-scale waves) model solutions in latitudes where the grid increment is the smallest.

The above discussion was in the context of using map projections to model limited areas of Earth's surface. However, there have also been methods developed for using combinations of map projections to model the entire sphere using a *composite grid*, where one of the objectives is to avoid the problems of latitude–longitude grids described later. For example, Phillips (1957a, 1962) used a combination of a Mercator projection for latitudes equatorward of a boundary latitude and stereographic projections for higher latitudes. And, Stoker and Isaacson (1975) and Dudhia and Bresch (2002) used two polar-stereographic projections that overlapped in equatorial regions. Calculations on these grids can communicate at the interfaces, thus avoiding the need for artificial lateral-boundary conditions, or the integrations can be separate. Figure 3.6 shows an example of two overlapping polar grids. The method of using two overlapping polar-stereographic projections for global simulations was compared with two spectral methods by Browning *et al.* (1989) in terms of memory requirements, execution time, and arithmetic-operation count. The results were mixed, but the conclusion was that the methods were generally competitive.

Another common approach to modeling the sphere with map projections is to circumscribe a regular polyhedron, such as a cube, by the sphere. On each face of the polyhedron is defined a regular Cartesian grid, and radials from the center of the sphere are projected through the grid points in order to map the grid to the surface of the sphere. In the case of the cube, the model equations are solved on each of the six grids, but the calculation of finite differences at the boundaries is challenging. Sadourny (1972), McGregor (1996), Rančić *et al.* (1996), Ronchi *et al.* (1996), and Purser and Rančić (1997, 1998) review the testing and properties of such polyhedral-gnomonic projections. Adcroft *et al.* (2004) describe the use of the expanded cube as the basis for the general-circulation model of the Massachusetts Institute of Technology, McGregor and Dix (2001) summarize the use of this approach in Australia's Commonwealth

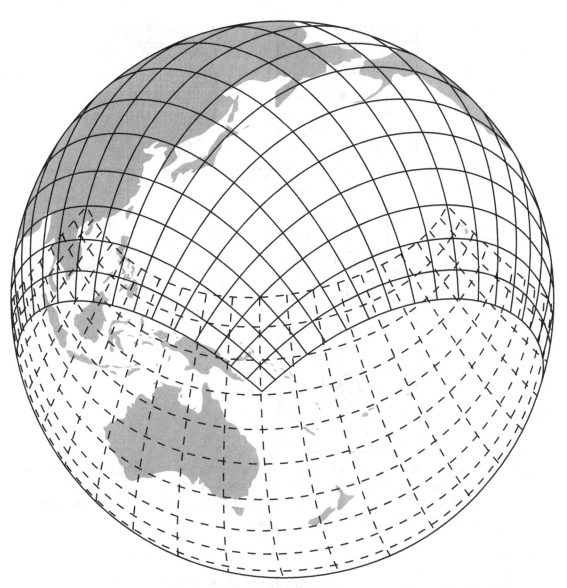

**Fig. 3.6**   A composite grid defined by overlapping North Polar and South Polar stereographic projections. From Williamson (2007).

Scientific and Industrial Research Organization (CSIRO) general-circulation model, and Zhang and Rančić (2007) apply the method to a version of the US National Weather Service's (NWS) Eta model. Figure 3.7 illustrates the projection of Earth's surface on the faces of an exploded cube, as well as an example of the relatively uniform distribution of grid points on the sphere. This approach produces a nonconformal projection, but additional transformations can convert it to one that has conformal properties.

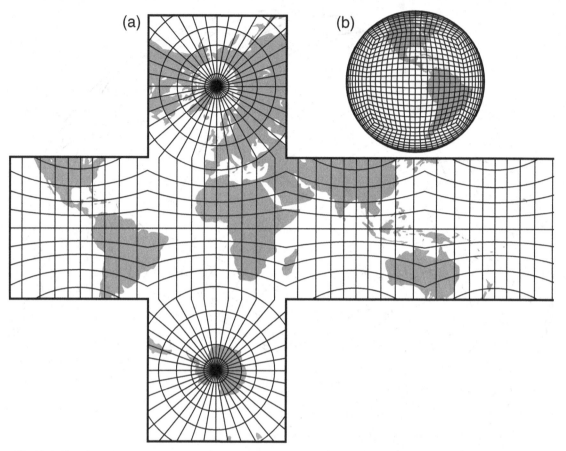

**Fig. 3.7**  A *cubic gnomonic projection*, used as the basis for a global model grid, is defined by establishing Cartesian grids on each face of a cube that is inscribed within a sphere. These grids are mapped to Earth's surface, producing the relatively evenly spaced grid points shown on the sphere in (b). The expanded cube with geographic and latitude–longitude references is shown in (a). Panel (b) is from Rančić *et al.* (1996).

## Latitude–longitude grids

In this approach, latitude and longitude are the horizontal coordinates, and the vertical coordinate is defined along the local radial from the center of Earth. On each vertical coordinate surface, the sphere is partitioned into grid cells using increments of latitude and longitude. If these intervals are constant over the entire sphere, the longitudinal distance between grid points becomes progressively smaller as the meridians converge at the poles (Fig. 3.8).

This requires that time steps be small in order to maintain stable solutions to the equations. In addition, the existence of the singularities at the poles, where the coordinate lines (the meridians) intersect, means that calculating horizontal derivatives can be problematic. Shortening the time step near the poles produces satisfactory results, but calculations for

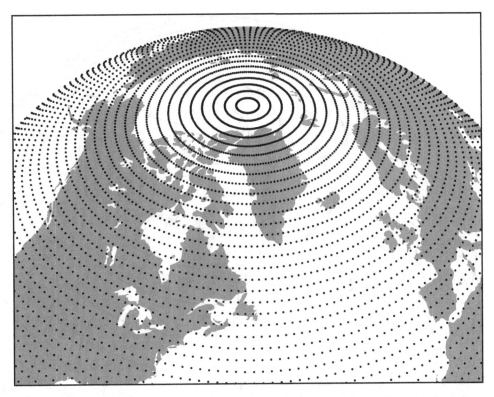

**Fig. 3.8**    A latitude–longitude grid shown for part of the sphere, where the points are defined at a uniform interval in each coordinate direction.

the small percent of Earth's surface area near the poles can consume more than half the computer time (Grimmer and Shaw 1967).

One approach to dealing with the small distance between grid points near the poles, and the associated impact on the time step and model performance, is to Fourier filter the variable fields in the east–west direction near the poles. This is accomplished by Fourier transforming a variable, filtering out the higher wavenumbers by truncating the series, and then inverse transforming back to physical/grid-point space to obtain a smoother field. This removal of small-scale information from the model solution effectively filters the faster modes, and permits the use of a longer time step in spite of the small longitudinal grid distance. Note that, in this approach, there is still the computational burden of solving the equations at the dense array of grid points near the poles, and the excessive density of points wastes memory. Williamson (2007) points out that this is an unsatisfying engineering approach, but it is in use today, for example in the optional finite-volume core of the National Center for Atmospheric Research's (NCAR) Community Atmosphere Model (CAM). Another method is to use larger increments of longitude as the pole is approached (Williamson and Rosinski 2000). An example of the resulting *reduced grid* is shown in Fig. 3.9.

**Fig. 3.9** A reduced grid, in which the longitudinal grid increment in degrees is increased with decreasing distance from the pole, where the objective is to maintain a relatively uniform physical distance between grid points. From Williamson (2007).

## Spherical geodesic grids

A desirable property of a grid is uniformity in the spacing of the grid points over the sphere, or a portion of the sphere. It has been shown that map projections can produce grids that have significant variability in the Earth-distance between points (Fig. 3.5), especially when the computational domain must span a large area. Similarly, latitude–longitude grids have inherently higher resolution where the meridians converge at high latitudes. However, geodesic grids have a nearly homogeneous distribution of points over the sphere.

In mathematics, a geodesic is the equivalent of a straight line, but on a curved surface. On a spherical surface, such as that of Earth, a geodesic is the shortest path between two points, specifically a segment of a great circle. A spherical geodesic grid is defined by spherical, equilateral triangles whose edges are geodesics. One way of defining this grid is to begin with an icosahedron, the geometric solid shown in Fig. 3.10a with 20 triangular faces (major triangles), 12 vertices, and 30 edges, where the vertices touch the surface of a sphere. The vertices may then be connected by geodesics on the sphere, producing spherical triangles. A grid may be created by dividing the major triangles into smaller ones (grid triangles) using a variety of approaches. For example, bisecting each edge of the icosahedron and connecting the bisection points produces four new equilateral triangles for each original one (Fig. 3.10b). The vertices of these new triangles can then be projected onto the sphere along a radial from the center (Fig. 3.10c), and then connected by geodesics to again produce spherical triangles (Fig. 3.10d). Even though the distances between adjacent points look uniform, they are not exactly so. A hint at the asymmetries from one part of the surface to the next can be seen in the fact that the "new" vertex facing the viewer in the upper-center (Fig. 3.10d) is surrounded by six adjoining triangles, while the "original" icosahedron vertex to its right is surrounded by only five. Williamson (1968) and Sadourny *et al.* (1968) describe another approach for dividing the major triangles into grid triangles, where the inequality in the distance between points is less than that resulting from the method just described. Figure 3.11 shows an example of the distribution of grid points over the sphere.

Some applications of spherical geodesic grids employ the above triangular cells, while others use a related grid with hexagonal cells. To obtain the latter, Voronoi cells are constructed based on the triangular grid, where such cells consist of the set of all points that are closer to a particular vertex than to any other vertex. For the twelve original vertices in the icosahedral grid (e.g., in Fig. 3.11), the Voronoi cells are pentagons. For all the rest, they are hexagons. Figure 3.12 illustrates the geometric relationship between the triangular

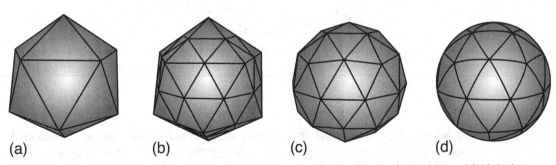

(a)                    (b)                    (c)                    (d)

**Fig. 3.10**  In the generation of a spherical geodesic grid, the major triangles of the icosahedron (a) are subdivided, where (b) shows one approach. The vertices of the new triangles are projected (c) onto the sphere that is coincident with the vertices of the icosahedron. Geodesic lines are then drawn between the new vertices to generate spherical grid triangles (d).

**Fig. 3.11**  An example of the relatively uniform distribution of grid points over the sphere, for a spherical geodesic grid, based on one method for dividing the major triangles of an icosahedron into grid triangles. Note that the horizontal resolution in this example is very coarse. Adapted from Williamson (1968).

and hexagonal grids. Randall *et al.* (2002) describe the relative merits of hexagonal, square, and triangular cells, and Weller *et al.* (2009) compare a few additional mesh-refinement methods.

Other advantages of the spherical geodesic grid include the fact that it is straightforward to selectively enhance the resolution in some areas (adding triangles), to better render fine-scale features in the vicinity of mountains or other types of small-scale local

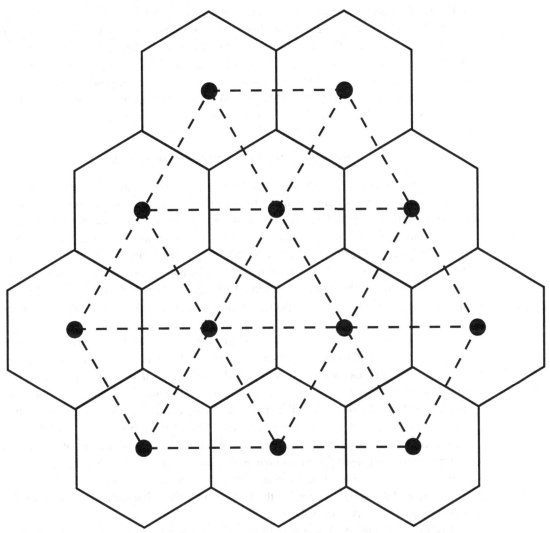

**Fig. 3.12**    Illustration of the geometric relationship between triangular cells and hexagonal cells, either of which can be used as the basis for a spherical geodesic grid.

forcing. A disadvantage is that the indexing of the grid points is more complex than for Cartesian or latitude–longitude grids, but Randall *et al.* (2002) show how the grid cells on the sphere can be separated into rectangular panels whose cell values can be logically organized in computer memory.

Examples of contemporary models that employ spherical geodesic grids are the Ocean–Land–Atmosphere Model (OLAM; Walko and Avissar 2008a,b), the Operational Multiscale Environment Model with Grid Adaptivity (OMEGA; Bacon *et al.* 2000), and the operational GME model of the German Weather Service (Majewski *et al.* 2002). Randall *et al.* (2002) describe the development of a coupled ocean–land–atmosphere spherical-geodesic-grid model for climate applications. Tomita and Satoh (2004) and Satoh *et al.* (2008) describe a model that uses this method for global, cloud-resolving simulations. And, the US National Oceanic and Atmospheric Administration (NOAA) is developing the Flow-following finite-volume Icosahedral Model (FIM) for operational use. Further discussions and literature summaries about spherical geodesic grids can be found in Sadourny *et al.* (1968), Williamson (1968), Baumgardner and Frederickson (1985), Nickovic (1994), Ringler *et al.* (2000), and Ringler and Randall (2002).

## Differential grid resolution across the sphere

For modeling global-scale processes, it is reasonable to desire somewhat uniform horizontal resolution over the sphere. However, in other model applications it is common to want greater resolution in certain regions. For example, in research settings a specific meteorological feature is often being studied, and the grid points and computational resources should be focussed there. For operational forecasting models, small-scale processes dominate in certain regions such as near complex terrain, making it desirable to have greater horizontal resolution there than elsewhere. Additionally, operational models are often set up to serve limited areas such as specific countries, so, again, greater resolution is needed there.

There is a variety of approaches that can be used to produce different horizontal resolutions over a three-dimensional computational volume. A common one is to embed a high-resolution limited-area model within a global model, with the global model providing lateral-boundary conditions to the limited-area model. The global model forecast or simulation is performed first. The limited-area model may consist of a single high-resolution grid, or a nest of multiple grids with grid spacings that change abruptly by a factor of perhaps three to five between adjacent grids. Figure 3.13 illustrates an example of a nest with multiple grids that is used for operational prediction over an area near the Chesapeake Bay in the eastern USA (Liu *et al.* 2008a). Section 3.5 on lateral-boundary conditions discusses the methods and limitations of this approach for obtaining higher horizontal resolution over specific areas of the sphere.

A property of some models is that all the grids in a horizontal nest must have the same vertical resolution (distribution of layers). In contrast, with others, not only can the vertical resolution vary among the grids in a nest, but vertical nesting is permitted. Figure 3.14c shows the common situation where the grids in a nest have the same vertical resolution. In contrast, Figs. 3.14a and b illustrate vertical nesting, where the inner grid not only has higher vertical resolution, but it can focus computational resources in certain vertical layers. For example, in the inner grid with higher horizontal resolution, enhanced vertical resolution can be used in the boundary layer, near the tropopause, or in layers with low-level jets, all regions with larger vertical gradients of variables. See Clark and Farley (1984) and Clark and Hall (1991, 1996) for additional information about vertical grid nesting in the Clark model.

Because wave reflections can sometimes occur at an abrupt variation in resolution, such as at the transition between grids in a nest (Davies 1983), there is a benefit to using a gradual variation in horizontal grid increment to achieve greater resolution in certain areas.

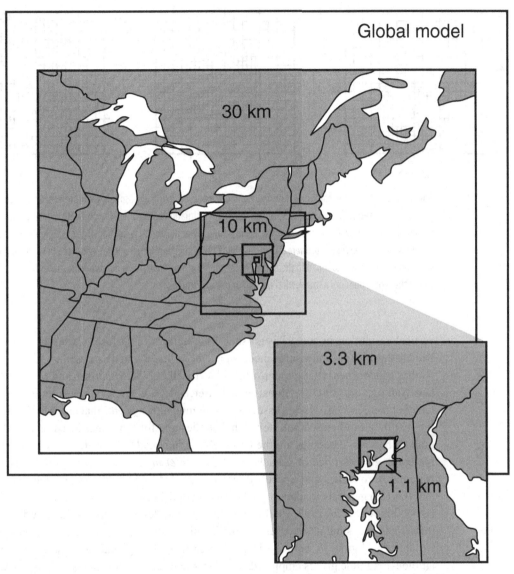

**Fig. 3.13** An example of a nested-grid model used for operational prediction. The model with the two-way interacting grids is embedded within a global-model prediction. Model grid increments are indicated. From Liu *et al.* (2008a).

Such grids with a gradual change in resolution are sometimes called stretched grids. Kalnay de Rivas (1972) and Fox-Rabinovitz *et al.* (1997) discuss the truncation error associated with the approximation of derivatives on a variable-resolution mesh. They point out that such errors with a nonuniform smoothly varying mesh are equivalent to those with a uniform mesh defined by a transformation, such as associated with map projections. That is, even though $\Delta X_G$ is constant, the effective grid increment on Earth's surface, $\Delta X_E$, varies smoothly. Virtually all models employ a stretched grid in the vertical, with higher

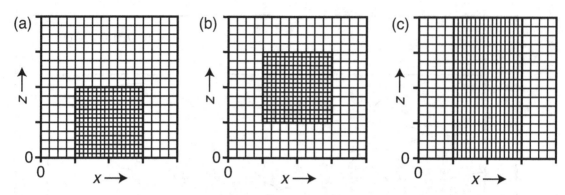

Fig. 3.14 Schematic examples of vertical grid nesting, where certain layers in the atmosphere are represented by a subgrid with greater vertical resolution. Three examples are shown for a situation where there is a grid with greater horizontal resolution that is horizontally nested within a grid having lower horizontal resolution. One example shows vertical nesting where the lower atmosphere is represented by the higher-resolution grid (a), another shows the enhanced vertical-resolution grid confined to the middle of the model atmosphere (b), and another illustrates the situation where the model does not allow vertical nesting or grid stretching (c).

vertical resolution in some layers, such as near Earth's surface. Cartesian, horizontally stretched grids for limited-area models have been investigated and used by Anthes (1970), Anthes and Warner (1978), Staniforth and Mitchell (1978), Staniforth and Mailhot (1988), and Walko *et al.* (1995b). Global-model stretched grids are sometimes motivated by the fact that there is a cost-saving associated with modeling systems that are sufficiently versatile that they can be used both for global studies and for those that focus on specific geographic regions. Examples are the Global Environmental Multiscale (GEM) model of the Meteorological Service of Canada (MSC) (Côté *et al.* 1998a,b; Yeh *et al.* 2002) and the National Aeronautics and Space Administration (NASA) Goddard Earth Observing System (GEOS) general circulation model (Suarez and Takacs 1995; Takacs and Suarez 1996; Fox-Rabinovitz *et al.* 1997, 2000). Côté *et al.* (1993) describe a global shallow-fluid model that employs a similar stretching method. This strategy uses spherical coordinates with variable resolution in both horizontal directions. For global studies, a conventional latitude–longitude grid is employed, with the singularities at the poles. When a concentration of grid points is needed for higher resolution in a particular region, the resolution is varied, as in the example shown in Fig. 3.15. The poles of the coordinate system do not necessarily coincide with Earth's poles. With increasing distance from the grid's poles, near the geographic poles in this example, the resolution in that direction increases, as shown, until becoming constant for the east–west belt around the sphere shown in gray. The resolution of the other spherical coordinate also varies, with uniform high resolution in the north–south belt through the Americas. Where the areas of highest resolution in the two horizontal dimensions overlap, a high-resolution regional grid with a uniform latitude–longitude grid increment of 0.04° exists in this example. Fox-Rabinovitz *et al.* (2006) discuss the Stretched-Grid Model Intercomparison Project (SGMIP), where a number of variable-resolution global models were used for regional-climate simulations.

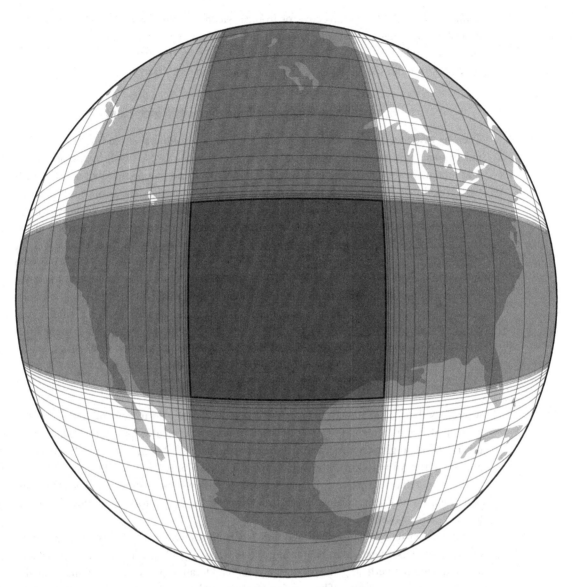

Fig. 3.15    An example of the variable-resolution horizontal grid of the operational regional configuration of the GEM model. See the text for details. Adapted from Yeh *et al.* (2002).

There are also various approaches in which the grid resolution changes during a model integration in order to better represent the evolving atmospheric processes. One method used with Cartesian limited-area models is called adaptive mesh, or grid, refinement, and is described in Berger and Oliger (1984) and Skamarock and Klemp (1993). This method uses the above concept of nested grids, but here the fine grids can change size, shape, location, and number (i.e., they can adapt) automatically based on estimates of the truncation error during a simulation or forecast. For example, fine-mesh grids are spawned automatically to follow convective or tropical storms as they move. Other methods are described in Dietachmayer and Droegemeier (1992) and Srivastava *et al.* (2000). Also, spherical geodesic grids easily accommodate adaptive mesh refinement methods because spherical triangles can be added or removed as needed (e.g., Bacon *et al.* 2000).

## Consistency of vertical and horizontal grid increments

There is evidence that model vertical and horizontal resolutions should not be specified independently. In particular, physical features that can be resolved well by the horizontal grid increment should also be resolvable by the vertical grid increment (and vice versa). If the vertical grid increment is too coarse to satisfy this criterion, the resulting truncation error will generate spurious gravity waves during the simulation and the features will be poorly rendered by the model. This problem has been most-commonly described in the context of sloping features in the atmosphere such as fronts (e.g., Snyder *et al.* 1993) or the slantwise convection resulting from conditional symmetric instability (e.g., Persson and Warner 1991). The mathematical relationship that defines consistency between the vertical and horizontal grid increments has been defined differently by different authors, but the expressions tend to be quite similar from a practical standpoint. For example, Pecnick and Keyser (1989) state that the optimal vertical grid increment is related to the horizontal grid increment by the expression

$$\Delta z_{opt} = s\Delta y, \tag{3.4}$$

where $s$ is the frontal slope, $\Delta z_{opt}$ is the optimal vertical grid increment, and $\Delta y$ is the horizontal grid increment. For synoptic-scale fronts with typical slopes from 0.005 to 0.02, this relationship gives optimal vertical grid spacings of 0.5–2.0 km for $\Delta y = 100$ km, and 50–200 m for $\Delta y = 10$ km. Alternatively, Lindzen and Fox-Rabinovitz (1989) suggest two consistency relationships, one for quasi-geostrophic flows and another for flows that contain gravity waves near a critical layer. In both cases, the $\Delta z_{opt}$ for midlatitudes is similar to that obtained from Eq. 3.4.

These consistency relationships and the associated research with hydrostatic and nonhydrostatic models suggest that decreasing the horizontal grid spacing of a model without also reducing the vertical grid spacing may not lead to an improvement, and may actually produce a worse simulation. The two-dimensional model simulations of frontogenesis presented by Pecnick and Keyser (1989) show that spurious gravity-wave structures and spurious large velocity and vorticity values result when $\Delta z > \Delta z_{opt}$. Also, Lindzen and Fox-Rabinovitz (1989) refer to instabilities, spurious amplitude growth, and other problems when an inconsistency exists between the vertical and horizontal resolutions. And, Gall *et al.* (1988) report that the intensity of erroneous waves generated at a front was diminished when the vertical grid spacing was reduced, to be consistent with the horizontal grid spacing.

Examples of the effects of using a vertical grid spacing that is insufficient to represent structures that are well resolved in the horizontal are shown in many of the above studies. As an illustration here, Fig. 3.16 shows the numerical noise in the vertical motion that results

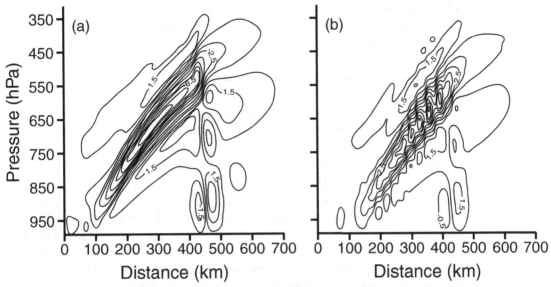

**Fig. 3.16** Vertical cross sections of vertical velocity, ω (solid lines, μb s⁻¹) after 24-h simulations of conditional symmetric instability with a 10-km horizontal grid increment and 75 layers (a) and 25 layers (b). From Persson and Warner (1991).

from the use of an inappropriately large vertical grid spacing in a simulation of slantwise convection. Figure 3.16a shows the smooth vertical-motion field from a 24-h simulation that employed 75 layers of equal depth, and a horizontal grid increment of 10 km, while Fig. 3.16b shows the same field for an analogous simulation that employed only 25 layers of equal depth. The truncation error associated with the poor vertical resolution of the feature has created a noisy vertical motion field and associated gravity waves (not shown). In the first experiment, $\Delta z = \Delta z_{opt}$, while in the second $\Delta z = 0.33 \Delta z_{opt}$. A third experiment used 25 layers, as in Fig. 3.16b, but the horizontal grid increment was increased from 10 km to 30 km. In this case, the use of both the coarser horizontal and vertical grid increments produced $\Delta z = \Delta z_{opt}$ and a smooth solution (not shown), but one with considerably less amplitude than in Fig. 3.16a because of the overall coarser resolution.

Even though these resolution-consistency studies clearly isolate the importance of this source of error for specific meteorological cases, this effect is not always responsible when model solutions appear to degrade with increasing horizontal resolution. For example, other problems such as the inappropriateness of resolution-dependent physical-process parameterizations may also be encountered as the horizontal resolution is increased in a model. And, it is well known that conventional objective measures of forecast skill, such as bias, mean-absolute error, and root-mean-square error, show lower skill for forecasts that have small-scale structures (i.e., that can result from high horizontal resolution), compared with forecasts that are smoother (Rife *et al.* 2004).

Many numerical models used for research and operational forecasting, especially those applied on the mesoscale, do not satisfy these consistency relations. For example, in the model simulations associated with Fig. 3.16, 750 vertical levels would have been required for consistency if the grid increment had been 1 km. Nevertheless, it is not clear what the practical consequences are of not satisfying the consistency relations, especially in light of the general success of numerical model studies and forecasts where the vertical resolution has presumably been insufficient. Lindzen and Fox-Rabinovitz (1989) suggest that this success is partly attributable to the use of horizontal numerical diffusion that limits the effective horizontal resolution of features. Complicating a desire to comply with the consistency constraint is the fact that, with operational modeling, in contrast to focussed case studies, there are many features with different slopes across the computational domain. And, the vertical grid spacing in most models varies considerably with distance above Earth's surface. Furthermore, two-way interacting model grids in a grid nest have different horizontal grid increments, but they typically must utilize the same vertical grid structure. Thus, $\Delta z_{opt}$ will vary considerable with time and place in the same model integration.

For a given amount of computational resources, modelers tend to maximize the horizontal resolution at the expense of the vertical resolution – the horizontal grid increment has tended to be the Holy Grail of modeling. However, for a given model application, even though there may not be computational resources available to completely satisfy the consistency criterion, the above experimental evidence suggests that modelers should not ignore this issue. Instead, a compromise should be made between the vertical and horizontal resolutions, where the sensitivity of the model solution to different choices should be evaluated using case studies of the prevailing meteorological processes in a given area. With the trend toward using an ensemble of coarser horizontal-resolution model runs, this consistency issue may, at least temporarily, become less critical.

### 3.2.2 Spectral methods

Early approaches to global grid-point modeling included the use of latitude–longitude grids with reduced time steps near the poles, quasi-homogeneous spherical-geodesic and cubed-sphere grids, and composite meshes. At the time that they were proposed, all of these approaches were problematic in some respect, and this resulted in the dominance of spectral modeling after the spectral-transform method (Machenhauer 1979) was introduced by Eliasen *et al.* (1970) and Orzag (1970), and implemented by Bourke (1974). This method dominated global modeling for decades, and still is widely used even though refinements to the above grid-point approaches and new computer architectures have resulted in the adoption of other options.

The spectral form of a series of meteorological equations is obtained by substituting finite expansions of the dependent variables, typically using double Fourier series or Fourier–Legendre functions (called *basis functions*) to represent the horizontal spatial variation. The orthogonality of these functions allows the derivation of a series of coupled, nonlinear, ordinary differential equations in the expansion coefficients, which are functions of time and the vertical coordinate. The equations are numerically integrated forward in time using conventional finite differencing in time and in the vertical dimension. Such models are initialized by forward transforming the standard dependent variables from physical space (grid-point values) to the transform space (expansion coefficients), and interpretable forecast fields are obtained by inverse transforming back to physical space.

Before further discussing the spectral method and its strengths and weaknesses, the one-dimensional shallow-fluid equations will be used to illustrate the approach. A Fourier series will be used as the basis function. In general, a one-dimensional field can be represented by the following series:

$$A(x) = \sum_{m=0}^{\infty} (a_m \cos mkx + b_m \sin mkx), \qquad (3.5)$$

where the Fourier coefficients $a_m$ and $b_m$ are real, $m$ is an integer wavenumber, $k = 2\pi/L$, and $L$ is the domain length. If we add and subtract the two exponentials in Euler's relations

$$e^{imkx} = \cos mkx + i \sin mkx \text{ and}$$

$$e^{-imkx} = \cos mkx - i \sin mkx,$$

where $i = \sqrt{-1}$, the following expressions are obtained:

$$\cos mkx = \frac{e^{imkx} + e^{-imkx}}{2} \text{ and} \qquad (3.6)$$

$$\sin mkx = \frac{e^{imkx} - e^{-imkx}}{2i}. \qquad (3.7)$$

Substituting Eqs. 3.6 and 3.7 into Eq. 3.5 produces

$$A(x) = \sum_{m=0}^{\infty} \left[ \left( \frac{a_m}{2} + \frac{b_m}{2i} \right) e^{imkx} + \left( \frac{a_m}{2} - \frac{b_m}{2i} \right) e^{-imkx} \right]. \qquad (3.8)$$

Doing some algebraic manipulation, and defining

$$C_0 = a_0,$$

$$C_m = \frac{a_m - ib_m}{2}, \text{ and}$$

$$C_{-m} = \frac{a_m + ib_m}{2},$$

allows Eq. 3.8 to be rewritten as

$$A(x) = C_0 + \sum_{m=1}^{\infty} C_m e^{imkx} + \sum_{m=1}^{\infty} C_{-m} e^{-imkx}.$$

The index in the last summation can be multiplied by $-1$, allowing the terms to be combined into

$$A(x) = \sum_{m=-\infty}^{\infty} C_m e^{imkx} = \sum_{|m| \leq \infty} C_m e^{imkx}.$$

Before substituting this solution into a set of meteorological equations, it will be assumed that the Fourier coefficients $C_m$ are a function of time, and a maximum value of the wavenumber $m$ will be defined. Recall that $m$ represents the number of waves over the domain length, $L$, so defining the highest wavenumber (or smallest wavelength) in the series determines the model resolution.

The shallow-fluid equations (Eqs. 2.36–2.38) will be converted to spectral form by representing the dependent variables in terms of truncated versions of the above Fourier series. Specifically,

$$u(x,t) = \sum_{|m| \leq K} U_m(t)e^{imkx},$$                      (3.9)

$$v(x,t) = \sum_{|m| \leq K} V_m(t)e^{imkx}, \text{ and}$$          (3.10)

$$h(x,t) = \sum_{|m| \leq K} H_m(t)e^{imkx},$$                      (3.11)

where $U_{-m}(t)$ is the complex conjugate of $U_m(t)$ and $K$ is the highest permitted wavenumber. Thus, the time dependence of the dependent variables will be represented through the complex Fourier coefficients $U$, $V$, and $H$, and the space dependence will be represented analytically in terms of the sinusoidal variation of the sine and cosine functions embodied by the exponential. To obtain the spectral equations, substitute Eqs. 3.9–3.11 into the differential equations Eqs. 2.36–2.38, multiply each equation by $e^{-ijkx}$, where $j$ is any arbitrary wavenumber, and integrate each equation with respect to $x$, from 0 to $L$. After using the following condition resulting from the orthogonality of the exponential,

$$\int_0^L e^{imkx}e^{inkx}dx = \begin{cases} 0; & m = -n, \\ L; & m = -n, \end{cases},$$

the shallow-fluid equations in spectral form are obtained. The original partial differential equations are now ordinary differential equations, where

$$\dot{U}_m = -ik \sum_{\substack{|p| \leq K \\ |m-p| \leq K}} (m-p)U_p U_{m-p} + fV_m - ikgmH_m,$$

$$\dot{V}_m = -ik \sum_{\substack{|p| \leq K \\ |m-p| \leq K}} (m-p)U_p V_{m-p} - fU_m - g\frac{\partial H}{\partial y}\delta_m, \text{ and}$$

$$\dot{H}_m = -ik \sum_{\substack{|p| \leq K \\ |m-p| \leq K}} (m-p) U_p H_{m-p} + V_m \frac{\partial H}{\partial y} - ik \sum_{\substack{|p| \leq K \\ |m-p| \leq K}} (m-p) H_p U_{m-p},$$

where $\delta_m = 1$ for $m = 0$ and $\delta_m = 0$ for $m \neq 0$. And,

$$\dot{U}_m = \frac{d}{dt} U_m(t), \text{ etc.}$$

The spatial derivatives have been evaluated analytically, leaving only the time derivatives to be approximated with finite differencing.

The basis functions used in most global spectral atmospheric models are *spherical harmonics*, a combination of sines and cosines in a Fourier series to represent the zonal variation of the dependent variables, and *associated Legendre functions* for the meridional variation. Such spherical harmonics take the form

$$\psi(\lambda, \phi) = \sum_{m=-K}^{K} \sum_{n=|m|}^{N(m)} \Psi_n^m Y_n^m(\lambda, \phi), \qquad (3.12)$$

where

$$Y_n^m(\lambda, \phi) = e^{im\lambda} P_n^m(\sin \phi),$$

$\psi$ is any dependent variable, $\lambda$ is longitude, $\phi$ is latitude, $m$ is zonal wavenumber, $K$ is the highest zonal wavenumber, $n$ is the order of the associated Legendre polynomial, $N(m)$ is the highest order of the associated Legendre polynomial, $\Psi_n^m$ are spectral coefficients, $Y_n^m$ are spherical harmonics, and $P_n^m(\sin \phi)$ are the associated Legendre functions of the first kind (which are polynomials). See, for example, Krishnamurti *et al.* (2006a) for the form of the associated Legendre polynomials. Other approaches include the use of two-dimensional Fourier expansions (Cheong 2000, 2006).

The relationship between the number of waves allowed in the meridional direction and the number of waves allowed in the zonal direction defines the type of truncation used in the model. Two types of truncation schemes are typically used in global spectral models – triangular and rhomboidal. The truncation defines the form of $N(m)$ in Eq. 3.12. For triangular truncation, $N(m) = K$ and the same number of waves is allowed in each direction. For rhomboidal truncation $N(m) = |m| + K$ and the number of meridional waves is greater than the zonal wavenumber by a constant factor. Figure 3.17 shows graphically the differences in these truncations. The triangular truncation is more commonly used today, for reasons discussed in Krishnamurti *et al.* (2006a). Triangular truncation provides a resolution that is uniform on the sphere and the same in the zonal and meridional directions. In contrast, rhomboidal truncation produces higher resolution near the poles.

Solving the nonlinear terms in models that are completely spectral is computationally intensive, and makes the process prohibitive. And, local-forcing processes (e.g., latent heat

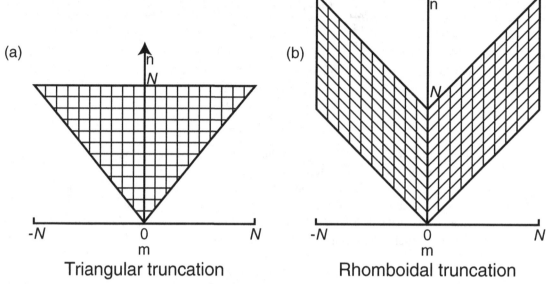

(a)

(b)

Triangular truncation

Rhomboidal truncation

**Fig. 3.17** Schematic showing the relationship between the permitted wavenumbers in the zonal (m) and meridional (n) directions for triangular and rhomboidal truncations.

release, differential surface heat fluxes), some of which are discontinuous, are only possible to represent in physical space. These problems have been resolved through the development of *pseudospectral* models that treat some processes in spectral space and others in physical, or grid-point, space. This approach is called the *transform method* because it involves transformations between spectral and physical space every time step. In particular, after extrapolations in time produce new expansion coefficients, the dependent variables are converted from spectral space to values defined on an appropriate grid using an inverse transform (e.g., Eqs. 3.9–3.11 for the one-dimensional Fourier expansion, or Eq. 3.12 for the two-dimensional expansion with spherical harmonics). The transform method works as follows, for a nonlinear term such as $u \partial u / \partial x$.

- Calculate $\partial u / \partial x$ in spectral space, and inverse transform it and $u$ to physical space on an appropriate grid.
- Calculate $u \partial u / \partial x$ on the grid by multiplication.
- Transform $u \partial u / \partial x$ back to spectral space, providing a value for each predicted wavenumber.
- Add this number to the tendency equations for the $u$ coefficients for each wavenumber, along with the contributions from other terms.
- Predict new values for the $u$ coefficients for each wavenumber.

No finite-difference approximations to the derivatives are required in this procedure. A number of particular spectral-transform methods exist. For example, Swarztrauber (1996) compared the accuracy of nine methods for solving the shallow-water equations.

For alias-free solutions to the nonlinear terms, with both triangular and rhomboidal truncations, the number of grid points in the zonal direction on the transform grid must be

$3N + 1$. In the meridional direction, $(3N + 1)/2$ points are needed for triangular truncation and $5N/2$ are needed for rhomboidal truncation. The points are equally spaced in the zonal direction, but are not in the meridional direction. Legendre polynomial solutions on this Gaussian grid are exact. However, the use of simple latitude–longitude transform grids for the above purpose has the previously described pole problem (a Gaussian grid would look similar to the one in Fig. 3.8). Constructing reduced Gaussian grids, of the same type shown schematically in Fig. 3.9 for a pure latitude–longitude grid, and described in Williamson (2007), is one way of addressing this. Figure 3.18 shows an example of a reduced Gaussian grid. Hortal and Simmons (1991) showed that the use of a reduced grid in

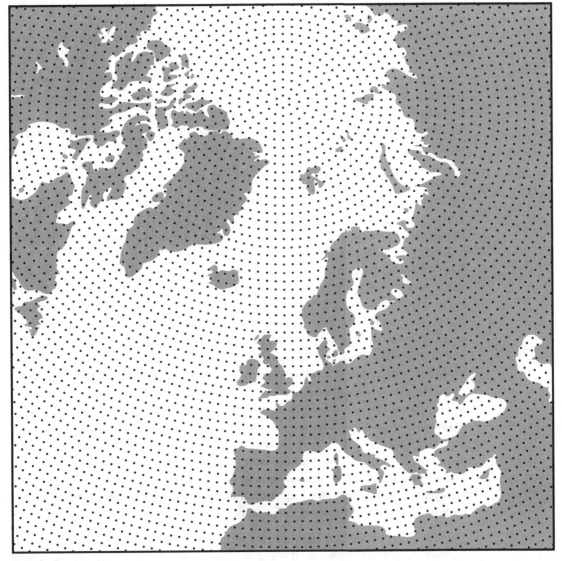

**Fig. 3.18**    An example of a reduced Gaussian T106 grid. From Hortal and Simmons (1991).

a spectral model produced no significant loss of accuracy relative to the use of a full grid. However, reducing the number of grid points leads to error in the calculation of nonlinear terms, and therefore some aliasing. One constraint for spectral applications of reduced grids is that the number of grid points on latitude circles must be consistent with the requirements of fast Fourier transform algorithms. The use of a relatively uniform grid increment across the sphere in reduced-grid, spectral-transform models is consistent with the fact that triangular truncation of the spherical harmonics produces a uniform resolution in spectral space.

Vertical derivatives in spectral models are typically approximated with standard finite differencing or with finite-element methods. Legendre polynomials and Laguerre polynomials have been used for vertical basis functions, but they both have significant disadvantages. Applications of the finite element approach are discussed in Béland and Beaudoin (1985), Steppeler (1987), and Hartmann (1988).

The following is a typical process by which forecasts are produced with global spectral models.

- Based on the desired resolution in physical space, a spectral truncation is chosen (e.g., triangular with $K = 42$ would be referred to as T42). The numbers of grid points in both the latitudinal and longitudinal directions are chosen, possibly based on a desire to avoid aliasing. For the chosen spectral truncation, the highest-degree Legendre polynomial is identified, and the latitudes where the roots of the polynomial occur are determined. These are the Gaussian latitudes, and are the basis for the Gaussian grid used in the transform.
- Observations of dependent variables are objectively analyzed, in physical space, to a grid.
- Gridded data representing model initial conditions are forward transformed to define expansion coefficients.
- Each time step, dependent variables are inverse transformed from spectral to physical space, and tendency contributions associated with local processes are calculated on the grid for the momentum, thermodynamic, and moisture equations. Such processes that are inherently local, with strong gradients or discontinuities that cannot be represented in spectral space include surface heat, moisture, and momentum fluxes; radiative flux divergence; latent-heat gains or losses; cloud-microphysical and convective processes; etc. Similarly, nonlinear terms are calculated on the grid according to the transform method described above. Terms with vertical derivatives may also be calculated in physical space. Tendency contributions are then transformed back to spectral space.
- Tendency contributions for terms calculated in spectral and physical space are added, and a time extrapolation is performed using standard time-differencing methods.
- At a desired frequency, dependent variables are reverse transformed to physical space and graphically displayed on a map projection to provide information on the state of the forecast and for comparison with observations.

Because modelers tend to think in terms of grid increments rather than wavenumbers for defining horizontal resolution, it is useful to have a simple conversion between global spectral resolution and the equivalent grid spacing. A few alternatives, based on different interpretations of the meaning of spectral resolution, were given by Laprise (1992).[2] The formulae are

$$L_1 = \frac{2\pi a}{3K + 1},$$

$$L_2 = \frac{\pi a}{K},$$

$$L_3 = \frac{2\sqrt{\pi}a}{K + 1},$$

$$L_4 = \frac{\sqrt{2}\pi a}{K},$$

where $K$ is as defined above, and $a$ is the radius of Earth. The spectral resolution is expressed in terms of $K$ and the type of truncation. If $K$ is 799, and the truncation is triangular, the resolution is given as T799. For a spectral resolution of T799, the estimates of the equivalent grid increments are $L_1 = 16.7$ km, $L_2 = 25.0$ km, $L_3 = 28.2$ km, and $L_4 = 35.4$ km. Some advantages and disadvantages of spectral models are summarized below.

Advantages

- There is generally no aliasing of quadratic nonlinear terms, and thus no nonlinear instability.
- There is no spatial truncation error because the derivatives are treated analytically, and thus there is no numerical dispersion of waves.
- Semi-implicit time-differencing schemes are easily implemented.
- There is almost no grid (computational) diffusion.

Disadvantages

- Local-forcing processes (e.g., latent-heat release, differential surface heat fluxes), some of which are discontinuous, are only possible to represent in physical space.
- When a linear combination of waves (e.g., spherical harmonics) is used to represent a large gradient or discontinuity, spurious waves can result (Gibbs phenomenon). In the case of specific humidity, this "spectral ringing" can result in negative values, which are physically impossible. And overshooting the correct solution can lead to spurious precipitation, called *spectral rain*.
- For higher resolutions, spectral models are computationally more demanding than grid-point models.
- Spectral models do not exactly conserve mass or energy.

---

[2] It should not be surprising that there is lack of agreement on the meaning of "horizontal resolution" in spectral space, given that there is even disagreement about the meaning of the term in the much more intuitive physical (grid-point) space. In addition to the fact that the scales represented by a grid-point model depend on the numerical smoothing and other factors, Pielke (1991) points out that modelers often erroneously use the term resolution to refer to the grid increment.

Lastly, a few regional, that is limited-area, spectral models have been developed and employed for research and operational prediction. One of the most widely used is the US National Centers for Environmental Prediction (NCEP) Regional Spectral Model (RSM; Juang and Kanamitsu 1994, Juang *et al.* 1997, Juang 2000, Roads 2000, Juang and Hong 2001) that has been used for operational weather prediction. The high-resolution RSM is typically used in a nest with a low-resolution global spectral model, with the two spectral models having the same vertical structure and physical processes. The regional model uses a double Fourier series as the basis function, and is defined on a map projection. Multiple spectral nests are possible within the global model. The Scripps Experimental Climate Prediction Center has used the RSM, coupled with the NCEP Global Spectral Model, for seasonal predictions (Roads 2004), and Han and Roads (2004) describe its use for 10-year climate simulations. There are numerous other applications of regional spectral models. The Florida State University nested regional spectral model has been used for weather and seasonal-climate simulations (Cocke 1998, Cocke *et al.* 2007), and the Japan Meteorological Agency has used such a model for operational prediction (Tatsumi 1986). Boyd (2005) provides a table of spectral limited-area models, and Krishnamurti *et al.* (2006a) provide a summary of the modeling process.

### 3.2.3  Finite-element methods

*Finite-element methods* were first developed for engineering applications, and have been since adopted for use in some models of ocean and atmospheric processes. They are analogous to spectral modeling methods, both being special cases of the Galerkin procedure in which the dependent variables are approximated by a finite sum of spatially varying basis functions with time-dependent coefficients. For spectral modeling, global (i.e., nonlocal) basis functions are employed, where for finite-element modeling the basis functions are low-order polynomials that are zero except in a localized region. In finite-element modeling, the computational domain is divided into a number of contiguous finite subregions called elements. On each element is defined a simple function, where continuity between functions on adjacent elements is required.

The finite-element method has been used in the operational Canadian Regional Finite Element (RFE) model (Staniforth and Mailhot 1988, Benoit *et al.* 1989, Tanguay *et al.* 1989, Belair *et al.* 1994), in the ECMWF model (Burridge *et al.* 1986), and elsewhere (Staniforth and Daley 1979). Finite-element representations are sometimes used only in the vertical, where finite-difference or spectral methods are employed for the horizontal (Staniforth and Daley 1977, Beland *et al.* 1983, Beland and Beaudoin 1985, Burridge *et al.* 1986, Steppler 1987). Good summaries of the application of finite-element methods in atmospheric models can be found in Cullen (1979), Staniforth (1984), and Hartmann (1988).

### 3.2.4  Finite-volume methods

In contrast to grid-point models where the prognostic quantity is the value of dependent variables at grid points, with *finite-volume models* it is the integrated value of a variable over a specific finite control volume. The control volumes are typically the traditional three-dimensional model grid cells, which leads to the fact that finite-volume methods are also referred to as cell-integrated methods. They are especially well suited for applications where it is important to conserve quantities such as mass, total energy, angular momentum, or entropy. Indeed, one of the reasons for the renewed interest in this approach is the significant lack of global mass conservation in models that use the semi-Lagrangian technique. There are two approaches in the finite-volume framework. One is the Departure Cell-Integrated Semi-Lagrangian (DCISL) scheme, and the other is the Flux-Form Semi-Lagrangian (FFSL) scheme. The DCISL and the FFSL differ in terms of how they estimate a property, for example the mass, of the cell at the trajectory's departure point in the semi-Lagrangian transport calculation. If mass conservation is the primary concern, the finite-volume method can be applied to the continuity equation, while conventional semi-Lagrangian grid-point methods are used for the other equations. Two recent examples of dynamical cores that use the finite-volume method are the European HIgh Resolution Limited Area Model (HIRLAM) and the NCAR global Community Atmospheric Model (CAM 3.0, Collins *et al.* 2006b). The HIRLAM employs the DCISL method and the CAM uses a flux-based scheme. The FIM model being developed by NOAA, mentioned earlier, also uses the finite-volume method. See Machenhauer *et al.* (2008) for a thorough summary of the use of finite-volume methods in atmospheric modeling.

## 3.3  Finite-difference methods

### 3.3.1  Time-differencing methods

Time-differencing methods can be explicit or implicit, or a combination of both. With *explicit methods*, the prognostic equation can be solved for the value of the dependent variable at the new (most-forward) time, with the new value on the left side of the equation, and the right side consisting of dependent variables defined at current or prior times. With *implicit methods*, dependent variables at the new time level appear on both sides of the equations, and the solution must be obtained iteratively. Semi-implicit techniques solve some terms in the equations explicitly, and some implicitly.

Unless the anelastic or Boussinesq equations described in Chapter 2 are used for a model, the solution contains acoustic waves and external gravity waves that both move with speeds at or near Mach number 1.[3] These meteorologically inconsequential waves can require the use of a small time step because of Courant-number constraints, and thus make the model computations inefficient. The next sections will explain how both explicit and implicit time-differencing methods deal with this problem. With *split-explicit* differencing, only the terms in the equations associated with the acoustic and external-gravity modes are computed with a short time step. The terms related to the meteorological processes use a longer time step, which is consistent with the relatively slow speed of those waves. With

---

[3]  The Mach number is the ratio of the phase speed of a wave to that of a sound wave.

semi-implicit differencing, implicit methods, whose time step is not constrained by the Courant number, are used for the fast acoustic and gravity modes, and explicit time differencing is used for the other terms associated with the slower meteorological waves.

## Explicit time differencing

There are two general types of explicit time-differencing approaches. One employs a single computational step, such as the forward-in-time or centered-in-time methods described earlier, to arrive at the new values of the dependent variables at the next time level. Another single-step scheme is the Adams–Bashforth method evaluated in Durran (1991). The other type uses multiple steps. When there are two steps, they are called *predictor–corrector schemes*. The first step is the predictor step and the second is the corrector step. Even though the multi-step methods involve a greater number of arithmetic operations, and therefore have greater computational expense, their numerical properties are superior in some respects to those of the single-step methods. In the following equations, $F$ represents the finite-difference approximation to all the terms on the right side of any of the model prognostic equations (the forcing), $\theta$ is any dependent variable, and $\theta^*$ is an intermediate solution. Equation 3.1 above shows an example of the centered-in-time, centered-in-space, single-step explicit scheme. Equation 3.13 is a representation for any such equation:

$$\theta_j^{\tau+1} = \theta_j^{\tau-1} + 2\Delta t F_j^{\tau}. \tag{3.13}$$

There are many multi-step schemes, and some have numerous variations with different accuracies. For example, one approach is the Lax–Wendroff scheme (Lax and Wendroff 1960):

$$\theta^*_j = \frac{1}{2}(\theta_{j+1}^{\tau} + \theta_{j-1}^{\tau}) + \frac{\Delta t}{2} F_j^{\tau} \qquad \text{(predictor step),} \tag{3.14}$$

$$\theta_j^{\tau+1} = \theta_j^{\tau} + \Delta t F^*_j \qquad \text{(corrector step).} \tag{3.15}$$

In the first step, a forward extrapolation is made from time $\tau$ for one-half the time step. In the second, these forecast values are used to calculate the tendency for the extrapolation from $\tau$ to $\tau+1$, which is a centered time step. Another two-step scheme is the Euler-backward (or Matsuno 1966) method. Here, we have

$$\theta^*_j = \theta_j^{\tau} + \Delta t F_j^{\tau}, \tag{3.16}$$

$$\theta_j^{\tau+1} = \theta_j^{\tau} + \Delta t F^*_j. \tag{3.17}$$

Another commonly used method, with many variations, is the Runge–Kutta scheme. One, described by Wicker and Skamarock (2002), is

$$\theta^*_j = \theta_j^{\tau} + \frac{\Delta t}{3} F_j^{\tau}, \tag{3.18}$$

$$\theta^{\dagger}_j = \theta_j^{\tau} + \frac{\Delta t}{2} F^*_j, \tag{3.19}$$

$$\theta_j^{\tau+1} = \theta_j^{\tau} + \Delta t F_j^{\dagger}. \tag{3.20}$$

This scheme is used in the community Advanced Research version of the Weather Research and Forecasting (WRF) model (Skamarock *et al.* 2008).

Another type of explicit time differencing is the so-called split-explicit method. The motivation for this approach, which is also called *time splitting*, is based on the fact that compressible, nonhydrostatic equations support sound waves (Mach number 1) and fast-moving external-gravity waves, as well as of course the meteorological waves (e.g., advective waves, Rossby waves, and internal-gravity waves that effect geostrophic adjustment) whose Mach numbers rarely exceed 0.3 even in the fastest jet-stream winds. Because the sound waves and external-gravity waves have small amplitudes and are not meteorologically significant, it is wasteful of computing resources to use the very-small time step that is needed to satisfy the Courant condition for these waves. There are a few methods to deal with this problem. One is to use split-explicit methods that integrate different terms in the equations using different time steps. The few terms associated with sound and external-gravity wave propagation are integrated with a small time step, and the rest of the terms that represent meteorological processes are integrated with a larger time step. All explicit methods have linear-stability criteria that are constrained by the Courant number. A number of nonhydrostatic models use this approach, including the WRF (Skamarock and Klemp 2008), the Mesoscale Model Version 5 (MM5, Dudhia 1993), the Lokal Modell (LM, Doms and Schättler 1997), the Coupled Ocean–Atmosphere Mesoscale Prediction System (COAMPS, Hodur 1997), and the Advanced Regional Prediction System (ARPS, Xue *et al.* 2000). Additional discussion of split-explicit time differencing can be found in Marchuk (1974), Klemp and Wilhelmson (1978), Wicker and Skamarock (1998), Klemp *et al.* (2007), Purser (2007), and Skamarock and Klemp (1992, 2008).

### Implicit and semi-implicit time differencing

An example of an explicit treatment of a one-dimensional linear advection equation would be

$$u_i^{\tau+1} = u_i^{\tau-1} - \frac{U\Delta t}{\Delta x}(u_{i+1}^{\tau} - u_{i-1}^{\tau}),$$

where there are no variables defined at $\tau+1$ on the right side of the equation. In contrast, the following form of the linear advection equation is implicit because $\tau+1$ values of dependent variables appear on the right:

$$u_i^{\tau+1} = u_i^{\tau} - \frac{U\Delta t}{2}\left(\frac{u_{i+1}^{\tau+1} - u_{i-1}^{\tau+1}}{2\Delta x} + \frac{u_{i+1}^{\tau} - u_{i-1}^{\tau}}{2\Delta x}\right). \tag{3.21}$$

The approximation to the space derivative is represented as a time average of the derivative evaluated at the forward time and at the current time, so that it applies at $\tau+1/2$. Such schemes applied to the full set of equations are typically unconditionally stable, and can use long time steps that are unconstrained by the Courant number. Because implicit equations need to be solved iteratively (e.g., for $u_i^{\tau+1}$), they require much more computation per time step than do explicit equations. Specifically, three-dimensional Helmholtz equations must be solved each time step. Unfortunately, for fully three-dimensional problems, with six or seven variables and complex equations, the computational savings from the longer time step are typically more than offset by the greater computational cost per time step. Motivated by this problem, Marchuk (1965) and many others since then have developed *semi-implicit* schemes, which do provide computational advantages compared with explicit methods. With semi-implicit approaches, some terms are treated explicitly and some are treated implicitly. That is, in the finite-difference equation, implicit terms use averaging operators similar to the one above, while explicit terms have the conventional formulation. Those terms treated implicitly are typically the ones associated with processes that motivate the use of the implicit method in general; that is, those terms associated with fast-moving acoustic and external-gravity waves that would normally demand the use of a short time step. The rest of the terms, which are related to the slower meteorological processes, are treated explicitly. The time step that is stable for the explicit terms is also stable for the implicit terms because they are stable for any time step. Robert (1979) analyzed the following semi-implicit form of the shallow water equations:

$$\frac{\partial u}{\partial t} = -u\frac{\partial u}{\partial x} - v\frac{\partial u}{\partial y} - \overline{\frac{\partial \phi}{\partial x}}^{t} \,,$$

$$\frac{\partial v}{\partial t} = -u\frac{\partial v}{\partial x} - v\frac{\partial v}{\partial y} - \overline{\frac{\partial \phi}{\partial y}}^{t} \,,$$

$$\frac{\partial \phi}{\partial t} = -u\frac{\partial \phi}{\partial x} - v\frac{\partial \phi}{\partial y} - (\phi - \phi_0)\left(\frac{\partial u}{\partial x} + \frac{\partial v}{\partial y}\right) - \overline{\phi_0\left(\frac{\partial u}{\partial x} + \frac{\partial v}{\partial y}\right)}^{t} \,.$$

Here, $\phi_0$ is a mean geopotential height and the overbarred height gradient and divergence terms are the implicit time-mean expression of Eq. 3.21. The rest of the terms are treated explicitly. Robert (1979) linearized this set of equations and showed that the gravity waves have no time-step restriction and that the advection has the standard CFL time-step restriction. The shallow-fluid equations are incompressible and have no acoustic mode.

### 3.3.2  Space-differencing methods

#### Eulerian space differencing

*Eulerian models* calculate the transport (advection) terms in the equations at points that have fixed horizontal and vertical coordinates. This is the approach described thus far in this chapter, regardless of whether the models are entirely grid-point based or use the spectral-transform method. Equations such as the following, for any independent variable $\alpha$, are solved for the values of the time derivatives at specific points, where the advection terms and the other forcing terms ($F$) apply at the same locations:

$$\left.\frac{\partial \alpha}{\partial t}\right|_{x,y,z} = -\vec{V} \cdot \nabla\alpha + F(x, y, z, t) \,.$$

In all cases, the advection (and other) terms require adherence to a stability criterion that is related to the time step, regardless of whether such a short time step is needed for accuracy (truncation-error control). Because of this inherent stability limitation of Eulerian methods, and the associated computational liability, the semi-Lagrangian approach described below has become widely used.

## Lagrangian and semi-Lagrangian space differencing

With purely *Lagrangian methods*, changes in the properties of individual moving parcels of air are calculated. That is, our reference frame is air parcels and not grid points. In this case, the following equation would apply, where the material derivative, or the derivative following the parcel, is on the left side:

$$\frac{d\alpha}{dt} = F(x, y, z, t). \tag{3.22}$$

For a perfectly conserved quantity,

$$\frac{d\alpha}{dt} = 0. \tag{3.23}$$

That is, the value of $\alpha$ does not change following the parcel. Integrating Eq. 3.23 in time, as part of a larger set of equations, would simply involve estimating how the conserved $\alpha$ field is redistributed by the wind. Such a Lagrangian forecast system could be initialized by beginning with regularly spaced parcels, and assigning values of $\alpha$ to parcels based on standard initialization techniques. However, after a short forecast period the parcels would become very unevenly distributed, providing unacceptable contrasts in spatial resolution. This problem motivated the development of *semi-Lagrangian* space differencing, where a completely new set of regularly spaced parcels is chosen each time step. One approach is, at each time step, to initially define the parcels at grid points, and move each parcel in space for one time step based on the prevailing velocity field. The new parcel positions will, of course, not be at grid points, but the parcels will have retained their original value of $\alpha$ so a new spatial distribution of $\alpha$ will be defined based on the irregularly spaced end-points of the trajectories. The values of $\alpha$ at trajectory end-points are then spatially interpolated to the original grid points, defining new regularly spaced parcels from which to begin the next time step. An alternative approach is to also begin with parcels at grid points, but calculate one-time-step back-trajectories using the same wind field as in the above process. Then the $\alpha$ at the back-trajectory's end-point is defined by spatial interpolation from the current grid-point values, and this value is assigned to the grid point (i.e., the value of $\alpha$ is the same at both ends of the trajectory). The latter approach is more common because it is more straightforward to interpolate from a regular grid to irregularly located points, than the opposite.

For the typical situation where there are forcing terms ($F$), and $\alpha$ is not conserved for each parcel (Eq. 3.22), a schematic one-dimensional finite-difference equation based on a trapezoidal integration approach would be the following, where $x_j = j\Delta x$ refers to the position at grid points, $t^n = n\Delta t$, and $\tilde{x}^n$ is the estimate of the starting position of the trajectory at time $t^n$ (notation based on Durran 1999):

$$\frac{\alpha(x_j, t^{n+1}) - \alpha(\tilde{x}_j^n, t^n)}{\Delta t} = \frac{1}{2}[F[x_j, t^{n+1}] + F(\tilde{x}^n, t^n)].$$

Here, the forcing is defined as an average of the values applicable at the beginning and end of the trajectory. If this is an equation for a chemical species, $\alpha$, the forcing term would represent sources or sinks of that material. If $\alpha$ is a meteorological variable, the forcing would represent all the standard terms on the right side of the equation, other than advection of course. These terms would be calculated using standard Eulerian differencing methods on the grid. The value of $F$ at grid point $x_j$ would be calculated directly, and the value of $F$ at $\tilde{x}^n$ would be defined by interpolation from adjacent grid points. Unfortunately, this form of the equation is implicit (variables defined at $t^{n+1}$ are on both sides of the equation) and would require additional computational work in order to solve it. Alternatively, the following explicit, centered-in-time approach could be solved explicitly. The time step is only limited by the constraints that trajectories cannot cross, and trajectory end-points need to be within the grid:

$$\frac{\alpha(x_j, t^{n+1}) - \alpha(\tilde{x}_j^{n-1}, t^{n-1})}{2\Delta t} = F[\tilde{x}_j^n, t^n].$$

As a result of the pioneering work of Robert (1981, 1982) and others, the semi-Lagrangian method has become an extremely popular approach for global and limited-area modeling, where some of the reasons are:

- it is often more efficient than competing Eulerian schemes because the CFL condition associated with advection terms is avoided,
- it can be used with both grid-point and spectral-transform methods,
- it may be combined with semi-implicit methods that are applied to the pressure-gradient and velocity-divergence terms, and
- the primary source of nonlinear instability, the nonlinear advection terms, does not exist.

Conversely, there are criticisms of semi-Lagrangian methods in that they generally fail to exactly conserve energy and mass. Summaries of semi-Lagrangian methods can be found in Staniforth and Côté (1991), Durran (1999), and Williamson (2007).

## Grid staggering methods

Grid staggering involves defining different dependent variables on different grids. Typically, the points in one mesh are offset from those in the other by $0.5\Delta x$. Figure 3.19 shows a one-dimensional schematic of an approach to staggering. For the unstaggered grid shown in Fig. 3.19a, calculation of an advection term such as $u\partial\theta/\partial x$ with a centered, three-point method would require differencing across a $2\Delta x$ interval. For the staggered grid in Fig. 3.19b, the derivative can be calculated by differencing across a $1\Delta x$ interval. This halves the effective grid increment for such terms, increasing the spatial resolution and decreasing the effects of truncation error on the solution. Also, a benefit of

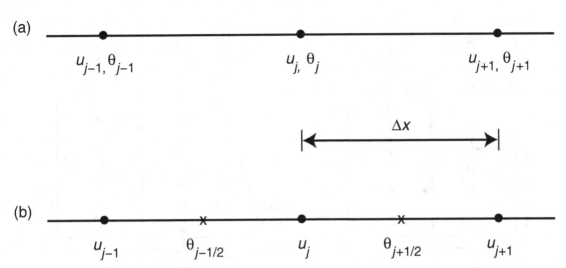

Fig. 3.19 Schematic of one-dimensional unstaggered (a) and staggered (b) grids. For the staggered grid, the mass-field variable ($\theta$) is offset by one-half grid increment from the momentum variable ($u$). Adapted from Durran (1999).

staggering the horizontal and vertical velocity in hydrostatic models, described by Pielke (2002a), is that, when the vertical velocity is diagnosed from the horizontal divergence by integrating the continuity equation, lateral boundary values for the horizontal velocity have no direct impact on the vertical velocity. This is illustrated in Fig. 3.20. For

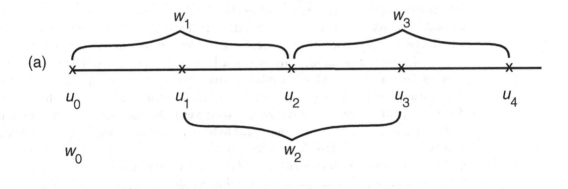

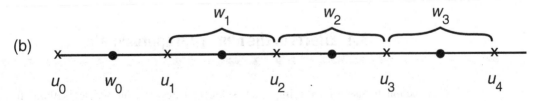

Fig. 3.20 Schematic showing horizontal and vertical velocities on a one-dimensional unstaggered grid (a) and on one type of staggered grid (b). The subscript shows the position of the grid points relative to the left boundary. Adapted from Pielke (2002a).

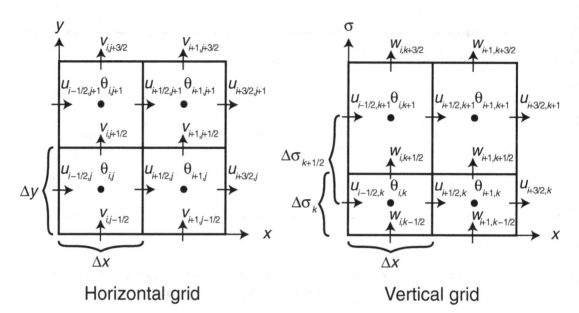

**Horizontal grid**                         **Vertical grid**

Fig. 3.21   The Arakawa-C grid-staggering method. Adapted from Skamarock *et al.* (2008).

unstaggered grids, where $u$ and $w$ are defined at the same points, Fig. 3.20a shows that the vertical velocity at the first interior point ($w_1$) is calculated using a lateral boundary value of $u(u_0)$. In contrast, for the staggered grid the vertical velocity at the first interior point is calculated using only nonboundary values (Fig. 3.20b). Figure 3.21 shows an example of a standard approach to horizontal and vertical grid staggering for a three-dimensional model. The wind components ($u, v$) are defined at the locations of the vectors in the figure, and the mass variables ($q, p, \theta$) are defined at the locations defined by the dots. This is called the Arakawa-C grid, where other alternative staggering methods are described in Arakawa and Lamb (1977) and Haltiner and Williams (1980). Unfortunately, the smaller distance over which the derivatives are calculated means that the effective grid increment is smaller, and therefore the time steps need to be smaller in order for the CFL condition to be satisfied. Nevertheless, the increase in effective resolution, relative to an unstaggered grid, is gained without the use of additional grid points, which would require more computer storage and more computations. Virtually all contemporary grid-point models use staggered meshes.

## 3.4  Effects of the numerical approximations

This section focusses on the important subject of how the numerical methods that are employed to integrate the equations can affect the model solution in various ways. The discussion of truncation error shows how the derivatives in the equations are incorrectly estimated by finite-difference approximations. Then is described how each term in the equations possesses criteria, based on model parameters and atmospheric conditions, that must be met in order for the model solution to be stable. The effects of the numerical methods on the phase and group speeds of meteorological waves in the model solution are described, illustrating that for some differencing schemes wave energy can even move in the wrong direction. The nonlinear interaction of waves on a grid can lead to the erroneous accumulation, through aliasing, of wave energy in small wavelengths, and this can lead to the problem of nonlinear instability. And, the concept of horizontal diffusion (the spread, and smoothing, of properties on the grid) is discussed because it is a process that can remove correct small-scale information in the model solution, and it can be used to control numerical problems in the model solution. Lastly, the strengths and weaknesses are summarized of the various vertical coordinates used in models.

### 3.4.1  Truncation error

Because the equations that govern atmospheric processes are differential equations, with derivatives in most of the terms, approximating the continuous space and time derivatives with finite-difference expressions represents a considerable potential source of error in the modeling process. It is straightforward to quantify this error by using Taylor's theorem, which defines a polynomial that approximates any function over an interval. A remainder term in the polynomial represents the difference between the values of the function and the approximation. The following polynomial is called *Taylor's series*, where $f$ is any meteorological variable in the derivative terms of Eqs. 2.1–2.6, and the series can be written for any independent variable:

$$f(x) = f(a) + (x - a)\frac{\partial f(a)}{\partial x} + \frac{(x-a)^2}{2!}\frac{\partial^2 f(a)}{\partial x^2} + \frac{(x-a)^3}{3!}\frac{\partial^3 f(a)}{\partial x^3} + \cdots$$

$$\cdots + \frac{(x-a)^n}{n!}\frac{\partial^n f(a)}{\partial x^n} + R(n, x). \tag{3.24}$$

It states that the value of a function, $f$, at any point, $x$, can be approximated by using the known value and derivatives at point $a$. In the case of an infinite series, the expression would be exact. For a series truncated at $n$ terms, there is a residual, $R$, that defines the error. The truncation error will be defined here for three finite-difference approximations to the derivative: two-point, three-point, and five-point formulae.

For a two-point approximation, let $x = a + \Delta x$ in Eq. 3.24, truncate the series by dropping second-order and higher terms, and solve for the derivative, to obtain the following:

$$\frac{\partial f(a)}{\partial x} = \frac{f(a + \Delta x) - f(a)}{\Delta x}. \tag{3.25}$$

This is called the forward-in-space differencing formula, which has first-order accuracy because second-order and higher terms in Taylor's series were dropped. An analogous backward-in-space formula results from letting $x = a - \Delta x$.

A three-point differencing scheme can be obtained by writing Taylor's series as

$$f(a + \Delta x) = f(a) + \Delta x \frac{\partial f(a)}{\partial x} + \frac{(\Delta x)^2}{2!} \frac{\partial^2 f(a)}{\partial x^2} + \frac{(\Delta x)^3}{3!} \frac{\partial^3 f(a)}{\partial x^3} + \cdots \qquad (3.26)$$

and

$$f(a - \Delta x) = f(a) - \Delta x \frac{\partial f(a)}{\partial x} + \frac{(\Delta x)^2}{2!} \frac{\partial^2 f(a)}{\partial x^2} - \frac{(\Delta x)^3}{3!} \frac{\partial^3 f(a)}{\partial x^3} + \cdots . \qquad (3.27)$$

Subtracting the two series produces

$$f(a + \Delta x) - f(a - \Delta x) = 2\Delta x \frac{\partial f(a)}{\partial x} + \frac{2(\Delta x)^3}{3!} \frac{\partial^3 f(a)}{\partial x^3} + \cdots .$$

Solving for $\partial f(a)/(\partial x)$ provides

$$\frac{\partial f(a)}{\partial x} = \frac{f(a + \Delta x) - f(a - \Delta x)}{2\Delta x} - \frac{(\Delta x)^2}{3!} \frac{\partial^3 f(a)}{\partial x^3} + \cdots . \qquad (3.28)$$

Truncating this equation after the first term on the right produces the following approximation, which we say has second-order accuracy because we ignore the third-order and higher terms in the series. This is called a three-point approximation to the derivative because it spans points $x - \Delta x$, $x$, and $x + \Delta x$ :

$$\frac{\partial f(a)}{\partial x} = \frac{f(a + \Delta x) - f(a - \Delta x)}{2\Delta x} .$$

One way of calculating the effect of truncating the series on the accuracy of the derivative is to compare the value of the derivative from this approximation with the exact value. Let $f = A \cos kx$, where $k = 2\pi/L$ and $L$ is the wavelength. The exact value of the derivative is

$$\frac{\partial f}{\partial x} = -kA \sin kx, \qquad (3.29)$$

where the approximation is

$$\frac{\Delta f}{\Delta x} = \frac{A \cos k(x + \Delta x) - A \cos k(x - \Delta x)}{2\Delta x} . \qquad (3.30)$$

Using trigonometric identities it can be shown that

$$\frac{\dfrac{\Delta f}{\Delta x}}{\dfrac{\partial f}{\partial x}} = \frac{\sin k\Delta x}{k\Delta x}, \qquad (3.31)$$

where, as $\Delta x / L \to 0$, $k\Delta x \to 0$ and $\sin k\Delta x \to k\Delta x$. That is, as the argument of the sine function approaches zero, so does the function itself, and the ratio approaches unity. This ratio defines the truncation error because it represents the error in the finite-difference approximation to the derivative that is associated with the truncation made in Taylor's series. Thus, for a wave of length $L$ that is defined by many grid points, the ratio of the approximation to the derivative and the exact derivative is near unity. Figure 3.22 shows this ratio for different wavelengths. For a given grid increment, the derivatives of long waves are clearly better represented than the derivatives of shorter waves. For example,

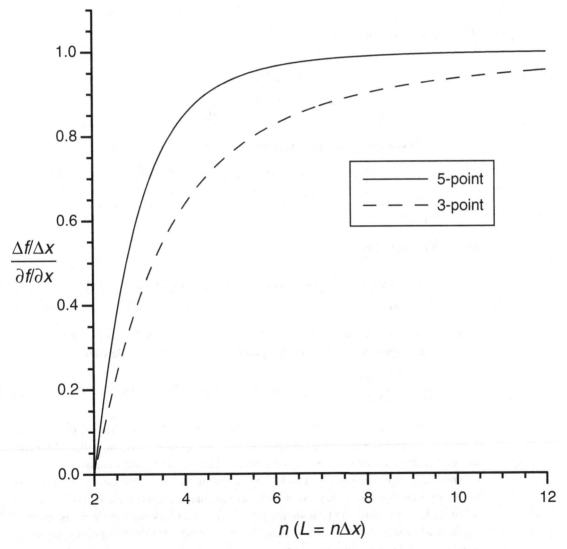

**Fig. 3.22** The ratio of the value of the numerical approximation to the derivative of the cosine function and the value of the true derivative, for different numbers of grid increments per wavelength (how well the wave is resolved), for the five-point (fourth-order) and three-point (second-order) approximations.

calculating the derivative of a wave of length $6\Delta x$ with the three-point approximation underestimates the derivative by 17%. For a $10\Delta x$ wave the error is only 6%. This leads to the common qualitative statement that at least ten grid increments are needed in order to properly represent a wave.

Taylor's series can also be used to estimate the truncation error for a five-point approximations to the derivative. For example, add Eqs. 3.26 and 3.27 and solve for the second derivative to obtain

$$\frac{\partial^2 f(a)}{\partial x^2} = \frac{f(a+\Delta x)+f(a-\Delta x)-2f(a)}{(\Delta x)^2} + \cdots$$

A third derivative can then be defined as

$$\frac{\partial^3 f(a)}{\partial x^3} = \frac{\dfrac{\partial^2 f(a+\Delta x)}{\partial x^2} - \dfrac{\partial^2 f(a-\Delta x)}{\partial x^2}}{2\Delta x}.$$

Using the above expression for the second derivative,

$$\frac{\partial^3 f(a)}{\partial x^3} = \frac{\dfrac{f(a+2\Delta x)+f(a)-2f(a+\Delta x)}{(\Delta x)^2} - \dfrac{f(a)+f(a-2\Delta x)-2f(a-\Delta x)}{(\Delta x)^2}}{2\Delta x},$$

and simplifying yields

$$\frac{\partial^3 f(a)}{\partial x^3} = \frac{f(a+2\Delta x)-f(a-2\Delta x)-2f(a+\Delta x)+2f(a-\Delta x)}{2(\Delta x)^3}.$$

Substituting this expression for the third derivative into Eq. 3.28 and simplifying produces the following expression, which is a five-point approximation to the derivative:

$$\frac{\partial f(a)}{\partial x} = \frac{1}{2\Delta x}\left[\frac{4}{3}(f(a+\Delta x)-f(a-\Delta x)) - \frac{1}{6}(f(a+2\Delta x)-f(a-2\Delta x))\right]. \quad (3.32)$$

The truncation error is calculated as before (Eqs. 3.29–3.31) and is shown in Fig. 3.22. This is a fourth-order accurate approximation to the derivative because the fifth-order and higher terms in the series were truncated, and has smaller error than the three-point/second-order approximation, as can be seen in the figure. It is interesting that Eq. 3.31 for the truncation error of the second-order approximation, and the analogous one for the fourth-order approximation (not shown), only depend on how well the wave is resolved on the grid ($n$ in $L = n\Delta x$) and not on $x$ itself. However, for the two-point approximation in Eq. 3.25, the truncation error also depends on $x$ (not shown). That is, the ratio $\Delta f/\Delta x \div \partial f/\partial x$ depends on position within the cosine wave. For a wave of length $8\Delta x$, this ratio is unity at $x = L/4$, but becomes very large in magnitude as $x \to 0, \pi$.

In summary, in a grid-point model (i.e., a nonspectral model), every time a derivative is calculated with a finite-difference approximation – which is in virtually every equation, at every grid point, at every time step – the derivative is imperfectly estimated, where the magnitude of the error depends on the sophistication of the differencing scheme and how well the wave is resolved by the grid. Such errors in the pressure-gradient terms in the momentum equations lead to errors in the geostrophic wind, errors in the divergence terms of the continuity equation result in incorrect vertical motions, errors in the gradients in advective terms lead to incorrect advective changes, etc. Also, the pressure gradient in equations expressed in terrain-following, sigma vertical-coordinate systems consists of two derivative terms (see Section 3.4.8), where each term is large and the small difference between them defines the pressure-gradient force. Over large terrain gradients, these terms become especially large, and truncation errors that do not cancel in the two terms create erroneous pressure gradients and accelerations. This problem is partly mitigated by using perturbation forms of the equations, where the derivatives apply to departures from a mean state.

### 3.4.2  Linear stability, and damping properties

The term *stability* in the context of atmospheric modeling is related to whether the amplitudes of waves in the numerical solution to Eqs. 2.1–2.7, or some other equation set that is the basis for a model, grow exponentially for numerical (i.e., nonphysical) reasons, quickly causing floating-point-overflow conditions that halt the integration of the equations. Many modelers will normally not need to worry about this problem because model codes often contain limits on the time step, and other parameters, that attempt to prevent the instability. Nevertheless, it is useful to know why these constraints exist, and what to expect if the linear-stability criteria are accidentally violated.

Different finite-difference approximations to the time and space derivatives in the equations of motion have different criteria for maintaining stable solutions. Some approximations are absolutely stable – that is, they are never unstable. Some are always unstable – they are called absolutely unstable – and cannot be used. Most are conditionally stable, meaning that stable solutions are obtainable for certain ranges of model parameters and meteorological conditions. Each term in Eqs. 2.1–2.7 contributes to the stability of the numerical solution of its respective equation, but the advection terms are often the most problematic. It is fortunate that the condition for stability of the linear advection equation is about the same as that of the nonlinear advection equation, allowing us to analytically calculate a useful stability criterion with the linear term. Because this kind of instability exists for linear advection, it is called a *linear instability*, contrasting it with the nonlinear instability problem that is described later. This section discusses the linear stability condition for both advection and diffusion terms.

### Linear stability of an advection term

The following linear equation will be used as the basis for our analysis of the stability of the advection equation. Assume that $h$ is a meteorological variable such as the height of a pressure surface or the depth of a shallow fluid, and that $U$ is a mean wind speed.

This notation indicates that the terms apply at grid point $j$ on the $x$ axis and at time step $\tau$:

$$\left.\frac{\partial h}{\partial t}\right|_j^\tau = -U\left.\frac{\partial h}{\partial x}\right|_j^\tau.$$

Assume harmonic solutions to this equation of the following form,

$$h = \hat{h}e^{i(kx - \omega t)}, \tag{3.33}$$

where $\hat{h}$ is the amplitude, $k = 2\pi/L$, $L$ is wavelength, $i = \sqrt{-1}$, and $\omega = Uk$ is wave frequency. Now assume that $\omega$ is complex, so that $\omega = \omega_R + i\omega_I$. The implication of this can be seen by substitution into Eq. 3.33, producing

$$h - \hat{h}e^{\omega_I t}e^{i(kx - \omega_R t)}. \tag{3.34}$$

The assumption of a complex frequency has allowed for a wave-amplitude variation with time, such that positive $\omega_I$ is associated with exponential wave growth as time ($t$) increases, negative $\omega_I$ is associated with wave damping, and $\omega_I = 0$ means that the amplitude remains constant at $\hat{h}$. The value of $\omega_I$ will determine which of these situations prevails, where wave growth is associated with an unstable model solution. The second exponential defines the phase of the wave in the $x$ direction.

For instructional purposes, we will first analyze the stability of the advection equation using forward differencing for the time derivative and backward differencing for the space derivative. The finite-difference expression is

$$\frac{h_j^{\tau+1} - h_j^\tau}{\Delta t} = -U\frac{h_j^\tau - h_{j-1}^\tau}{\Delta x}, \text{ or} \tag{3.35}$$

$$h_j^{\tau+1} - h_j^\tau = -\frac{U\Delta t}{\Delta x}(h_j^\tau - h_{j-1}^\tau), \tag{3.36}$$

where $\tau$ is the time-step number and $j$ is the grid-point number. Expressing the assumed form of the solution in Eq. 3.34 in finite-difference form by letting $x = j\Delta x$ and $t = \tau\Delta t$ produces

$$h_j^\tau = \hat{h}e^{\omega_I \tau\Delta t}e^{i(kj\Delta x - \omega_R \tau\Delta t)}. \tag{3.37}$$

Substitute this form of the solution into the finite-difference expression Eq. 3.36, producing

$$e^{\omega_I \Delta t}e^{-i\omega_R \Delta t} - 1 = -\frac{U\Delta t}{\Delta x}[1 - e^{ik\Delta x}]. \tag{3.38}$$

Using Euler's relations

$$e^{ix} = \cos x + i\sin x \text{ and} \tag{3.39}$$

$$e^{-ix} = \cos x - i\sin x \tag{3.40}$$

in Eq. 3.38 yields

$$e^{\omega_I \Delta t}(\cos\omega_R \Delta t - i\sin\omega_R \Delta t) = 1 + \frac{U\Delta t}{\Delta x}(\cos k\Delta x - 1 + i\sin k\Delta x).$$

In order to obtain information about whether the solution damps or amplifies, the complex equation is separated into its real and imaginary parts:

$$e^{\omega_I \Delta t} \cos \omega_R \Delta t = 1 + \frac{U \Delta t}{\Delta x} (\cos k \Delta x - 1), \tag{3.41}$$

$$e^{\omega_I \Delta t} \sin \omega_R \Delta t = -\frac{U \Delta t}{\Delta x} \sin k \Delta x. \tag{3.42}$$

Squaring both sides of each equation and adding them eliminates the real part of the frequency, leaving

$$e^{\omega_I \Delta t} = \sqrt{1 + 2\left(\frac{U \Delta t}{\Delta x}\right)(\cos k \Delta x - 1)\left(1 - \frac{U \Delta t}{\Delta x}\right)}. \tag{3.43}$$

Recall that Eq. 3.34 shows that the value of this exponential in the assumed form of the model solution controls whether the solution increases or decreases in amplitude with increasing time. That is

$$e^{\omega_I t} = e^{\omega_I \tau \Delta t} = \left(e^{\omega_I \Delta t}\right)^\tau,$$

so the value of the solution amplifies or damps exponentially as the time-step value $\tau$ increases with the model integration.

For this particular combination of space and time differencing schemes, Eq. 3.43 shows the dependence of the exponential on wavelength and the ratio $U \Delta t / \Delta x$. Figure 3.23 indicates that the model solution damps exponentially with time for $U \Delta t / \Delta x < 1$, it does not change amplitude when this ratio is unity, and it amplifies exponentially for ratios greater than unity. Shorter wavelengths are damped more severely than are longer wavelengths. Thus, the stability criterion for this differencing scheme is $U \Delta t / \Delta x \leq 1$. The ratio $U \Delta t / \Delta x$ is the previously defined CFL condition. It is also called the *Courant number*, and is described in Courant *et al.* (1928). Thus, given the chosen grid increment, and the largest wave speed that is likely to exist anywhere on the grid during the forecast ($U$), the time step required for stability is chosen. Note that such selective damping of short waves that are poorly resolved on a grid is sometimes considered to be an advantageous property of a differencing scheme.

A similar procedure can be used to analyze the stability criterion for the forward-in-time, centered-in-space advection equation

$$\frac{h_j^{\tau+1} - h_j^\tau}{\Delta t} = -U \frac{(h_{j+1}^\tau - h_{j-1}^\tau)}{2 \Delta x},$$

obtaining

$$e^{\omega_I \Delta t} = \sqrt{1 + \left(\frac{U \Delta t}{\Delta x}\right)^2 \sin^2 k \Delta x}.$$

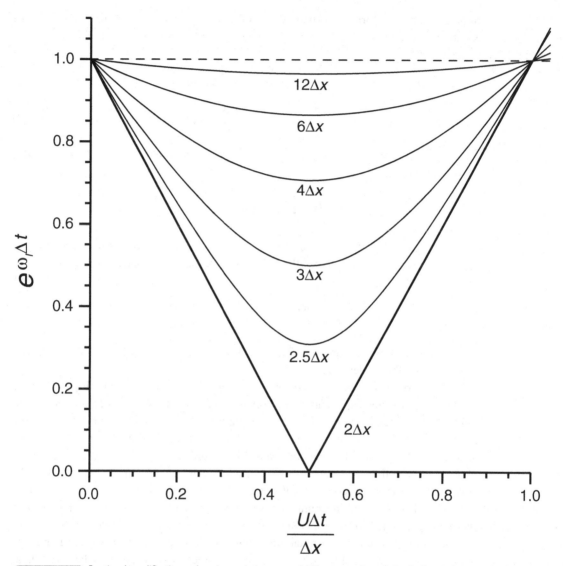

**Fig. 3.23** Fractional amplification or damping each time step, of different wavelengths for the forward-in-time, backward-in-space linear advection equation, as a function of the Courant number.

The only time step that will allow the exponential to be less than or equal to unity, and guarantee a stable solution, is zero. Thus, this differencing scheme is absolutely unstable.

Now consider the stability of the three-point centered-in-space, centered-in-time, linear advection equation:

$$\frac{h_j^{\tau+1} - h_j^{\tau-1}}{2\Delta t} = -U\frac{(h_{j+1}^{\tau} - h_{j-1}^{\tau})}{2\Delta x}. \qquad (3.44)$$

From the assumed form of the solution in Eq. 3.37, it is easy to obtain

$$h_{j+\beta}^{\tau} = e^{i\beta k\Delta x}h_j^{\tau} \quad \text{and} \tag{3.45}$$

$$h_j^{\tau+\upsilon} = e^{\omega_I \upsilon\Delta t}e^{-i\omega_R \upsilon\Delta t}h_j^{\tau} . \tag{3.46}$$

The substitution of Eq. 3.45 into the right side of Eq. 3.44, and then the use of Euler's relations (Eqs. 3.39 and 3.40), provides

$$h_j^{\tau+1} = h_j^{\tau-1} - \frac{U\Delta t}{\Delta x}(2i\sin k\Delta x)h_j^{\tau} .$$

Defining $\alpha = (2U\Delta t/\Delta x)\sin k\Delta x$ produces

$$h_j^{\tau+1} = h_j^{\tau-1} - i\alpha h_j^{\tau} \quad \text{and} \tag{3.47}$$

$$h_j^{\tau+2} = h_j^{\tau} - i\alpha h_j^{\tau+1} . \tag{3.48}$$

Using Eq. 3.47 to eliminate $h_j^{\tau+1}$ in Eq. 3.48 yields

$$h_j^{\tau+2} = (1-\alpha^2)h_j^{\tau} - i\alpha h_j^{\tau-1} . \tag{3.49}$$

The matrix form of Eqs. 3.47 and 3.49 is

$$\begin{bmatrix} h_j^{\tau+1} \\ h_j^{\tau+2} \end{bmatrix} = \begin{bmatrix} 1 & -i\alpha \\ -i\alpha & 1-\alpha^2 \end{bmatrix} \begin{bmatrix} h_j^{\tau-1} \\ h_j^{\tau} \end{bmatrix} . \tag{3.50}$$

The time step for this differencing scheme is $2\Delta t$ (Eq. 3.44), so we can represent the phase and amplitude change in the solution during that period by substituting 2 for $\upsilon$ in Eq. 3.46, which becomes

$$h_j^{\tau+2} = \lambda h_j^{\tau} , \tag{3.51}$$

where $\qquad \lambda = e^{2\omega_I \Delta t}e^{-i2\omega_R \Delta t} .$

Because $\lambda$ represents the change between any two time steps, we can also write

$$h_j^{\tau+1} = \lambda h_j^{\tau-1} . \tag{3.52}$$

Substitution of Eq. 3.52 into Eq. 3.47 and Eq. 3.51 into Eq. 3.49 yields

$$0 = (1-\lambda)h^{\tau-1} - i\alpha h_j^{\tau} ,$$

$$0 = (1-\alpha^2-\lambda)h_j^{\tau} - i\alpha h^{\tau-1} .$$

Thus, Eq. 3.50 becomes

$$
\begin{bmatrix} 0 \\ 0 \end{bmatrix} = \begin{bmatrix} 1-\lambda & -i\alpha \\ -i\alpha & 1-\alpha^2-\lambda \end{bmatrix} \begin{bmatrix} h^{\tau-1} \\ h_j^{\tau} \end{bmatrix}.
$$

This linear system of equations has a nontrivial ($h \neq 0$) solution if the determinant of the matrix of coefficients equals zero. That is

$$
\begin{vmatrix} 1-\lambda & -i\alpha \\ -i\alpha & 1-\alpha^2-\lambda \end{vmatrix} = 0 \text{, yielding}
$$

$$
\lambda^2 + (\alpha^2-2)\lambda + 1 = 0 \text{, with the two solutions being}
$$

$$
\lambda = 1-\frac{\alpha^2}{2}\pm\frac{\alpha}{2}\sqrt{\alpha^2-4}. \tag{3.53}
$$

Recall that $\lambda$ represents the amplitude and phase change of the wave over a centered time step, $2\Delta t$. Thus, the magnitude of $\lambda$ represents the amplification or damping of the wave during the $2\Delta t$ period:

$$
|\lambda| = \left| 1-\frac{\alpha^2}{2}\pm\frac{\alpha}{2}\sqrt{\alpha^2-4} \right|.
$$

If $\alpha^2-4 = 0$, $|\lambda|= 1$ for both solutions. For $\alpha^2-4 > 0$, $|\lambda| > 1$ for the solution with the negative sign. For $\alpha^2-4 < 0$, $\lambda$ is complex, and solving for $|\lambda|$ using

$$
|\lambda| = \sqrt{|\lambda_R|^2 + |\lambda_I|^2}
$$

also yields $|\lambda|=1$. Thus $|\lambda| =1$ for $\alpha^2 \leq 4$, where the latter is the requirement for stability. Note that it is sufficient that only one solution for $|\lambda|$ be greater than unity when $\alpha^2 > 4$ in order to make this condition unstable. Substituting the definition for $\alpha$ into the stability requirement leads to

$$
\frac{U\Delta t}{\Delta x}\sin k\Delta x \leq 1, \tag{3.54}
$$

and because the sine can equal unity, the stability requirement is $U\Delta t/\Delta x \leq 1$. Note that if this condition is marginally violated, the wave of length $4\Delta x$ for which the sine function in Eq. 3.54 is unity will be the first to become unstable. Using a similar approach, it can be shown that the stability requirement for the five-point centered-in-space (see Eq. 3.32), and centered-in-time, linear advection equation

$$
\frac{h_j^{\tau+1}-h_j^{\tau-1}}{2\Delta t} = -U\frac{\frac{4}{3}(h_{j+1}^{\tau} - h_{j-1}^{\tau})-\frac{1}{6}(h_{j+2}^{\tau} - h_{j-2}^{\tau})}{2\Delta x} \tag{3.55}
$$

is $U\Delta t/\Delta x \leq 0.73$. These centered-in-space and centered-in-time schemes, unlike the forward-in-time and backward-in-space method (Fig. 3.23), do not damp for any stable value of the Courant number.

As noted earlier, the advection term has one of the most restrictive linear-stability criteria of all the terms in the model equations. That is, the time step must be sufficiently small to guarantee that the Courant number ($U\Delta t/\Delta x$) is always less than unity at any time in the integration and at all locations on the grid. The grid increment in this ratio is chosen to be sufficiently small so that processes or meteorological features that are being simulated or forecasted can be defined on the grid with acceptably small truncation error. The velocity scale is a function of the prevailing meteorology, and cannot be controlled. That leaves the time step as the only free parameter that can be adjusted to maintain a stable solution. A useful geometric way of visualizing this stability criterion is that $U\Delta t$, the numerator in the ratio, is the distance traveled by an advecting feature in one time step. If this distance is greater than one grid increment (the denominator), the ratio is greater than unity and the solution is linearly unstable. Unfortunately, further analysis of this term in the context of the full equations, which contain more than advection effects, reveals that the speed in the stability criterion that must be accommodated when choosing a time step is $U + C_P$. This is the advective speed plus the phase speed of the fastest wave on the grid. If the model equations admit sound waves or external-gravity waves, this phase speed can be $300\ \text{ms}^{-1}$. So, in choosing a stable time step, this largest wave speed must be estimated, as well as the magnitude of the advective speed in the strongest jet on the grid. Many models will make these estimations internally, and choose a "safe" time step, allowing also for the fact that a horizontally varying map-scale factor will cause the Earth distance between grid points to depart from $\Delta x$. Nevertheless, sometimes the estimates will not be sufficiently conservative, and a linear instability will occur in extreme circumstances. Such occasional stability problems may be acceptable given the fact that an extremely conservative small time step would waste computer resources.

The linear stability of the vertical-advection term also has a potentially serious constraint on the time step. This condition parallels that of the horizontal-advection term, and for the three-point approximation is $w\Delta t/\Delta z \leq 1$ for $z$ as the vertical coordinate. Analogous to the horizontal-advection problem, the velocity in this expression is the maximum wave velocity in the vertical, which could be simply the advective velocity, or it could be the sum of the advective velocity and that of vertically propagating gravity waves and sound waves. The vertical grid increment, $\Delta z$, typically varies significantly across the depth of the model atmosphere, where smaller values are often used in the boundary layer and near the tropopause in order to resolve large vertical gradients. The vertical advective wave speed is often a maximum near the level of nondivergence, but can be locally large when convective circulations are explicitly represented in the model. Where especially small vertical grid increments and large vertical velocities coexist, the constraint on the time step may be greater than that associated with the horizontal-advection term. For example, recall the discussion in the previous chapter of techniques (Boussinesq, anelastic, hydrostatic approximations) for filtering sound waves from the model solution. The combination of sound waves propagating in the vertical at $300\ \text{m s}^{-1}$, and the fact that vertical grid increments are typically much smaller than horizontal grid increments for most models, would require an extremely small time step. Thus, there is great motivation to remove the sound waves, or to deal with them numerically in such a way that they do not represent a severe constraint on the time step for the entire equation (for example, the split-explicit time differencing described previously in Section 3.3.1).

## Linear stability of an explicit horizontal-diffusion term

Even though much more will be discussed later in this chapter about explicit numerical diffusion terms, in the overall context of the diffusion or damping of model solutions, the linear stability of the following low-order diffusion term will be evaluated here. The strength of the diffusion effect is controlled by the specified magnitude of the positive diffusion coefficient, $K$:

$$\frac{\partial h}{\partial t} = K\frac{\partial^2 h}{\partial x^2}. \tag{3.56}$$

The term on the right side of this equation is added to the standard physical-process terms in a prognostic equation for the variable $h$, where the purpose is to damp poorly resolved and sometimes-erroneous small space-scale features in the model solution. If the damping is sufficiently strong, some of the problems described later related to the nonphysical behavior of small-scale energy on the grid can be mitigated. As in the analysis of the advection equation, the amplitude change of the solution depends on the value of the imaginary part of the wave frequency in the assumed form of the wave solution in Eq. 3.34. Approximating this equation with the centered-in-time, centered-in-space approach used for the advection equation,

$$h_j^{\tau+1} - h_j^{\tau-1} = \frac{2K\Delta t}{(\Delta x)^2}(h_{j+1}^{\tau} + h_{j-1}^{\tau} - 2h_j^{\tau}),$$

yields

$$e^{\omega_I t} = \frac{2K\Delta t}{(\Delta x)^2}(\cos k\Delta x - 1)\pm\sqrt{4\left(\frac{K\Delta t}{(\Delta x)^2}\right)^2(\cos k\Delta x - 1)^2 + 1}, \text{ for } \omega_R = 0.$$

Unless $K = 0$, the exponential is more-negative than $-1$, for the negative sign on the radical, and thus that solution is absolutely unstable, amplifying and changing sign (i.e., phase) each time step. If a forward-in-time, centered-in-space approximation is used,

$$h_j^{\tau+1} - h_j^{\tau} = \frac{K\Delta t}{(\Delta x)^2}(h_{j+1}^{\tau} + h_{j-1}^{\tau} - 2h_j^{\tau}), \text{ and}$$

$$e^{\omega_I t} = 1 + \frac{2K\Delta t}{(\Delta x)^2}(\cos k\Delta x - 1), \text{ again for } \omega_R = 0. \tag{3.57}$$

For infinitely long waves, the exponential equals unity, so there is no damping or amplification regardless of the value of $K$ or $\Delta t$. For waves of length $2\Delta x$,

$$e^{\omega_I t} = 1 - \frac{4K\Delta t}{(\Delta x)^2}.$$  (3.58)

For $0 < K\Delta t/\Delta x^2 \leq 1/4$, $0 \leq e^{\omega_I t} < 1$ and the $2\Delta x$ solution damps;

for $1/4 < K\Delta t/\Delta x^2 \leq 1/2$, $-1 \leq e^{\omega_I t} < 0$ and the solution damps while changing phase every time step;

and for $1/2 < K\Delta t/\Delta x^2$, $e^{\omega_I t} < -1$ and the solution amplifies while changing phase each time step.

Thus, for physically realistic solutions, the linear stability criterion is $K\Delta t/\Delta x^2 \leq 1/4$. Section 3.4.7 discusses other types of horizontal-diffusion terms. Note that models also typically have vertical-diffusion terms with analogous stability criteria. For thin model layers (small $\Delta z$) and a large diffusion coefficient, for example in the boundary layer, this criterion may be easily violated.

## Maintaining linear stability with multiple terms in an equation

The individual analyses of the linear stability conditions for the finite-difference approximations to the advection and diffusion terms provided quantitative information about the combinations of values of model parameters (e.g., time step) and meteorological conditions required for stable, realistic solutions. For the advection term, the stability constraint was $U\Delta t/\Delta x \leq 1$ for centered-in-time, and second-order centered-in-space differencing. For the second-order diffusion term, realistic solutions with forward-in-time, centered-in-space, differencing required that $K\Delta t/\Delta x^2 \leq 1/4$. With both of these terms in the same equation, there are questions about how we choose our parameters appropriately to maintain stability, and how we accommodate the fact that one term employs centered-in-time differencing and the other uses forward-in-time differencing. The latter issue can be addressed by evaluating the diffusion term at the $\tau - 1$ time, and extrapolating over a $2\Delta t$ interval to the $\tau + 1$ time. Equation 3.59 shows the prognostic finite-difference equation for the two terms combined.

$$\frac{h_j^{\tau+1} - h_j^{\tau-1}}{2\Delta t} = -u_j^{\tau} \frac{(h_{j+1}^{\tau} - h_{j-1}^{\tau})}{2\Delta x} + \frac{K}{(\Delta x)^2} (h_{j+1}^{\tau-1} + h_{j-1}^{\tau-1} - 2h_j^{\tau-1}).$$  (3.59)

The previous separate analysis of the stability of the individual linear versions of the advection and diffusion terms was necessary for mathematical reasons, but the actual operative constraint is based on the combination of all the terms in an equation. Nevertheless, in practice, the constraints associated with the *individual* terms are considered, and the time step from the most restrictive one is employed for the integration. For example, in the case of

Eq. 3.59, let the grid increment be 25 km, and the estimated largest speed on the grid be 50 m s$^{-1}$. Because unity is the largest stable value of the Courant number, the maximum time step would be 500 s. In practice, the chosen time step used is typically about 20–25% less than this limit to account for the facts that (1) the estimate of the largest stable Courant number is based on a linear analysis, (2) the distance on Earth between grid points is less than $\Delta x$ in some areas because of the map projection, and (3) the maximum wind speed may be incorrectly estimated. So, the actual time step used in this case would perhaps be 400 s. For the diffusion term, assume that we would like to damp 25% of the amplitude of the $2\Delta x$ wave each time step, which means that $e^{\omega_I t}$ must equal 0.75. Because Eq. 3.58 shows that the damping rate is a function of both the time step and the value of the diffusion coefficient, $K$, the time step from the advection equation can be used, and the value of $K$ chosen to achieve the desired damping. Thus $K \approx 10^5$ would produce the desired damping.

### 3.4.3 Phase/group-speed errors

In this section it will be shown how finite-difference approximations can lead to physically unrealistic phase and group speeds. Consider first the forward-in-time and backward-in-space approximation to the advection equation discussed earlier (Eq. 3.35). The speed of an advecting feature should simply be $U$. To define the advective speed in the numerical solution, divide Eq. 3.42 by Eq. 3.41 to eliminate $\omega_I$, take the inverse tangent of the resulting equation, and use the fact that $\omega_R = C_R k$, obtaining

$$C_R = \frac{1}{k\Delta t} \operatorname{atan} \left[ \frac{(U\Delta t/\Delta x)\sin k\Delta x}{1 + (U\Delta t/\Delta x)(\cos k\Delta x - 1)} \right].$$

This phase speed of the advective wave on the grid is a function of wavelength (in terms of the wavenumber $k$) and the Courant number. Figure 3.24 illustrates this relationship. For all wavelengths, Courant numbers of less than 0.5 cause the waves to move more slowly than they should, and Courant numbers of greater than 0.5 cause the waves to move at an erroneously high speed. In general, waves whose phase speed is a function of wavelength are called *dispersive* (e.g., Rossby waves). Even though the advective wave is not dispersive in nature, in the numerical solution it is, so this process is called *numerical dispersion*.

For the three-point, centered-in-space and centered-in-time advection equation, for the situation with a stable solution where $\alpha^2 - 4 \leq 0$ in Eq. 3.53,

$$\lambda = e^{2\omega_I \Delta t} e^{-i2\omega_R \Delta t} = 1 - \frac{\alpha^2}{2} \pm \frac{i\alpha}{2}\sqrt{4 - \alpha^2}.$$

For nondamping, stable solutions, the first exponential is equal to unity. Employing Euler's relation, Eq. 3.40, to rewrite the second exponential, separating the real and imaginary parts of the complex equation, recalling that $\omega_R = C_R k$, substituting the definition of $\alpha$, and employing trigonometric identities yields

$$C_R = \frac{1}{k\Delta t} \operatorname{asin}\left( \pm \frac{U\Delta t}{\Delta x} \sin k\Delta x \right). \tag{3.60}$$

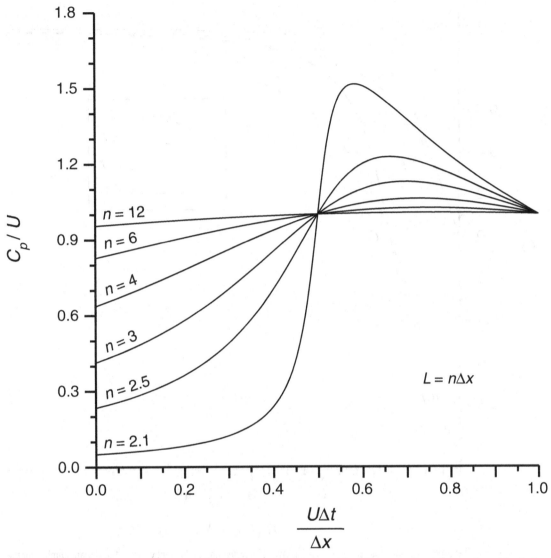

**Fig. 3.24**  Ratio of the numerical phase speed and the true advective speed for different wavelength features ($n\Delta x$) for the forward-in-time, backward-in-space linear advection equation, as a function of the Courant number.

There are two waves defined here. The one corresponding to the positive argument of the inverse sine is an approximation to the physical wave, moving in the correct direction but slower than the true feature being advected. The other wave, called the *computational mode* or *ghost mode*, is entirely fictitious, with no counterpart in nature, and moves in the opposite direction. The amplitude of the computational mode is typically much smaller than that of the physical mode. The phase speed of the physical mode is shown in Fig. 3.25, as a function of wavelength and the Courant number. This differencing scheme

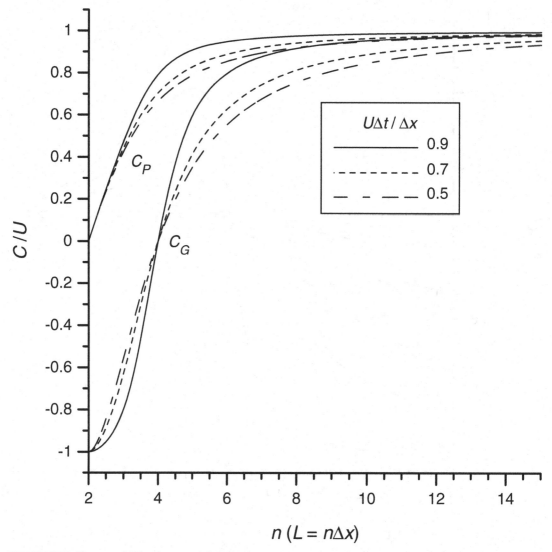

**Fig. 3.25** Phase speed ($C_P$) and group speed ($C_G$) as a function of wavelengths for different values of the Courant number, for the three-point, centered-in-space and centered-in-time approximation to the linear advection equation. The phase and group speeds pertain to the physical, not the computational, mode.

is also dispersive, with the phase speeds of the longer waves better approximating the true speed, and the speed of the $2\Delta x$ wave being zero. Also, the wave speeds are more realistic for Courant numbers closer to unity.

Examples of model solutions for the three-point (second-order accuracy) centered-in-space and centered-in-time approximation to the linear advection equation (Eq. 3.44) are shown in Fig. 3.26. For the "no-diffusion" curve, the model represents only the linear advection equation defined on a one-dimensional grid that has periodic lateral-boundary

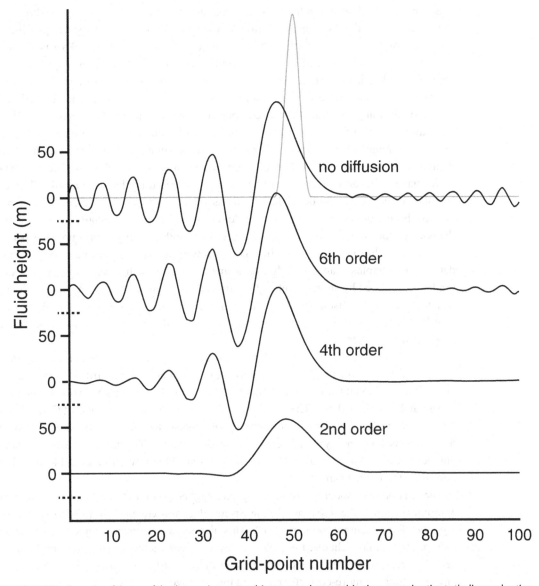

**Fig. 3.26**  Examples of the use of the three-point, centered-in-space and centered-in-time approximation to the linear advection equation. The initial condition was a symmetric wave in the height field, centered on grid-point 50, and is shown by the light line at the top. The exact solution is also shown by the light line, after the wave has advected over about 100 grid increments from left to right across the grid, exiting at the right boundary and entering at the left. Numerical solutions are shown with heavier lines, for no diffusion and diffusion operators of different order. The Courant number was 0.1.

conditions. That is, a wave exiting one end of the computational domain enters on the opposite end. The initial conditions were defined by a wave in the height field centered on grid-point 50 and with the shape of the lighter curve in the figure. The advective velocity was from left to right in the figure, with a magnitude of 10 m s$^{-1}$. The lighter line in the figure also shows the theoretical solution for the wave after it has traversed approximately 100 grid points (100 km), exiting on the right and entering on the left. That is, the wave moved at exactly the advective speed, $U$, and there was no damping or amplification. In contrast, the black lines (focus on the upper one in this discussion) show the numerical solution at the same time for a Courant number of 0.1 ($\Delta t = 10$ s). The original wave, even though it appears smooth, was composed of many Fourier components with different wavelengths, which, as we have seen, have different numerical phase speeds. The longer wavelengths have speeds closer to the correct value (see Fig. 3.25), but the shorter components move at speeds that are proportional to their wavelength. The very short waves have not even exited the grid on the right at this time, having moved at less than half the correct speed. Some of the erroneous wave energy might be associated with the previously mentioned computational mode, but it is difficult to visually separate it from the poorly represented short waves in the physical mode. Clearly, this is not an especially satisfactory solution for representing the advection process. In particular, when model dependent variables change rapidly over a small distance, such as across fronts, the sharp gradient is defined by short-wavelength Fourier components. Thus, even when such physical features are realistically rendered in model initial conditions, as the short wavelengths become separated from the longer wavelengths as the feature propagates, the gradient will weaken.

To illustrate how the choice of the Courant number can affect numerical dispersion, Fig. 3.27 depicts model solutions analogous to the upper one (no diffusion) in Fig. 3.26, but for the additional Courant numbers of 0.5 ($\Delta t = 50$ s) and 0.9 ($\Delta t = 90$ s). The differences can be explained by referring to Fig. 3.25, which shows that the use of Courant numbers closer to unity produces more-correct phase speeds for the 3–10$\Delta x$ wavelengths, and thus there is less energy in the erroneously slow waves. The influence of the Courant number is dependent on the specific approximation to the advection term, but this shows how great the impact can be.

The analogous model solution using the higher-accuracy, fourth-order (five-point) approximation for the derivative in the linear advection equation (Eq. 3.55) is shown in Fig. 3.28 for a Courant number of 0.1. This also shows erroneous numerical dispersion, like the three-point scheme, but it is less severe. Nevertheless, significant amplitude has been lost relative to the correct solution, and there is still wave energy in erroneous features that trail the more-correctly rendered longer wave.

In nature, the advective wave is nondispersive, and the phase speed and the group speed are equal. Given that these centered-in-space and centered-in-time solutions exhibit numerical dispersion, it is revealing to calculate the group speed ($C_G$) of the waves, or in other words the speed at which the wave energy propagates. In general,

$$C_G = \frac{\partial}{\partial k}(C_P k).$$

(3.61)

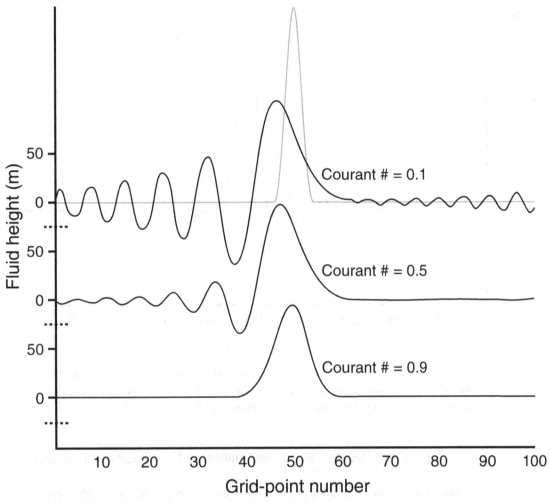

**Fig. 3.27** Model solutions for the linear advective wave, without diffusion, for the different indicated values of the Courant number. Second-order, centered-in-space and centered-in-time differencing was used. The exact solution is shown by the light line at the top.

For the three-point, centered-in-space and centered-in-time approximation to the advection equation, Eq. 3.60 represents the phase speed, $C_P$. Substituting this expression into Eq. 3.61 and evaluating the derivative gives

$$C_G = \frac{U \cos k\Delta x}{\left[1 - \left(\frac{U\Delta t}{\Delta x}\sin k\Delta x\right)^2\right]^{1/2}}.$$

This group speed is plotted in Fig. 3.25 as a function of wavelength and Courant number. For a wave of length $4\Delta x$, the group speed is zero. The $2\Delta x$ wave energy travels at the correct speed, but in the wrong direction. Thus, the energy propagation properties of the shorter waves are severely mishandled by this finite-difference approximation.

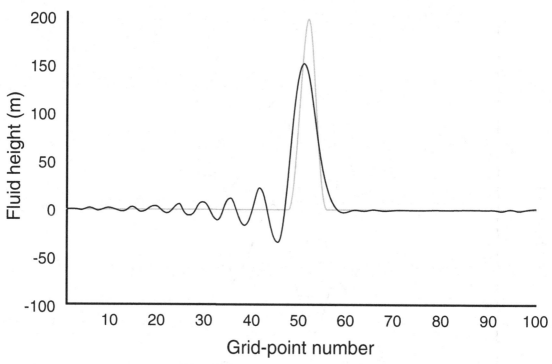

**Fig. 3.28** Analogous to the no-diffusion curve in Fig. 3.26, except that the model used the fourth-order-accurate, five-point, approximation to the spatial derivative in the advection equation. No explicit diffusion was employed, and the Courant number was 0.1. The exact solution is shown by the light line at the top.

### 3.4.4  Properties of some example, multi-step, time-differencing schemes

Section 3.3.1 defines only a few of the many multi-step time-differencing schemes that have been used in atmospheric models. Even though Durran (1999) provides a thorough discussion of their numerical properties, a few schemes will be mentioned here. In general, these methods are popular because their stability criteria are often not as stringent as for the single-step methods that are the focus above, they can have relatively high orders of accuracy, and some very selectively damp the smaller, poorly resolved wavelengths. Figure 3.29a shows the fractional damping each time step associated with the Lax–Wendroff and Euler-backward time-differencing schemes applied to the linear-advection equation. In each case, the second-order, centered-in-space approximation is used for the space differencing (the variable $F$ in Eqs. 3.14–3.20). Because of the desirability of selectively damping short-wavelength, poorly resolved waves, while leaving the well-resolved waves relatively undamped, the Lax–Wendroff scheme is superior in this respect. The numerical dispersion caused by these schemes is shown in Fig. 3.29b, where the Euler-backward method produces more-correct phase speeds for the better-resolved waves. An example of an advective-wave solution using a multi-step time-differencing scheme with high-order space differencing is seen in Fig. 3.30. The third-order Runge–Kutta time-differencing scheme,

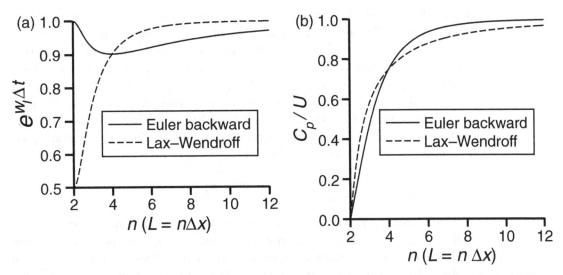

Fractional damping each time step (a) and the ratio of the numerical phase speed and the correct phase speed (b), as a function of wavelength for the Lax–Wendroff and Euler-backward time-differencing schemes, with second-order, centered space differencing.

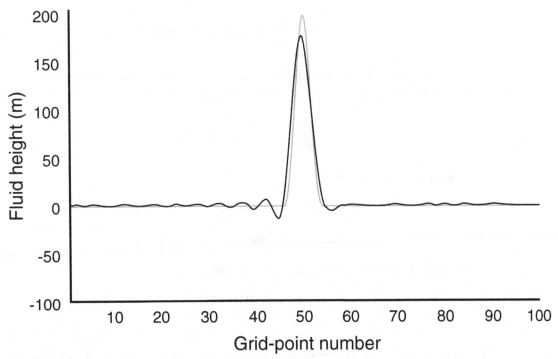

Analogous to the no-diffusion curve in Fig. 3.26, except that third-order Runge–Kutta time differencing and sixth-order space differencing were used. There was no explicit diffusion employed. The exact solution is shown by the light line at the top.

shown in Eqs. 3.18–3.20 and employed in the dynamic core of the community Advanced Research WRF (ARW) model (Skamarock *et al.* 2008), is combined with sixth-order space differencing. Compared to the wave solution shown in Fig. 3.28 resulting from single-step, centered-in-time differencing and fourth-order centered-in-space differencing, there is less numerical dispersion and the amplitude of the primary wave is better preserved. The ARW offers second- through sixth-order options for the space differencing, where the default is the fifth-order option because the odd-order schemes have desirable implicit damping properties, whereas the even-order schemes do not. But, regardless of the approach for the space-differencing, the Runge–Kutta time differencing contributes some damping of its own.

### 3.4.5  Aliasing

*Aliasing* is a process by which two waves represented on a model grid interact through a nonlinear term in the equations to produce fictitious waves, resulting in an erroneous redistribution of energy (amplitude) in the wave spectrum, and possibly even leading to an instability that is fatal to the model integration. This process can be illustrated with a simple, nonlinear advection term:

$$\frac{\partial u}{\partial t} = -u\frac{\partial u}{\partial x}.$$

For simplicity, assume that $u$ can be represented mathematically as a sum of cosine waves, such as

$$u = \sum_{m=0}^{\infty} a_m \cos k_m x, \text{ where } k_m = \frac{2\pi}{L}m,$$

$L$ is the length of the computational grid, and $k_m$ is a wavenumber.[4] Differentiating this expression with respect to $x$, as in the advection term, we obtain

$$\frac{\partial u}{\partial x} = -\sum_{m=0}^{\infty} a_m k_m \sin k_m x,$$

and multiplying by $-u$ gives

$$-u\frac{\partial u}{\partial x} = -(a_0 + a_1 \cos k_1 x + a_2 \cos k_2 x + \ldots + a_m \cos k_m x + \ldots) \times$$
$$(a_1 k_1 \sin k_1 x + a_2 k_2 \sin k_2 x + \ldots + a_n k_n \sin k_n x + \ldots).$$

The result of any two waves, $k_m$ and $k_n$, interacting is

$$a_n a_m k_n \sin k_n x \cos k_m x.$$

----

[4] Note that both $k_m$ and $m$ are wavenumbers. The $m$ represents the number of waves on the domain of length $L$, and is nondimensional. The $k_m$ is $2\pi$ divided by the wavelength ($L/m$), has dimensions of inverse distance, and is sometimes referred to as a rotational wavenumber.

But

$$\sin x \cos y = \frac{\sin(x + y) + \sin(x - y)}{2},$$

and therefore the interaction product is

$$a_n a_m k_n [\sin(k_n + k_m)x + \sin(k_n - k_m)x] =$$

$$a_n a_m k_n \left[ \sin\frac{2\pi}{L}(n + m)x + \sin\frac{2\pi}{L}(n - m)x \right].$$

Thus, when wavenumbers $m$ and $n$ interact, they produce two waves, one with wavenumber $n+m$ and one with wavenumber $n-m$. This is no problem in continuous space where all wavenumbers are possible, but it can be in the discrete (grid-point) space of a model. For example, assume a one-dimensional grid having $j_{max}$ intervals (see Fig. 3.31), where $j_{max}$ is an even number. Table 3.1 shows the range of wavenumbers and wavelengths on this grid, where the longest complete wave that can be represented is defined by the domain length, $L$, and the shortest wave is defined in terms of the grid increment, $\Delta x$.

Now consider what happens when resolvable wavenumbers $m$ and $n$ interact to yield a wavenumber that is larger than what is permitted by the grid (i.e., a wavelength of less than $2\Delta x$). This would result from the $m+n$ interaction product rather than the $m-n$ product, so $m+n > j_{max}/2$. A way of defining $m+n$ without the inequality is $m+n = j_{max} - s$, where $s < j_{max}/2$. So the wave resulting from the problematic $m+n$ interaction would be

$$\sin\frac{2\pi}{L}(m + n)x = \sin\frac{2\pi}{j_{max}\Delta x} \cdot (j_{max} - s) \cdot (j\Delta x)$$

$$= \sin 2\pi \frac{j_{max} - s}{j_{max}}j = \sin\left(2\pi j - \frac{2\pi s j}{j_{max}}\right).$$

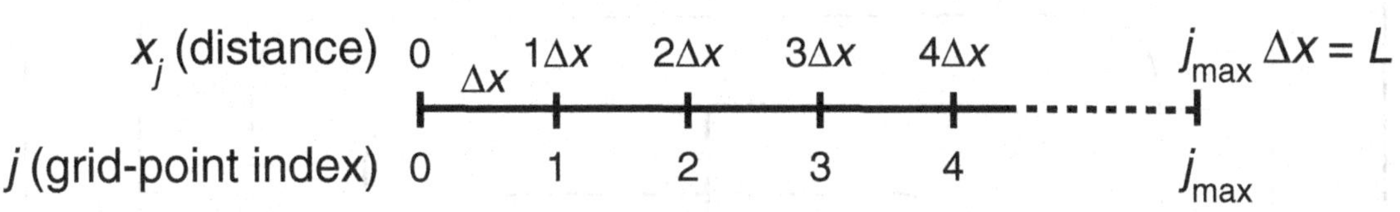

Fig. 3.31   A one-dimensional grid with $j_{max}$ grid intervals and a length of $L$.

| Table 3.1 Corresponding wavenumbers and wavelengths on a grid with $j_{max}$ points | |
|---|---|
| Wavenumber | Wavelength |
| 1 | $j_{max} \Delta x$ (longest) |
| 2 | $j_{max} \Delta x/2$ |
| . | . |
| . | . |
| . | . |
| $j_{max}/4$ | $4\Delta x$ |
| $j_{max}/2$ | $2\Delta x$ (shortest) |

But $\sin (x-y) = \sin x \cos y - \cos x \sin y$, so that

$$\sin\left( 2\pi j - \frac{2\pi sj}{j_{max}}\right) = \sin 2\pi j \cos\frac{2\pi sj}{j_{max}} - \cos 2\pi j \sin\frac{2\pi sj}{j_{max}},$$

where the sine function in the first term on the right side is equal to zero, and the cosine function in the second term is equal to unity. Thus,

$$\sin\frac{2\pi}{L}(m+n)x = \sin\frac{2\pi sj}{j_{max}} = \sin\frac{2\pi s}{L}x.$$

So the unresolvable wavenumber shows up on the grid as wavenumber $s$, such that $s = j_{max} - (m+n)$. For example, say $m = \frac{1}{2}j_{max}$ (a $2\Delta x$ wave) and $n = \frac{1}{4}j_{max}$ (a $4\Delta x$ wave), so $m+n = \frac{3}{4}j_{max}$ (a $\frac{4}{3}\Delta x$ wave) and $m-n = \frac{1}{4}j_{max}$ (a $4\Delta x$ wave). But the $\frac{4}{3}\Delta x$ wave is unresolvable, and the aliasing produces energy in the $4\Delta x$ wavelength ($s = j_{max} - \frac{3}{4}j_{max} = \frac{1}{4}j_{max}$).

To illustrate all possible interactions, assume $j_{max} = 24$. Any interaction that produces a wavenumber greater than 12 will result in aliasing. The erroneous redistribution of energy on this grid is illustrated in Fig. 3.32.

Not only does this aliasing process cause energy to be incorrectly located in the wrong scales, resulting in errors in the model solution, it can also result in the model solution becoming unstable and stopping the numerical integration process. This is called a

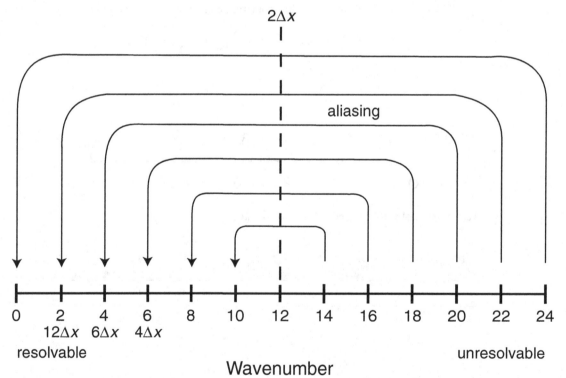

Fig. 3.32 Illustration of how nonlinear interactions on a 24-point grid produce aliasing, resulting in energy being erroneously placed in the wrong wavelengths. Interactions that produce unresolvable wavelengths on the right result in the energy being "folded" over the smallest resolvable wavenumber 12 ($2\Delta x$) to the resolvable side of the scale.

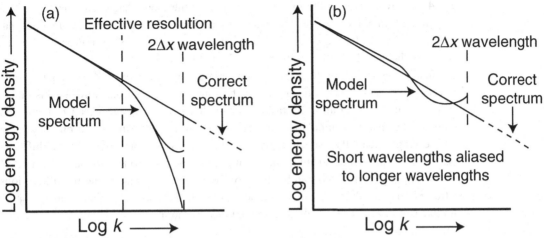

**Fig. 3.33** Schematic of the correct (straight line) and model kinetic-energy spectra. In (a), two examples are shown of the normal damping of the kinetic energy at the high-wavenumber end of the model spectrum by the model diffusion (see Section 3.4.7). In this situation, any energy aliased to the high wavenumbers has been controlled by the diffusion. In (b) is shown a model solution where aliasing has added erroneous energy that remains in the high wavenumbers. Adapted from Skamarock (2004).

nonlinear instability and will be discussed in the following section. Aliasing is also mentioned in the discussion of spectral modeling, wherein each wave interaction is analytically treated, and those interactions that would result in unresolvable products are not permitted. This is one of the advantages of the spectral-modeling approach.

Figure 3.33 shows the impact of aliasing on a model's kinetic-energy spectrum. The plot on the left shows a normal spectrum, where aliasing is not a significant problem. The filtering of the high wavenumber (short wavelength) part of the model solution, through processes described in Section 3.4.7 below, causes a loss of kinetic energy in the segment of the spectrum between the $2\Delta x$ wavelength and the effective resolution of the model. This damping is desirable because these wavelengths are poorly represented on the grid. The right panel shows the spectrum when aliasing has added erroneous energy in the resolved part of the spectrum, overwhelming the desirable damping illustrated in the left panel, and impacting the model's representation of physical processes.

### 3.4.6 Nonlinear instability

The discussion of computational instability in Section 3.4.2 was based on a linear differential equation, and adhering to the appropriate stability criterion is sufficient to avoid problems of that type. For nonlinear equations, there is a similar criterion, but even when this condition is satisfied it is possible for another type of instability to develop in the numerical solution. As shown in Chapter 2, primitive-equation models are based on nonlinear equations, and the source of the problem is the aliasing that was just described. The symptom of *nonlinear instability* that results from aliasing is a rapid buildup of energy in the 2–4$\Delta x$ wavelengths in the model solution, after the model has been integrated for a long period of time. The cause of this can be inferred from the list of wave interactions that are associated with aliasing on the 24-point grid shown in Fig. 3.32 (see Problem 1 at the end of this chapter). In particular, there are 42 combinations of wavenumbers $m$ and $n$ that result in aliasing, and 30 of these interactions produce energy in the 2–4$\Delta x$ range. Couple this with the fact that every aliasing interaction involves at least one wave in this range, and it is clear that such an uncontrolled accumulation of energy can lead to numerical problems. The energy accumulation in these short wavelengths can be controlled by using a differencing scheme that selectively damps the short waves, or scale-selective diffusion (dissipation) terms can be added to the equations (see next section). Alternatively, spectral or semi-Lagrangian methods can be used for the nonlinear terms so that the interactions are treated analytically. Continued integration of a model that has a kinetic-energy spectrum like that in Fig. 3.33b could lead to nonlinear instability.

### 3.4.7 Diffusion: real, explicit numerical, implicit numerical, grid

The diffusion processes described here all have the effect of spatially spreading features in the heat, moisture, and momentum fields of the modeled fluid. This can have the effect of damping the amplitude of perturbations in the variables, so sometimes they are referred to as damping processes. Because the diffusion or damping is scale selective, the methods may also be considered filters. There is obviously real (physical) diffusion, or mixing, in the atmosphere, caused by turbulence, and this needs to be represented in some realistic way. In addition, however, a nonphysical, scale-selective, diffusion or damping process is incorporated in all models, through explicit terms in the predictive equations or implicitly through the use of damping differencing schemes. The purpose is to "clean up" unrealistic features in the model solution associated with lateral-boundary noise, computational modes, and erroneous shortwave energy from numerical wave dispersion. Lastly, even if none of the above diffusion processes were incorporated in the model, there would still be the spatial spread of information about the model variables through the vertical and horizontal finite differencing. The terminology used in the literature for referring to the different types of diffusion is not standard, so the reader should be cautious.

### Physical diffusion

The atmosphere contains turbulence that smooths out, or diffuses, structures in the momentum, thermal, and moisture fields, in all three coordinate directions. Where gradients exist, turbulent fluxes transport properties, such that the amplitudes of maxima or minima in physical fields are reduced. Because of the intensity of shear- and buoyancy-driven turbulence in the boundary layer, and the typically strong gradients there, planetary boundary-layer parameterizations are needed to represent this important physical process. Likewise, turbulent mixing can be important elsewhere, in the free atmosphere, such as near jets in the wind field and in the vicinity of moist convection, and models need to be able to treat the associated mixing in a realistic way. This is the "real" diffusion, or mixing, that must be represented in a physically faithful model of the atmosphere. Chapter 4 discusses the representation of this *physical diffusion* by turbulence parameterizations.

## Explicit numerical diffusion

In the previous section on aliasing, it was mentioned that one way of controlling the artificial accumulation of energy in short wavelengths, and the resulting nonlinear instability, is through the use of diffusion terms, on the right side of the predictive equations, that are explicitly designed to damp these short wavelengths. In addition to controlling this instability, damping the short wavelengths also improves model solutions like those shown in Figs. 3.27 and 3.28, where numerical dispersion causes short wavelengths to be erroneously separated from the physical solution. A challenge is sufficiently damping the erroneous component of the model solution while not damping the physically realistic part.

There are a few different mathematical forms, shown in Eq. 3.62, that are used for the term that explicitly controls shortwave amplitudes:

$$\frac{\partial h}{\partial t} = (-1)^{n/2+1} K_n \nabla^n h. \tag{3.62}$$

Here, $K$ is the *diffusion coefficient*, $h$ is any dependent variable, and $n = 0, 2, 4, 6$ indicates the order of the term. The right side for zero-order ($n = 0$) damping is $-K_0 h$. This produces a non-scale-selective relaxation, and is typically applied near lateral and upper boundaries. The second-order term has the form of the Laplacian, and is the equivalent of Eq. 3.56 except that it has two horizontal space dimensions. Recall that a term with this form appears in the physical heat-diffusion equation, where higher values are always transferred down gradient toward regions with smaller values. The fact that the change in the property depends solely on the sign and magnitude of the curvature (the second derivative) means that new extrema are not added to a field. This second-order term is less scale-selective than the higher-order ones. Equation 3.57 represents the amount of damping per time step for the forward-in-time, centered-in-space finite-difference scheme for the one-dimensional problem. Equation 3.63 shows analogous equations for the fourth- and sixth-order diffusion as well. The upper equation is the same as Eq. 3.57, and represents the damping per time step for second-order diffusion, and the middle and lower equations apply to fourth- and sixth-order diffusion, respectively:

$$e^{\omega_I t} = 1 - K\Delta t \begin{bmatrix} (2 - 2\cos k\Delta x)/(\Delta x)^2 \\ (6 - 8\cos k\Delta x + 2\cos 2k\Delta x)/(\Delta x)^4 \\ (20 - 30\cos k\Delta x + 12\cos 2k\Delta x - 2\cos 3k\Delta x)/(\Delta x)^6 \end{bmatrix}. \tag{3.63}$$

The amount of damping at different wavelengths for these three diffusion-operator options is important because it is necessary to filter the poorly resolved, small scales, especially in the $2$–$4\Delta x$ range, without greatly damaging the amplitudes of the better-resolved length scales. Figure 3.34 shows the amount of damping per time step for the second-, fourth-, and sixth-order terms as a function of wavelength. For each curve, the values of K and $\Delta t$ have been chosen so that the $2\Delta x$ wave is completely removed, each time step. In

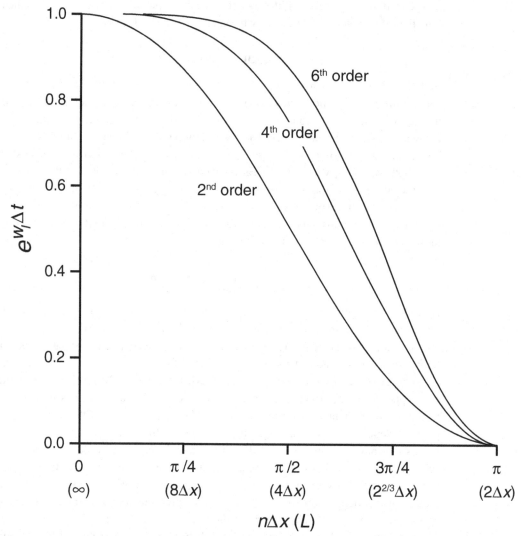

Fig. 3.34 Damping rate per time step (Eq. 3.63) for second-, fourth-, and sixth-order diffusion, for $K\Delta t$ chosen such that there is 100% damping of the $2\Delta x$ wave each time step. The abscissa is wavelength ($n\Delta x$, bottom) and wavenumber (top).

practice, it would be unusual to use such a large diffusion coefficient, but such an assumption allows us to normalize the curves to reveal relative damping rates. For example, for the reasonably well-resolved $8\Delta x$ wave, second-order diffusion removes ~15% of the amplitude each time step, whereas the higher-order approaches remove 1–2% or less. For the $4\Delta x$ wave, second-order diffusion removes about twice the amplitude per time step as does the fourth-order diffusion.

Figure 3.35 shows the result of the damping in Eq. 3.62, over multiple time steps, by the second- and sixth-order diffusion terms ($n = 2$ and $n = 6$, respectively). A square-wave, even though it possesses first-order discontinuities that would typically not exist for most variables in the atmosphere, is chosen for the initial shape of the feature to be diffused.

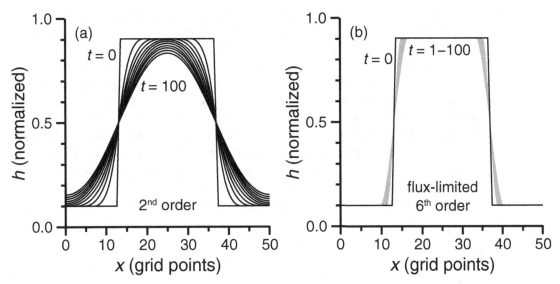

**Fig. 3.35**   Damping of a square wave by second-order diffusion (a, $n = 2$ in Eq. 3.62), and sixth-order diffusion with a flux limiter to prevent added noise from Gibbs phenomenon (b, $n = 6$ in Eq. 3.62). The curves in panel (a) correspond to the solution for the wave at intervals of $10\Delta t$. The gray area in panel (b) defines the envelope of the ten $10\Delta t$-interval curves. As with Fig. 3.34, $K\Delta t$ was chosen such that there is 100% damping of the $2\Delta t$ wave each time step. Adapted from Xue (2000).

Figure 3.35a shows that 100 time steps of the second-order diffusion, in addition to damping the small scales that make up the corners of the square wave, suppresses the amplitude of the $25\Delta x$ main wave by ~15%. Higher-order diffusion terms, even though they are more scale-selective, can introduce new local extrema, or noise, near large gradients in the model solution. This effect is known as the Gibbs phenomenon. In these schemes, the diffusive flux is not necessarily down gradient, leading to the nonphysical artifacts. One remedy to this problem is described in Xue (2000), wherein diffusive fluxes are set to zero whenever they are in the same direction as the gradient. Figure 3.35b is analogous to 3.35a, but pertains to the sixth-order diffusion with this flux limiter.

Figure 3.26 provides an additional illustration of the effect of diffusion operators of different order. Here, the one-dimensional, shallow-fluid, second-order, advection equation, with a Courant number of 0.1 that produced the "no diffusion" solution, is rerun with different-order diffusion terms. Even though the fourth-order-advection term used for Fig. 3.28 is more realistic, the second-order approach is employed in this illustration because it is easier to visualize the effects of the diffusion on the different wavelengths. For each experiment, the diffusion coefficient was chosen such that 10% of the $2\Delta x$ wave amplitude was damped each time step. The second-order diffusion damps all wavelengths, including the main wave. Fourth-order diffusion is more selective in its damping, affecting shorter waves the most. The sixth-order term only touches the very-shortest waves, especially those on the right side of the domain that have not yet exited the grid. Further discussions and examples of the use of diffusion terms or filters to remove small-scale wave energy can be found in Shapiro (1970, 1975), Raymond and Garder (1976, 1988), Raymond (1988), Durran (1999), Xue (2000), and Knievel *et al.* (2007).

It is worth noting that the diffusion term should be calculated on a quasi-horizontal surface, and not on a model's constant sigma or potential temperature vertical-coordinate surfaces. For example, the temperature on a terrain-following sigma surface will typically be a minimum over a mountain. Thus, diffusion of temperature (from high to low values) in the thermodynamic equation, if calculated on the sigma surface, will produce temperature increases over the mountain. This temperature rise over the elevated terrain will result in the development of an erroneous thermally direct wind circulation. Thus, for each grid point, the variable being diffused should be vertically interpolated from the model-coordinate surface to the horizontal surface passing through the grid point. The value of the diffusion term calculated on the horizontal surface should then be used in the tendency equation.

To illustrate the effect of this diffusion on the spatial spectrum of a model variable, Fig. 3.36 shows the kinetic-energy spectrum for a WRF-model forecast having a 10-km grid

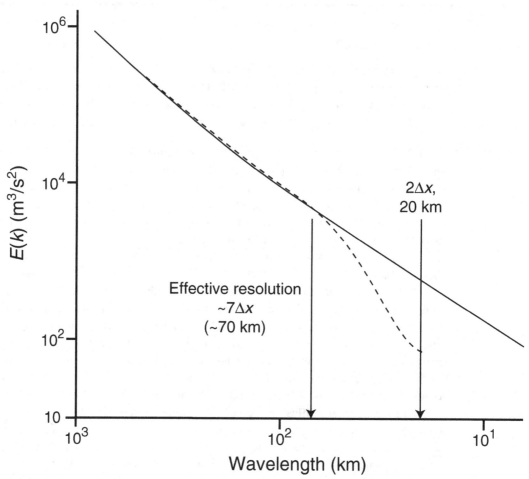

**Fig. 3.36**     The effect of diffusion on the kinetic-energy spectrum for a WRF-model forecast having a 10-km grid increment. The expected slope of $k^{-5/3}$ is shown as a reference, and is reproduced by the model for wavelengths above $7\Delta x$. But the energy between the $2\Delta x$ and $7\Delta x$ wavelengths has been damped by the diffusion, resulting in an effective resolution of 70 km, not 20 km. Adapted from Skamarock (2004).

increment. The expected slope of $k^{-5/3}$ is shown as a reference, and is reproduced by the model for wavelengths above $7\Delta x$. But the energy between the $2\Delta x$ and $7\Delta x$ wavelengths has been damped by the diffusion, resulting in an effective resolution of 70 km, not 20 km. See Frehlich and Sharman (2008) for an additional analysis of effective model resolution.

## Implicit numerical diffusion

Some finite-difference schemes, such as the forward-in-time and backward-in-space method described in Section 3.4.2, and the odd-order Runge–Kutta time-differencing schemes mentioned in Section 3.4.4, selectively damp certain wavelength bands (e.g., Fig. 3.23). If the damping is controllable and sufficiently scale-selective, this is a desirable property of the differencing scheme, and an explicit numerical diffusion term may not be needed to damp poorly resolved shortwave energy.

## Grid diffusion

This process results from the fact that the model variables at each grid point are affected each time step by the variables at neighboring grid points, through terms with spatial derivatives. The grid increment and the time step thus define the rate at which grid diffusion causes the spread of atmospheric properties through every nonzero term in the equations that involves a spatial finite-difference expression. Naturally, processes in the real atmosphere such as advection, turbulent diffusion, and inertia–gravity wave motion cause information to spread spatially, but grid diffusion is nonphysical, ubiquitous, and can act rapidly. For example, consider a model with a 25-km grid increment. If we assume a maximum wave speed of 50 m s$^{-1}$ anywhere on the grid, and if we conservatively require that the Courant number be less than 0.7, the time step would be 350 s. Also assume that the wind speed in the boundary layer is $5\,\mathrm{m\,s^{-1}}$. With a three-point finite-difference approximation to an advection term, where the tendency at each grid point uses information that is one grid increment away, the information propagates at a speed of $\Delta x/\Delta t$, or over 70 m s$^{-1}$. This fictitious propagation is over 10 times faster than the speed of the transfer of information by advection in the boundary layer. Even though this process is nonphysical, the resulting smoothing or mixing is sometimes used to represent the real diffusion in the atmosphere. Unfortunately, the effect is not controllable in terms of its overall strength and its ability to selectively damp small scales.

## 3.4.8 Numerical implications of the choice of the model vertical coordinate

The following sections provide a brief summary of the numerical implications of the use of the historically most-common choices for the vertical coordinate in NWP models. See Sundqvist (1979) for more information.

### Height above sea level

At face value, this would seem like an intuitively appealing coordinate. In particular, the coordinate surfaces are fixed relative to Earth's surface, and, unlike some other options, the pressure-gradient force in the momentum equations is represented by one term, the gradient of pressure. There are significant problems, however. Because the height surfaces are penetrated by orographic features at low levels, there are areas of grid points on the coordinate surfaces where atmospheric properties are undefined. This makes it virtually impossible to properly calculate derivatives on the constant-$z$ surfaces at the grid points located next to these voids, and it is impossible to employ spectral methods to define the horizontal variation of model variables. Lastly, grid-point-model codes typically scan systematically through the rows and columns of grid points in the matrix of points, and it becomes very cumbersome to interrupt this process with breaks in the pattern where the points do not exist.

There is an approach to the use of $z$ coordinates, called the volume-fraction or shaved-grid-cell method (Adcroft *et al.* 1997, Steppeler *et al.* 2002, Walko and Avissar 2008b), that avoids some of the above disadvantages. Computations for cells that are partially embedded below the terrain surface are modified to account for the kinematic effect of the barrier. Even with this approach, a disadvantage relative to terrain-following vertical-coordinate systems is that employing high vertical resolution immediately above the surface requires the use of many thin model layers when there is a large variation in topographic height, thus increasing the computational expense. Another disadvantage is that the grid cells that intersect topography have different properties than the rest, and thus they must be treated differently in the numerical algorithms.

## Pressure

Pressure is the variable that radiosondes use to define vertical position when observed values of dependent variables are transmitted to the ground station. So, in some sense, it may be reasonable to use this as the vertical coordinate in a model in which the radiosonde observations must be assimilated. However, this coordinate has virtually the same problems as does the height system. But the difficulty with orography interrupting the surfaces is even more problematic because the heights of the pressure surfaces change with time, and thus does the pattern of the grid points that are masked. During the integration, grid points will appear and disappear, and it is very difficult to assign realistic physical properties to grid points that are only temporarily part of the calculations.

## Potential temperature

Under hypothetical adiabatic conditions, the potential temperature ($\theta$) of a parcel does not change as it moves, and the parcel remains on $\theta$ surfaces. That is, $\theta$ surfaces are *material surfaces*, and when these surfaces are the vertical coordinate surfaces, the vertical motion ($d\theta/dt$) is zero. Even though both real and model atmospheric processes are close to being adiabatic, outside of the boundary layer and where phase changes are not consuming or releasing latent heat, they are never perfectly adiabatic because radiative flux divergences are never zero. So, over those large volumes where $d\theta/dt$ is small, vertical advection is also small, and artificial grid diffusion in the vertical is small. This reduced artificial vertical spread of moisture and other scalar quantities in θ coordinates leads to their more-realistic transport.

Because this variable is obviously linearly related to temperature, where temperature gradients are largest the model potential-temperature surfaces are more-tightly packed. This means that there is more vertical resolution in the model where it is needed most in order to represent large gradients, for example along quasi-horizontal frontal surfaces and near the tropopause. Figure 3.37 illustrates a cross section of a front in both pressure (a) and θ (b) coordinates. In θ coordinates (Fig. 3.37b), the strong wind shear in the frontal zone (shaded area) spans one-quarter of the vertical extent of the cross section, whereas in pressure coordinates (Fig. 3.37a) this shear is concentrated within a narrow region in the vertical. Also, the fact that the coordinate surfaces approximate material surfaces implies that horizontal gradients will be smaller than when the coordinate surfaces cut across fronts. Thus, the truncation error associated with horizontal and vertical derivatives will be smaller. Because isentropic surfaces intersect Earth's surface, with or without orography, potential temperature has disadvantages similar to those of the pressure and height coordinates. Also, near the strongly heated surface of Earth, potential temperature can decrease with height (superadiabatic lapse rates) in a shallow layer, below where it displays its normal increase with height. Thus, in a model with this vertical coordinate, any lapse rate that approaches the adiabatic value during the forecast must be artificially adjusted to a

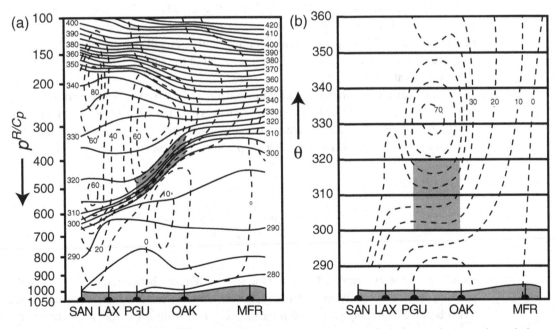

**Fig. 3.37**  Cross section of a front in pressure (a) and isentropic (b) coordinates, where the horizontal axis is north–south along the coast of the western USA. The gray area in the two cross sections spans the same volume of atmosphere. The solid lines are isentropes and the dashed lines are isotachs. From Benjamin (1989), based on Shapiro and Hastings (1973) and Bleck and Shapiro (1976).

more-stable one to avoid the situation where potential temperature is the same at two places on the vertical scale. Another problem that is in common with sigma coordinates (see below) is that the pressure-gradient term appears as the horizontal derivative of the Montgomery potential, which consists of the sum of two terms ($C_pT + gz$). The horizontal derivatives of these individual terms can be large, and the pressure gradient is represented by a small difference between the two large derivatives. Thus, noncancelling truncation errors in the derivatives can produce large errors in the pressure gradient.

## Sigma-p

The so-called sigma coordinate systems are terrain following and thus avoid the above noted problems of the height, pressure, and potential-temperature coordinates that intersect the land or water surface. The pressure-based sigma coordinate (Phillips 1957b, Gal-Chen and Somerville 1975) is defined as

$$\sigma = \frac{p - p_t}{p_s - p_t},$$

where $p_t$ is a constant pressure chosen for the top of the model, $p_s$ is the surface pressure, and $p$ is local pressure at any point in the column. If the top of the model is defined to be at the top of the atmosphere, we simply have $\sigma = p/p_s$. For $p = p_s$, the condition at the surface, $\sigma = 1$ everywhere. For $p = p_t$, $\sigma = 0$ everywhere. Thus, over any column of model atmosphere, $0 < \sigma < 1$. Because surface pressure and local pressure are functions of time, the vertical positions of sigma-coordinate surfaces will change. Figure 3.38 shows a vertical cross section of sigma surfaces in a model of the eastern USA, where the model top is located at 500 hPa.

As noted earlier, the pressure gradient in equations expressed in sigma vertical-coordinate systems consists of two terms. In the sigma-p system, one contains the derivative of the surface pressure, $p^*$, and the other contains the derivative of the geopotential height of the sigma surface, as shown in the following pressure-gradient term from the first equation of motion.

$$\frac{\partial p^* u}{\partial t} \propto -mp^* \left( \frac{RT}{p^* + \dfrac{p_t}{\sigma}} \frac{\partial p^*}{\partial x} + \frac{\partial \phi}{\partial x} \right).$$

Each term is potentially very large, and the small difference between them represents the pressure gradient force. Where there are large terrain-elevation gradients, these individual terms become especially large, and truncation errors that do not cancel in the two terms create erroneous pressure gradients and accelerations. This issue is partly addressed by defining a base state condition, and by using perturbation forms of the equations where the derivatives apply to departures from a mean state. See Mesinger *et al.* (1988) for a discussion of the history and shortcomings of this coordinate.

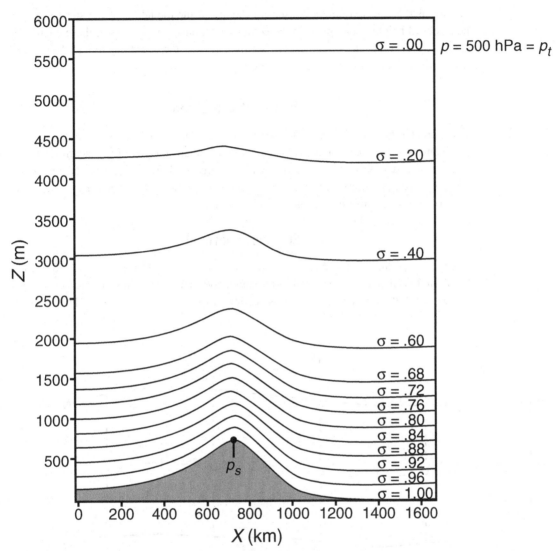

Fig. 3.38 Cross section of sigma surfaces for a model grid over the eastern USA. The model top is at 500 hPa. Adapted from Warner *et al.* (1978).

## Sigma-z

Where the above sigma-p coordinate system is normalized by the pressure depth of the model atmospheric column, the sigma-z system (Kasahara 1974) is normalized by the physical depth of the atmosphere. Specifically

$$\sigma = \frac{z_t - z}{z_t - z_s},$$

where $z_t$ is the constant height chosen for the top of the model, $z_s$ is the surface height, and $z$ is local height at any point in the column. Obviously the heights of these coordinate surfaces do not vary with time. Like the sigma-p system, the coordinate ranges from 0 to 1 through the depth of the model atmosphere.

### Hybrid isentropic-sigma

This approach involves the use of terrain-following sigma coordinates in the lower troposphere and isentropic coordinates above. It retains the advantages of the isentropic representation, but avoids the previously mentioned major shortcoming that occurs in the boundary layer. A variety of these hybrid schemes has been developed. Benjamin *et al.* (2004b) should be consulted for more information.

### Step-mountain (eta)

A vertical coordinate described by Mesinger *et al.* (1988), Black *et al.* (1993), Black (1994), and Wyman (1996) is the step-mountain coordinate, also known as the eta coordinate. Figure 3.39 shows a vertical cross section of the coordinate surfaces. The approach

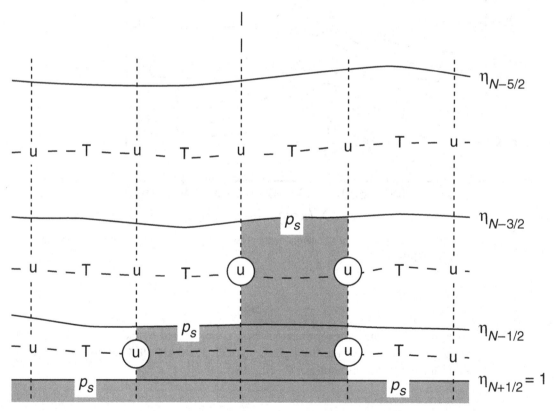

**Fig. 3.39** Cross section of the three lower model levels for the step-mountain coordinate system, showing where variables are defined. The shaded area represents the land surface. From Mesinger *et al.* (1988).

was motivated in order to avoid the problems of the sigma system that are associated with steep orographic slopes. As seen in the figure, the orography is constructed from the three-dimensional grid boxes of the model, with the surface elevation being defined with a discrete set of values. At the vertical surfaces, which are essentially internal hard boundaries, the normal component of the velocity (circled in the figure) is zero. The coordinate surfaces are quasi-horizontal. The eta coordinate is defined as

$$\eta = \frac{p - p_t}{p_s - p_t}\eta_s = \eta_s\sigma,$$

where $\sigma$ is the definition of the sigma coordinate provided above, $p_t$ is the pressure at the model top, $p_s$ is the pressure at the surface, and

$$\eta_s = \frac{p_{rf}(z_s) - p_t}{p_{rf}(0) - p_t},$$

which is the value of eta at Earth's surface. The reference pressures, $p_{rf}$, correspond to the pressures at the interface between model layers. For $p_t = 0$, the eta-coordinate's definition simplifies to

$$\eta = \eta_s\sigma = \frac{p_{rf}(z_s)}{p_{rf}(0)}\sigma.$$

For flat terrain ($z_s = 0$), the eta coordinate is identical to the sigma coordinate. See the above references for example simulations using this coordinate.

### 3.4.9  Time smoothers and filters

Propagating disturbances can be damped by both space and time smoothers. The explicit numerical diffusion operators described above in Section 3.4.7 are intended to damp or smooth small-scale disturbances in terms of the spatial variability. Other operators smooth in the time dimension. Again, propagating disturbances may be smoothed in either way. In particular, there are situations where, with centered time-differencing methods, the model solutions at odd and even time steps can depart from each other. This *separation of the solution* results from the fact that, after the initial forward time step, the leapfrog differencing allows even and odd time steps to affect each other only through the derivative. That is, the leap is from even-to-even and odd-to-odd time steps. This $2\Delta t$ oscillation can easily be damped with a time smoother. One of the most popular is described by Asselin (1972),

$$\alpha^\tau = (1 - \beta)\alpha^\tau + \frac{\beta}{2}(\alpha^{\tau+1} + \alpha^{\tau-1}),$$

where $\alpha$ is any dependent variable and a typical value of $\beta$ is 0.1. It can be applied at every time step, or intermittently.

# 3.5  Lateral-boundary conditions

The values of dependent variables must be specified (i.e., not internally calculated) at the lateral edges of the computational grids of Limited-Area Models (LAMs; that is, everything but global models). Even though some global models used for weather or climate prediction are capable of resolving mesoscale processes, for the foreseeable future there will be a need to embed even-higher resolution LAM grids within the coarser models. Thus, the challenges of dealing with Lateral-Boundary Conditions (LBCs) will need to be addressed. The LBCs should have the following properties.

- Meteorological features should propagate from coarse- to fine-mesh grids without significant distortion.
- Inertia–gravity waves should propagate through the boundary, especially longer-length waves that are related to important physical processes such as geostrophic adjustment. Shorter waves may be damped on outflow, but they should not be reflected.
- The LBCs should not allow artificial dynamic/numerical feedbacks between grids that can cause a catastrophic termination of the model integration.

Note that there are numerous references that describe various kinds of evidence of the potentially serious effect of LBC error on LAM forecasts (e.g., Miyakoda and Rosati 1977, Oliger and Sundstrom 1978, Gustafsson 1990, Mohanty *et al.* 1990, and Warner *et al.* 1997). Much of the following analysis of LBC effects is based on Warner *et al.* (1997).

## 3.5.1  Sources of LBC error

Because the negative impacts of LBCs on LAM solutions are inevitable, our objective should be to understand the nature of the problems well and learn how to mitigate their effects. The LBC's negative influences can be attributed to at least six factors.

- *Low resolution of LBC data* – Open LBCs (see Section 3.5.3) are defined based on forecasts from coarser-resolution models or analyses of observations, depending on whether the LAM is being used for operational or research applications. In either case, the horizontal, vertical, and temporal resolution of the boundary information is generally poorer than that of the LAM, and thus the boundary values interpolated to the LAM grid at every time step have the potential of degrading the quality of the solution.
- *Errors in the meteorology of the LBCs* – Even if the LBC-data resolution is hypothetically similar to that of the LAM, and there is little interpolation error, the quality of the LBC data may be erroneous for other reasons, especially if they are based on other model forecasts. That is, the forecast that provides the LBCs may simply be wrong in some important respect having nothing to do with its resolution. In any case, these errors will be transmitted to the LAM domain at the grid interface.
- *Lack of interactions with larger scales* – Specified LBCs determine the computational-domain-scale structure of the meteorological fields. But, these longer wavelengths

cannot interact with the model solution on the interior. This limited spectral interaction can affect the evolution of the LAM forecast because the LAM solution cannot feed back to the large scales.

- *Noise generation* – The specific LBC formulation used can produce transient, nonmeteorological, inertia–gravity modes on the LAM domain. Even though these modes are thought to not interact strongly with the meteorological solution, they are superimposed on the physically realistic fields and can complicate the interpretation of the forecast.
- *Physical-process parameterization inconsistencies* – The physical-process parameterizations may, sometimes out of necessity, be different for the LAM and the coarser-resolution model providing the LBCs. The resulting inevitable differences in the solution at the boundary may cause spurious gradients and feedbacks between the two grids, which can influence the solution on the LAM domain.
- *Phase- and group-speed contrasts* – Earlier in this chapter it was shown that some differencing schemes can cause phase- and group-speed errors whose magnitude depends on how well a wave is resolved on the grid. Thus, as a wave passes between computational areas with different grid increments, waves can be stretched or compressed. Browning *et al.* (1973) refer to a numerical refraction effect resulting from the phase-speed differences, that causes "unexpectedly large errors" on the coarse mesh of two-way interacting grids.

### 3.5.2  Examples of LBC error

At least four general types of studies have been performed, from which we can gain insight into LBC error. One involves the application of model computational domains of different size, and from these simulations a direct determination is made of the effect of the proximity of the lateral boundaries on some measure of the quality of the simulation. Another type can be grouped into the general category of mesoscale predictability studies wherein a control simulation is first performed with a LAM. Then, perturbations are imposed on the model initial conditions or LBCs, and the differences between the model solutions with and without the perturbations are analyzed and ascribed to specific factors, including the LBCs. A third category of study uses an adjoint model from which actual LBC-sensitivity fields are produced directly. Relevant studies from which we can gain insight are described here. And a fourth type is the Big-Brother–Little-Brother experiment, which is discussed in more detail in Chapter 10.

#### Domain-size sensitivity studies

One of the first studies of the effects of defining LAM LBCs with a coarser-resolution forecast was that of Baumhefner and Perkey (1982). A LAM (Valent *et al.* 1977) with a 2.5° latitude–longitude (lat–lon) grid was embedded within, and obtained its LBCs from, a 5° lat–lon hemispheric model (Washington and Kasahara 1970). Both models used the same vertical grid structure and physical-process parameterizations. The LBC "error" was first assessed by comparing the solution from this nested system with that from a non-nested, 2.5° lat–lon version of the hemispheric model. Figure 3.40 shows the midtropospheric pressure error (difference between the LAM and hemispheric-model solutions) associated with the LBCs for a 48-h forecast period. Large pressure errors with amplitudes of 5–10 hPa propagate rapidly onto the forecast domain at middle and high latitudes, primarily from the west and north boundaries, with speeds of 20°–30° lon day$^{-1}$. Comparison of this error distribution with the location of synoptic disturbances (not shown) shows that the error maxima are associated with areas in which significant changes are taking place at the boundaries. The fairly inactive large-scale meteorological conditions in the subtropics and tropics generate very little LBC error. For LAM simulations in which the LBCs were provided by a 2.5° lat–lon hemispheric model (i.e., the LAM and hemispheric models had the same horizontal resolution), errors were also large and had a similar distribution, indicating that significant LAM errors in these regions resulted from

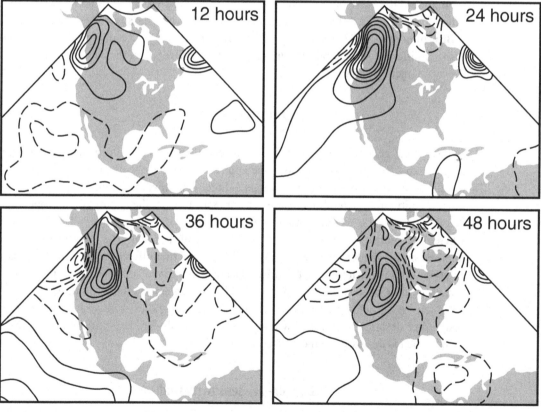

**Fig. 3.40** For different simulation lead times, pressure difference at 6 km ASL (about 500 hPa) between simulations from a 2.5° lat–lon hemispheric model and a limited-area model with the same resolution embedded within a 5° lat–lon hemispheric model. The differences are associated with boundary-condition effects. The area delineated is that of the LAM domain. The isobar interval is 1 hPa, and negative values are dashed. Adapted from Baumhefner and Perkey (1982).

the LBC formulation itself and not just the quality of the LBC data. Figure 3.41 summarizes the Root-Mean-Square (RMS) error growth in 500-hPa heights on the limited-area domain associated with the use of LBCs from the 2.5° (dotted curve) and 5° (dashed

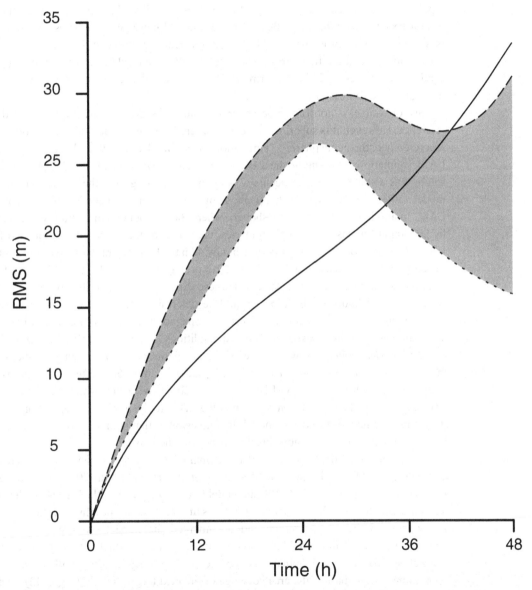

**Fig. 3.41** RMS 500-hPa height differences (m), where the solid line shows the difference between 5° and 2.5° hemispheric simulations over the area of the LAM grid, the dashed line shows the difference between the 2.5° hemispheric simulation and that from the 2.5° LAM whose LBCs are provided by the 5° hemispheric simulation, and the dotted line shows the difference between the 2.5° hemispheric simulation and that from the 2.5° LAM whose LBCs are provided by the 2.5° hemispheric simulation. The abscissa is forecast hours. From Baumhefner and Perkey (1982).

curve) lat–lon hemispheric models. It is revealing that the error growth in the LAM is similar whether or not the LBC information was defined by a model of the same or worse resolution. The solid curve shows the difference between the 2.5° and 5° hemispheric simulations over the area of the LAM domain, and represents the error that is associated with the use of the 5° unbounded grid compared to the 2.5° unbounded grid. The most rapid error growth is during the first 24 h for both the 2.5° and 5° LBCs. The fact that the error associated with the 2.5° LBCs decreases after 24 h probably indicates that some of it is related to rapidly propagating and damped transients generated at the lateral boundaries early in the simulation. In contrast, when the 5° LBCs are used there is a continuing propagation of coarse-resolution information that causes the error to generally increase throughout the forecast.

This, of course, is not true forecast error because observations are not being used as a reference. However, it is sobering to see that, when the hemispheric 2.5° simulation is used as a reference, the hemispheric 5° simulation shows smaller error than do either of the 2.5° LAM simulations containing the LBC error. That is, when using the 2.5° hemispheric solution as a standard, higher accuracy is obtained by using only the coarse hemispheric model rather than the coarse hemispheric model with an embedded higher-resolution LAM. In another experiment (not shown), where the computational domain was extended by 20° lon at the east and west boundaries, the center of the domain was protected from LBC contamination for a longer period, but by 48 h the high central latitudes were contaminated from both the east and the west by error propagating inward at about $30°$ lon day$^{-1}$. Baumhefner and Perkey (1982) state that "these experiments lead to the not too surprising conclusion that boundary locations should be determined from the forecast time frame selected and the typical boundary error propagation rate." Comparison of model-simulation error defined based on observed conditions for the 2.5° hemispheric model and the 2.5° LAM embedded within the 5° hemispheric model revealed that the LBCs increased the total simulation error at high latitudes by up to 50% after 24 h. That is, the total error growth from all non-LBC sources is about twice that which is related to the LBCs. Naturally, the relative contribution of the LBCs to the total error depends greatly on the overall predictive skill of the model. It is noteworthy that similar results were obtained using two totally different algorithms for specifying the LBCs.

A well-controlled demonstration of the domain-size problem is described by Treadon and Petersen (1993), who performed a series of experiments with 80- and 40-km grid-increment versions of the US NWS Eta model (Black *et al.* 1993, Black 1994) with a winter and summer case. While maintaining the same resolution and physics, they progressively reduced the area coverage and documented the impact on forecast skill. The "control simulation" utilized the full computational domain of the Eta model, while experimental simulations used domains that were progressively smaller, with each having approximately one-half of the area coverage of the next larger domain (Fig. 3.42). In each case, US NWS global spectral, T126, previous-cycle forecasts were used for LBCs. For a winter cyclogenesis case, the 80- and 40-km grid-increment models with the full domain produced reasonably accurate forecasts. However, the forecast on the smallest domain, which had its lateral boundaries close to the area affected by the storm, had 500-hPa RMS

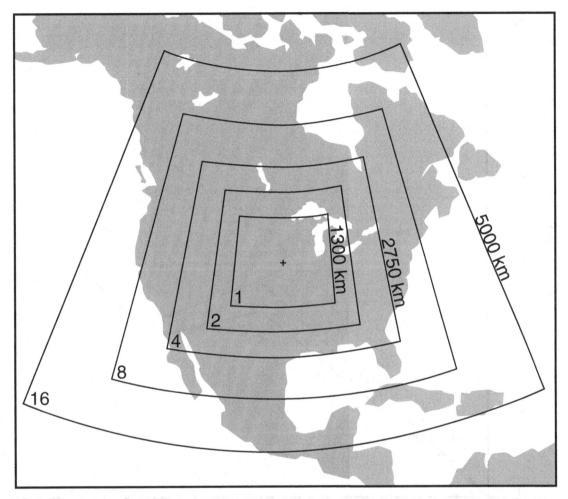

**Fig. 3.42**  Five integration domains of the 80-km grid-increment Eta model used in the domain-size sensitivity study. The grid number corresponds to the factor by which the grid area is larger than that of the smallest grid. From Treadon and Petersen (1993).

height errors that were twice as large (relative to data analyses) as those of the forecast on the full domain, by only 12 h into the forecast period. In addition, the surface low-pressure center was much weaker than observed, and was erroneously placed, in the forecast on the smallest domain. For a summer case, with much weaker flow over the small domains, the error growth was qualitatively similar to that of the winter case. Again, RMS 500-hPa height errors were more than twice as large on the smallest domain than they were on the largest domain by the 36-h forecast time (Fig. 3.43). An example is shown in Fig. 3.44 of the rapid influence that the LBCs can have at upper levels, even when the cross-boundary flow is weak to moderate. For this summer case, Fig. 3.44 illustrates two 12-h simulations of 250-hPa isotachs from the 40-km grid-increment Eta model. Figure 3.44a shows a

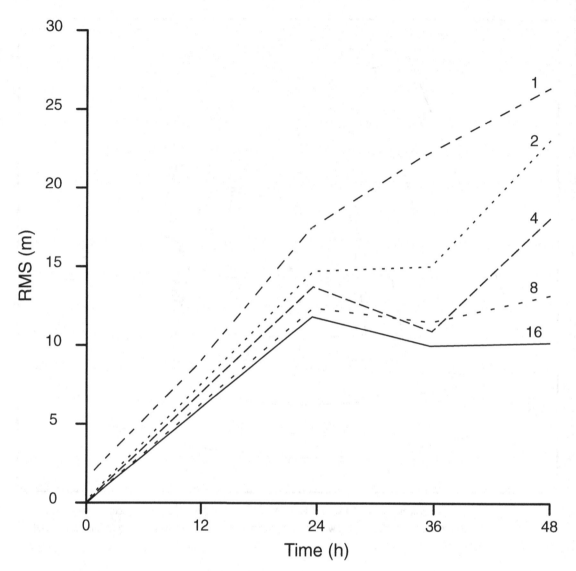

**Fig. 3.43** Temporal evolution of the RMS 500-hPa height errors (relative to data analyses) associated with the use of each of the five computational grids in Fig. 3.42, for a forecast initialized at 0000 UTC 3 August 1992. For each experiment, the errors were calculated for the same area of the innermost grid (number 1). The abscissa is forecast time. The grid numbers correspond to those defined in Fig. 3.42. From Treadon and Petersen (1993).

strong narrow jet streak simulated on the largest domain, while Fig. 3.44b shows that the same feature on the smallest domain (with the same resolution) has been considerably smoothed.

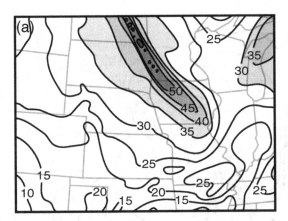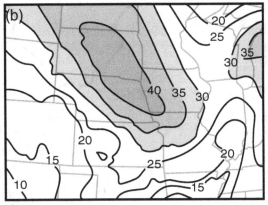

**Fig. 3.44** Twelve-hour simulations of 250-hPa isotachs (m s$^{-1}$) from the 40-km grid-increment Eta model initialized at 1200 UTC 3 August 1992, based on experiments that used the largest computational domain (a) and the smallest (b). Both maps apply for the area of the smallest grid. The isotach interval is 5 m s$^{-1}$. From Treadon and Petersen (1993).

## Mesoscale predictability studies

Predictability studies with mesoscale LAMs have demonstrated that error growth is much different than what has been documented for global models (Anthes *et al.* 1985, Errico and Baumhefner 1987, Vukicevic and Paegle 1989, Warner *et al.* 1989). When small perturbations (errors) are added to the initial conditions (but not the boundary conditions) of a LAM, the simulation from the perturbed initial state and that from the unperturbed (control) initial state do not diverge as they would with an unbounded model. The perturbed atmosphere on the domain interior is advected out of the domain at the outflow boundaries, and the use of identical LBCs in the two simulations causes unperturbed atmosphere to be swept in at the inflow boundaries.

In a predictability study that is revealing of LBC effects, Vukicevic and Errico (1990) used a relatively coarse resolution version of The Pennsylvania State University–NCAR Mesoscale Model Version 4 (MM4) with a grid increment of 120 km for a 96-h simulation of Alpine cyclogenesis. The LBCs were defined for MM4 using data analyses, and simulations from the NCAR global Community Climate Model Version 1 (CCM1) that was initialized at the same time as the LAM. In one experiment, a control simulation was first performed with MM4, and then the initial conditions were perturbed and the model was again integrated. The LBCs were based on analyses of data and were thus "forecast-error free" and the same for both simulations. Figure 3.45 shows the 96-h 500-hPa geopotential-height differences between the two simulations. The largest differences between the two simulations are on the eastern, downwind half of the domain because the identical LBC data strongly influence the model solutions on the western half.

To gain further insight about LBC effects on LAM solutions, an additional experiment used normal (control) and perturbed-initial-condition CCM1 forecasts to define the LBCs of a corresponding pair of MM4 forecasts that had initial conditions that were identical

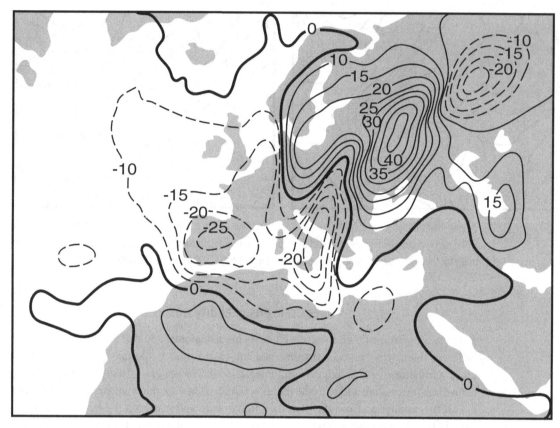

**Fig. 3.45**  The 500-hPa 96-h geopotential-height difference between a control simulation with MM4 and a parallel one with per-turbed initial conditions. The contour interval is 5 m. The LBCs were identical and based on analyses of observations. From Vukicevic and Errico (1990).

and equal to those of the control CCM1 simulation. The perturbed CCM1 initial conditions were defined so as to emulate expected operational measurement errors. Thus, this experimental design has considerable relevance to operational forecasting with a LAM because it isolates the effects of normal errors in a coarse-mesh forecast on the dynamical evolution of a LAM forecast for which it provides LBCs. Figure 3.46 shows the 500-hPa geopotential-height difference in the two 6-h LAM solutions, where differences of over 10 m appear near the domain center over Europe. During this short time, high-frequency transient modes resulting from the LBC formulation have contaminated the entire domain. It is important to recognize that the LAM domain employed here has perhaps four times the area of many LAMs, and thus the LBC error effects would normally be felt on considerably shorter time scales. Based on these results, Vukicevic and Errico (1990) state that "medium range forecasts with nested limited-area models may not significantly reduce RMSEs relative to the same forecasts performed with global models."

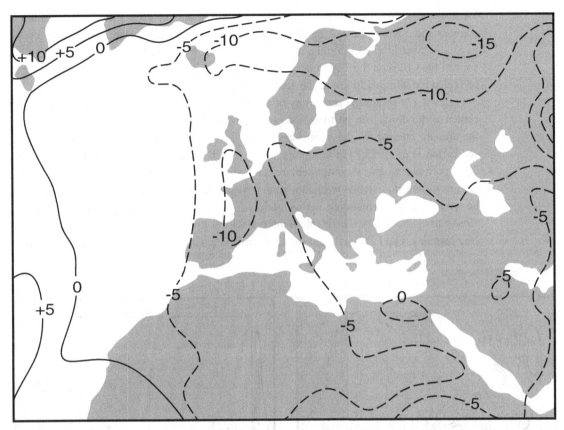

**Fig. 3.46**   The 500-hPa geopotential height difference between a 6-h control simulation with MM4 and a parallel one with perturbed LBCs. The contour interval is 5 m, where negative values are dashed. Adapted from Vukicevic and Errico (1990).

## Adjoint sensitivity studies

Variational techniques employing an adjoint model have been used to investigate the sensitivity of LAM forecasts to initial conditions and boundary conditions. The adjoint operator produces fields that indicate the quantitative impact on a particular aspect of the forecast of any small, but arbitrary, perturbation in initial conditions, boundary conditions, or model parameters. This approach has an advantage over the traditional types of predictability studies discussed above in that the resulting dependencies are not sensitive to the specific perturbations applied to the initial or boundary conditions. For a more in-depth discussion of this technique, the reader should consult Hall and Cacuci (1983), Errico and Vukicevic (1992), and Errico (1997).

Errico *et al.* (1993) applied this approach to investigate the sensitivity of LAM simulations to conditions on the domain interior and LBCs. A dry version (no moisture variables) of the MM4 model and its adjoint were employed, where the model had a grid increment of 50 km and 10 computational layers. The LBCs were provided by linear temporal interpolation between 12-h T42 analyses from the European Centre for Medium-Range

Weather Forecasts (ECMWF). The sensitivity was tested in 72-h simulations of both a summer and a winter case. A number of aspects of the simulations were investigated relative to their sensitivity to initial and boundary conditions. We will concentrate on the influence of the LBCs on the 72-h relative vorticity at the 30 grid points on each computational level that are within 150 km of the center of the domain.

Figure 3.47a shows the sensitivity of the 72-h relative vorticity in a small column in the center of the domain to perturbations of the initial 400-hPa $v$-component of the wind on the domain interior for the winter case. (For further discussion of the sensitivity metric, see Errico *et al.* 1993.) For comparison, Fig. 3.47b illustrates the sensitivity of the same 72-h vorticity to the $v$-component of the wind on the lateral boundaries. The LBC-sensitivity metric extends over four rows and columns of grid points near the boundary because the LBC formulation in this model is such that LBCs are defined at all four points closest to the boundary. The isopleth intervals differ greatly between Figs. 3.47a and 3.47b (see caption). The LBC and grid-interior sensitivities are only in the upwind directions to

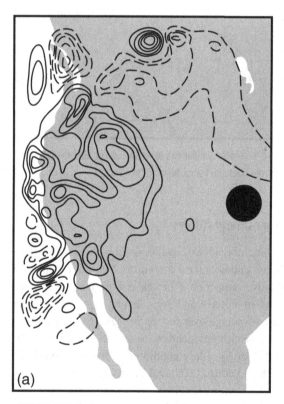

**Fig. 3.47**  Sensitivity of the 72-h relative vorticity in a limited volume in the center of the domain (black circle, panel a) to perturbations of the 400-hPa $v$-component of the wind on (a) the domain interior and (b) the lateral boundaries at the initial time for the winter case. For panel "a" ("b"), the maximum absolute value is 1.4 units (8 units), and the isopleth interval is 0.25 unit (1 unit). Only the western one-half of the computational domain is shown. Adapted from Errico *et al.* (1993).

**Table 3.2** Maximum values of the metric of the sensitivity of the 72-h relative vorticity near the center of the domain to the 400-hPa $v$-wind component on the lateral boundaries and on the domain interior. Values are shown pertaining to the sensitivity of the 72-h vorticity to the $v$-component perturbations at four times during the simulation.

|                              | Simulation time (h) | | | |
| --- | --- | --- | --- | --- |
|                              | 0   | 24  | 48  | 60  |
| Lateral boundary sensitivity | 8   | 40  | 150 | 52  |
| Interior sensitivity         | 1.4 | 18  | 76  | 93  |

*Source:* From Warner *et al.* (1997).

the west and north. Table 3.2 summarizes the maximum value of the sensitivity metric on the domain interior and on the lateral boundaries at four times during the simulation, and indicates that, as expected, the sensitivity of the 72-h vorticity to conditions on the domain interior is less for early times of the simulation. That is, the 72-h vorticity simulation tends to "forget" the impact of the perturbations as these conditions become more temporally removed. In terms of the effect on the 72-h simulation, the 48-h LBCs are more important than those at other times because the 24-h difference (between 48 h and 72 h) is the time required for the LBC signal to propagate to the center of the domain at this level. It is interesting that the 72-h forecast is less sensitive to initial condition ($t = 0$) perturbations (1.4 units) than it is to LBC perturbations at any time (8–150 units). The results for lower levels in the model (i.e., perturbations below 400 hPa) with weaker winds are qualitatively similar except that it naturally requires more time for LBC effects to penetrate to the center of the domain. For the summer case, the weaker wind speeds cause a factor-of-two slower propagation of the sensitivity.

## Big-Brother–Little-Brother experiments

In these experiments, a high-resolution model whose grid spans a large area is used to generate a reference simulation. This is the *Big-Brother simulation*. Then, using the identical model, another simulation is performed for a sub-area within the reference-simulation's grid. Lateral-boundary conditions are provided based on a data set that results from filtering all but the larger scales from the Big-Brother solution. This is the *Little-Brother simulation*. Because the experimental conditions in the two simulations are exactly the same, except for the presence of the LBCs in the Little-Brother simulation, differencing the two model solutions over the area of the smaller grid isolates the effect of the LBCs. See the discussion in Section 10.4 for additional information and references about this type of experiment. An example illustrating LBC effects that have been isolated using this method is shown in Fig. 3.48. Shown is the computational-domain-averaged precipitation rate for the area of the small grid, based on both the Big-Brother and Little-Brother simulations. The Canadian Regional Climate Model (Caya and Laprise 1999) was employed here in a test of regional climate modeling methods. The existence of the LBCs in this case had very little effect on the average precipitation rate.

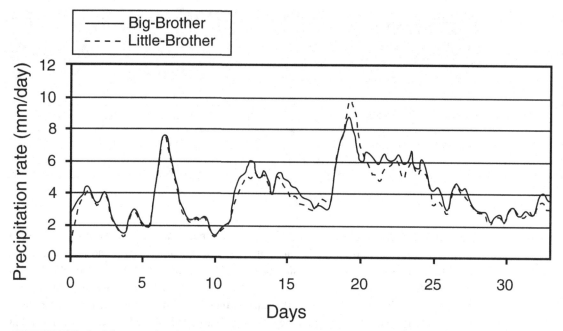

**Fig. 3.48** Spatially averaged precipitation rate for the area of the small grid in a Big-Brother–Little-Brother-experiment, for both the Big-Brother and Little-Brother simulations. Adapted from Denis *et al.* (2002).

### 3.5.3 Types of LBC formulations

Open, or free, LBCs allow values of variables to be externally specified based on forecasts from a larger model grid (e.g., a global model), or from gridded analyses of data. There are two approaches for defining LBCs from a coarser-resolution grid. One involves the simultaneous integration of the LAM and a coarser-mesh model within which it is embedded, where the information flow between the domains is in both directions. See Harrison and Elsberry (1972), Phillips and Shukla (1973), and Staniforth and Mitchell (1978) for a discussion of such techniques. In the other approach, LBCs are prescribed based on the output from a previous integration of a coarser-mesh model or an analysis of observations. The development of these techniques is described in Shapiro and O'Brien (1970), Asselin (1972), Kesel and Winninghoff (1972), and Anthes (1974). The first approach is called *two-way interactive nesting*, and the latter is called *one-way*, or *parasitic, nesting*. In both cases, meteorological information from the coarser-mesh domain must be able to enter the fine-mesh domain, and inertia–gravity and other waves must be able to freely exit the fine-mesh domain. With the two-way interacting boundary conditions, the information from the fine mesh can affect the solution on the coarse mesh, which can feed back to the fine mesh. An example of the desirability of this approach is provided in Perkey and Maddox (1985), who use numerical experiments to show that a convective-precipitation system can influence its large-scale environment, which can then feed back to the mesoscale. Note that LAMs that employ a two-way interacting nested grid system must generally obtain LBCs for their coarsest-resolution domain from a previously run global model or from analyses of observations. Thus, whether or not a two-way interacting nesting strategy is employed, the use of a one-way interacting interface condition is almost always necessary.

For the interface condition between domains of a two-way interacting nest, a variety of approaches are successfully used for interpolating the coarser-grid solution to the finer grid, and for filtering the finer-grid solution that is fed back to the coarser grid (Clark and Farley 1984, Zhang *et al.* 1986, Clark and Hall 1991). For one-way interacting grids, techniques are common that filter or damp small scales in the fine-mesh solution near the boundary (Perkey and Kreitzberg 1976, Kar and Turco 1995). For example, in the Perkey and Kreitzberg (1976) approach, a wave-absorbing or sponge zone near the lateral boundary prevents internal reflection of outward-propagating waves through an enhanced diffusion as well as truncation of the time derivatives. In these approaches, the fine grid is forced with large-scale conditions through a relaxation or diffusion term (Davies 1976, 1983, Davies and Turner 1977).

It is intuitive that two-way interactive nesting should provide for better model solutions on the finest grid than does one-way, parasitic nesting, simply because upscale effects can feed back to the fine mesh. This has, in fact, been demonstrated, for example by Clark and Farley (1984) for forced gravity-wave flow, and as noted earlier by Perkey and Maddox (1985) for convection. However, there are sometimes practical reasons for one-way nesting. For example, for operational nested modeling systems, one-way nesting allows the coarse-grid forecast to be completed first, and the products made available quickly to forecasters while the more computationally intensive calculations are taking place for the finer grids. And, in situations where significant computer-memory limitations exist, it is sometimes essential to limit calculations to one grid at a time.

For simple research or educational models, *periodic*, or *cyclic*, LBCs may be employed. Here, the grid points near one edge of the domain are coupled with those near the opposite edge, so that features that exit at one boundary enter at the other. This is illustrated in Fig. 3.49 for a three-point horizontal differencing scheme applied on a one-dimensional grid. At each time step, after the extrapolation in time is performed for grid points 2 through $j_{max-1}$, the values of the variables at the penultimate points are used to redefine the values at the corresponding edge points. For a model that uses five-point horizontal

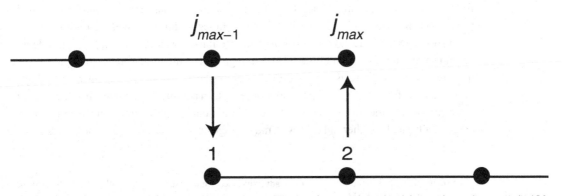

**Fig. 3.49**   Transfer of information between grid points at the edges of a computational grid that employs cyclic, or periodic, LBCs.

differencing, one more overlap point would be required near each edge. In a model with two horizontal dimensions, LBCs can be periodic in both directions. Or, a periodic LBC can be employed in only one direction, and an impervious wall boundary condition can be assumed on the other two edges, in which case the model is called a channel model.

### 3.5.4  Some practical recommendations

The studies described in the previous section, and others, are synthesized into the following recommendations for how LBC effects can be minimized in any LAM application.

#### (1) Utilize a lateral-boundary *buffer zone*

The LBC errors that reach the central part of a LAM grid are sometimes so large as to render the LAM forecast to be of no more value than that of the coarser-mesh model that produces the LBCs. In this situation, if enough computational resources are available, the lateral boundaries can be distanced from the central part of the LAM grid so that LBC errors do not penetrate to this region during a forecast with the desired duration. Alternatively, a standard domain area can be employed and the forecast duration can be limited so as to prevent penetration of the LBC errors into the central area of meteorological interest. Table 3.3 illustrates the domain sizes and forecast-duration limits that are necessary in order to minimize LBC impacts for different meteorological regimes and forecast-area length scales.

To illustrate the implications of the need for an LBC "buffer zone", a typical LAM configuration will be assumed, and the useful length of the forecast will be calculated. It will be assumed that the lateral boundaries are removed in each direction from the area of meteorological interest (having length scale $L$) by a distance equal to one-half $L$. For example, if the computational domain has 100 grid points in each direction, the inner protected area of meteorological interest on the model domain is represented by the central subset of $50 \times 50$ points. Most modelers would agree that this is a reasonable compromise, even though there are three times as many computational points in the buffer-zone region outside the area of interest than there are in it. This seemingly large computational "overhead" is generally accepted as unavoidable. The useful period of the forecast is defined here as the time required for LBC influences to advect to the central forecast area.[5] Table 3.3 shows the useful forecast periods (entry a) for four different computational areas with different scales (rows), and for four different meteorological regimes (columns). Average midtropospheric wind speeds ($S$ in Table 3.3) are used in the advection-time calculation for midlatitude winter and summer regimes, and for the tropical regime. For the midlatitude-uncoupled regime, it is assumed that there is weak vertical coupling and that the dominant meteorological processes are forced by lower-tropospheric effects. The smallest domain has the size of a large city (row 1, metropolitan area), the next larger one spans an area equivalent to the coverage of a typical weather radar (row 2, radar-range area), the next larger one covers about a

---

[5]  For simplicity, it is assumed here that the advective speed represents the speed with which LBC error penetrates inward on the LAM domain. However, LBC errors may be propagated by nonadvective waves such as inertia–gravity or Rossby waves.

**Table 3.3** For four different computational areas (rows) and four different meteorological regimes (columns): [a]useful duration of forecasts for a standard domain; [b]width of buffer zone required (in units of $L$, defined in column 2) for forecasts of "standard" duration (defined in column 3); and [c]ratio of buffer-zone grid points to central forecast-area grid points for forecasts of "standard" duration.

| Forecast domain size | Interior forecast-area length scale ($L$) | "Standard" forecast duration | Meteorological regimes | | | |
|---|---|---|---|---|---|---|
| | | | Winter mid lat $S = 30$ m s$^{-1}$ (~60 kt) | Summer mid lat $S = 15$ m s$^{-1}$ (~30 kt) | Tropical $S = 8$ m s$^{-1}$ (~15 kt) | Mid lat uncoupled $S = 5$ m s$^{-1}$ (~10 kt) |
| Metropolitan area | 50 km | 6 h | [a]14 min [b]13.0 $L$ [c]724 | [a]28 min [b]6.5 $L$ [c]194 | [a]52 min [b]3.5 $L$ [c]63 | [a]1.4 h [b]2.2 $L$ [c]27 |
| Radar-range area | 500 km | 18 h | [a]2.3 h [b]3.9 $L$ [c]76 | [a]4.6 h [b]1.9 $L$ [c]23 | [a]8.7 h [b]1.0 $L$ [c]8 | [a]13.9 h [b]0.6 $L$ [c]4 |
| Regional area | 2000 km | 36 h | [a]9.3 h [b]1.9 $L$ [c]23 | [a]18.5 h [b]1.0 $L$ [c]8 | [a]34.7 h [b]0.5 $L$ [c]3 | [a]55.6 h [b]0.3 $L$ [c]1.7 |
| Continental area | 5000 km | 72 h | [a]23.1 h [b]1.6 $L$ [c]16 | [a]46.3 h [b]0.8 $L$ [c]6 | [a]86.8 h [b]0.4 $L$ [c]2 | [a]138.9 h [b]0.3 $L$ [c]1.3 |

*Source:* From Warner *et al.* (1997).

quarter of a typical continent (row 3, regional area), and the largest one covers an entire continent (row 4, continental area). For the metropolitan-area domain, the forecast is hardly more than a "nowcast", regardless of the regime (entry a, useful forecast length). The radar-range and regional domains are of a scale that might be appropriate for regional weather prediction for small to moderate size countries, but unless they are in the tropics the forecast period is generally limited to considerably less than one day. Only for continental domains can useful forecasts have durations beyond a day.

Also shown in Table 3.3 is the lateral boundary displacement (entry b), in units of $L$, required to produce a forecast of "standard" duration (column 3) without LBC-error penetration to the domain interior. In addition, for each of these extended domains is computed the ratio of the number of buffer-zone grid points to the number of interior forecast-area grid points (entry c), which serves as a metric of the computational overhead resulting from the need for a buffer zone. If the buffer-zone width is increased for the small domains to allow for forecasts with a longer, more operationally useful, duration, the computational overhead generally becomes quite large. For example, to obtain a 6-h forecast in winter with the metropolitan area domain could require an overhead factor of between 500 and 1000. Often it is possible to take advantage of an asymmetry in the speed/direction climatology of the prevailing advecting wind, and increase the width of the buffer zone in the direction of stronger prevailing flow. Using available computational resources wisely by asymmetrically protecting the domain interior is recommended, but this will likely only permit an increase in the useful duration of the forecast by less than 50% compared to the use of a symmetric buffer zone with the same number of grid points. It has been implied that the LBC error is sufficiently large that it overwhelms the forecast accuracy when the error penetrates to the domain interior. However, there are measures that can be taken to control the amplitude of the LBC errors, and some current LBC formulations may not be especially damaging to the model solution.

### (2) Minimize interpolation error with the lateral-boundary data

The actual magnitudes of LBC errors will depend on a number of factors including the quality of the coarse-mesh forecast that is producing the LBCs and the magnitude of the error associated with the spatial and temporal interpolation from the coarse mesh to the LAM domain at the lateral boundaries. The interpolation error can be reduced through the frequent passing of LBC information from the coarse-mesh model to the LAM. For example, passage of a fast-moving mesocyclone through the boundary may be missed entirely if LBC data are updated only every six hours.

### (3) Use compatible numerics and physics with the LAM and the model providing the LBCs

The use of reasonably consistent physical-process parameterizations (convection, cloud microphysics, turbulence, and radiation) on the two grids will minimize unrealistic gradients that can develop at the interface and propagate onto the LAM domain through advection and inertia–gravity waves. For example, Warner and Hsu (2000) show how parameterized convection on an outer grid can strongly influence resolved convection on an inner grid through LBC-forced mass-field adjustments.

### (4) Employ well-tested and effective LBC formulations

Many LBC formulations for meteorological models are inherently ill-specified mathematically, and thus engineering approaches have been devised to minimize the potentially serious numerical problems that can develop. The LBC formulation used should be sufficiently well tested and designed so that it does not generate significant-amplitude, inertia–gravity waves that can move toward the central area of the domain at much greater than advective speeds. Even though some of the examples presented earlier demonstrate that this error can be significant, the use of appropriately engineered LBC algorithms can often limit the amplitude of this mode of error propagation to acceptable levels.

### (5) Allow for effects of data assimilation on LBC impact

The use of a preforecast FDDA period can have both a positive and negative effect on the LBC influence, whether continuous or intermittent assimilation techniques are utilized. On the one hand, the preforecast integration period will allow LBC error to propagate closer to the domain center by the start of the forecast. Conversely, the data assimilated during the period will partially correct for errors of LBC origin that are within the influence region of the observations.

### (6) Account for importance of local forcing

If strong local forcing mechanisms prevail within the fine mesh, and dominate the local meteorology, the forecast quality may not be as strongly affected by LBC errors as it would otherwise be. For example, the time of onset of a coastal-breeze circulation is more strongly correlated with local thermodynamic effects than with specific characteristics of the large-scale flow field and its LBC-related errors.

### (7) Avoid strong forcing at the lateral boundaries

Strong dynamic forcing at the lateral boundaries can create numerical problems with many LBC formulations. Even though it is not possible to avoid the passage of transient high-amplitude meteorological phenomena through the boundaries, it is possible to avoid collocating lateral boundaries with known regions of strong surface forcing such as associated with steep orography and surface temperature gradients. Locating large terrain gradients near or at lateral boundaries is one of the most common ways in which LBCs can cause the catastrophic failure of a model integration.

### (8) Utilize interactive grid nests when possible

When a LAM cannot influence the solution of the coarser-mesh model that provides its boundary values, the scale interactions of the LAM-resolved waves and those on the large scale are prevented. In addition, the use of a two-way interactive interface can, but will not necessarily, reduce the development of spurious gradients at the boundaries. Thus, interactive boundaries should be employed where possible, rather than one-way-specified boundaries.

### (9) With any new model application, perform sensitivity studies to determine the LBC influences

After considering the experiences described in the last section, it should be clear that LBC sensitivity studies should be performed for any new application of a LAM, especially if the aforementioned recommendations regarding the buffer-zone width are not taken literally. These sensitivity studies should include the testing of the dependence of forecast accuracy on buffer-zone width, the sensitivity of the forecast quality to different LBC formulations, and a comparison of the LAM skill to that of other operational modeling systems that have unbounded domains. A practical test for any LAM application is to compare the solution over the limited area with that from a model with equivalent resolution integrated over a much larger domain (Yakimiw and Robert 1990). If the LAM is to be used operationally, the forecasts naturally should be evaluated for LBC sensitivity over a wide range of events within all seasons.

# 3.6 Upper-boundary conditions

Artificial upper-boundary conditions are required in all atmospheric models because the model atmospheres do not extend to infinity. Indeed, for some historical applications the upper boundary has been located within the troposphere in order to save computational resources. An example of this approach is that Lavoie (1972) placed the upper model boundary, the "lid", at the top of the boundary layer. Pielke (2002a) describes the location of the upper boundary in various historical model applications.

Upward-propagating internal-gravity waves, for example generated by mountains or by deep convective storms, can extend to great heights in the atmosphere. Commonly used upper-boundary conditions (e.g., rigid lid, free surface) completely reflect these waves, which is a problem because no such reflection happens in nature, and erroneous downward-propagating waves contaminate the model solution. There are a number of approaches for preventing this from happening. One involves the use in the model of a gravity-wave absorbing layer, or sponge layer, immediately below the model top, to prevent the wave from reaching the top and reflecting. Such wave absorption can be produced by employing a greatly enhanced, artificial horizontal and/or vertical diffusion (viscosity), where the viscosity increases from the standard value at the bottom of the layer to a maximum at the top boundary. A particular disadvantage of this approach is that the absorbing layer may need to be thick, spanning a large number of model layers and thus involving a large computational overhead. The overall effectiveness of the absorption depends on the wavelength of the gravity wave, the thickness of the absorbing layer, and the distribution of viscosity in the layer. Note that using a shallow absorbing layer with a very large, but computationally stable, viscosity will not be effective because large gradients in viscosity will also cause wave reflections. Klemp and Lilly (1978) defined the entire upper half of their computational domain as the absorbing layer. Figure 3.50 shows a two-dimensional model solution for idealized flow over a maximum in the orography, with and without the use of a viscous damping layer. The Gaussian obstacle had a 5-km half-width, and an amplitude of 1 km. Shown is the vertical motion field in the lowest 10 km of the 50-km-deep model. The model is described in Sharman and Wurtele (1983). The damping spanned the 20 km below a rigid lid that defined the model top. Without the damping, the reflected waves produce considerable noise in the troposphere, over 40 km below the model top. The waves in the solution for the experiment with the absorbing layer could be a result of imperfect damping, or more likely they could be a consequence of wave reflections from the lateral boundaries. An alternative approach for damping the waves before they reach the upper boundary is to use a Raleigh damping layer, again below the model top, where model variables are relaxed toward a predetermined reference state. For example, the Rayleigh damping term in a prognostic equation would be like

$$\frac{\partial \alpha}{\partial t} = \tau(z)(\alpha - \bar{\alpha}),$$

where $\alpha$ is any dependent variable, $\bar{\alpha}$ is the reference value of that variable, and $\tau(z)$ increases upward within the damping layer and defines its vertical structure (e.g., see

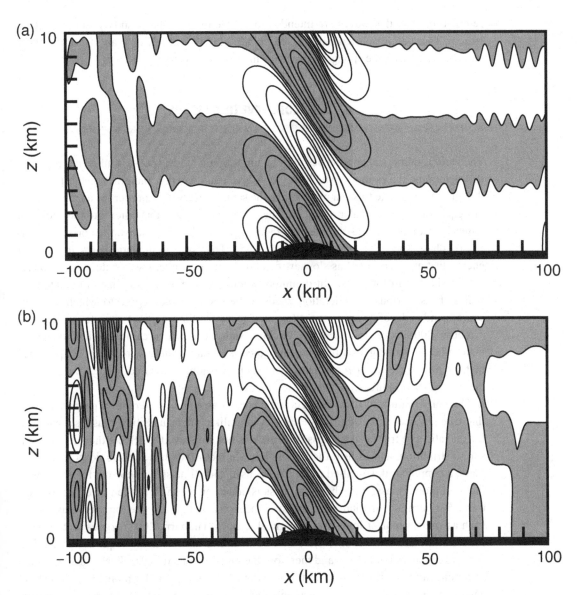

**Fig. 3.50** Vertical motion over the lowest 10 km of the atmosphere from two-dimensional model simulations of flow over elevated orography (see the black shading at the bottom), with a viscous layer at the top of the model (a) and without a viscous layer (b). The flow was from left to right, and downward motion is shaded. The damping layer spanned the upper 20 km of the 50-km-deep model. Provided by Robert Sharman.

Durran and Klemp 1983). Israeli and Orzag (1981) compare the Rayleigh-damping and viscous-damping methods.

A completely different approach, which does not rely on an absorbing layer, involves the use of a radiation boundary condition. Here, values of variables at the boundary are modified during the integration to minimize wave reflection. Clearly the term radiation refers to the fact that waves are intended to radiate through the boundary, and not reflect from it. These approaches are discussed in Durran (1999) and Klemp and Durran (1983), and compared with the sponge approaches in Israeli and Orzag (1981).

## 3.7  Conservation issues

The various numerical approaches used in atmospheric models possess inherent properties that determine the extent to which they conserve mass, energy, and other quantities. Even though we might like to see a model have the same conservation properties as the continuous equations and the real atmosphere, there are many factors that enter into the choice of numeric methods, such as the inherent damping of small-scale energy, the correct rendering of phase speeds, and numerical efficiency. That said, systematic leaks in mass or energy that are manifested as slow artificial drifts in the model mean state may be tolerable for short-term forecasts, but definitely would not be for integrations on climate time scales. Thus, serious consideration needs to be given to the degree to which spurious sources and sinks of physical quantities are acceptable for a particular model application. Thuburn (2008) contains a summary of conservation issues for weather-prediction and climate models, and suggests that we can expect accurate solutions from models provided that the time scale for artificial numerical sources is long compared to the time scale for the true physical sources.

The conservation of mass is arguably the most absolute conservation property, given that true physical sources are irrelevant. And, unlike other quantities, mass is conserved for diabatic and frictional processes. If mass is not conserved, it affects the surface pressure distribution, and in turn the circulations. Sometimes when models do not conserve mass, a nonphysical, so-called mass-fixer is used each time step to correct for changes in the total global mass, but where the mass is added or removed in the correction is arbitrary. Furthermore, if total mass is not conserved, neither are the various constituents such as water vapor or long-lived chemical species. Thuburn (2008) argues that, at least for long climate simulations, there is a very strong argument for requiring dynamical cores to conserve total mass, and therefore the mass of constituents. He also discusses the situations in which it is important to conserve momentum, angular momentum, potential enstrophy, energy (kinetic, and available and unavailable potential), entropy, and potential vorticity.

## 3.8  Practical summary of the process for setting up a model

This section is meant to summarize how our knowledge of the numerical processes discussed in this chapter should serve as a guide for setting up a model. There are additional factors that must be considered, such as the appropriateness of physical-process parameterizations, but these are reviewed in other chapters. It is assumed in the following that the time step is internally determined by the model, and that there is no choice in the methods used to solve the equations (e.g., spectral versus grid-point approaches, explicit versus semi-implicit time differencing, etc.).

- Based on a knowledge of the purpose for using the model and the prevailing meteorology in the geographic area to be modeled, determine the physical processes that must be simulated or forecasted.
- Choose a horizontal grid increment that is sufficiently small to resolve all the processes to be represented on the grid.
- Define a vertical distribution of grid points that adequately defines anticipated important vertical structures (e.g., boundary-layer gradients, low-level jets, the tropopause) and, if possible, ensure reasonable compatibility of the vertical grid increment with the horizontal increment.
- For limited-area models, choose the map projection that is most suitable for the range of latitudes represented by the model grid. View a graphic of the map-scale factor at each grid point when setting up the model grid to confirm the degree to which it departs from unity.
- Compare the model solution with observations, and quantify the skill. If the model is to be used as a general research or operational-forecasting tool, numerous cases should be chosen from all seasons. Just because the model has been reported to be accurate for other locations and configurations, do not assume that this step can be avoided.
- For limited-area models, perform tests to define the sensitivity of the accuracy of the model solution to different locations of the boundaries and different domain sizes.
- Perform tests to determine the sensitivity of the model accuracy to the vertical and horizontal grid increments.

Section 10.1 provides additional practical guidance for applying models to perform research case studies.

## SUGGESTED GENERAL REFERENCES FOR FURTHER READING

Durran, D. R. (1999). *Numerical Methods for Wave Equations in Geophysical Fluid Dynamics*. New York, USA: Springer.

Krishnamurti, T. N., H. S. Bedi, V. M. Hardiker, and L. Ramaswamy (2006). *An Introduction to Global Spectral Modeling*. New York, USA: Springer.

Staniforth, A., and N. Wood (2008). Aspects of the dynamical core of a nonhydrostatic deep-atmosphere, unified weather and climate-prediction model. *J. Comput. Phys.*, **227**, 3445–3464.

Williamson, D. L. (2007). The evolution of dynamical cores for global atmospheric models. *J. Meteor. Soc. Japan*, **85B**, 241–269.

World Meteorological Organization (1979). *Numerical Methods Used in Atmospheric Models, Volume II*. Global Atmospheric Research Programme, GARP Publication Series No. 17. Geneva, Switzerland: World Meteorological Organization.

## PROBLEMS AND EXERCISES

1. For the 24-point grid referenced in Fig. 3.32, list all the combinations of interacting wavenumbers that produce aliasing, and the erroneous wavenumbers that results from the interactions.

2. Derive an expression for the ratio of the five-point numerical approximation to the derivative and the analytic solution for the derivative, analogous to what is shown for the three-point approximation in Section 3.4.1.

3. Explain graphically, or in words, why the truncation error for the forward-in-space differencing formula in Eq. 3.25 is dependent on position within the wave, in addition to how well the wave is resolved on the grid.

4. Given that Fig. 3.27 shows that the use of Courant numbers close to unity produces more-realistic solutions than do smaller values, for the centered-in-space and centered-in-time approximation to the advection term, why can't we use sufficiently large time steps to ensure the prevalence of these large Courant numbers?

5. Prove the orthogonality of the exponential function.

6. Using the programming language of your choice, construct a one-dimensional model based on the shallow-fluid equations (Chapter 2) with three-point time and space differencing and no explicit diffusion. Assume periodic lateral-boundary conditions, and perform the following experiments.
   - Simulate an advective wave and a gravity wave.
   - Choose a time step that violates the linear stability criterion, and output the model solution each time step.
   - Add an explicit diffusion term and show its effect on the model solution for different diffusion coefficients.
   - Alter the time step to evaluate how the use of different Courant numbers affects the model solution.
   - For the same initial conditions, evaluate the effect of horizontal resolution on the model solution.

7. Some studies with LAMs (e.g., Alpert *et al.* 1996) suggest that the quality of simulations decreases as lateral boundaries become too distant from or too close to the area of meteorological interest. Provide possible explanations for this situation.

8. Given the typical spacing of radiosonde soundings (~400 km), explain the types of meteorological processes that can be adequately resolved by them in the initial conditions.

9. Regarding Fig. 3.44, explain why the jet streak simulated on the small domain is so much smoother than the one simulated on the larger domain.

# 4 Physical-process parameterizations

## 4.1 Background

The parameterization problem involves algorithmically or statistically relating the effects of physical processes that cannot be represented directly in a model to variables that are included. Physical processes are parameterized for a few reasons.

- The small scales involved make it too computationally expensive to represent a process directly.
- The complexity of a process makes it too computationally expensive to represent directly.
- There is insufficient knowledge about how a process works to explicitly represent it mathematically.

The representation of atmospheric processes in models takes place within the dynamical core as well as through the so-called model "physics". The dynamic processes include the propagation of various types of waves (e.g., advective, Rossby, inertia–gravity). Even though the physics processes are parameterized to a large degree, their correct rendering by a model is nevertheless essential for the prediction of virtually all of the dependent variables. The parameterized processes that are discussed in this chapter include cumulus convection, cloud microphysics, turbulence, and radiation. Land-surface processes are also parameterized because they occur on too small a scale to be represented directly, but they are discussed separately in Chapter 5.

Even though parameterizations are typically developed and discussed independently from each other, and from the dynamical core, this is artificial and should be avoided. This is because parameterizations do interact, and the realism of this interaction determines the accuracy of the model. For example, the parameterized spectral solar radiation represents an energy flux at the land surface, and the land-surface parameterization partitions some of it to the sensible heating of the ground. The resulting land–atmosphere fluxes provide lower-boundary conditions to the surface-layer and boundary-layer parameterizations of turbulence, which distribute heat and moisture throughout the lower atmosphere. And, when the water vapor condenses, parameterizations of convection are relied upon to represent all aspects of the associated subgrid-scale processes. And, microphysical processes that are related to the development of hydrometeors are parameterized for stable precipitation. In turn, convective-cloud and stable-cloud effects on radiation must be parameterized, where this radiation attenuation strongly influences the surface temperature through the land-surface parameterization. Thus, because of these interdependencies, parameterizations should not continue to be developed in isolation. A more holistic approach is greatly needed if we are to reduce model error.

An issue that will be discussed below is that the performance of some parameterizations can depend on season and the meteorological processes that prevail in specific geographic regions. For example, some convective parameterizations are more appropriate for midlatitudes, while others perform better in the tropics. And, models employed for polar applications will use parameterizations that are different from those that are applicable for midlatitude, coastal-zone simulations, etc. And, the same parameterization is sometimes tuned for specific needs. An obvious related issue is that global models must use the same parameterizations for all geographic areas, thus eliminating the option of choosing ones that best suit a particular region.

Figure 4.1 illustrates how parameterizations fit within the overall framework of a model. The term "resolved" in the upper box refers to grid-scale processes that do not need to be parameterized. An important aspect of this figure is that the primary inputs for parameterizations of any type are the resolved-scale structures of the atmosphere that control the process that is being parameterized. As a very simple example, the resolved static stability near the ground can be used to infer the strength of subgrid-scale turbulence in the boundary layer, which controls the grid-box-average vertical fluxes in the tendency equations for temperature, humidity, and wind. Similarly, layer-average relative humidity from the model can be used to infer the fractional-area coverage of subgrid-scale cumulus clouds, which can be used in the equation that calculates the radiation reaching the surface. Thus, a parameterization relates the resolved-scale input variables to the resolved-scale effects ($\partial \Phi_P / \partial t$ in Fig. 4.1) of the parameterized process. It will be seen that parameterizations can have a wide range of complexities. The middle box in

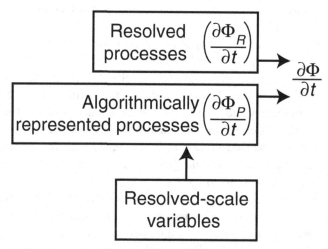

Fig. 4.1 Schematic showing how a predictive equation for a dependent variable $\Phi$ has contributions from terms that correspond to resolved processes (subscript $R$) and parameterized processes (subscript $P$). The inputs to the parameterizations are the resolved-scale atmospheric variables.

Fig. 4.1 can be a simple look-up table, or it can be so computationally intensive that $\partial\Phi_P/\partial t$ cannot be recalculated at every time step

Refer to Eqs. 2.1–2.6 to see how the effects of parameterized processes are included in the prognostic equations. The momentum equations, Eqs. 2.1–2.3, contain friction terms (*Fr*), that are separated into viscous and turbulent stresses in Eq. 2.16. The rest of the terms are solved using the methods described in Chapter 3, and their numerical form is part of the dynamical core of the model. But the friction, for example in the boundary layer, results from the existence of the turbulent eddies that cannot be resolved. The boundary-layer parameterization will define these friction terms. Similarly, the diabatic heating–cooling term (*H*) in the thermodynamic equation, Eq. 2.4, contains contributions from phase changes of water that are defined in the microphysical and convective parameterizations, from heat transport by the parameterized turbulence fluxes near the surface, and from the radiation parameterization.

It is important to be aware that parameterizations are generally developed with certain grid increments in mind. That is, only those aspects of the physical system that are not resolved by the model need to be parameterized. Thus, the modeler should be aware of such assumptions when deciding on parameterizations to employ for a given purpose. There is another general parameterization issue that is related to model resolution: As model grid increments continue to decrease as computing power increases, the models begin to partially (i.e., poorly) resolve some processes that are being parameterized. Thus, there is the risk of "double counting" processes. The consequence of this situation is that there is a range of grid increments for which a process is too poorly resolved to represent explicitly, but there is not a sufficient separation between resolved and parameterized scales for the parameterization assumptions to be valid.

Stensrud (2007) represents the best review available of all the types of parameterizations used in atmospheric models. The discussions in this chapter often follow those in that reference.

## 4.2 Cloud microphysics parameterizations

Cloud microphysics encompasses all cloud processes that occur on the scales of the cloud droplets and the hydrometeors, rather than on the scale of the cloud itself. The correct modeling of these processes determines the skill with which precipitation type, amount, and spatial distribution are forecast. Similarly, microphysical processes are the cause of the potentially destructive straight-line winds of convective outflow boundaries. And, cloud horizontal and vertical distributions must be modeled well in order for the radiation and surface energy budgets to be predicted with accuracy. Microphysical processes are also critically important in climate modeling. For example, the physical system may respond to increases in greenhouse gases with an alteration in global cloud properties (and albedo), which can have a potential positive or negative feedback relative to the original temperature increase. And, the microphysical impacts of increased natural or anthropogenic aerosols in the atmosphere of a modified climate must be represented in a model because this can change precipitation efficiency.

Historically, stratiform clouds have been explicitly represented in NWP models, with the microphysics parameterized, because their large horizontal extent has allowed them to be resolved by most grids. In contrast, the small horizontal size of most convective clouds relative to typical grid increments has meant that they have been subgrid-scale phenomena, and thus their effects are represented through parameterizations. Thus, the same model parameterizes one type of cloud and explicitly represents another. The two components of the model compete for water vapor, and model output files generally include separate variables for convective precipitation and stable precipitation. This situation still prevails for most operational NWP models, and for climate models.

For research applications, and for some operational LAMs, the horizontal grid increments can be sufficiently small so that the models, called cloud-resolving models, can explicitly represent moist convection on the grid. This allows the same model code, including microphysics parameterizations, to represent all moist processes – a much more appealing situation than the one noted above where different parts of the model apply to convective and stratiform cloud. The horizontal grid increment below which cloud-generating circulations can be explicitly represented by a model is very situation dependent. Weisman *et al.* (1997) suggests that a grid increment of 4 km is sufficient to resolve squall-line convection. However, there will never be a grid increment below which microphysical process will not need to be parameterized. This is because such processes, as we will see, exist on the cloud-droplet and rain-drop scales of micrometers to millimeters; indeed, even the molecular scales are relevant.

## 4.2.1  Microphysical particles and processes

The following summary is provided for the reader who has not had the benefit of a course in cloud microphysics. The particle types and the microphysical processes that they undergo are important in the context of the generation of precipitation in its various forms, and thus they should be parameterized in some way in atmospheric models. Further information can be obtained from Fletcher (1962), Rogers (1976), Cotton and Anthes (1989), Rogers and Yau (1989), Houze (1993), Pruppacher and Klett (2000), and Straka (2009). The particle types that are involved in microphysical processes are listed below.

- *Cloud droplets* – These are liquid drops, with a typical radius of 10 μm, that form through the condensation of water vapor in the presence of a cloud-condensation nucleus (CCN, small particles that have an affinity for water).
- *Rain drops* – Cloud droplets can grow to rain drops through the accretion mechanism described below, or rain drops can result from the melting of snow crystals. Rain drop radii range from 100 to 1000 μm.
- *Ice crystals* – Water droplets freeze in the presence of an ice nucleus (IN, similar to a CCN) at temperatures below the normal freezing point. Larger droplets freeze at higher temperatures.

- *Aggregates of ice crystals, snow flakes* – These are clusters of ice crystals formed when ice crystals with different terminal velocities collide and coalesce. Snow flakes are formed by this process.
- *Rimed ice particles* – These form when ice crystals collide and coalesce with cloud droplets at temperatures below freezing. If the features of the ice crystal can be distinguished, it is called a rimed ice particle.
- *Graupel particles* – If the crystal features of a rimed ice particle are not recognizable, it is called a graupel particle. Graupel also results from the instantaneous freezing of rain drops, when the sub-freezing drops collide with ice crystals.
- *Hail stones* – As graupel particles fall through the cloud of sub-freezing liquid, they grow by riming. Hail stones result from cases of extreme riming.

Some of the microphysical processes are as follows.

- *Condensation* – Liquid droplets form when water saturation is exceeded at temperatures from –40 °C to above freezing. The condensation takes place on CCN that are natural or anthropogenic, typically submicrometer-sized, particles.
- *Accretion* – In the warm-cloud process, droplets with different masses have different terminal velocities, and the resulting collisions between droplets can result in coalescence and droplet growth. As a droplet grows, so does its vertical velocity relative to the smaller cloud droplets, thus increasing the rate of collisions.
- *Evaporation* – Cloud droplets and rain drops evaporate.
- *Ice and snow aggregation* – When ice crystals and snow flakes collide and coalesce, it is called aggregation.
- *Accretion by frozen particles* – Snow, graupel, or hail collect other solid or liquid particles as they fall.
- *Vapor deposition* – The saturation vapor pressure with respect to liquid water is higher than the saturation vapor pressure with respect to ice. Thus, if a cloud that contains both droplets and ice crystals is saturated with respect to water, it is supersaturated with respect to ice. As the ice crystals grow by vapor deposition, the air becomes subsaturated with respect to the liquid surface, and cloud droplets evaporate. This process is called the Bergeron–Findeisen mechanism.
- *Melting* – As snow flakes fall into the lower troposphere, below the freezing level, they may melt and form rain drops. Similarly, hail and graupel begin to melt as they fall below the freezing level.
- *Freezing* – Water droplets freeze in the presence of IN, riming involves the freezing of water droplets that collide with ice crystals, and rain drops can freeze to form graupel.

Figure 4.2 illustrates the microphysical processes that must be represented in some form in a model in order to predict the types of precipitation shown at the bottom of the figure. A similar diagram can be found in Cotton and Anthes (1989).

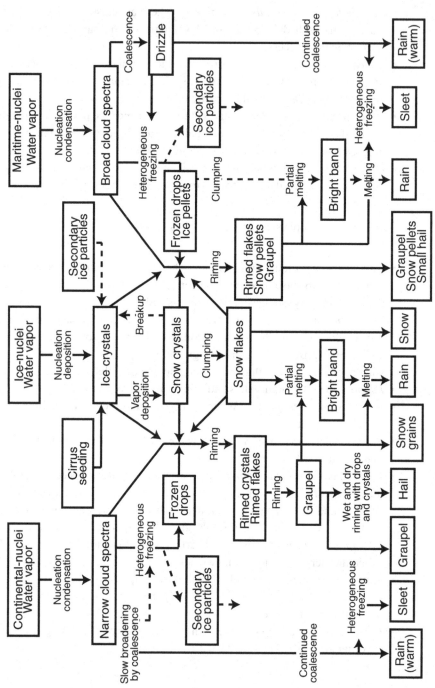

**Fig. 4.2**    Schematic showing the microphysical processes that are important for defining the spatial and temporal distribution of the types of precipitation shown at the bottom of the figure. The purpose of the figure is to emphasize the complexity of the processes that must be represented in a model. See Stensrud (2007) for details, and Cotton and Anthes (1989) for a similar diagram. Adapted from Braham and Squires (1974).

## 4.2.2 Microphysical parameterizations

Microphysical parameterizations aim to represent, as thoroughly as possible, the processes described in the previous section. The parameterizations are divided into two categories, based on how the size distributions of particle types are represented. In *bin models*, the particle size spectrum is divided into intervals, and the particle concentrations are predicted for each interval, or bin. Changes for each bin can result from conversions between particle types, and from the increase or decrease of particle sizes. This requires a predictive equation for each particle type and size bin, which must be solved at each grid point. Thus, the use of bin models is very computationally intensive and is presently limited to research activities, and not operational weather and climate prediction. In contrast, *bulk microphysical parameterizations* assume a prescribed analytic form for the size spectrum of each particle type – e.g., exponential (Kessler 1969) or gamma (Walko *et al.* 1995a) distributions – and the evolutions of the size spectra are obtained by solving predictive equations for the moments. Single-moment, bulk parameterizations only involve prediction of the particle mixing ratio or specific humidity; that is, the ratio of the mass of a particular particle type to the volume or mass, respectively, of the dry air in which the particles are distributed. Double moment schemes predict both the particle mixing ratio and the particle number concentration. Triple-moment schemes add radar reflectivity to the predictive equations, allowing the shape parameter in the gamma distribution to vary independently.

To illustrate an example of how bulk microphysical parameterizations are represented in a single-moment model, the following are predictive equations for the specific humidity of five different forms of water: water vapor ($q_v$), cloud water ($q_{cw}$), cloud ice ($q_{ci}$), snow ($q_s$), and rain ($q_r$). Tensor notation is used for notational brevity, such that when a subscript appears twice in the same term it is assumed that the term is summed over all possible subscript values. For example, the first term to the right of the equal sign represents advection in the three space directions, such that $u_i$ equals $u$, $v$, and $w$, and $x_i$ equals $x$, $y$, and $z$ for $i = 1, 2, 3$, respectively. The next term to the right is the turbulent mixing term (see Chapter 2) in tensor notation, and is interpreted in the same way. The third terms on the right side in the equations for snow and rain represent the fact that these two types of hydrometeors have significant terminal velocities ($V_T$). Where there are vertical derivatives in the mass of a species, there will be a contribution to the tendency in proportion to the terminal velocity. The rest of the terms on the right ($S$) represent various sources and sinks associated with conversions from one type of particle to another.

$$\frac{\partial q_v}{\partial t} = -u_i\frac{\partial q_v}{\partial x_i} - \frac{1}{\rho_0}\frac{\partial}{\partial x_i}\rho_0\overline{u_i'q_v'} - S_{deps} - S_{depci}$$

$$+ S_{evapr} - S_{vcondtocw}$$

$$\frac{\partial q_{cw}}{\partial t} = -u_i\frac{\partial q_{cw}}{\partial x_i} - \frac{1}{\rho_0}\frac{\partial}{\partial x_i}\rho_0\overline{u'_i\,q'}_{cw} + S_{vcondtocw} - S_{freezcw}$$

$$- S_{cwtor} - S_{acccwbyr} - S_{acccwbys}$$

$$\frac{\partial q_{ci}}{\partial t} = -u_i\frac{\partial q_{ci}}{\partial x_i} - \frac{1}{\rho_0}\frac{\partial}{\partial x_i}\rho_0\overline{u'_i\,q'}_{ci} + S_{freezcw} + S_{depci} - S_{citos} - S_{acccibys}$$

$$\frac{\partial q_s}{\partial t} = -u_i\frac{\partial q_s}{\partial x_i} - \frac{1}{\rho_0}\frac{\partial}{\partial x_i}\rho_0\overline{u'_i\,q'}_s - V_{Ts}\frac{\partial q_s}{\partial z} + S_{citos}$$

$$+ S_{acccibys} + S_{acccwbys} + S_{deps} - S_{smelttor}$$

$$\frac{\partial q_r}{\partial t} = -u_i\frac{\partial q_r}{\partial x_i} - \frac{1}{\rho_0}\frac{\partial}{\partial x_i}\rho_0\overline{u'_i\,q'}_r - V_{Tr}\frac{\partial q_r}{\partial z} - S_{evapr} + S_{acccwbyr} + S_{cwtor} + S_{smelttor}$$

The sources and sinks are defined as follows.

$S_{evapr}$ evaporation of rain drops

$S_{acccwbyr}$ accretion of cloud-water droplets by rain drops

$S_{cwtor}$ growth of cloud-water droplets to rain drops by cold-cloud (Bergeron–Findeisen) process

$S_{smelttor}$ melting of snow to produce rain drops

$S_{citos}$ growth of cloud ice to snow

$S_{acccibys}$ accretion of cloud ice by snow

$S_{acccwbys}$ accretion of cloud water by snow

$S_{deps}$ growth of snow by vapor deposition

$S_{freezcw}$ freezing of cloud water to produce cloud ice

$S_{depci}$ growth of cloud ice by vapor deposition

$S_{vcondtocw}$ condensation of vapor to form cloud-water droplets

It is within these "S" terms that the parameterizations of microphysical processes are represented. Some schemes are simple, and only a few of the particle types and interactions (conversions) are operative. Others are more complex, with many more interactions. Figure 4.3 illustrates the microphyical processes that are represented in three different example parameterizations, and emphasizes the fact that the number of interactions allowed among particles varies greatly. Shown are the interactions for the Dudhia (1989), Reisner *et al.* (1998), and Lin *et al.* (1983) schemes. The reader should refer to Stensrud (2007) for examples of different approaches for actually representing the different processes.

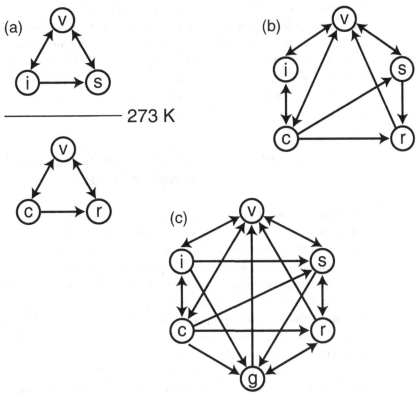

Fig. 4.3 Microphysical processes represented in three different parameterizations: (a) Dudhia (1989), (b) Reisner *et al.* (1998), and (c) Lin *et al.* (1983). Particle types are abbreviated as vapor (v), cloud ice (i), snow (s), cloud water (c), rain water (r), and a combination of graupel and hail (g). The arrows indicate the direction of the particle interactions. The line in panel (a) separates the processes above the freezing level (above the line) from those below the freezing level. From Stensrud (2007).

### 4.2.3 Initialization of microphysical variables

Ideally it would be possible to initialize the microphysical variables just as we do the other dependent variables. However, there are a couple of impediments to success. First, the existence of specific types of microphysical particles in the atmosphere can only be roughly inferred, at best, based on satellite cloud imagery (cloud ice and water) and various sources of precipitation observations (rain and snow). The vertical distributions of the particles, and their horizontal spatial detail at the cloud scale, are unknown, thus making their initialization very problematic. In addition, the microphysical variables respond quickly to forcing by atmospheric circulations on the cloud scale and larger, so initializing the variables without also including corresponding compatible circulations would be futile. For example, if cloud and precipitation observed along a front are used to initialize microphysical variables, the variables will only be retained by the model if it has the front and associated vertical circulation in the correct location. Without the frontal lifting, the cloud and precipitation will dissipate. Because of the above issues, microphysical variables are sometimes assumed to have zero concentrations at the initial time of a forecast, where the expectation is that they will spin up to realistic values within the initial 3–6 h. When sequential data-assimilation methods are employed (see Chapter 6), the model simulation that provides the first-guess for the analysis can be used to define the microphysical variables. Or, when continuous data-assimilation, such as Newtonian relaxation, is used, model-generated microphysical variables are automatically part of forecast initial conditions.

## 4.2.4 Modeling the effects of anthropogenic and natural aerosols on microphysical processes

The atmospheric CCN represent the subset of the general population of aerosols that can nucleate a cloud droplet at a particular water saturation. The ability of a particle to act as a CCN depends primarily on chemical composition and size. Thus, an important issue related to the inclusion of microphysical processes in NWP and climate models is the correct representation of the details of the CCN from natural and anthropogenic sources (Rosenfeld *et al.* 2008). This is challenging from a number of respects. One is that atmospheric aerosol properties are not systematically observed at all. Another is that the complex interactions among aerosol particles, hydrometeors, and cloud dynamics, including the dynamic competition for water vapor among nuclei of different sizes and composition, mean that predicting the specific response of the system to the types and amounts of available aerosols (even if we had that information) can be challenging. An example of the importance of knowing simply the approximate amount of CCN follows. If there are many CCN available, cloud-droplet concentrations can be large. For a given liquid-water content, this means that droplets are smaller, the cloud optical thickness and albedo are higher, and precipitation efficiency is reduced. The lower precipitation efficiency leads to higher cloud liquid-water content, cloud lifetime, and cloud thickness (Albrecht 1989). We thus can have the situation where an increase in CCN availability can lead to a reduction in rain-drop concentration. Complicating the situation are the facts that some aerosols can decrease cloud albedo (Kaufman and Nakajima 1993) and their chemical compositions influence their activation as cloud droplets (Raymond and Pandis 2002). Thus, the lack of our ability to operationally predict aerosol properties has implications for the predictability of microphysical processes and clouds, and weather in general. For example, Taylor and Ackerman (1999) found that the elevations of cloud tops and the microphysical structure of stratus clouds were significantly affected by aerosols emitted by ships into an otherwise clean maritime environment.

Because of long-term trends in anthropogenic aerosols from pollution, and mineral aerosols from desertification, and the fact that aerosols, clouds, and precipitation are critical components of the climate system, it is especially important in climate modeling for aerosols to be simulated in terms of their sources, sinks, and transport. See Levin and Cotton (2009) for a complete summary of aerosol effects on microphysics, and Heintzenberg and Charlson (2007) for their role in climate.

## 4.3  Convective parameterizations

It is important to be able to accurately simulate moist convection[1] with models for a variety of practical reasons. Intense moist convection can lead to flash flooding, gust fronts, and tornadoes. And, the aggregate effect of individual convective elements is an important component of monsoon circulations, the Hadley and Walker circulations, and the ENSO. These large-scale processes need to be simulated properly in climate predictions and long-range weather predictions, and thus models need to be able to accurately represent the effects of the convection on their resolved scales. And, shallow cumulus clouds dominate the tropics, and are common in other latitudes, greatly impacting the global albedo. The effects of these clouds thus need to be represented in weather and climate models, which must reasonably render the radiation budget.

In general, convective parameterizations activate moist convection at relative humidities lower than water-vapor saturation at a grid point. This is because the convective columns are subgrid scale, so the grid-box-average relative humidity will be sub-saturated even though there are saturated regions within the grid box. In addition to generating grid-box-average values of convective precipitation, the schemes also define the effects of the subgrid-scale convection on other grid-scale variables. The overall objective is for these parameterizations to define convection in the right place and at the right time (with the correct diurnal cycle if applicable), and with the correct evolution and intensity. And, the parameterization should define the appropriate modification by the convection to the large-scale environment, so that subsequent convection can be accurately predicted. Figure 4.4 shows the overall concept, even though Mapes (1997) points out how it is a simplification. In general, the large-scale processes (e.g., low-level convergence, destabilization of a deep layer) control the moist convection, and, in turn, the convection will

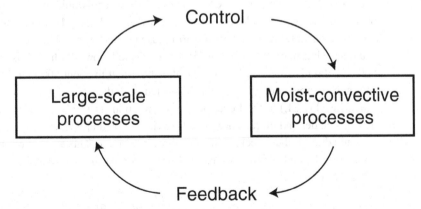

**Fig. 4.4**  Schematic of the interaction between large-scale processes and moist convection. Adapted from Arakawa (1993).

---

[1]  The term moist convection refers to convection that leads to cloud, and possibly precipitation, formation.

modify the large scale, possibly with the aggregate latent heating from the convective cells maintaining the larger-scale circulation. The purpose of the parameterization is to emulate this process.

Moist convection in the atmosphere can be classified into two types. Deep convection extends vertically over a large fraction of the depth of the troposphere, and is associated with (1) low-level convergence that exists on a scale larger than the individual updrafts and (2) deep conditional instability. In contrast, shallow convection spans only a small fraction of that depth, with cloud tops perhaps a few kilometers above the surface. Precipitating deep convection dries the environment by removing water vapor, and warms it as a result of the compensating subsidence. However, nonprecipitating shallow moist convection has no direct net influence on the environment. Its existence does indirectly impact the environment, however, because the clouds reflect solar energy and the resulting shading of the ground means a cooler boundary layer.

Cloud-resolving models, that are capable of explicitly resolving convective-scale circulations, employ grid increments of perhaps 1 km and are commonly used in research (see Wu and Li (2008) for a review and a comprehensive list of references). For example, Weisman *et al.* (1997) use a model with a grid increment of 4 km that they state explicitly resolves squall-line convection. However, because moist convection consists of a mix of updrafts and downdrafts that often have scales of a few hundred meters to a few kilometers, it will be years before operational global and limited-area weather-forecast models are capable of resolving them. And, it will be decades before global climate models have sufficient horizontal resolution to resolve moist convection. Thus, convective parameterizations will be needed well into the foreseeable future.

## 4.3.1 Types of convective parameterizations

A common feature of most convective parameterizations is that they calculate the Convective Available Potential Energy (CAPE) and the Convective INhibition (CIN) of the environment in order to estimate the characteristics of convection. Figure 4.5 graphically illustrates these two variables in terms of a typical warm-season sounding on a thermodynamic chart. In this sounding (thin line), the lapse rate is dry adiabatic below 800 hPa, it is isothermal between 800 and 700 hPa, and nearly dry adiabatic above that to about 500 hPa. The heavy solid line shows the temperature of a parcel that is lifted from the surface, through the Lifting Condensation Level (LCL) to the Level of Free Convection (LFC). Between the LCL and LFC the parcel is colder and more dense than its environment, and is thus negatively buoyant. Energy is required to lift the parcel against this downward force. This amount of energy is, by definition, the CIN, and is proportional to the striped area in the figure. Mathematically, CIN is defined as follows:

$$CIN = -g \int_{SL}^{LFC} \frac{\theta(z) - \bar{\theta}(z)}{\bar{\theta}(z)} dz \,,$$

where $\theta$ is the potential temperature of a parcel rising dry or moist adiabatically from its Starting Level (SL) to the LFC, $\bar{\theta}$ is the potential temperature of the environment, and the

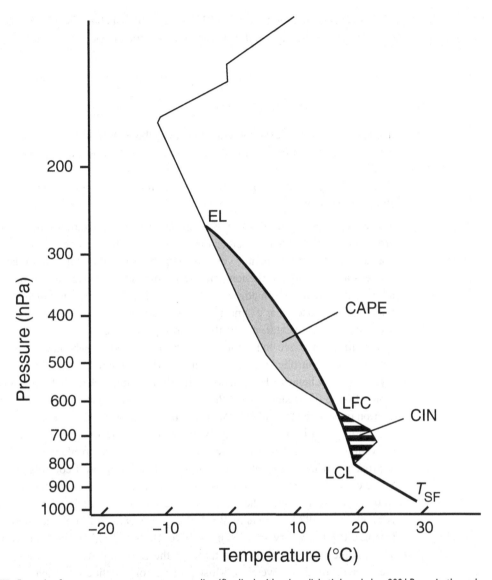

Fig. 4.5 Example of a warm-season temperature sounding (fine line) with a dry adiabatic layer below 800 hPa, an isothermal layer from 800 to 700 hPa, and another near-dry-adiabatic layer up to about 500 hPa. The heavy solid line defines the temperature of a parcel that is lifted from the surface through the Lifting Condensation Level (LCL) and the Level of Free Convection (LFC), to the Equilibrium Level (EL). The areas are shown that define the Convective Available Potential Energy (CAPE) and the Convective INhibition (CIN). The surface temperature is $T_{SF}$.

negative sign exists so that CIN is positive when energy is required to lift the parcel to its LFC. If there is sufficient energy to lift the parcel above the LFC, it will rise buoyantly along the heavy line until it reaches the Equilibrium Level (EL) where the parcel is neutrally buoyant. It is assumed with this simple parcel theory that there is no mixing between the parcel and its environment. The CAPE is the buoyant energy available to an ascending parcel as it rises from the LFC to its EL where it loses its buoyancy, and is proportional to the shaded area in the figure. It is defined mathematically as

$$CAPE = g \int_{LFC}^{EL} \frac{\theta(z) - \bar{\theta}(z)}{\bar{\theta}(z)} dz \, .$$

Thus, for convection to exist there must be available CAPE to provide the buoyant energy to accelerate parcels upward, and there must be a method by which parcels overcome the prevailing CIN.

There are a large number of ways of categorizing convective-parameterization schemes, including the following.

- There are general approaches to the convective-parameterization problem, and schemes are identified in terms of whether they follow a particular one. For example, schemes that adjust the environmental vertical temperature profile when the relative humidity exceeds a threshold and the temperature profile is unstable are called moist-convective adjustment schemes. Sometimes schemes are identified in terms of the first author to publish an approach. For example, methods that produce convection based on resolved-scale moisture convergence are called Kuo-type schemes.
- A classification suggested by Mapes (1997) is based on whether the development of convection is controlled by the creation of CAPE or the removal of CIN. So-called deep-layer-control schemes, also termed equilibrium-control schemes, tie the development of convection to the creation of CAPE by large-scale processes. In these methods, the convection is assumed to maintain the instability in the large-scale environment in a state of equilibrium that is near neutrality. Alternatively, low-level-control schemes, also called activation-control schemes, relate convection to the removal of CIN. In fact, many approaches include elements of both low-level and deep-layer controls.
- Some convective parameterizations represent the effects of only deep moist convection (most schemes discussed here), while others apply to only shallow convection (e.g., Albrecht *et al.* 1979, Deng *et al.* 2003, Bretherton *et al.* 2004). Some apply to both types (Tiedtke 1989, Gregory and Rowntree 1990, Betts and Miller 1993, and Kain 2004).
- The schemes can be classified in terms of the environmental, grid-scale, variables that are affected by the convection. Most schemes only define the impact on the environmental temperature and humidity, but some also treat effects on the momentum (e.g., Fritsch and Chappell 1980, Han and Pan 2006).
- Some methods directly define the final state of the environment after the convection has effected the change, while others attempt to simulate the process by which the change takes place. The former, generally more simple, approaches are called static schemes, while the latter are referred to as dynamic schemes.
- A distinction among methods is the nature of the so-called trigger function. This is the set of criteria in the parameterization that prescribes where and when the parameterized convection will be activated. The importance of this component of a convective parameterization was demonstrated by Kain and Fritsch (1992), who tested five different trigger functions in the same model and in the same parameterization, the

Kain–Fritsch scheme (Kain and Fritsch 1993), for the same meteorological case. There were substantial differences in the simulated parameterized convection for the different trigger functions. A similar dependence was found by Stensrud and Fritsch (1994).

- The scales resolved by the models are a way of classifying these schemes, such that there are mesoscale-model parameterizations and coarse-grid-model parameterizations. The particular distinction is that the mesoscale models (grid increments of 5–50 km) have sufficient horizontal resolution to explicitly resolve mesoscale circulations associated with the convection, where examples include thunderstorm outflow boundaries, mesohighs and mesolows, rear-inflow jets, and midlevel vortices associated with mesoscale convective systems. Thus, the mesoscale models need to only parameterize the convective-scale processes (e.g., Stensrud and Fritsch 1994, Zheng *et al.* 1995), whereas coarser-grid models must parameterize both the mesoscale and the convective-scale processes, and their many interactions (Frank 1983).

A given convective parameterization can be classified in terms of a number of the methods in the above incomplete list.

## 4.3.2 Scale considerations

In the above discussion is mentioned the relationship between model resolution and the nature of the convective parameterization. But, the parameterization problem also depends on other scale issues. In particular, Frank (1983) elaborates on a discussion in Ooyama (1982) regarding the relationships between convective parameterizations and the scale of the convective process. Figure 4.6 illustrates the scale regions that have relevance to the parameterization of convection. The abscissa is the physical length scale ($L$) of the process, and the ordinate is the dynamic length scale expressed in terms of the Rossby radius of deformation ($R$). The latter length scale is defined as

$$R = \frac{NH}{(\zeta+f)^{1/2}(2Vr^{-1}+f)^{1/2}},$$

where $N$ is the Brunt–Väisälä frequency, $H$ is the scale height of the circulation, $\zeta$ is the relative vorticity, $f$ is the Coriolis parameter, $V$ is the rotational component of the wind, and $r$ is the radius of curvature of streamlines. Within region I, with length scales of less than 10 km, are individual convective cells and clouds. Regions I and II pertain to phenomena, referred to as dynamically small, where $L < R$, and region III pertains to dynamically large phenomena where $L > R$. Frank (1983) states that the parameterization problem is somewhat simpler for region-III processes because there is a stronger relationship between the large-scale flow and convection. For example, if a cold-front's strength and position are simulated correctly, the associated convective rainfall will be straightforward to represent with a parameterization. A parallel argument is that, when latent heating from convection affects the mass field in dynamically large systems, the system will adjust through changes

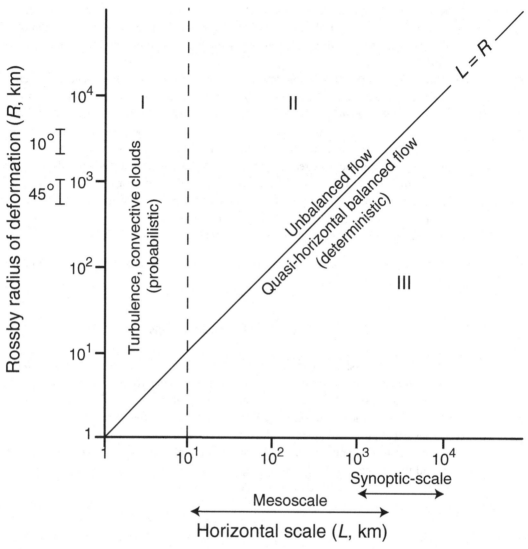

Fig. 4.6 Schematic showing three regimes (I, II, III) of atmospheric circulation, defined in terms of the prevailing physical length scale (abscissa) and the dynamic length scale (ordinate), where the latter is represented by the Rossby radius of deformation. The intervals defined on the ordinate indicate the values of $R$ that are typical at the noted latitudes. Adapted from Frank (1983), and originally from Ooyama (1982).

in the rotational component of the wind.[2] Because the rotational wind component is not strongly related to convection, there are no links, or only weak links, to secondary convective processes. In contrast, for dynamically small systems, adjustment of the mass field to

---

[2] On dynamically large scales, the mass field will "dominate" the geostrophic-adjustment process, and the imbalance caused by the latent heating will produce an adjustment toward a geostrophic value in the rotational component of the wind. See Section 6.10.1 for a discussion of the geostrophic-adjustment process.

the latent heating will cause divergent circulations that will influence the future evolution of convection, making the process more difficult to parameterize.

### 4.3.3  Relationship between the subgrid-scale (convective) precipitation parameterization and the resolved-scale precipitation

For all but very-high-horizontal-resolution models that can explicitly represent convective cells, models generally employ both convective and microphysical (Section 4.5.1) parameterizations. This means that precipitation can be produced by the model both when the convective parameterization is triggered and when the explicit processes represented in the microphysical parameterization produce resolved-scale precipitation that reaches the ground. In the former case, subgrid-scale precipitation is represented by the parameterization at the model-resolved grid scale, for sub-saturated grid-box conditions. In the latter case, grid-box saturation is required at some point in the column. There are two precipitation variables defined on the model grid at the surface; one is the parameterized convective precipitation and one is the resolved-scale precipitation. Even though these are summed to produce a total-rainfall field, model developers often look at both output fields separately to help them better understand internal model processes. This dual treatment of precipitation processes, by two generally distinct components of the model, leads to some conceptual and real difficulties. For example, the convective parameterization often does not produce cloud water and ice on the grid scale, even though precipitation has been generated, and thus no radiative effects of the clouds are rendered in the model. And, precipitation generated in one geographic region of a meteorological event will be produced by the parameterization, and by the explicit microphysics code in another geographic area. For example, in a mesoscale convective system, the microphysics parameterization may produce precipitation in the trailing stratiform-precipitation region, while the convective parameterization represents the precipitation elsewhere. Or, in an extratropical cyclone the microphysics and convective parameterizations could predominate in the precipitating regions of the warm and cold fronts, respectively. Figure 4.7 illustrates how the partitioning of precipitation from these two sources can depend on the meteorological event and the convective parameterization. Shown in both panels is the ratio (percentage) of the convective precipitation to the total precipitation, for 36-h simulations that employed four different convective parameterization schemes in a model that was otherwise the same. The left panel applies to simulations of a mesoscale convective system that occurred in the spring season, and the right one pertains to an Arctic front in the winter, with some convection in the warm air mass. There were clearly great differences among the simulations in terms of the partitioning of the precipitation between the resolved and subgrid components. For example, for the mesoscale convective system (a), the use of the Anthes–Kuo parameterization caused virtually all of the precipitation to be produced by the subgrid-scale mechanism, whereas when the Grell parameterization was used a large percentage of the precipitation was produced by the resolved-scale mechanism. Relationships between the two parts of the model that simulate precipitation processes are clearly not simple, nor is it always easy to anticipate which one will dominate for a particular case. And, these results make it clear that it is not reasonable to equate

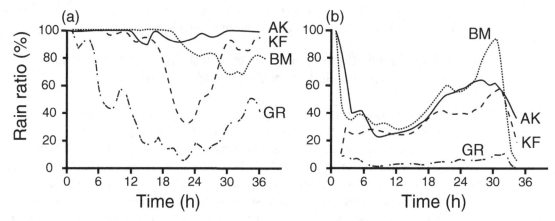

**Fig. 4.7** Ratio (percentage) of subgrid-scale precipitation to total precipitation (sum of subgrid and resolved precipitation) for simulations of two meteorological cases, where four different convective parameterizations were used for each case. The horizontal grid increment was 36 km, and the ratios are based on totals for the computational area. Panel (a) pertains to simulations of a mesoscale convective system that occurred in May, and panel (b) pertains to simulations of an Arctic front in February, with some convection in the warm air mass. The four convective parameterizations were the Grell (GR; Grell 1993, Grell *et al.* 1994), Kain–Fritsch (KF; Kain and Fritsch 1993), Betts–Miller (BM; Betts and Miller 1986), and Anthes–Kuo (AK; Anthes 1977, Grell *et al.* 1994) schemes. Adapted from Wang and Seaman (1997).

resolved and subgrid precipitation produced in the model to stratiform and convective precipitation in the atmosphere.

In spite of the common lack of a direct link between the convective and resolved-scale precipitation code in a model, there are exceptions. For example, hybrid methods partition a fraction of the parameterized precipitation from the convective scheme to the grid-scale precipitation defined in the microphysics scheme (e.g., Frank and Cohen 1987).

### 4.3.4 Summary of example convective-precipitation parameterizations

Many different convective parameterizations have been developed and used in models of various scales (e.g., Arakawa and Schubert 1974; Kuo 1974; Kreitzberg and Perkey 1976; Anthes 1977; Brown 1979; Fritsch and Chappell 1980; Molinari and Corsetti 1985; Betts and Miller 1986; Frank and Cohen 1987; Tremback 1990; Grell 1993; Kain and Fritsch 1993; Janjić 1994, 2000; Grell and Dévényi 2002; Kain 2004).

The following are commonly used schemes. The very brief descriptions are meant only to illustrate some of the high-level properties of the methods. Stensrud (2007) and Wang and Seaman (1997) should be consulted for more-lengthy summaries.

### Grell scheme

This is a variant of the Arakawa–Schubert parameterization (Arakawa and Schubert 1974), and is a deep-layer-control scheme. In the Arakawa–Schubert method, shallow and deep cumulus clouds, with a spectrum of sizes in each grid box, are idealized as plumes. In contrast, the Grell scheme only uses one cloud size, which is justified given that the applications are on the mesoscale. The subgrid precipitation is calculated by

$$P = Im(1 - \beta),$$

where $I$ is the condensate in the updraft, $m$ is the updraft mass flux at cloud base, and $(1 - \beta)$ is the precipitation efficiency. The latter is assumed to be a function of the resolved environmental wind shear in the lower troposphere. Convective downdrafts are parameterized.

## Anthes–Kuo scheme

This parameterization (Anthes 1977, Grell *et al.* 1994) is a variation of one of the earliest convective schemes (Kuo 1965, 1974), and uses a column-integrated moisture convergence ($M$) to determine the location and intensity of convection. When conditional instability exists and the moisture convergence exceeds a threshold, convection is initiated. Because the parameterized convection is based on the source of buoyant energy, it is classified as a deep-layer-control scheme. The moisture that is converging in the column is partitioned into convective precipitation and moistening of the column. The precipitation rate ($P$) is calculated by

$$P = (1 - b)M,$$

where

$$b = 2(1 - \overline{RH})$$

and $\overline{RH}$ is the column-mean relative humidity. This is a computationally undemanding scheme, which is one reason for its somewhat enduring popularity. It is, however, not especially well founded because moisture convergence does not necessarily result in convective activity. There are better schemes now available.

## Betts–Miller scheme

The Betts–Miller scheme (Betts and Miller 1993, Janjić 1994) is another deep-layer-control scheme that, upon initiation of convection, adjusts the model profiles of temperature and moisture in each grid column toward specified reference profiles that correspond to a quasi-equilibrium condition that is associated with deep convection (Betts 1986). The parameterized precipitation is calculated by

$$P = \int_{P_B}^{P_T} \frac{q_R - q}{\tau g} dp,$$

where $q$ is the grid-point specific humidity, $q_R$ is based on the deep-convection reference profile for specific humidity, $\tau$ is the time scale over which the adjustment occurs, and the $P_T$ and $P_B$ are the pressures at the top and bottom of the cloud.

## Kain–Fritsch scheme

The Kain–Fritsch scheme (Kain and Fritsch 1993) is an updated version of the Fritsch–Chappell scheme (Fritsch and Chappell 1980). Here, the activation of convection is defined by low-level forcing, and is also a function of the CAPE at a grid point. So it is both a low-level- and deep-layer-control scheme. The convective precipitation is calculated as

$$P = ES,$$

where $E$ is the precipitation efficiency and $S$ is the sum of the vertical fluxes of vapor and liquid at about 150 hPa above the LCL.

### 4.3.5  The choice of convective parameterization, and its impact on the simulation

Convective parameterizations employ a wide variety of approaches to the problem, and assumptions, and they inevitably perform best for those situations where the assumptions are better satisfied. This can make them dependent on the geographic area and the prevailing meteorological process. For example, some parameterizations seem to work best in the tropics, or in midlatitudes, or in high latitudes. Unfortunately, the parameterizations in global models must perform adequately for all climates and weather scenarios.

The point was made earlier that parameterization methods used for coarse-resolution models must represent both the convective-scale processes as well as the mesoscale processes that are related to the convection. In contrast, schemes used in mesoscale models only need to parameterize the convective scales. In a nested system of grids in a LAM, which may span resolutions from the synoptic scale to the mesogamma scale, it is not unreasonable to use different parameterizations on the different grids. On a very-high-resolution grid in a nest, which can explicitly resolve convection, it would be appropriate to not use any parameterization. Unfortunately, most convective parameterizations were designed for models with grid increments of 20–30 km or larger. Even though there is evidence that some can still be used with grid increments as small as 10 km, there currently is no good solution to the problem of how to represent convection between that resolution and those that are needed to explicitly resolve convection.

To illustrate the potential sensitivity of the accuracy of the precipitation forecast to the choice of the convective parameterization, see Fig. 4.8. The average total (convective and resolved) rain rate is plotted for a spring-season convective event (panel a), based on observations, and for five simulations that used different treatments for the convection – four different parameterizations, and no parameterization. For all simulations, the model grid increment was 12 km. At specific times in the simulations, the rain rate varied by as much as a factor of three or four among the different parameterizations. Also depicted is the bias score averaged for three warm-season convective events (panel b), again for each of the four parameterizations and for the use of no parameterization. The horizontal grid increment was 36 km, representing an appropriate resolution for the use of any of the parameterizations tested. Both the simulation-average scores on the right, as well as the

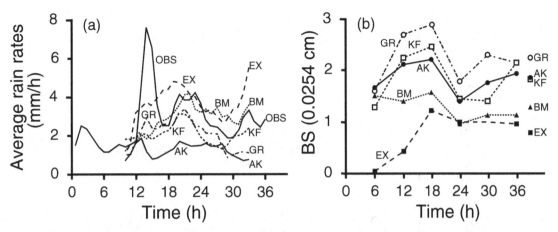

Fig. 4.8    Average rain rate for a spring-season convective event (a), based on observations (OBS), and for five simulations that used different treatments for the convection – four different parameterizations, and no parameterization (EX). For all simulations the model grid increment was 12 km. Also depicted is the bias score averaged for three warm-season convective events (b), again for each of the four parameterizations and for the use of no parameterization. The horizontal grid increment was 36 km. The four convective parameterizations were the Grell (GR), Kain–Fritsch (KF), Betts–Miller (BM), and Anthes–Kuo (AK) schemes. Adapted from Wang and Seaman (1997).

time-dependent curves, show a substantial dependence of the precipitation amount on the parameterization that was employed. The Betts–Miller scheme produced a simulation-average bias of close to unity, whereas the Grell scheme had a bias that exceeded two. The simulation that used no convective parameterization (explicit – EX) severely under-predicted the early precipitation amounts because of the time required to develop grid-scale saturation. This is one of hundreds of examples in the literature of the dependence on many factors of convective-parameterization performance.

Not only is model-simulated precipitation sensitive to the parameterization scheme employed on that grid, it has been shown that precipitation from cloud-resolving simulations on a fine grid in a nest is sensitive to the convective parameterization that is employed on the surrounding coarser grids. For example, Warner and Hsu (2000) describe tests with an operational LAM having three grids in a nest. The two coarser grids had grid increments of 10 km and 30 km, and thus required the use of a convective parameterization. In contrast, the innermost grid, with a grid increment of 3.3 km, explicitly represented the convection (i.e., no parameterization was used). For a model simulation of summer convection in the southwestern USA, Fig. 4.9 shows the grid-average hourly rain rate produced on the convection-resolving grid of this model when three different convective parameterizations were used on the surrounding grids. Also shown is the rainfall on that grid when no parameterization was used on the outer grids, as well as the rainfall estimated by the reflectivity from the WSR-88D radar. Even though the model used on the inner grid was identical in all four simulations, there clearly was a large impact of the choice of the convective parameterization used on the other grids. Through LBC effects, the parameterizations on the coarser grids caused stabilization and drying to various degrees on the convection-resolving grid, resulting in substantial differences in the simulated precipitation.

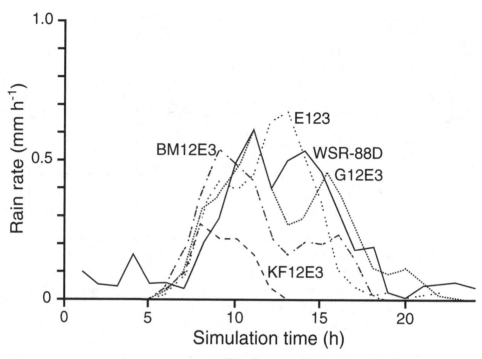

**Fig. 4.9** Grid-average hourly rain rates on the inner, convection-resolving grid (grid 3) of a nested LAM, when three different convective parameterizations were used on the surrounding grids (grids 1 and 2). Also shown is the rainfall on that grid when no parameterization was used on the outer grids (E123), as well as the rainfall estimated by the reflectivity from the WSR-88D radar. The lines are labelled in terms of the convective parameterization used: BM is Betts–Miller, G is Grell, KF is Kain–Fritsch, and E is explicit. The model configuration used on the inner grid was identical in all four simulations. Adapted from Warner and Hsu (2000).

## 4.4 Turbulence, or boundary-layer, parameterizations

### 4.4.1 Boundary-layer structure

At the lower boundary of the troposphere is the turbulent layer through which the influence of the surface is directly transmitted to the free atmosphere above. Through this boundary layer, or mixed layer, turbulent eddies transport water vapor and heat upward from their source at the surface. Also, the frictional stress exerted by the surface on the atmospheric fluid is transmitted by the turbulence. There are two causes of turbulence, or sources of turbulent energy. One is the buoyancy that creates rising parcels of air, or convection, and the compensating subsidence, when the land surface is heated during the day. The other source is related to the rate of change of the horizontal wind speed with height – i.e., the vertical shear of the horizontal wind. When this shear is small, and there is no buoyancy, the flow is nonturbulent, or laminar. When the shear exceeds a threshold, the flow becomes turbulent, with the turbulent energy derived from the mean wind. The buoyancy-driven convective turbulence is dominant during the day, while the shear-driven turbulence is more common at night. Because the wind speed perpendicular to any surface, including the ground, must be zero, turbulence cannot exist at the surface and cannot transport heat or moisture there. Thus, in a very shallow layer of a few molecules to a few millimeters above the surface, called the laminar sublayer (also, microlayer), transfers of heat, moisture, and surface-frictional effects are through molecular processes. Thus, the laminar sublayer is the nonturbulent interface between the ground and the turbulent mixed layer, and the mixed layer is the turbulent interface between the laminar sublayer and the free atmosphere above the mixed layer. The lower 50–100 m of the mixed layer, where the turbulent transport of heat, moisture, and momentum vary relatively little (compared to the situation in the mixed layer above), is called the surface layer.

The vertical extent of the turbulent mixing defines the daytime (convective) boundary-layer depth. At night, the atmosphere near the surface cools, and the source of the buoyant energy is eliminated. Any new turbulent energy in this stable layer near the ground must now be derived from the vertical shear in the horizontal wind. Unless the horizontal wind is exceptionally strong, this nocturnal boundary layer is much more shallow than the daytime one. The turbulence and the well-mixed profiles of the different meteorological variables penetrate progressively upward during the daytime heating cycle. Figure 4.10 shows measured profiles of the vertical structure of potential temperature at a few times during the daytime heating cycle in the Great Basin Desert in the USA. As the heating continued during the day, the depth of the approximately constant potential-temperature layer progressively increased. At night, as the cooling land surface in turn cools the lowest layer of the atmosphere, a temperature inversion forms.

Figure 4.11 illustrates distinctions between nocturnal turbulence that is generated from wind shear alone, and daytime turbulence that results from both buoyant motion as well as wind shear. Both curves show the variation with time of the vertical inclination of the wind at 29 m Above Ground Level (AGL), based on bivane measurements.[3] The lower curve, with relatively small-amplitude and high-frequency excursions from the horizontal direction, shows the effect of nocturnal turbulence that results only from the vertical shear of the horizontal wind. In contrast, the upper daytime curve shows similar high-frequency variability, but it is superimposed on a lower-frequency change with a period of perhaps 15–60 seconds. The longer-period changes during the day are associated with larger horizontally moving turbulent eddies.

In purely laminar flow, the layers of air slide over each other without much mixing between them. The only mixing that occurs is through the exchange of molecules between layers. Molecules from a slower-moving layer nearer the ground enter a faster-moving layer above, and the slower speed of the molecules represents a drag that slows down the upper layer. Analogously, faster-moving molecules move downward, with the drag effect causing the lower layer to speed up. Molecules of water vapor move between layers, causing a net transfer from moist layers to dry layers, and heat is transferred by virtue of the

---

[3]  A bivane is a wind vane with two axes of rotation, one horizontal and one vertical.

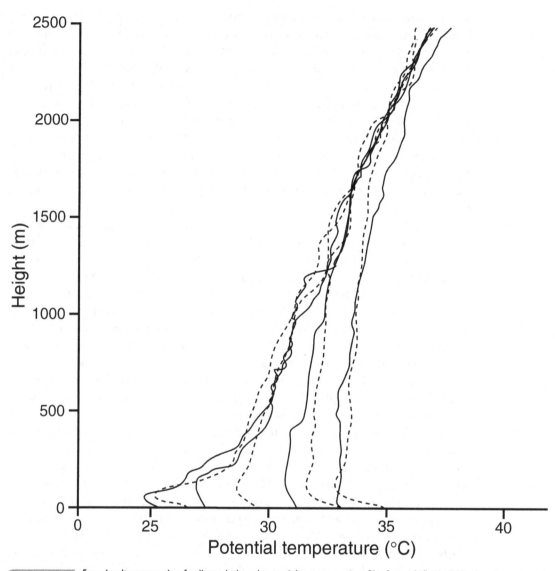

**Fig. 4.10** Four simultaneous pairs of radiosonde-based potential-temperature profiles for a salt flat (solid line) and a vegetated, sandy site (dashed line) in the Great Basin Desert in September 1997 on a relatively clear calm day. There were four pairs of radiosonde ascents during the day at 0830 LT (0850 for the salt-flat sounding), 1000 LT, 1200 LT, and 1400 LT (left to right in the figure). The two locations are about 20 km apart. Provided by Elford Astling, US Army Dugway Proving Ground.

different kinetic energies of the molecules exchanged between layers. In nonturbulent flow, this is how the layers of air "feel" each other. With turbulent flow, there are eddies that mix the air between the layers, with this type of mixing being much more efficient than the molecular mixing of laminar flow. One can imagine the vertical exchange of properties with turbulent mixing in the same way as with molecular mixing. For example,

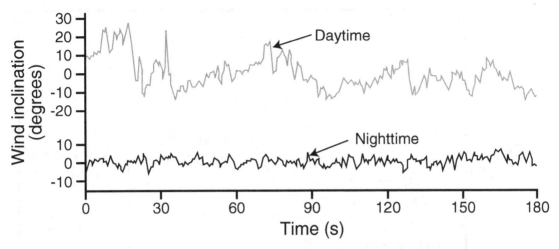

Fig. 4.11 The variation with time of the vertical inclination of the wind at 29 m AGL, based on bivane measurements, for daytime and nighttime conditions. The mean horizontal wind speed was 3–4 m s$^{-1}$. From Priestley (1959).

because water-vapor content near the surface generally decreases with height, upward-moving air in the turbulent eddies will contain more water vapor and downward-moving air will contain less water vapor.

Figure 4.12a is a schematic of the geometry of the daytime (convective) and nocturnal (stable) mixed-layer structure, and of the transitions between the two regimes. During the daylight hours, the convective mixed layer will increase in depth as the surface heating generates buoyancy-driven turbulence that erodes upward into the troposphere. The depth will typically reach about 1 km, but may span the entire troposphere in strongly heated deserts. After sunset, the ground and the lower atmosphere cool, and the buoyant source of turbulent energy diminishes. The nocturnal, or stable, mixed layer derives most of its turbulent energy from the wind shear, with the depth of the layer being considerably less than in the daytime. In contrast to the daytime, at night the mixing can be intermittent. The shear will develop to a critical value; mixing will abruptly ensue and decrease the shear to a subcritical value, shutting off the mixing; the shear will then increase again; etc. At the ground, this process is manifested as periods of calm that are occasionally interrupted when moderate or strong winds are briefly mixed downward from above. Above the stable, nocturnal boundary layer there exists residual turbulence from the daytime mixed layer, with the intensity decaying with time as a result of internal friction within the fluid.

Figure 4.12b shows typical vertical daytime profiles of wind speed ($u$), water-vapor density ($\rho_v$), and potential temperature ($\theta$). Parcels of unsaturated air that are mixed upward or downward by turbulence, cool and warm, respectively, at the dry adiabatic lapse rate. Thus, the well-mixed vertical profile of temperature is the dry adiabatic lapse rate, which is known as a "neutral" temperature profile. In these conditions, the potential temperature is uniform with height. The temperature itself decreases at about 10 °C km$^{-1}$ in this region. Within the surface layer, closer to the ground, the temperature decreases even more rapidly

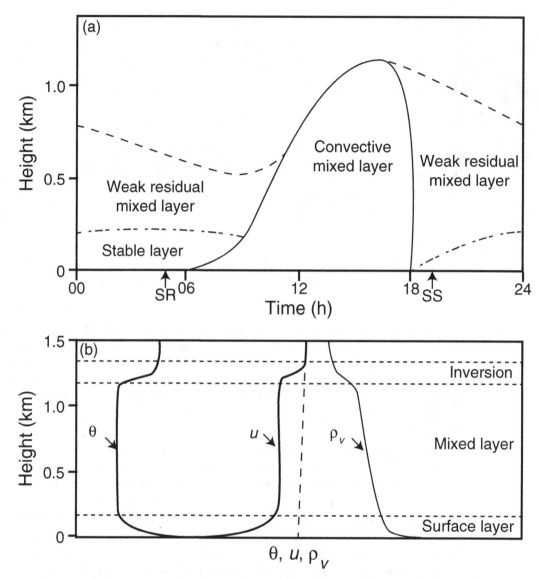

**Fig. 4.12** Schematics of (a) the typical diurnal variation of the boundary-layer structure and (b) typical vertical profiles within the daytime boundary layer of potential temperature ($\theta$), horizontal wind speed ($u$), and water-vapor density ($\rho_v$). The times of sunrise (SR) and sunset (SS) are shown in (a). The dashed line in (b) represents the wind speed that would exist without friction between the atmosphere and ground. Adapted from Oke (1987).

with height at a superadiabatic rate. At the top of the mixed layer is often a potential temperature inversion, within which potential temperature increases with height. The transition from uniform potential temperature to the inversion above is often used to define the depth of the mixed layer based on radiosonde soundings. Wind speed increases rapidly with height within the surface layer, from zero at the ground, and remains relatively uniform within the mixed layer. Throughout the midlatitude troposphere above the mixed layer, the climatological north–south temperature contrast causes the wind speed to increase with height up to the tropopause. The dashed line represents the value that the wind speed would attain without the retarding effect of friction, which is transmitted through the boundary layer by turbulence. Above the mixed layer in the free atmosphere, where turbulence does not transmit the frictional stress of Earth's surface, the wind speed is greater. The water-vapor content, here defined in terms of the density of the water vapor, is fairly uniform within the mixed layer, but it does decreases somewhat with height because the source is at the surface and entrainment mixes in drier air from above the boundary layer.

### Internal structures within boundary layers

Daytime convective boundary layers and the nighttime residual mixed layers are represented in Fig. 4.12 as simple structures with smooth temperature lapse rates. However, various factors can cause a considerable amount of internal structure to exist. First, when the boundary layer contains layers of dust, the radiative heating and cooling effects of the dust impact the vertical temperature profile. Even if the dust appears to be uniformly distributed throughout the boundary layer, the vertical sorting of different particle sizes and mineral types (having different optical properties) can produce vertical differences in heating and cooling rates. Another factor is the development of internal boundary layers that result from air flowing over surfaces with contrasts in properties such as the heat flux or roughness. For example, if the horizontal wind transports boundary-layer air from a smooth, hot surface to a cooler, rougher one, an internal boundary layer that is forced by the rougher surface develops within the original boundary layer. That is, the different surface roughness and heat flux would produce an internal boundary, within the mixed layer, across which the wind and temperature profiles would differ. This boundary would intersect the surface at the edge of the temperature and roughness contrast, and rise with increasing distance downstream. There are always subtle to major contrasts in surface properties, so whenever the boundary-layer air moves horizontally these internal boundary layers will complicate the structure. For example, the potential-temperature profiles in Fig. 4.10 show considerable variability within the boundary layer.

### Aerodynamic roughness and the vertical wind profile

The near-surface turbulent fluxes of heat, moisture, and momentum are influenced by the structure and spacing of surface-roughness elements such as rocks, vegetation, and soil grains. In general, rougher surfaces cause more-intense turbulence. Expressions representing the effect of the turbulence on the vertical wind profile in the surface layer employ a parameter called the roughness length ($z_0$) to describe the roughness characteristics of the surface. In particular, it can be shown that, for conditions of neutral stability (strong convective mixing),

$$u(z) = \frac{u_*}{k} \ln\frac{z}{z_0}, \tag{4.1}$$

where $u$ is the speed of the mean wind at height $z$, $u_*$ is the friction velocity, $k$ is the von Karman constant with a value that is thought to be 0.35–0.40, and $z$ is the height above the ground. The friction velocity represents the drag of the atmosphere against Earth's surface, or the frictional stress. Recall that the surface layer, where this equation applies, is the lower 50–100 m of the mixed layer where the turbulent fluxes of heat, moisture, and momentum vary relatively little in the vertical.

If $z$ in Eq. 4.1 is set to $z_0$, $u$ is equal to zero. Thus, $z_0$ is the height above the ground at which the mean wind speed goes to zero in neutral conditions, and is proportional to the roughness of the surface. Because $u_*$ is not a function of height in the surface layer and $k$ is a constant, $u$ increases logarithmically with increasing $z$. Figure 4.13 shows a typical vertical profile of horizontal wind speed for neutral conditions (i.e., a solution to Eq. 4.1). If the equation is solved for $u_*$, it is clear that $u_*$ is linearly related to the slope of the line ($u/\ln(z/z_0)$). The $y$-intercept of the line is $z_0$. Also shown are profiles of $u$ with height for unstable and stable conditions; that is, with temperature lapse rates greater than and less than, respectively, the neutral (dry adiabatic) value. For stable conditions the profile is concave downward, and for unstable conditions it is concave upward.

Obviously it is important to be able to estimate the roughness length in order to apply this equation. Bagnold (1954) determined that the roughness length over a bare flat sand surface is approximately equal to the mean diameter of the sand grains, divided by 30 (roughly $10^{-5}$ m for typical sand). When vegetation is present, the roughness lengths are larger and more difficult to obtain (Driese and Reiners 1997). Even though studies that estimate roughness lengths over agricultural fields with regularly spaced plantings are abundant, there are fewer studies of less-homogeneous natural environments.

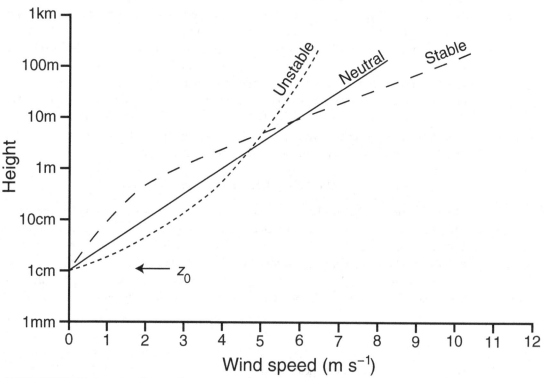

**Fig. 4.13**   Wind speed as a function of height within the surface layer, for neutral, stable, and unstable conditions. From Stull (1988).

## Boundary layers that are detached from the surface

All boundary layers originate because of the influence of a surface with which they are in contact. However, boundary layers do not necessarily remain in contact with the surface over which they form, with important consequences for the weather. An example of one such situation is shown in Fig. 4.14, where a boundary layer first develops over the high desert plateau of northern Mexico. When southwesterly lower-tropospheric winds cause this heated layer of air to move toward the lower terrain elevations to the northeast, the boundary layer becomes detached from the surface. When this happens, a temperature inversion, i.e., a stable layer of air, forms at the base of the Elevated Mixed Layer (EML). This stable layer can inhibit the

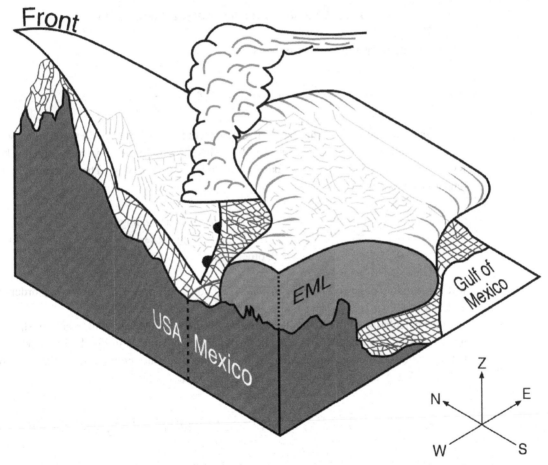

**Fig. 4.14** Schematic of an EML-related severe-weather environment over the southern Great Plains of North America. In the foreground is the high plateau of Mexico, from which a mixed layer is flowing to the northeast and becoming elevated over the Great Plains of the USA. The northwest quadrant of the area is the southern Rocky Mountains. Convective clouds are seen on the northwest edge of the EML, associated with low-level southeasterly flow causing moist, unstable air to run out from under the convection-inhibiting inversion at the base of the EML. The meteorological case discussed in the text had a front positioned to the east of the Rocky Mountains, as illustrated. Adapted from Lakhtakia and Warner (1987).

development of convective rainfall because parcels of air that are buoyant in the neutral lapse rates near the surface are no longer buoyant when they encounter this warm elevated mixed layer. The large CIN associated with this thermal "lid", represented by the hot boundary-layer air, allows moisture and heat to build up in the layer between it and the ground. If the convective motions erode through the inversion, or leak out around it, the consequence is a sudden release of energy that can lead to severe convective storms. This process produces severe convective weather in the semi-arid Great Plains of the USA in the spring. There are many other areas in the world where mixed layers become elevated, with important potential consequences for downstream precipitation. These situations must be represented in models in order for precipitation forecasts to be accurate.

### 4.4.2 Boundary-layer parameterization closures

The momentum equations in tensor notation are

$$\frac{\partial u_i}{\partial t} = -u_j\frac{\partial u_i}{\partial x_j} - \delta_{i3}g + f\varepsilon_{ij3}u_j - \frac{1}{\rho}\frac{\partial p}{\partial x_i},  \tag{4.2}$$

where the viscous-stress terms have not been included. As noted earlier, we sum over repeated indices that appear with variables that are multiplied by each other. The Kronecker delta, $\delta_{mn}$, equals 1 when $m = n$ and is zero otherwise. The alternating unit tensor is defined as

$$\varepsilon_{ijk} = \begin{array}{l} +1, \quad \text{if } i, j, k \text{ are in ascending order} \\ -1, \quad \text{if } i, j, k \text{ are in descending order} \\ \phantom{-}0, \text{ otherwise,} \end{array}$$

where ascending order means that $i, j, k$ are 1, 2, 3 or 2, 3, 1 or 3, 1, 2. Descending order is 3, 2,1 or 2, 1, 3 or 1, 3, 2. The value is zero if any of the $i, j, k$ are the same.

In the calculation of the Reynolds' equation for the $u$ component of momentum, shown in Eq. 2.15, the first step was to separate each dependent variable in Eq. 2.1 into a perturbation and a mean component, where it was assumed that the perturbation is associated with turbulence. Making this same substitution into Eq. 4.2, produces

$$\frac{\partial}{\partial t}(\bar{u}_i + u_i') = -(\bar{u}_j + u_j')\frac{\partial}{\partial x_j}(\bar{u}_i + u_i') - \delta_{i3}g + f\varepsilon_{ij3}(\bar{u}_j + u_j') - \frac{1}{(\bar{\rho} + \rho')}\frac{\partial}{\partial x_i}(\bar{p} + p').$$

Expanding this, and making the assumption that $\rho' \ll \bar{\rho}$, produces

$$\frac{\partial \bar{u}_i}{\partial t} + \frac{\partial u_i'}{\partial t} = -\bar{u}_j\frac{\partial \bar{u}_i}{\partial x_j} - u_j'\frac{\partial u_i'}{\partial x_j} - \bar{u}_j\frac{\partial u_i'}{\partial x_j} - u_j'\frac{\partial \bar{u}_i}{\partial x_j} - \delta_{i3}g + f\varepsilon_{ij3}\bar{u}_j + f\varepsilon_{ij3}u_j' - \frac{1}{\bar{\rho}}\frac{\partial}{\partial x_i}\bar{p} - \frac{1}{\bar{\rho}}\frac{\partial}{\partial x_i}p'.$$

$$\tag{4.3}$$

Now, applying the Reynolds' postulates as well as the continuity equation, as in Section 2.2, produces

$$\frac{\partial \bar{u}_i}{\partial t} = -\bar{u}_j \frac{\partial \bar{u}_i}{\partial x_j} - \delta_{i3} g + f \varepsilon_{ij3} \bar{u}_j - \frac{1}{\rho} \frac{\partial}{\partial x_i} \bar{p} - \frac{\partial}{\partial x_j} \overline{u'_i u'_j}. \tag{4.4}$$

The Reynolds' stress term on the right represents the effects of turbulence on the mean motion, where the quantity $\overline{u'_i u'_j}$ is called a double correlation or a second statistical moment. Given that the model equations only predict the mean quantities, these covariances need to be defined in some way. There are two approaches for accomplishing this. One involves linking the magnitude of this (second-moment) term to the resolved-scale, or mean (first-moment), variables. The other requires the development of a predictive equation for the covariances.

To obtain a predictive equation for the covariances, begin by subtracting Eq. 4.4 from Eq. 4.3, producing a predictive equation for the turbulent velocity components:

$$\frac{\partial u'_i}{\partial t} = -u'_j \frac{\partial u'_i}{\partial x_j} - \bar{u}_j \frac{\partial u'_i}{\partial x_j} - u'_j \frac{\partial \bar{u}_i}{\partial x_j} + f \varepsilon_{ij3} u'_j - \frac{1}{\rho} \frac{\partial}{\partial x_i} p' + \frac{\partial}{\partial x_j} \overline{u'_i u'_j}. \tag{4.5}$$

Multiply this by $u'_k$, apply an averaging operator and Reynolds' postulates, to produce an equation for the second term on the right in the following predictive equation for the covariance:

$$\frac{\partial}{\partial t} \overline{u'_i u'_k} = \overline{u'_i \frac{\partial}{\partial t}(u'_k)} + \overline{u'_k \frac{\partial}{\partial t}(u'_i)}. \tag{4.6}$$

Then change all $i$ indices to $k$ in Eq. 4.5, multiply every term by $u'_i$, and again apply an averaging operator and Reynolds' postulates, producing an equation for the first term on the right in Eq. 4.6. Add the two equations to produce the following predictive equation for the covariance:

$$\frac{\partial}{\partial t} \overline{u'_i u'_k} = -\bar{u}_j \frac{\partial}{\partial x_j} \overline{u'_i u'_k} - \overline{u'_i u'_j} \frac{\partial}{\partial x_j} \bar{u}_k - \overline{u'_k u'_j} \frac{\partial}{\partial x_j} \bar{u}_i - \frac{\partial}{\partial x_j} \overline{u'_i u'_k u'_j}$$

$$+ f(\varepsilon_{kj3} \overline{u'_i u'_j} + \varepsilon_{ij3} \overline{u'_k u'_j}) - \frac{1}{\rho}\left[ \frac{\partial}{\partial x_i} \overline{p' u'_k} + \frac{\partial}{\partial x_k} \overline{p' u'_i} - \overline{p'\left(\frac{\partial u'_i}{\partial x_k} + \frac{\partial u'_k}{\partial x_i}\right)} \right].$$

This represents six different predictive equations, for $\overline{u'u'}$, $\overline{u'v'}$, $\overline{u'w'}$, $\overline{v'v'}$, $\overline{v'w'}$, and $\overline{w'w'}$. Unfortunately, a triple-correlation or third-moment term now exists on the right side of the equation. If predictive equations are derived for these moments, quadruple correlations will appear on the right side. This situation where there are always more unknowns than equations, requiring that the unknown terms be represented as some function of the known variables, defines the turbulence-parameterization closure problem. Table 4.1 summarizes the equations for the different statistical moments.

**Table 4.1** Example of the prognostic equations for the first three statistical moments, indicating the number of equations and the number of unknowns.

| Example prognostic variable | Statistical moment | Equation | Variable parameterized | Number of equations | Number of unknowns |
|---|---|---|---|---|---|
| $\bar{u}_i$ | First | $\dfrac{\partial \bar{u}_i}{\partial t} = \ldots - \dfrac{\partial}{\partial x_j}\overline{u'_i u'_j}$ | $\overline{u'_i u'_j}$ | 3 | 6 |
| $\overline{u'_i u'_j}$ | Second | $\dfrac{\partial}{\partial t}\overline{u'_i u'_j} = \ldots - \dfrac{\partial}{\partial x_k}\overline{u'_i u'_j u'_k}$ | $\overline{u'_i u'_j u'_k}$ | 6 | 10 |
| $\overline{u'_i u'_j u'_k}$ | Third | $\dfrac{\partial}{\partial t}\overline{u'_i u'_j u'_k} = \ldots - \dfrac{\partial}{\partial x_m}\overline{u'_i u'_j u'_k u'_m}$ | $\overline{u'_i u'_j u'_k u'_m}$ | 10 | 15 |

When there are predictive equations for the first moments of the state variables ($u$, $v$, $w$, $T$, etc.), and the covariance terms (e.g., $\overline{u'v'}$ ) are parameterized in terms of the first moments, it is called a first-order closure. With second-order closure methods, there are predictive equations for both the state variables and the covariances, and the triple correlations are parameterized in terms of the first and second moments. Thus, the order of the closure is defined in terms of the highest-order prognostic equations that are retained. Stull (1988) illustrates the use of correlation triangles to summarize the unknowns for different orders of closure (Table 4.2). Note that these two tables apply only to the momentum

**Table 4.2** Correlation triangles illustrating the unknowns associated with the different levels of the turbulence closure, for the momentum equations only

| Order of closure | Correlation triangle of unknowns |
|---|---|
| Zero | $\bar{u}$ <br> $\bar{v}$    $\bar{w}$ |
| First | $\overline{u'u'}$ <br> $\overline{u'v'}$    $\overline{u'w'}$ <br> $\overline{v'v'}$    $\overline{v'w'}$    $\overline{w'w'}$ |
| Second | $\overline{u'u'u'}$ <br> $\overline{u'u'v'}$    $\overline{u'u'w'}$ <br> $\overline{u'v'v'}$    $\overline{u'v'w'}$    $\overline{u'w'w'}$ <br> $\overline{v'v'v'}$    $\overline{v'v'w'}$    $\overline{v'w'w'}$    $\overline{w'w'w'}$ |

*Source:* From Stull (1988).

equations, and thus there are more unknowns when considering the full set of equations. A motivation for higher-order closures is the assumption that the more moments that are predicted, the more accurate will be the solution for the state variables. That is, the higher the moment that is defined by a parameterization, the less the approximation will affect the principal forecast variables, the first moments.

There also are closure methods wherein some of the terms in a particular moment category are parameterized and some are explicitly predicted. For example, in the prognostic equations for the first moments, some second moments on the right side may be parameterized while others are predicted. If all the second moments are predicted, the closure would be second order. If they are all parameterized, it would be first order. Thus, in this case it would be referred to as a 1.5 order method. There are other noninteger closures.

Regardless of the order of the closure, there are two different approaches that can be used in the parameterization. In one, the unknown quantity at a grid point is defined in terms of known quantities, or their vertical derivatives, at the same grid point. Of course the derivative would have to be calculated using adjacent grid points in the vertical. This is called a *local closure*, where such methods have been used through the third order. Alternatively, the unknown variable at a grid point can be estimated using known quantities that are defined at locations that are a significant distance away in the vertical. This is a *nonlocal closure*. The nonlocal closures and the higher-order local closures generally yield more-accurate solutions than do the lower-order local closures, but the former involve greater computational expense and model-code complexity (Stull 1988). An advantage of the second-order or higher closure methods is that the second moments of the wind components can be used to quantify the total Turbulent Kinetic Energy (TKE), such that

$$\frac{TKE}{m} = \frac{1}{2}\overline{u_i u_i},$$

where $m$ is mass. The TKE variable is useful for any application that requires knowledge of the turbulence intensity (such as dispersion of air pollution and turbulence loading on structures).

### 4.4.3 Local closures

There are many, many different local and nonlocal closure approximations in use in models, and this section and the next one will only provide a couple of illustrations. A common local closure approximation is referred to as K-theory or gradient-transport theory. In a first-order closure, we must parameterize the second moments. Assume the following generic predictive equation for a variable $\xi$ :

$$\frac{\partial}{\partial t}\bar{\xi} = \dots - \frac{\partial}{\partial x_j}(\overline{\xi' u'_j}).$$

A closure approximation for the flux $\overline{\xi' u'}_j$ is

$$\overline{\xi' u'}_j = -K\frac{\partial}{\partial x_j}\bar{\xi}, \tag{4.7}$$

where the parameter $K$ is a scalar with units of $\text{m}^2\text{s}^{-1}$. For positive $K$, the above equation states that the flux $\overline{\xi'u'_j}$ is down the local gradient of $\bar{\xi}$. Combining the above equations leads to

$$\frac{\partial}{\partial t}\bar{\xi} = \ldots + \frac{\partial}{\partial x_j}K\frac{\partial}{\partial x_j}\bar{\xi},$$

which now has no undefined variables on the right. The entire system of equations would be closed because there is a prognostic or diagnostic equation for each variable. The coefficient $K$ is referred to by a variety of names, including the eddy viscosity, the eddy diffusivity, the eddy-transfer coefficient, the turbulent-transfer coefficient, and the gradient-transfer coefficient. Different $K$ values are sometimes associated with the transfer of different variables, so that we sometimes see the $K$s written as $K_m$, $K_E$, and $K_H$, for momentum, moisture, and heat, respectively. Obviously the values of $K$ control the turbulent flux, and intuition tells us that a parameterization of $K$ in terms of the first moments would link it to the wind shear (Richardson number) and the static stability, quantities that are related to the mechanical and buoyant production of turbulence, respectively. Stull (1988) reviews many methods for the parameterization of $K$ in the context of this local closure. That reference also reviews many other local closures.

Because the turbulent flux is defined by local conditions, this simple closure works best for situations when the turbulent eddies are small and locally generated. For large eddies, the nonlocal closures described in the next section provide better results.

### 4.4.4 Nonlocal closures

Nonlocal-closure methods are motivated by the recognition that much of the mixing in the boundary layer can be associated with large eddies whose vertical dimension is approximately that of the boundary-layer depth, and that these eddies are not related to the local static stability or wind shear at some point in the middle of the boundary layer. Rather, these eddies are driven by the deeper mean stability that spans the boundary layer, which in turn responds to the surface heat flux. Figure 4.15 is a schematic that illustrates the distinction between local closures and a couple of different types of nonlocal closures. In Fig. 4.15a, which applies to local closures, the unknown higher-moment terms at the middle grid point in the column are parameterized through the use of known variables at that point, or through derivatives of the known variables whose calculation requires values at adjacent grid points.

One framework for treating the nonlocal-closure problem is called transilient turbulence theory (Stull 1988, 1993). Imagine a column of model grid boxes, where a small subset of the boxes is shown in Fig. 4.15, and identify a single reference grid box within the column. And, assume that we can define the grid boxes above and below from which turbulent eddies mix air into the reference grid box, as well as the grid boxes that receive air from the reference grid box. Figure 4.15b shows mixing between a reference grid box at the midpoint of the column, and all the surrounding boxes. The same process can be used to define the vertical turbulent mixing between all the other grid points and their

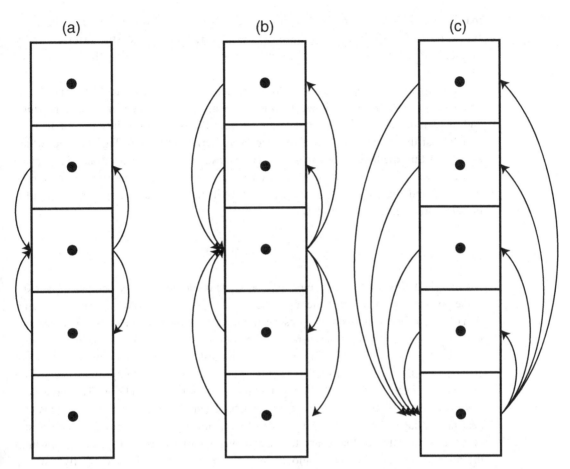

**Fig. 4.15** Schematic of the distinction between local closures and two different types of nonlocal closures. Panel (a) applies to local closures, and (b) and (c) pertain to nonlocal closures. See the text for details.

surroundings. Given this conceptual view of the situation, the process by which heat, water vapor, or a passive tracer are transferred among boxes can be quantified using the transilient-turbulence concept and the notation of Stull (1988). Let $\bar{\xi}_i$ represent the concentration of a passive tracer in a reference grid box $i$, let $c_{ij}$ be the fraction of air in box $i$ that is transported by turbulence from the donor box $j$ during a time period $\Delta t$, and let $\bar{\xi}_j$ be the concentration of the tracer in donor box $j$. To find the new concentration in box $i$ after the elapsed time, we simply sum the contributions from all $N$ grid boxes in the column, such that

$$\bar{\xi}_i(t + \Delta t) = \sum_{j=1}^{N} c_{ij}(\Delta t)\bar{\xi}_j(t).$$

This equation defines the exchange between each box $i$ and all the other boxes $j$. It is a matrix equation, where $c_{ij}$ is an $N \times N$ matrix of mixing coefficients called a transilient matrix, and $\bar{\xi}_i$ and $\bar{\xi}_j$ are $N \times 1$ matrices (vectors). Because the transilient matrix is a function of the turbulence, it is the same for all variables $\xi$. See Stull (1993) for discussions of many nonlocal parameterizations in the context of transilient-turbulence theory.

An example of a simple nonlocal closure is described in Blackadar (1978) and Zhang and Anthes (1982). Here, the intensity of the vertical convective transfer of heat, moisture, and momentum during the day is determined from the surface heat flux and the thermal structure of the entire mixed layer (not the local thermal structure). Figure 4.15c, which applies to this method, shows that the vertical exchange is visualized as taking place between each model layer in the boundary level and the lowest model layer. That is, both the small and the large eddies have their roots in the surface layer. An example of a turbulent-transfer term is as follows:

$$\frac{\partial \theta}{\partial t} = \dots m(\theta_a - \theta),$$

where $\theta$ is the local potential temperature at some model level in the boundary layer, $\theta_a$ is the potential temperature at the top of the surface layer ($\sim 10$ m AGL), and $m$ is a function of the surface heat flux and represents the fraction of the mass in a grid box in the column that is exchanged with the surface grid box during a specific time interval.

Figure 4.16 shows another way of viewing the distinction between nonlocal and local closures. On the left (a) is a vertical profile of the mean potential temperature within and above a forest canopy during the daylight hours (a convective mixed layer). There is a shallow inversion within the canopy, with an unstable layer above that, which transitions to a near-neutral layer. Above this, near the top of the boundary layer, the profile becomes stable. The vertical dashed lines show the deep movement of air parcels (open circles) within the boundary layer. The demarcations along the three vertical lines to the right of the sounding, also in panel (a), show how the layer is divided into turbulent and nonturbulent (laminar) flow regimes, and into stability regimes according to whether local or nonlocal methods are used in the definition. In panel (b) are shown heavy vertical arrows (heat flux) that indicate the directions and magnitudes of the vertical fluxes of heat that are observed under these conditions. Also shown are the heat fluxes estimated using local and nonlocal closures. A local-closure approximation for the vertical heat flux, such as Eq. 4.7 with $\bar{\theta} = \bar{\xi}$ and $j = 3$, would be

$$\overline{\theta' w'} = -K_H \frac{\partial}{\partial z} \bar{\theta},$$

such that the direction and magnitude of the flux is defined by the vertical gradient of the potential temperature. Use of this approach results in the "Local static stability" layering (a) and the "Local interpretation" of the heat flux directions (b). These heat fluxes are clearly inconsistent with the observed fluxes shown with the large arrows (b). In contrast, when the static stability is defined across the depth of the boundary layer by comparing the potential temperature of the near-surface parcel with those of the parcels above, to reflect the deep-layer stability, the more-realistic nonlocal static stability and heat-flux profiles

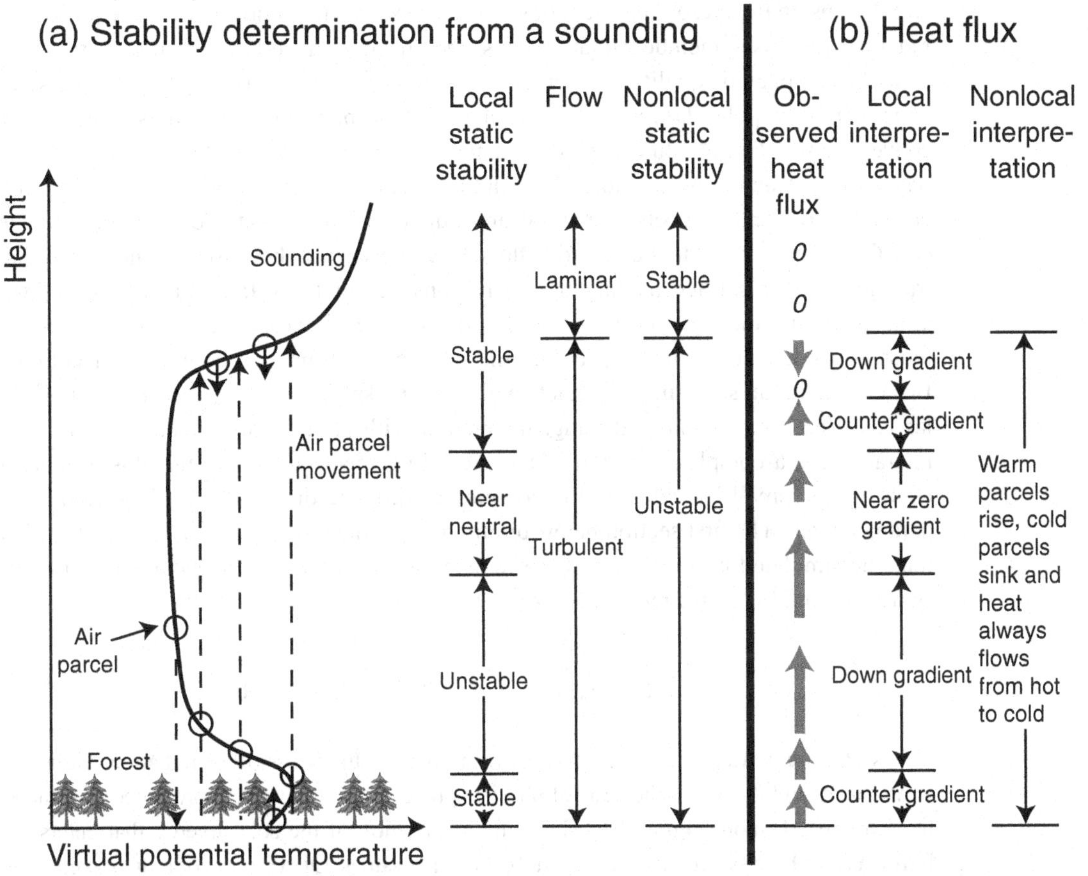

## (a) Stability determination from a sounding | (b) Heat flux

Fig. 4.16    Schematic showing stability defined based on local and nonlocal methods, and associated heat fluxes for each type of method. See the text for details. Adapted from Stull (1991).

result. Equations such as the above one for the local-closure approximation to the vertical heat flux can incorporate a correction term to the local gradient that incorporates the effects of the large eddies, such that

$$\overline{\theta' w'} = -K_H\left(\frac{\partial}{\partial z}\bar{\theta} - \gamma\right).$$

This correction defines an effective vertical lapse rate that is more unstable than the actual one, maintaining an upward heat flux through a deeper layer.

## 4.5 Radiation parameterizations

Electromagnetic radiation from the Sun is responsible for the existence of all processes in the atmosphere, including the midlatitude westerlies and cyclones, monsoons, tropical cyclones, and the Hadley circulation on the global scale, to convection and coastal circulations on the mesoscale. Less obvious, but nevertheless important, consequences of radiative processes include radiation fogs, the strong surface-based temperature inversions that impact air quality, evaporation of water at Earth's surface, the extreme katabatic winds in polar latitudes, the development of buoyant instabilities that lead to convective weather, etc. In order to simulate these processes correctly, models must represent the interaction of the radiation with the land, oceans, vegetation, clouds, air molecules, and mineral aerosols of natural and human origin. A critical component of the calculations is to provide the radiative flux at Earth's surface, because it is the spatial distribution of this surface heating that is responsible for the differential heating of the atmosphere. In addition, the radiative flux divergence of energy within the atmosphere must be calculated in order to define the radiative heating and cooling in a column. Because radiation sometimes interacts with the atmosphere and the surface at a molecular level (e.g., molecular scattering, interaction with cloud droplets) and because this interaction is a complex function of the prevailing spectrum of wavelengths, the processes are too small in scale and too complex to simulate directly. Thus they need to be parameterized. The first section below briefly reviews the relevant radiative processes that must be simulated by models, and the remaining ones describe the parameterization of shortwave and longwave radiative fluxes.

## 4.5.1  Processes that must be represented

This section reviews some basic concepts regarding the transmission of radiation within the atmosphere, and illustrates the sorts of physical processes that must be represented by a radiation parameterization. Figure 4.17 shows the disposition of the solar energy that enters the Earth–atmosphere system, in terms of global-average values. Of 100 units of radiation entering the atmosphere annually from the Sun, 31 units are both reflected and scattered back to space. This includes 6 units that are reflected from Earth's surface (land and ocean), 17 units that are reflected and scattered from clouds, and 8 units that are reflected and scattered from molecules and dust in the atmosphere. Twenty units are absorbed by the atmosphere and clouds. The remaining 49 of the 100 units are direct and diffuse solar radiation that are absorbed by Earth's surface. The partitioning of radiant energy at Earth's surface between the absorbed (49 units) and reflected (6 units) components is typically calculated within the land-surface model (Chapter 5) that is part of the atmospheric model. The other processes – absorption by the atmosphere, clouds, and dust; scattering by air molecules and dust; and reflection by clouds and dust – are represented in the radiation parameterization. Figure 4.18, again based on a global annual average, summarizes what happens to the 69 units of energy that are shown in Fig. 4.17 to be absorbed by both Earth's surface and atmosphere. Energy gains at the surface, totaling 144 units, are 49 units from direct and diffuse solar radiation and 95 units of infrared from the clouds and gas of the atmosphere. Surface energy losses, which also must total 144 units, include 114 units of infrared emitted to space and the atmosphere, 23 units lost by evaporation, and 7 units lost to the atmosphere through sensible heat fluxes. The net radiation is the radiative gain of 144 units minus the radiative loss of 114 units. Also, the 155 units of energy gains by the atmosphere from the various sources on the left must equal the losses from

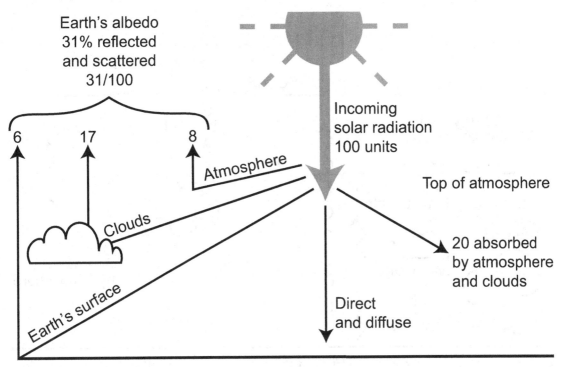

Earth's albedo
31% reflected
and scattered
31/100

6          17          8

Atmosphere

Incoming
solar radiation
100 units

Top of atmosphere

Clouds

20 absorbed
by atmosphere
and clouds

Earth's surface

Direct
and diffuse

49 absorbed at surface

**Fig. 4.17** The estimated atmospheric energy budget in terms of the global annual-average components. The estimated partitioning of energy is based on various sources.

the processes on the right. The land-surface component of the modeling system is responsible for calculating the evaporation rate, and the emission and absorption of infrared energy; the boundary-layer parameterization calculates the turbulent fluxes of heat by convection; the precipitation processes (both parameterized and resolved-scale) define the latent heating resulting from water condensation; and the radiation parameterization defines the remaining processes. These estimates of the various components of the budgets are being routinely revised, and there are some important differences among the estimates (e.g., Mitchell 1989, Kiehl and Trenberth 1997). It is important to remember that the magnitudes of these components are global averages, and thus they represent a large contribution from conditions over the oceans. Some energy-budget diagrams of this sort show an actual energy flux at the top of the atmosphere, rather than an arbitrary base value of 100 units. The global-average, top-of-the-atmosphere energy flux on a horizontal surface is about 342 W m$^{-2}$. The individual components of the budget thus represent partitions of this base value.

All substances with a temperature greater than absolute zero emit radiation. The rate at which energy is radiated at a given temperature, summed over all wavelengths, is proportional to the fourth power of the absolute temperature, which is the basis for the Stefan–Boltzmann law. With an added dimensionless coefficient, $\varepsilon$, the law states that

$$\text{Energy emitted} = \varepsilon \sigma T^4, \tag{4.8}$$

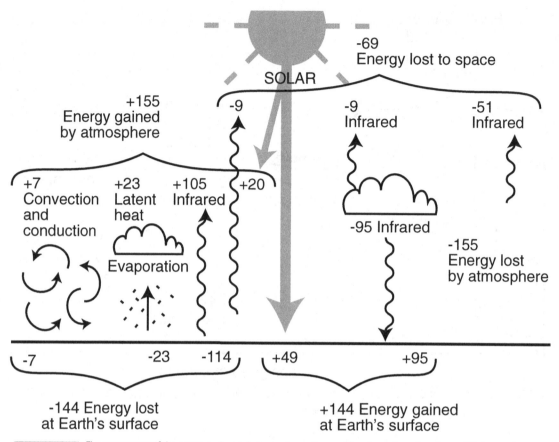

Fig. 4.18 The components of the energy exchange between the surface and the atmosphere, based on a global annual average. The estimated partitioning of energy is based on various sources.

where $\sigma$ is the Stefan–Boltzmann constant, equal to $5.67 \times 10^{-8}$ W m$^{-2}$ K$^{-4}$. If the radiating body emits the maximum possible radiation per unit area per unit time, the emissivity, $\varepsilon$, is equal to unity, and the emitter is called a black body. Less efficient radiators have emissivities between zero and unity. The intensity of the radiation emitted by a black body at different wavelengths is a function of temperature, and is prescribed by Planck's law. This intensity distribution with wavelength has a very similar shape for emitters of any temperature (see the examples of Fig. 4.19), and has a single maximum. The particular wavelength composition of the emitted energy depends on the temperature of the emitter, such that a temperature increase not only increases the total amount of energy emitted (Eq. 4.8), but it also increases the fraction from the shorter wavelengths. That is, the curve in Fig. 4.19 shifts to the left as temperature increases, and the wavelength of the peak emission, $\lambda_{max}$, moves accordingly, such that

$$\lambda_{max} = 2.88 \times 10^{-3}/T,$$

where $\lambda_{max}$ is in meters and $T$ is in kelvin.

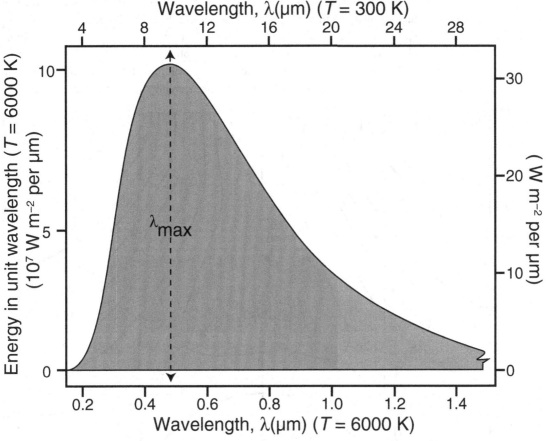

**Fig. 4.19** The spectral distribution of radiant energy emitted from a black body at (a) 6000 K, left vertical and lower horizontal axes and (b) 300 K, right vertical and upper horizontal axes. The $\lambda_{max}$ is the wavelength at which the energy output per unit wavelength is a maximum. Approximately 10% of the energy is emitted at wavelengths longer than those shown. Adapted from Monteith and Unsworth (1990).

In most atmospheric applications, we are only concerned with wavelengths in the ultra-violet, visible, and infrared portions of the full electromagnetic spectrum. Much of the Sun's energy is in the visible part of the spectrum, from 0.36 µm (violet) to 0.75 µm (red), as shown in Fig. 4.19, which depicts the energy intensity for the different wavelengths emitted by the Sun which has a surface temperature of about 6000 K. There is also significant solar energy emitted in the wavelength bands that are shorter than the violet (the ultraviolet) and longer than the red (the infrared), with the total solar ultraviolet–visible–infrared band extending from about 0.15 µm to about 3.0 µm. The infrared energy in this solar spectrum is referred to as the solar infrared, to contrast it with the longer wavelength infrared that is emitted by the cooler Earth and its atmosphere. Figure 4.19 also shows the spectrum of the energy emitted by the Earth and its atmosphere, which have a temperature of roughly 300 K. The wavelength band is all within the infrared, and extends from about 3 µm to 100 µm. Thus, the solar spectrum from 0.15 to 3.0 µm is referred to as shortwave radiation, while Earth's spectrum in the range 3.0–100 μm is longwave radiation. Note that the maximum intensity of the Sun's radiation is about a factor of $10^7$ greater than that for Earth. The peak intensity of the solar spectrum is at 0.48 μm ($\lambda_{max}$ is green, in the middle of the visible spectrum), whereas for the Earth–atmosphere system it is at about 10 μm.

Figure 4.20 shows the seasonal and latitudinal variation of the possible daily total (without atmospheric attenuation) solar radiation receivable on a horizontal surface. The

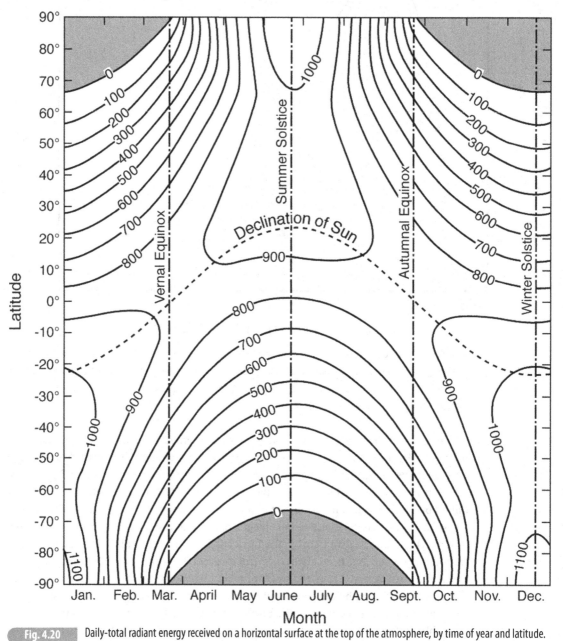

**Fig. 4.20**   Daily-total radiant energy received on a horizontal surface at the top of the atmosphere, by time of year and latitude. The isopleths are labelled in cal cm$^{-2}$. From List (1966).

values represent the time integral of the unattenuated direct solar energy flux during day-light hours, which is based on the intensity of the Sun's radiation, and the Earth–Sun geometry that controls the daily evolution of the Sun angle. No meteorological effects are accounted for (reflection by clouds, scattering, etc.), so this is the maximum possible energy receivable. It is the effect of the atmosphere on this radiation that must be parameterized.

Because of the existence of clouds, dust, and optically active (absorbing and emitting) gases, Earth's atmosphere is far from transparent to the passage of the radiant energy emitted by Earth and the Sun. Each of the constituents has its own unique effect on each individual wavelength band of the radiation. In terms of the bulk effect of the atmosphere, part of the radiation is reflected or scattered, part is absorbed, and the remainder is transmitted. The fraction of the incident radiation that is absorbed (absorptivity) individually by the atmosphere's major gaseous components, and by the total atmospheric mixture of gases, is shown in Fig. 4.21. It is clear that the gaseous medium totally absorbs some wavelength

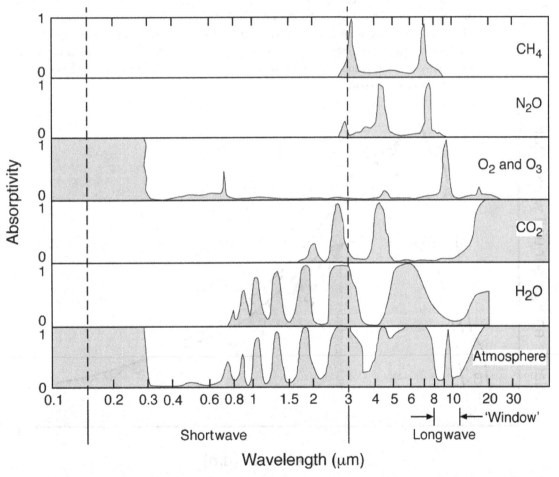

**Fig. 4.21** The absorptivity (fraction absorbed) of the atmosphere's major gaseous components, and of the total atmospheric mixture of gases. Adapted from Fleagle and Businger (1963).

bands of radiant energy, while it is relatively transparent to others. Within the visible wavelengths of the solar spectrum, from 0.36 to 0.75 μm, where the greatest energy is represented, not much absorption occurs from gases. At ultraviolet wavelengths shorter than 0.30 μm, ozone effectively absorbs virtually all of the energy, and water vapor becomes an important absorber at wavelengths longer than 0.80 μm. There is a "window" in the longwave absorption from about 8 to 11 μm, which encompasses the wavelength of the peak emission at 10 μm from the Earth–atmosphere system that emits at about 300 K (Fig. 4.19). Clouds absorb and emit radiation in this window, and thus their correct parameterization is important to the modeled radiation balance. Figure 4.22 shows the effects of both scattering and absorption on the attenuation of the solar beam as it penetrates the cloud-free atmosphere. The outer curve, A, is the solar spectrum at the top of the atmosphere, which differs from the smooth blackbody spectrum of a 6000 K radiator shown in Fig. 4.19. Here, the degree to which the emissivity departs from unity depends on wavelength. The remainder of the curves show the progressive modification of the top-of-the-atmosphere spectrum by ozone absorption, molecular scattering, aerosol scattering, and water vapor and oxygen absorption. The lower curve (E) shows the energy that survives the downward transit of the solar beam through the atmospheric medium. It is the purpose of radiation parameterizations to estimate the three-dimensional time-of-day-, seasonal-, weather-, and climate-dependent details of these processes whose annual-average effects are shown in these figures.

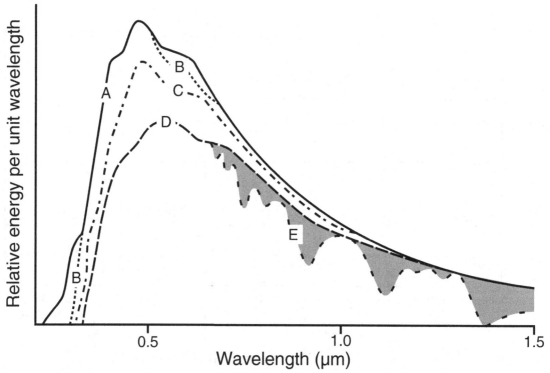

**Fig. 4.22**  Attenuation of the extraterrestrial solar spectrum (A) by ozone absorption (B), molecular scattering (C), aerosol scattering (D), and water vapor and oxygen absorption (E). Curve E is terrestrial sunlight. From Henderson (1977).

## 4.5.2 The general framework for representing radiation in models

The above physical effects of absorption, emission, scattering, and reflection are accounted for in calculations of the vertical fluxes of longwave and shortwave radiation at every grid point in a model. The vertical convergence of these energy fluxes is then employed in the thermodynamic energy equation (Eq. 2.4), such that

$$\frac{\partial T}{\partial t} \propto \frac{1}{\rho c_p}\frac{\partial}{\partial z}(F_D - F_U),$$

where $F_D$ is the downward flux and $F_U$ is the upward flux of radiant energy of all wavelengths. This equation is converted to finite-difference form, and solved at every grid point with the other terms in the equation. In addition, the downward flux at Earth's surface is also calculated and used as input to the model's land-surface parameterization (Chapter 5). The challenge is to calculate the longwave and shortwave fluxes with sufficient accuracy, especially for climate models where a small percentage error could be damaging, yet with sufficient efficiency so that the models can execute in the required period of time. This latter issue can be critical because radiation parameterizations can be very time consuming, and thus a variety of alternative approaches and approximations have been developed for solving the radiative-transfer equations. This is a sufficiently important issue that these parameterizations are often not called every time step, in order to reduce the computational demand.

Stephens (1984) lists a few factors that make the accurate modeling of radiation processes challenging, including the following.

- Radiation can simultaneously influence the atmospheric dynamics in multiple ways, and the required accuracy of the approximations in the parameterization can be dependent on the prevailing meteorological process. This makes it difficult to choose a set of approximations that is satisfactory in all situations.
- The atmospheric dynamics respond to diabatic heating from radiative flux convergence, phase changes, and sensible heating. However, these processes are sometimes coupled, for example when radiative cooling leads to condensation, leading to complex nonlinear interactions that may be difficult to approximate.

Another challenge is the following one. Required for calculations of shortwave and longwave radiative transfer are vertical profiles of temperature, pressure, and water-vapor mixing ratio from the model. In addition, concentrations of many other optically active gases, as well as natural and anthropogenic aerosols, are needed. In some forecasting situations, climatological distributions of the gases and aerosols are assumed. But, it is well known that there is considerable temporal and spatial (horizontal and vertical) variability in these quantities, and this is motivating efforts to include estimates of their concentrations in model initial conditions. For simulation of future climate, experimental approaches involve the specification of future increases in carbon dioxide and other gases according to different scenarios (see Chapter 16).

Note that there will be no attempt in this text to go into detail about radiative-transfer calculation methods, and associated approximations used in various parameterizations. Readers interested in the details should consult Liou (1980), Stephens (1984), and Stensrud (2007).

### 4.5.3 Parameterization of longwave fluxes

The simplest, least computationally demanding, and generally least accurate approach relates bulk properties of the resolved-scale atmosphere to the radiative flux. For example, there are simple, empirical approaches to estimating the downwelling longwave radiation ($F_{LD}$) at the ground using near-surface values of temperature. For example, Unsworth and Monteith (1975) suggest that

$$F_{LD} = c + d\sigma T_a^4,$$

where $T_a$ is the 2-m air temperature, $c = -119 \pm 16$ W m$^{-2}$, and $d = 1.06 \pm 0.04$ for a location in the UK. A similar relationship was proposed by Anthes *et al.* (1987) to calculate the net longwave flux at the surface:

$$F_{Lnet} = \varepsilon_g \varepsilon_a \sigma T_a^4 - \varepsilon_g \sigma T_g^4,$$

where $\varepsilon_g$ is the substrate, or ground, emissivity, $T_g$ is the ground temperature, and $T_a$ and $\varepsilon_a$ are the temperature and the emissivity, respectively, at 40 hPa above the surface.

The general solution of the radiative transfer problem involves estimation of the following integral equations:

$$F_U(z) = \int_0^\infty \pi B_\upsilon(z=0) \tau_\upsilon^f(z, z=0) d\upsilon + \int_0^\infty \int_0^z \pi B_\upsilon(z') \frac{d\tau_\upsilon^f}{dz'}(z,z') dz' d\upsilon \qquad (4.9)$$

and

$$F_D(z) = \int_0^\infty \int_z^\infty \pi B_\upsilon(z') \frac{d\tau_\upsilon^f}{dz'}(z,z') dz' d\upsilon, \qquad (4.10)$$

where $F_U(z)$ and $F_D(z)$ are the upward and downward longwave fluxes through level $z$, respectively, $\upsilon$ is frequency, $B_\upsilon$ is Planck's function, $\tau_\upsilon^f$ is the diffuse transmission function defined by the hemispheric integral

$$\tau_\upsilon^f(z,z') = 2\int_0^1 \tau_\upsilon(z,z',\mu)\mu d\mu,$$

where $\mu$ is the cosine of the zenith angle and

$$\tau_\upsilon(z,z',\mu) = \exp\left[-\frac{1}{\mu}\int_{u(z)}^{u(z')} k_\upsilon(p, T) du\right].$$

The quantity $k_\upsilon(p,T)$ is the absorption coefficient and $u$ is the concentration of the attenuating gas defined along the path from $z$ to $z'$. The first term on the right in Eq. 4.9 represents the attenuation of the radiation emitted by Earth's surface. The second term in Eq. 4.9 and the single term in Eq. 4.10 correspond to the analogous emittance of longwave radiation by the atmosphere. The various parameterizations used for calculation of the longwave fluxes employ a variety of approximations for calculating the integrals in the above four equations. See Liou (1980), Stephens (1984), and Stensrud (2007) for discussions of specific techniques, and of how the effects of clouds on longwave fluxes are parameterized.

### 4.5.4 Parameterization of shortwave fluxes

The transfer of solar radiation in the atmosphere is less complex than that for longwave radiation because there is not the complexity of the simultaneous absorption and emission from layer to layer in the vertical. That is, the atmosphere does not emit in these wavelengths. However, shortwave radiative transfer has additional complexity in the sense that molecular scattering is important, unlike the situation with longwave transfer.

As with longwave parameterizations, there are relatively simple empirical methods for estimating the shortwave fluxes at Earth's surface. Examples are described in Anthes *et al.* (1987), Savijärvi (1990), and Carlson and Boland (1978). For example, Anthes *et al.* (1987) employ the following expression for the shortwave flux absorbed by the ground surface:

$$H_S = S_0(1-\alpha)\tau\cos\zeta,$$

where $S_0$ is the solar constant, $\alpha$ is the albedo, $\zeta$ is the solar zenith angle, and $\tau$ is the shortwave transmissivity. The transmissivity calculation is based upon Benjamin (1983), and accounts for absorption and scattering by direct and diffuse radiation, and the effects of multiple cloud layers.

For more-complex calculations, the direct irradiance at a level $z$ is defined by Beer's law as

$$F_D(z,\mu_0) = \mu_0\int_0^\infty S_\upsilon(\infty)\tau_\upsilon(z,\infty,\mu_0)d\upsilon,$$

where $F_D(z,\mu_0)$ is the downward radiant flux through level $z$ for a beam of radiation with a zenith angle $\theta_0$ ($\mu_0 = cos\theta_0$), the integral is over frequency ($\upsilon$), $S_\upsilon(\infty)$ is the solar irradiance at the top of the atmosphere, and the monochromatic transmittance function is

$$\tau_\upsilon(z,\infty,\mu_0) = \exp\left(-\frac{1}{\mu_0}\int_{u(z)}^{u(\infty)} k_\upsilon du\right).$$

See Stensrud (2007) and Stephens (1984) for a discussion of approximations to these integrals used in parameterizations, as well as to the expressions that represent shortwave absorption and scattering. These references also discuss the parameterization of cloud effects on the shortwave fluxes.

# 4.6 Stochastic parameterizations

The influence of unresolved scales on the resolved scales is typically parameterized through the use of deterministic formulae. In contrast, stochastic parameterizations recognize that there are multiple possible states of the unresolved processes corresponding to a given state of the resolved variables, and that these can feed back differently to the resolved scales. These stochastic approaches take various forms. One example is described in Lin and Neelin (2000), wherein a random component is added to the CAPE that is calculated from the temperature and moisture profiles in the deep-layer-control scheme of the Betts and Miller (1986) convective parameterization. Addition of this random component improved the tropical intraseasonal variability of modeled convection. Similarly, Grell and Dévényi (2002) developed a parameterization that can use a large ensemble of closure assumptions and parameter values, and statistical techniques are then used to define the proper feedback to the resolved model variables. This method has been used operationally in the NCEP Rapid Update Cycle (RUC) model. Other examples of the many applications of stochastic parameterizations are described in Palmer (2001), Jung *et al.* (2005), and Plant and Craig (2008).

# 4.7 Cloud-cover, or cloudiness, parameterizations

With high-resolution cloud-resolving models, e.g., with grid increments of 1 km, it is possible to reasonably assume that an entire grid box is either cloudy or cloud-free. This is clearly not a reasonable assumption for global weather and climate models having horizontal grid increments of 10–100 km. Thus, for such coarser-resolution models there is a need to parameterize the cloud geometry in order to properly allow for the effects of the clouds on the radiation and surface-energy budgets. See Tompkins (2002, 2005) and Tompkins and Janisková (2004) for background and additional references.

Geometric cloud properties that might be estimated include:

- the horizontal fractional coverage of the grid box by cloud,
- the vertical fractional coverage of the grid box by cloud, and
- the overlap of the clouds in each vertical column.

Because the microphysics and convection parameterizations do not directly yield the above information, it must be independently estimated. Two approaches are used: relative-humidity-based methods and statistical methods.

Imagine a single grid volume whose horizontal area is partially filled with clouds, where such clouds extend in the vertical through the model layer. In the subvolume filled with cloud, the air is saturated. Elsewhere, the region is cloud-free and unsaturated. In this case, the grid-volume-mean relative humidity is obviously less than 100%. Thus, if a model predicts the grid-resolved relative humidity to be 100%, it can be assumed that the grid-box area is filled with cloud. If the model grid-resolved relative humidity is less than 100%, it remains to be inferred by the parameterization whether there is any cloud in the grid box whose effects on the radiation and surface-energy budget should be accounted for.

All parameterization approaches assume the existence of fluctuations of humidity and/ or temperature on the subgrid scale. Figure 4.23 shows an example distribution of mixing ratio and saturation mixing ratio over a grid increment, where the mean grid-box condition is unsaturated. Without the fluctuations, there would be no regions where $q_s < q$, leading to saturation. There are a number of different approaches for relating subgrid cloud cover to resolved-scale variables. One set of methods is based on diagnostic relationships between subgrid-scale fractional cloud cover and relative humidity. They generally apply to only the first property in the above list – the horizontal fraction of the grid box that is covered by cloud. A common mathematical relationship between relative humidity and fractional cloud cover, proposed by Sundqvist *et al.* (1989), is

$$C = 1 - \sqrt{\frac{1 - RH}{1 - RH_{crit}}},$$

where $C$ is cloud fraction and $RH_{crit}$ is the critical Relative Humidity (RH) above which cloud is assumed to form. Here $0.0 \le C \le 1.0$ and $RH_{crit} \le RH \le 1.0$, so that $C$ increases monotonically from 0.0 to 1.0 as $RH$ increases from $RH_{crit}$ to 1.0. Because such a simple relationship is not likely to be equally suitable over a wide range of cloud types and climates and weather regimes, many alternative approaches have been suggested. For example, Slingo (1980, 1987) proposes separate RH-based relationships for convective clouds and high-, medium-, and low-level stratiform clouds. And Xu and Randall (1996) include both the total cloud-water and cloud-ice mixing ratio, and RH as predictors.

Another general approach to the problem involves specification of the subgrid-scale probability distribution function for the humidity, as well as possibly that of the temperature. Various symmetrical and asymmetrical distributions have been assumed, where Fig. 4.24 shows a

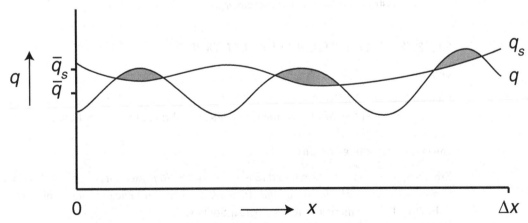

**Fig. 4.23**    Example of the variation of mixing ratio (*q*) and saturation mixing ratio (*q*ₛ) over a grid increment, where regions of saturation exist in spite of the fact that the grid-average condition is unsaturated.

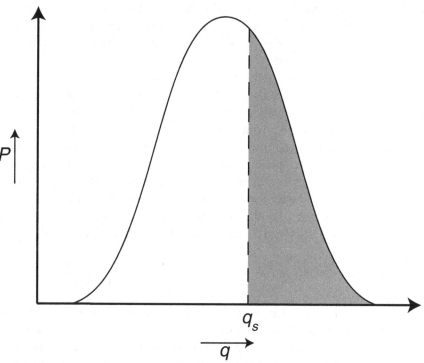

**Fig. 4.24**  Schematic example of the probability distribution function of the mixing ratio on subgrid scales. Shown is the
saturation mixing ratio ($q_s$), which is based on the assumption of a uniform temperature. The region is shaded for
which saturation implies the existence of cloud cover.

schematic example of how this can be used to estimate the cloud fraction. In the figure, it is
assumed, as a simplification, that the temperature does not fluctuate on these scales, so that
there is a constant saturation mixing ratio. The cloud cover is defined by the integral over the
part of the distribution for which $q$ exceeds $q_s$.

### SUGGESTED GENERAL REFERENCES FOR FURTHER READING

### General

Stensrud, D. J. (2007). *Parameterization Schemes: Keys to Understanding Numerical
Weather Prediction Models*. Cambridge, UK: Cambridge University Press.

### Convective parameterizations

Emanuel, K. A., and D. J. Raymond (eds.) (1993). *The Representation of Cumulus Convection in Numerical Models of the Atmosphere*. Meteorological Monographs, No. 46,
Boston, USA: American Meteorological Society.

Frank, W. M. (1983). Review: The cumulus parameterization problem. *Mon. Wea. Rev.*,
**111**, 1859–1871.

Smith, R. K. (ed.) (1997). *The Physics and Parameterization of Moist Atmospheric Convection*. Dordrecht, the Netherlands: Kluwer Academic Publishers.

### Turbulence parameterizations

Stull, R. B. (1988). *An Introduction to Boundary Layer Meteorology*. Dordrecht, the Netherlands: Kluwer Academic Publishers.

### Radiation parameterizations

Liou, K.-N. (1980). *An Introduction to Atmospheric Radiation*. London, UK: Academic Press.
Stephens, G. L. (1984). The parameterization of radiation for numerical weather prediction and climate models. *Mon. Wea. Rev.*, **112**, 826–867.

### Cloud microphysics parameterizations

Cotton, W. R., and R. A. Anthes (1989). *Storm and Cloud Dynamics*. London, UK: Academic Press.
Fletcher, N. H. (1962). *The Physics of Rain Clouds*. Cambridge, UK: Cambridge University Press.
Houze, R. A., Jr (1993). *Cloud Dynamics*. London, UK: Academic Press.
Pruppacher, H. R., and J. D. Klett (2000). *Microphysics of Clouds and Precipitation*. Dordrecht, the Netherlands: Kluwer Academic Publishers.
Rogers, R. R. (1976). *A Short Course in Cloud Physics*. Oxford, UK: Pergamon Press.
Rogers, R. R., and M. K. Yau (1989). *A Short Course in Cloud Physics*. 3rd edn. Oxford, UK: Butterworth-Heinemann.
Straka, J. M. (2009). *Cloud and Precipitation Microphysics: Principles and Parameterization*. Cambridge, UK: Cambridge University Press.

### Cloud-cover parameterizations

Tompkins, A. M. (2005). *The Parameterization of Cloud Cover*. ECMWF Technical Memorandum.

## PROBLEMS AND EXERCISES

1. Why might it be problematic for global models to use different parameterizations for different geographic regions, even though this could mean that the parameterizations perform better?
2. At about what horizontal grid increment do you imagine that parameterizations for sub-grid-scale cloud cover will not be needed? How might the answer to this be weather-regime dependent?
3. Suggest types of sensing systems that might be useful for initializing microphysical variables.

4. Describe how the existence of mineral aerosols of natural and human origin can influence microphysical and radiative processes in the atmosphere.

5. Given that mineral aerosols have important impacts on microphysical and radiative processes, how can their effect be practically represented in model forecasts?

6. What physical mechanisms might be responsible for convective parameterizations on the outer grids of an interacting nest influencing the resolved convective precipitation on the inner grid (e.g., as in Fig. 4.9)?

# 5 Modeling surface processes

## 5.1 Background

The surface processes whose numerical simulation is discussed here occur near both the land–atmosphere and the water–atmosphere interfaces. Over land, the movement of heat and water within the plant canopy and the ground beneath it must be represented in both weather- and climate-prediction models. Through this movement of heat and water across the land–atmosphere interface, properties of the land surface such as temperature and wetness are felt by the atmospheric boundary layer and the free atmosphere above. The atmosphere, in turn, affects the substrate and vegetation properties through radiation, precipitation, and controls on evapotranspiration. The effect of the surface on the frictional stress felt by the air moving over it is more the subject of boundary-layer meteorology and parameterizations rather than land-surface physics, so most of the discussion of this topic is found in Chapter 4. Over water, the interaction is complicated by the fact that the wind stress causes currents, waves, and vertical mixing of the water, which affect surface temperature and evaporation.

The skillful numerical prediction of atmospheric processes of many types and scales depends on the proper representation of surface–atmosphere interactions. For example, the prediction of convection relies on the accurate calculation by the model of surface fluxes of heat and water vapor. And, direct thermal circulations on the mesoscale, forced by horizontally differential heating at the surface, can dominate the local weather and climate near coastlines and sloping orography. On larger scales, monsoon circulations respond to seasonal variations in surface-heating differences between continents and oceans. And the modification of air masses by surface–atmosphere heat fluxes is an important factor that controls near-surface temperatures on the synoptic scale. On the global scale, the processes associated with the ENSO cycle involve ocean–atmosphere interaction. Indeed, of course the entire general circulation of the atmosphere is driven by differential surface heating, and a critical link in the global hydrologic cycle is the evapotranspiration at the surface. There are many more examples of atmospheric phenomena whose accurate simulation or forecasting depends on skillfully modeling the surface–atmosphere interaction, and the subsurface processes, over land and water.

Although many similar land-surface processes prevail on both weather and climate time scales, there are some that are only important for climate forecasting. Examples of these include interactions related to the carbon cycle, as well as changes in plant species associated with drought and other climate change. Modeling both categories of processes will be considered in this chapter, although the climate-related ones are discussed in greater detail in Chapter 16.

Because most students and professionals who are studying NWP will not have had the same background in land and ocean processes as they have had in atmospheric dynamics and thermodynamics, a summary is offered here in order to illustrate what must be included in a complete model. More detail will be provided below about land processes relative to ocean processes because, on weather-prediction time scales of a week or two, it is typical to assume that water conditions (e.g., temperature) are constant. In contrast, because of the fast response of the land conditions to precipitation, diurnal and day-to-day variations in solar radiation, frontal passages, etc., land-surface and subsurface processes are modeled explicitly or parameterized in virtually all models.

There are two ways in which Land-Surface Models (LSM) are used in NWP. They are integral components of the atmospheric model, and are run simultaneously with the rest of the code to predict or simulate surface fluxes of heat, moisture, and momentum. And, they are used as the basis of Land Data-Assimilation Systems (LDAS), which are run as stand-alone systems that ingest observed meteorological variables in order to diagnose current substrate temperature, moisture, and vegetation conditions for use in model initialization. That is, the LDAS is not coupled with the atmospheric model when it is run.

There is a hierarchy of LSMs that are appropriate for different applications, and all the processes discussed here do not need to be represented in every LSM. Nevertheless, regardless of the level of complexity of the treatment, the surface-process representations are an integral part of the atmospheric-model code. Thus, it is important for the model user to understand the strengths and weaknesses of this part of the model, just as with any other aspect of the model – just because it is not atmospheric science does not mean that LSMs can be treated as black boxes.

## 5.2  Land-surface processes that must be modeled

Land-surface models use atmospheric information (wind speed, temperature, etc.) from the atmospheric model's surface-layer representation, precipitation forcing from the convective and microphysics parameterizations, and radiative forcing from the radiation scheme. This forcing is used with information about the land's state variables to calculate the surface fluxes of heat and moisture to the atmosphere, reflected shortwave radiation, and longwave radiation emitted to the atmosphere and space. An appreciation for the breadth of the topic of land-surface processes and modeling can be developed through Fig. 5.1, which shows the various prevailing physical processes. The processes involve heat transfers, and the movement and transformation of water in its various forms. Within the substrate, there are the following processes.

- Liquid water is transported downward through gravity drainage and in all directions through capillary effects. Water also can rise and fall through changes in the water table.
- Water vapor moves vertically through the air spaces by convection and molecular diffusion.

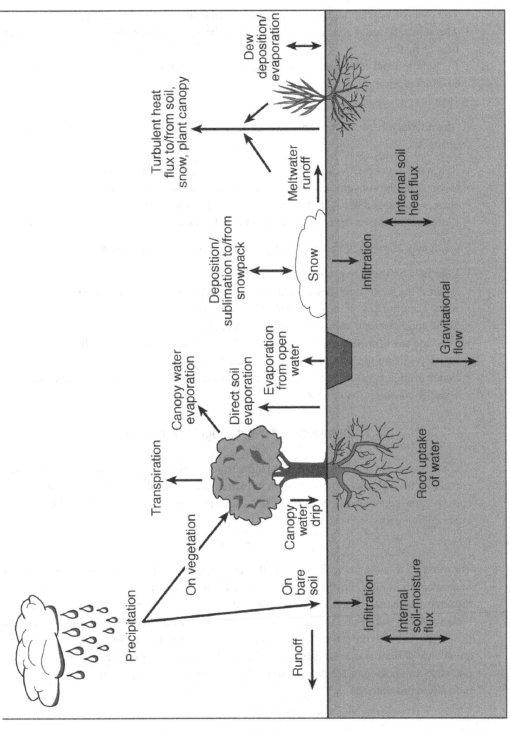

**Fig. 5.1** Schematic showing physical processes that are associated with the movement of heat and water within the substrate and at the surface. Adapted from Chen and Dudhia (2001).

- The roots of vegetation draw water from the substrate within the root zone.
- The substrate water freezes and thaws, with the release and consumption of the latent heat of fusion.
- Evaporation and condensation occur, with the release and consumption of the latent heat of condensation.
- Heat is conducted.

At the interface between the substrate surface and the atmosphere, there are the following exchanges.

- Rainwater, snowmelt water, irrigation water, and dew enter the substrate.
- Water from the substrate evaporates and sublimates into the atmosphere.
- Heat is exchanged between the atmosphere and the substrate.
- Liquid water passes from the underground roots to the above-ground stems and leaves.

Immediately above the interface, the following processes are important.

- Rain falls on the bare ground and vegetation.
- Water drips from vegetation onto the bare ground or onto other vegetation.
- Snow accumulates on the bare ground and vegetation.
- Snow and frost melt and sublimate, consuming heat.
- Dew and frost form on the bare ground and vegetation, releasing latent heat.
- Fog deposits on the bare ground and vegetation.
- Water evaporates from the leaf surfaces of vegetation, and transpires from vegetation, with the consumption of heat.

## 5.2.1 The energy and water budgets of the land surface

The energy and water budgets at the land surface control the temperature and moisture content of the substrate and vegetation, which interact with the atmosphere. The energy-conservation equation can be written for a unit mass or unit area of the surface that is experiencing gains or losses of energy:

$$R = LE + H + G \tag{5.1}$$

The variable $R$ is the net radiation, $L$ is the latent heat of evaporation, $E$ is the evaporation or condensation rate, $H$ is the sensible-heat exchange between the substrate/vegetation and the atmosphere, and $G$ is the sensible-heat exchange (conduction) between the surface and the subsurface substrate. The quantity $LE$ is the latent-heat flux, and $H$ and $G$ are heat fluxes also, all with the same dimensions as $R$. The net radiation represents the rate of radiant energy gain or loss at the surface of Earth, after accounting for all the various sources and sinks of short- and longwave radiation. The surface radiative energy balance is symbolically represented as

$$R = (Q + q)(1 - \alpha) - I{\uparrow} + I{\downarrow}, \tag{5.2}$$

where $R$ is the net radiation, $Q$ is the direct-solar and $q$ is the diffuse-solar radiation incident on Earth's surface, $\alpha$ is the surface albedo, $I{\uparrow}$ is the outgoing longwave radiation from the surface, and $I\!\downarrow$ is the absorbed downwelling longwave radiation that has been emitted by the atmosphere (gas, particulates, and clouds). Equation 5.1 simply states that the radiative energy gain or loss at the surface, $R$, must equal the sum of the other three terms. During the day, the rate of energy gain at the surface must equal the loss of energy associated with evaporation, the loss associated with heat conduction away from the surface into the substrate, and the loss of sensible heat to the atmosphere.

The water budget for a shallow soil layer at the surface can be represented as

$$\frac{\partial \Theta}{\partial t} = P - ET - RO - D, \tag{5.3}$$

where $\Theta$ (dimensionless) is the volumetric soil water content; $P$ is the rate of input through precipitation, snowmelt, dew and fog deposition, and irrigation water; $ET$ is the rate of loss through evapotranspiration; $RO$ is the rate of loss through lateral runoff; and $D$ is the rate at which water is lost through drainage to deeper layers.

## 5.2.2 Vertical heat transport within the substrate

Vertical heat transport within substrates is mostly through conduction (i.e., molecular diffusion), even though convective and advective movement of air can transport heat when the porosity (percentage air space) is high. This subsurface transport is important because it strongly modulates the thermal-energy budget at the surface. For example, heat gained and stored by the substrate during the day can be released to the atmosphere at night, affecting the boundary-layer structure and moderating the nocturnal minimum temperatures.

Because the substrate is a medium potentially consisting of solid, liquid, and gas phases, the thermal conductivity will depend on the proportions and characteristics of these components. The direction of the conductive heat transfer is from higher temperature to lower temperature, and the magnitude of the heat flux is proportional to the temperature gradient. Mathematically,

$$H_s = -k_s \frac{\partial T_s}{\partial z}, \tag{5.4}$$

where $H_s$ is the heat flux in the soil (positive upward), $k_s$ is the soil thermal conductivity, $z$ is distance on the vertical axis (positive upward), and $T_s$ is the substrate temperature. That is, the heat flux is proportional to the temperature gradient multiplied by a factor that reflects the ability of the substance to transfer heat. This is called a flux-gradient form of equation. The negative sign indicates that the flux is in the direction of lower temperature. The thermal conductivity is formally defined as the quantity of heat that flows through a unit cross-sectional area per unit time, when there exists a temperature gradient of one degree per unit distance perpendicular to the cross section. In general, a soil consists of solid substrate particles, liquid water, ice, and air spaces, and the relative contributions from the conductivity of these four components determines the total soil conductivity. The existence of water in the soil dramatically increases the thermal conductivity, not only because the conductivity of water is high, but because the water displaces the air which has an especially low conductivity (i.e., air is a good thermal insulator). Specifically, the conductivity of air is about two orders of magnitude lower than that of rock or wet soil. Thus, tabulated values of conductivity should specify the soil-moisture content and the porosity of the soil. Figure 5.2 shows qualitatively how the conductivity and other thermal properties of soil depend on the soil-moisture content.

Another important physical property of a substrate is the thermal, or heat, capacity, which describes how much heat is required to raise the temperature of a unit volume by one degree. As with the conductivity, the value of the soil heat capacity ($C_s$) depends on the fractions of the soil solids, liquid water, ice, and air. Air has a low heat capacity, so displacing air with water, as soil is moistened, raises the heat capacity of the air–water–solid mixture (Fig. 5.2b). Thus, more heat is required to raise the temperature of a moist soil than a dry soil. A related quantity is the specific heat ($c$), which is the amount of heat required to raise the temperature of a unit mass by one degree. Thus, it is equal to the heat capacity divided by the density of the substrate ($c = C/\rho$). The heat capacity and the specific heat are sometimes referred to as the soil's thermal sensitivity.

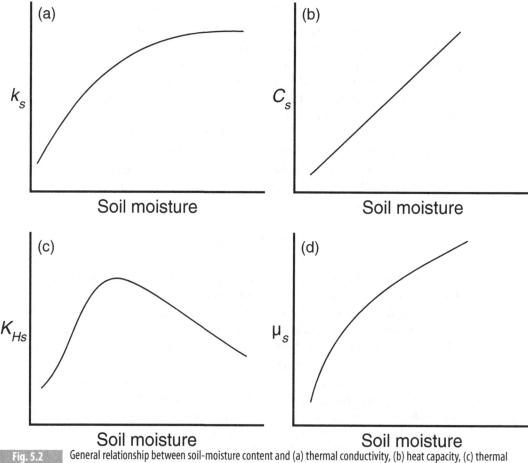

**Fig. 5.2**  General relationship between soil-moisture content and (a) thermal conductivity, (b) heat capacity, (c) thermal diffusivity, and (d) thermal admittance for most soils. From Oke (1987).

A quantity that is related to the heat capacity and the thermal conductivity is the thermal diffusivity ($K_{Hs} = k_s/C_s$). We will see that it determines the speed with which a temperature change propagates through a medium such as soil. Imagine the surface of the substrate heating up during the daytime heating cycle, which creates an upward temperature gradient immediately below the surface. Equation 5.4 predicts a downward heat flux that is directly proportional to the soil's thermal conductivity. Because the temperature gradient and therefore the heat flux are still small some distance below the surface, there will be a heat-flux convergence between the surface and that level. That is, the downward heat flux into the layer from the top is greater than the downward heat flux out of the layer at the bottom. This will raise the temperature of the soil in the layer in inverse proportion to its heat capacity, as shown in Eq. 5.5:

$$\frac{\partial T_s}{\partial t} = -\frac{1}{C_s}\frac{\partial H_s}{\partial z}. \tag{5.5}$$

In the above example, $H_s$ just below the surface has a large negative value, and is less negative with greater depth. The vertical derivative is thus negative, providing for a temperature increase. Combining Eqs. 5.4 and 5.5 gives

$$\frac{\partial T_s}{\partial t} = \frac{1}{C_s}\frac{\partial}{\partial z}\left(k_s\frac{\partial T_s}{\partial z}\right) = \frac{k_s}{C_s}\frac{\partial^2 T_s}{\partial z^2} = K_{Hs}\frac{\partial^2 T_s}{\partial z^2}, \tag{5.6}$$

where the simplification has been made that $k_s$ does not vary with depth. Here we see that the rate of temperature change is proportional to the diffusivity and the second derivative of temperature with respect to depth. Figure 5.3 shows a schematic of an idealized temperature distribution immediately above and below the air–ground interface, for both nighttime and daytime conditions. During the day, the curvature of the temperature profile below the surface is such that the second derivative in Eq. 5.6 is positive and the temperature increases. When the curvature reverses at night, the substrate cools. Note that a constant rate of temperature change with depth (a straight, sloping line in Fig. 5.3) would produce the same flux everywhere (Eq. 5.4), and no temperature change (Eq. 5.5).

Figure 5.2 shows that diffusivity is directly proportional to soil moisture, for low soil moisture, because the conductivity increases faster with increasing soil moisture than does the heat capacity. When the conductivity curve develops less slope than does the heat-capacity curve at higher soil moistures, the diffusivity begins to decrease.

Another way of understanding how the diurnal temperature wave at and below the surface is related to substrate properties is through the concept of thermal admittance, which is a property of the interface between two media (e.g., the substrate and the atmosphere). It is defined as $\mu_s = (k_sC_s)^{1/2}$, and is a measure of the ability of a surface to accept or release heat. Consider your feet in contact with a tile floor that has a temperature that is lower than your skin temperature. The temperature gradient at the skin–tile interface causes a flux from you to the tile, and a critical factor that determines whether the surface "feels" cold to you is whether the tile surface warms very rapidly to your skin temperature, or whether it remains colder. In the case of the tile, the conductivity is large so that the heat it gains

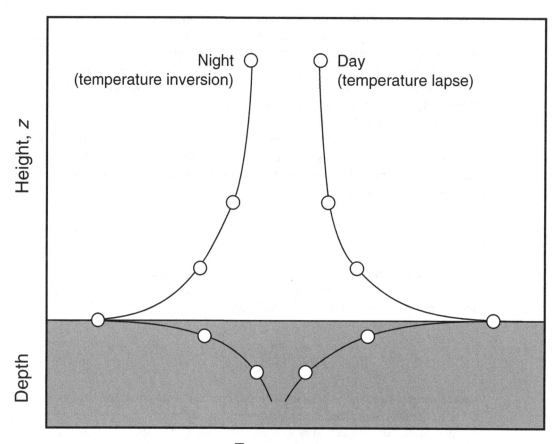

Fig. 5.3 Idealized vertical profiles of temperature in the soil and the atmosphere near the surface. Adapted from Oke (1987).

from your body is conducted rapidly from the surface to the tile material below. Thus, the surface temperature does not rise rapidly, as it would if you were standing on a wood floor having a lower conductivity (i.e., wood is a better thermal insulator). Analogously, if the material has a high heat capacity, its temperature is not going to rise rapidly in response to heat input from your skin. Thus, high conductivity and high heat capacity (i.e., large admittance) contribute to sustaining a large heat transfer across an interface because the temperature contrast is maintained. Of course, the temperature response of the second medium is equally important, so in the case of the surface–air interface, the admittance of the air must also be considered.

Thus, when the surface in contact with the atmosphere has high admittance (i.e., high conductivity and/or high heat capacity), the temperature of the surface does not increase as much during the daytime as it would for a low-admittance surface. This has implications for the daytime surface-energy budget in terms of smaller sensible-heat fluxes to the atmosphere (a cooler boundary layer) and weaker longwave emission from the ground.

The amplitude and time lag of the temperature wave that propagates downward into the substrate during the day also depend on the conductivity and the specific heat. The amplitude of the daily temperature oscillation is greater for large conductivities and for small heat capacities. That is, the amplitude is proportional to the diffusivity, $K_{Hs} = k_s/C_s$, such that

$$(\Delta T_s)_z = (\Delta T)_0 e^{-z(\pi/K_{Hs}P)^{1/2}}, \tag{5.7}$$

where $z$ is the depth below the surface, $P$ is the wave period (24 h is the dominant period in this discussion of the diurnal temperature wave), $(\Delta T)_0$ is the amplitude of the temperature wave at the surface ($z = 0$), and $(\Delta T_s)_z$ is the amplitude at depth $z$. That is, the amplitude of the diurnal temperature oscillation decreases exponentially with depth. Figure 5.4a shows

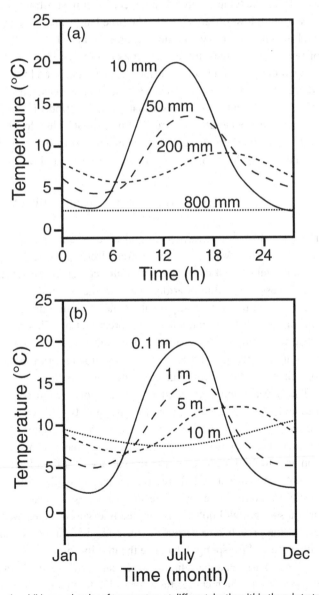

**Fig. 5.4**  Idealized (a) diurnal and (b) annual cycles of temperature at different depths within the substrate. From Oke (1987).

idealized diurnal temperature variations at different depths on a cloudless day. The time lag with depth shown in the figure is defined by

$$(t_2 - t_1) = \frac{(z_2 - z_1)}{2}(P/\pi K_{Hs})^{1/2}, \tag{5.8}$$

where $t_1$ and $t_2$ are the times that a temperature-wave maximum or minimum reaches levels $z_1$ and $z_2$, respectively. In other words, the difference is the time required for the temperature wave to pass from one depth to the other. Other variables have the same definition as before. The temperature wave travels faster in substrates with higher thermal conductivity and with lower heat capacity. The time lag means that the near-surface substrate can be cooling while warming continues a short distance below. At some depth, the curves are out of phase so that the time of the temperature maximum at that depth corresponds to the temperature minimum at the surface. Equations 5.7 and 5.8 apply also to the annual cycle, where the period corresponds to that of the seasonal rather than the diurnal cycle (Fig. 5.4b). Equation 5.7 shows that, with a period of 365 days, the depth to which the thermal wave penetrates is about 14 times greater than for the diurnal period (for the same $(\Delta T)_0$ ). If the diurnal thermal wave penetrates to 0.5 m with a particular amplitude, the annual wave will penetrate to 7 m with the same amplitude.

### 5.2.3 Vertical water transport within the substrate

There are two general ways for liquid water to enter the substrate between the water table and the surface. Water can move upward from the water table through capillary action (or the water table itself can rise). And, water can enter the substrate from the surface, where this process is called infiltration. The efficiency of infiltration depends on a number of factors such as rainfall intensity, total rainfall amount during a storm, the physical composition of the substrate, and antecedent precipitation. This rate of water movement into the substrate is important because it determines the potential for runoff and flooding, the amount of water near the surface that is available for evaporation, the availability of water for plants, and the extent of groundwater replenishment.

The upward and downward water transport within soils can take place through five mechanisms: Three apply to liquid water and two to water vapor. For liquid water, there are two forces that operate. One is gravity and the other is related to the surface tension between the soil particles and the water. It perhaps seems unusual to refer to surface tension as a force, but molecules at the surface of a liquid experience molecular forces that are not symmetrical and, therefore, not balanced like those experienced by molecules in the fluid's interior, away from the surface. Consider that a volume of soil (or, more familiarly, a sponge) will not drain completely dry after being wetted. Eventually, the drainage rate will approach zero because surface-tension forces, which promote retention of water within the soil (or sponge), balance the gravity force.

In the first mechanism, liquid water can move vertically as a result of a change in the pressure head, which causes the water table to change. That is, a water surface must be in dynamic equilibrium with its surrounding fluid, so local water excesses or deficits compared to surroundings are reconciled through water movement that equalizes the pressure. A simple illustration of this effect is found in desert environments where a hydrologically closed basin, perhaps containing a salt flat, is surrounded by mountains. Rainfall and snowmelt over the mountains refresh the water table below, which increases the water pressure and causes the groundwater to move laterally (from high to low pressure) into the central basin until equilibrium is attained with a higher water table there. In these situations, frequently the water table reaches the surface, creating seasonal lakes over the salt flats, which should be represented in a model. Analogously, extraction of water at a well site decreases the pressure there and causes inflow from the surroundings, lowering the water table over a wider area. This movement of water in response to pressure differences is forced by gravity; that is, the force of gravity is responsible for static pressure within a fluid. (Static pressure is related to the weight of the fluid above a point, whereas dynamic pressure is a result of fluid movement.)

Second, liquid water can move through soils by capillary action. This movement results from surface-tension effects between the water and soil particles. For example, water can move from the water table into the dry layer above through capillary effects. The capillary-rise layer has a lower bound at the water table and an upper bound that depends on soil properties. In general, capillary effects contribute to the spread of water from wetter to dryer soil. These surface-tension forces that bind the water to the soil particles are determined by the soil porosity and the soil moisture itself. The more porous the soil and the more dry the soil, the weaker are the surface-tension effects. For example, capillary movement of water can be blocked by a layer of open-textured soil (e.g., coarse sand or gravel) or by a dry layer of soil.

Third, downward liquid-water transport between the substrate particles is forced by gravity, where this water is supplied through infiltration – the entry at the surface of water from rain, snowmelt, or irrigation. The infiltration rate is limited by the rate of soil-water movement below the surface, called percolation, with any excess running off laterally or ponding at the surface. The correct modeling of this type of soil-water movement is important to the prediction of groundwater recharge; the evaporation rate; runoff, erosion, and flood production; the availability of water for plant uptake; and chemical changes such as salinization. The rate of this downward water movement, defined as the hydraulic conductivity, is controlled by the surface-tension effects between the soil particles and the water. The force of gravity draws the water downward, but surface-tension forces between the water and soil particles promote retention of the water. This latter effect of water retention is quantified in terms of the soil-moisture potential, which can be visualized as the amount of energy necessary to extract water from the soil matrix. Tight soils, such as clay, have a high potential, or water-retention capacity, compared to sand. Also, dry soils have a higher potential than wet soils: i.e., it takes less energy to extract a unit of moisture from a wet soil than from the same soil after it has become drier. Thus, the hydraulic conductivity is greater when the soil is wet and porous.

The amount of liquid water in soils is generally defined in terms of the soil-moisture content, which is the percentage of the volume of a soil that is occupied by water. The upper limit of the soil-moisture content is determined by the porosity. Coarse-textured soils, such sand, tend to be less porous than fine-textured soils, even though the mean pore size is greater in the former. The temporal change in the soil-moisture content at any point in a soil can be represented by an equation that is similar to Eq. 5.5, which expresses the local temperature change in terms of the difference in the vertical heat fluxes into and out of a layer (as represented for a point by the vertical derivative of the flux). Analogously, for soil moisture the vertical derivative of liquid-water fluxes must be represented. If the soil-moisture flux toward a point is greater than the flux away from it, the soil moisture increases, and vice versa. The following equation expresses local changes with time in volumetric soil-water content ($\Theta$) as a result of vertical variations in the vertical volume flux of liquid water ($q$). Note that this expression applies for vertical water transport within subsurface layers, and thus there are no direct effects of sources and sinks of water from precipitation, evaporation, and runoff (as there are for Eq. 5.3, which applies to the surface layer). However, the loss by canopy transpiration of water taken from the root zone is represented by $E_t$. The term $D_\Theta \partial \Theta / \partial z$ is associated with capillary (i.e., surface tension) water movement and $K_\Theta$ represents gravity-forced water movement. These are the second and third mechanisms described above, respectively.

$$\frac{\partial \Theta}{\partial t} = -\frac{\partial q}{\partial z} + E_t = \frac{\partial}{\partial z}\left(K_\Theta + D_\Theta\frac{\partial \Theta}{\partial z}\right) + E_t = \frac{\partial K_\Theta}{\partial z} + \frac{\partial}{\partial z}\left(D_\Theta\frac{\partial \Theta}{\partial z}\right) + E_t. \quad (5.9)$$

Here, $K_\Theta$ is hydraulic conductivity and $D_\Theta$ is soil-water diffusivity. The subscripts on $K$ and $D$ refer to their dependence on $\Theta$. The terms conductivity and diffusivity have been borrowed from the equations for the molecular diffusion and conduction of heat, which have terms similar in form to those above. Unfortunately, this terminology does not reflect the actual physical processes that are represented in the equation. The hydraulic conductivity and soil-water diffusivity are highly nonlinearly dependent on the soil moisture (Chen and Dudhia 2001), and have been calculated using various mathematical expressions. For example, Ek and Cuenca (1994) use

$$K_\Theta = K_{\Theta s}(\Theta/\Theta_s)^{2b+3}, \text{ and} \quad (5.10)$$

$$D_\Theta = -(bK_{\Theta s}\Psi_s/\Theta)(\Theta/\Theta_s)^{b+3}, \quad (5.11)$$

where $K_{\Theta s}$ is the saturation hydraulic conductivity, $\Theta_s$ is the saturation volumetric soil-moisture content, $\Psi_s$ is the saturation soil-moisture potential (a negative number), and $b$ is an empirically defined coefficient. All of these quantities are functions of the soil type.

The above mechanisms apply to liquid-water movement. However, there are also two mechanisms by which water vapor can move vertically above the water table through porous, dry soil: convection and vapor diffusion. Convection requires that the temperature of the soil, and that of the air within the soil, decrease upward in the soil more rapidly than the dry adiabatic lapse rate that is required for the triggering of buoyant motion. The flux of water by vapor diffusion is proportional to the gradient of the water-vapor content of the air within the soil, and results in water vapor transport from areas of higher to lower concentration, without the need for any movement of air on scales larger than the molecular.

## 5.2.4  Liquid-water transport within vegetation, and transpiration

Vegetation is important to the moisture budget because the roots access shallow and deep moisture that is not otherwise directly available for evaporation at the surface. This moisture is transferred through the xylem up the stems to the leaves, where it evaporates within the intercellular spaces of the leaves and is released through the stomata into the atmosphere. The latent heat consumed in this process is provided by the foliage, in contrast to evaporation from bare ground where the latent heat comes from the substrate. In either case, the energy loss is part of the surface-energy budget, which affects the atmosphere.

The rate of loss of water by transpiration from vegetation depends on many factors, including vegetation type and density, atmospheric humidity, time of day, season, and the degree of heat and water stress to which the vegetation has been subjected. There has been considerable historical controversy about the dependence of the transpiration rate on soil moisture. A wilting-point value of soil-moisture content has been defined as the limit below which the vegetation permanently wilts and transpiration ceases. This is a convenient concept, but ignores the fact that the moisture content within the root zone is not uniform, and that different coexisting vegetation types have greatly different tolerances for soil dryness. Field capacity is another threshold on the soil-moisture scale, with implications for vegetation. It is defined as the moisture value below which internal drainage ceases. That is, for a soil-moisture content that is less than the field capacity, the soil will retain the moisture and none will drain downward. This is another concept that has gained popularity because of its simplicity rather than its strict accuracy. Some have used the assumption that water is equally available to vegetation for any moisture value above the wilting point. Others have assumed that the vegetation is under stress for wetnesses between the wilting point and field capacity, with a transpiration rate that is dependent on soil moisture, and that only for wetnesses above field capacity is there no longer a stress.

## 5.2.5  Heat and water-vapor exchange between the surface and the atmosphere

It was mentioned previously that the vertical transfer of sensible heat at the substrate–atmosphere interface occurs through conduction. This takes place within a very shallow layer of atmosphere, called the laminar (nonturbulent) sublayer, having a depth of a few molecules to, at most, a few millimeters. Above this layer, the transfer is through turbulent eddies of air. This turbulence does not contribute to the flux at the surface because the eddies cannot exist there, where the velocity normal to the surface must be zero.

Because all nonradiative transfer of heat at the surface is through conduction, the heat flux can be represented (Eq. 5.12) by the same sort of flux-gradient relationship employed to represent heat transport by conduction within the substrate (Eq. 5.4). A similar expression (Eq. 5.13) can be used for the vapor flux. If it is assumed that the same type of equations can be applied for turbulent transfer as are used for molecular transfer, the equations can be rewritten as follows with the molecular diffusivities replaced with eddy diffusivities:

$$H = -C_a K_{Ha} \frac{\partial T}{\partial z}\bigg|_0 = -\rho c_p K_{Ha} \frac{\partial T}{\partial z}\bigg|_0 \text{ and} \qquad (5.12)$$

$$LE = -C_a K_{Wa} \frac{\partial q}{\partial z}\bigg|_0 = -\rho c_p K_{Wa} \frac{\partial q}{\partial z}\bigg|_0 . \qquad (5.13)$$

Here, $C_a$ is the heat capacity of the atmosphere, $c_p$ is the specific heat at constant pressure of the atmosphere, $K_{Ha}$ and $K_{Wa}$ are the diffusivities of heat and water vapor in the air, respectively, $q$ is specific humidity, and the vertical derivatives are evaluated within the laminar sublayer near the surface.

A challenge to applying these equations is that the exchange coefficients are functions of distance from the surface and static stability, varying by over three orders of magnitude from day to night. Alternative expressions for $H$ and $LE$ can be obtained if we vertically integrate these equations with the assumption that the fluxes do not vary much with height within the first couple of meters. The resulting expressions are

$$H = \rho c_p D_H (T_g - T_a) \text{ and} \qquad (5.14)$$

$$LE = \rho L D_W (q_{s,sat}(T_g) - q_a), \qquad (5.15)$$

where $D_H$ and $D_W$ are transfer coefficients that are integral functions of $K_{Ha}$ and $K_{Wa}$, $T_g$ is the temperature of the surface, and $T_a$ and $q_a$ are the temperature and specific humidity, respectively, of the air at a specified level near the surface. The value of the specific humidity at the surface, $q_s$, is equal to the saturation value, $q_{s,sat}$, at the temperature $T_g$, of any surface at which evaporation is occurring – water bodies, damp soil, leaf stomata.

Thus, the sensible- and latent-heat fluxes between the substrate and the atmosphere can be represented in terms of the differences between the temperature and humidity at and immediately above the surface. The direction of the fluxes depends on the sign of the difference, and the magnitude of the fluxes depends on the degree of the contrast between the conditions at the two levels. The transfer coefficients are functions of factors that affect the intensity of the turbulence, such as the roughness of the surface, the vertical shear of the horizontal wind from which turbulent energy can be derived, and the vertical lapse rate of atmospheric temperature, which determines whether turbulent energy is available from buoyancy. For example, evaporation rates ($LE$) are high when the atmosphere is dry (small $q_a$), the surface is warm (large $q_{s,sat}$) and wet, and the near-surface wind speed is high (producing a large shear, and thus a large transfer coefficient, $D_W$).

Another way of visualizing the controls on the surface heat flux is through the concept of thermal admittance. Most of the earlier discussion about admittance was in the context of the substrate properties; however, it was pointed out that the admittance of the atmosphere on the other side of the interface is equally important in determining the heat fluxes. This atmospheric admittance is defined as $\mu_a = (k_a C_a)^{1/2} = C_a K_{Ha}^{1/2}$, where $K_{Ha}$ is the eddy diffusivity of Eq. 5.12. For example, suppose that the surface is receiving solar radiation during the day. This heating of a thin layer of substrate and air at the interface will produce a temperature gradient within both the air and the substrate (see Fig. 5.3). The energy not lost by longwave emission and evaporation will be partitioned between sensible-heat fluxes into the atmosphere and the substrate in proportion to the relative admittances of the two media. Say, for example, that the substrate has a very low admittance because of poor thermal conductivity, but the boundary layer has a large admittance because the turbulence is well developed and thus the eddy diffusivity is large. The heat flux into the soil (Eq. 5.4) will thus be small in spite of the large $\partial T_s / \partial z$ , but the heat flux into the atmosphere (Eq. 5.12) will be large because the eddy diffusivity, $K_{Ha}$, is large. Thus, more of the radiant-energy input to the surface will be partitioned to the sensible-heat flux to the atmosphere rather than to the substrate. Alternatively, at night the radiative cooling of the surface draws heat from the air and the substrate in proportion to their admittances. Because calm, near-surface winds and a stable vertical profile of temperature mean that turbulence is weak and the eddy diffusivity is small, the atmospheric admittance is small at night and most of the surface heat lost to space by radiation is provided by the substrate rather than by the atmosphere. Methods of estimating atmospheric and substrate admittances are discussed in Novak (1986).

### 5.2.6  Horizontal water movement at and below the surface

Water from rainfall or snowmelt that accumulates on the surface too rapidly to infiltrate downward, ponds in the low spots of the substrate and eventually runs off laterally at the surface. When this occurs, the runoff is channeled across the landscape until reaching streams and rivers. The rate of infiltration, which determines the partitioning to runoff, of course depends on the soil type, the density and type of vegetation, the amount of organic litter on the surface, and the soil-moisture content. Some of the excess that runs off laterally will possibly infiltrate at another location. Because runoff is caused by the potential energy of the water, the horizontal redistribution is greatest and occurs most rapidly over steeply sloping terrain.

## 5.3  Ocean or lake processes that must be modeled

This section will provide only a brief summary of processes at and below a water surface that can affect the atmosphere in model simulations. The reader should be able to distinguish, based on scale, discussions of those processes that are associated with oceans and seas, in contrast to lakes. More-comprehensive discussions of ocean processes can be found in Miller (2007) and Haidvogel and Beckmann (1999). Chapter 16 on climate modeling discusses ocean and sea-ice processes that must be included for simulations on seasonal and longer time scales. As with the land surface, it is not necessary to include all of these processes for all model applications. Section 5.5 will clarify the level of complexity that must be included for different modeling situations.

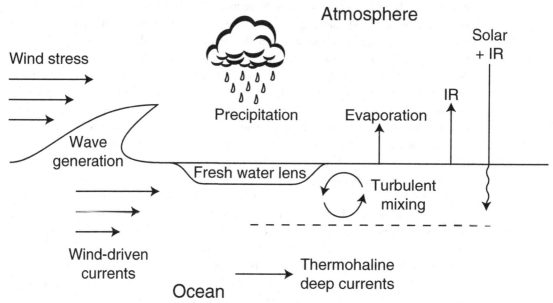

**Fig. 5.5** Schematic showing physical processes that are associated with the movement of heat and mass within the ocean. The IR notation refers to infrared radiation.

Figure 5.5 illustrates some of the processes that may be represented, explicitly or through parameterization, in models of the coupled atmosphere–water system. The wind at the surface causes waves, where the wave height is a function of the wind speed and the fetch. In turn, the stress between the atmosphere and the water is a function of the wave height. The surface waves and the subsurface turbulence, caused by the wind stress, mix the water through a layer that is tens to hundreds of meters deep (the mixed layer). The density, which is a function of the temperature and salinity, is relatively uniform in this well-mixed layer. In addition to the depth and intensity of the mixing being functions of the waves and wind speed, they also depend on the density stratification, or stability, of the near-surface water. The more stable the surface water, the weaker is the mixing and the shallower is the mixed layer (analogous to the atmospheric mixed layer). The near-surface stability depends on the vertical distribution of heating from incoming atmospheric radiation, the flux of fresh water from precipitation at the surface, and the mixing of warmer surface water with cooler water below through the turbulence. Precipitation is less dense than saline ocean water, and can remain on the top as a fresh-water lens. This increases the stratification, making it more difficult to mix the less-dense, heated, surface water downward by turbulence. This, in turn, results in higher Sea-Surface Temperatures (SST) associated with the fresh-water lens.

The radiation budget differs from that over a land surface in a number of respects, one of the most important being that the incoming solar and infrared radiation penetrate the medium and distribute the energy over a depth that depends on the water's turbidity. The greatest amount of radiant energy is absorbed near the surface, before the beam loses intensity through attenuation. The turbulence in the ocean distributes this warmer, near-surface water through the mixed layer. The diurnal variation of surface temperature is significant (Kawai and Wada 2007), but less than that over land because the energy is distributed through a layer by virtue of the penetration of the radiation and the turbulent mixing.

The wind also drives basin-scale near-surface ocean circulations, called gyres. Near coastlines, the Ekman drift associated with the Coriolis force can deflect the water movement away from the coast, causing upwelling that greatly influences water temperatures and the coastal climate.

# 5.4 Modeling surface and subsurface processes over land

As noted earlier, LSMs can be run as integral components of LAMs and global models, or they can be used autonomously as part of a LDAS, with input from observations instead of the atmospheric-model output. Modeling land-surface processes begins with mapping the types of substrate (e.g., rock, sand, loam) and vegetation (e.g., shrubs, coniferous trees, deciduous trees) over the computational area. A look-up table is then used to provide base values for physical variables, corresponding to the different substrate and vegetation types. Such variables would include substrate thermal conductivity, heat capacity, porosity, albedo, etc., where adjustments are made to variable values that are functions of surface or subsurface moisture. Figure 5.6 shows a schematic of the overall land-surface-modeling process. The landscape-related input variables are provided by the module in the upper left. In the upper right is the input of initial estimates of time-dependent variables such as substrate moisture content and temperature, which vary in both time and space, and are predicted by the LSM. Because these variables are not observed operationally, it is necessary to adjust the estimated values through the use of a LDAS, which requires observed atmospheric variables for forcing. It is necessary for this LDAS to run for months to years, in order to spin up the correct current conditions for the soil-temperature and -moisture profiles. After this spinup period, the LSM and soil state are ready to use for research or operational prediction. See Section 5.4.2 for further discussion of the concept of the LDAS. At the bottom of the figure is an LSM, coupled with an atmospheric model, that has used input from the LDAS to define the land-surface and subsurface conditions at the beginning of the model integration. Once the land-variable profiles are spun up, the LDAS is integrated forward on an hourly basis, as input observations become available, and output is used to initialize the land variables for operational forecasts.

The next subsection reviews land-surface-modeling methods, the second one describes the use of LSMs in LDASs, and the last one discusses how the LSMs are coupled with atmospheric models for weather and climate prediction.

## 5.4.1 Land-surface models

Now that we have reviewed the array of land-surface and subsurface physical processes that can affect surface–atmosphere interaction, this section will describe how the processes can be represented in a model. There are dozens of different specific formulations for

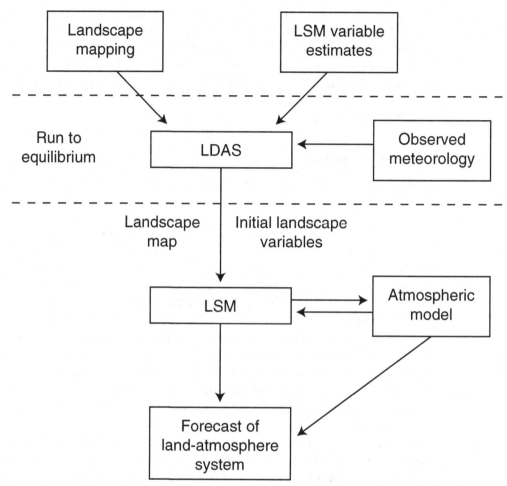

Schematic of the overall land-surface modeling process. At the upper left, the landscape properties (e.g., soil and vegetation type) are defined, and in the upper right a first guess of the vertical profiles of soil moisture and temperature are generated. These data sets are input to a LDAS that is driven by observed atmospheric forcing, and that is integrated forward in time for a historical period of months to years in order to allow the spinup of realistic, current, three-dimensional soil temperature and moisture fields. These conditions can then be used as input to a coupled atmosphere–land modeling system for production of weather forecasts, or they can be used directly for hydrological or other applications.

LSMs. For example, the Project for Intercomparison of Land-surface Parameterization Schemes (PILPS) included 23 schemes (Henderson-Sellers *et al.* 1995, Shao and Henderson-Sellers 1996, Chen *et al.* 1997). These references should be consulted for a summary of the varied computational approaches used in the participating LSMs, and for documentation of the models. The more-complete LSMs represent all the processes depicted in Fig. 5.1, and sometimes more when simulations are on interseasonal and climate time scales. One of the outcomes of the intercomparison was that sophisticated LSMs do not consistently outperform the simpler schemes. The reason for this is that it is virtually impossible to accurately define values on the local scale for the vast number of physical quantities that are used by some vegetation–soil–hydrology models.

Essentially, the LSM solves numerical (finite-difference) forms of Eqs. 5.1, 5.2, 5.3, 5.6, 5.9, 5.14, and 5.15. The net radiation defined in Eq. 5.2 is calculated as follows.

- The flux of direct solar radiation, $Q$, at the surface is calculated using the astronomical equations that define the Earth–Sun relationship during the diurnal and seasonal cycles (the elevation and azimuth angles of the incoming solar beam), the slope of the terrain, the attenuation of the solar beam by gases, clouds, and particles in the atmosphere (calculated by a radiation parameterization), and the solar spectral flux at the top of the atmosphere. For climate–time-scale simulations, the solar flux is varied based on known periodicities in the solar output and changes in Earth's orbital parameters.
- The indirect-solar radiation, $q$, is obtained from a radiation parameterization, using the above information about the direct solar beam, and information about atmospheric particulate and liquid scatterers.
- The albedo, $\alpha$, is based on the tabulated substrate and vegetation properties of the grid box, as well as the substrate wetness (which affects the albedo). Strictly speaking, the albedo is dependent on the wavelength, the incident angle of the radiation, and the viewing angle, but a single value is usually used as an approximation.
- The upward-propagating infrared energy at the surface is calculated from the time-dependent skin temperature ($T_g$) of the vegetation and soil, using $I{\uparrow} = \varepsilon\sigma T_g{}^4$, where $\varepsilon$ is the emissivity and $\sigma$ is the Stefan–Boltzmann constant.
- The downwelling infrared energy at the surface is calculated from $I{\downarrow} = \varepsilon I$ (incident), where the incident longwave flux at the surface is provided by the radiation parameterization.

The resulting net radiation is used in Eq. 5.1. The terms in this equation for the sensible ($H$) and latent ($LE$) heat-flux exchanges between the surface and the atmosphere are calculated with Eqs. 5.14 and 5.15, respectively. The heat flux between the surface and the uppermost substrate layer ($G$) is computed with Eq. 5.4, with the surface (skin) temperature ($T_g$) and the temperature of the uppermost soil layer used to calculate the vertical temperature derivative. Each of these terms in Eq. 5.1 is a function of $T_g$, which is obtained by iteration.

Within the substrate, temperature change is computed by integrating Eq. 5.6 in time. The water budget for the surface is computed by integration of Eq. 5.3, where the precipitation rate ($P$) is obtained from the atmospheric model, the loss by evapotranspiration ($ET$) and lateral runoff ($R$) are computed using various approaches, and the drainage to deeper layers is computed with flux terms such as in Eq. 5.9. Changes in the soil-moisture content of subsurface layers are calculated by integrating Eq. 5.9.

Some of the major ways in which land-surface parameterizations differ from each other are listed below.

- Land- and sea-ice process modeling
- Vegetation canopy representation
- Runoff calculation and surface routing (results from merger of hydrologic models and LSMs)

- Grid-box partitioning (whether model grid boxes contain a mixture of surface types, or whether the dominant surface type is applied to the entire grid box)
- Groundwater modeling
- Snowpack, snow-cover, snow-albedo treatments
- Dynamic vegetation, multi-layer vegetation canopy representation
- Urban canopy modeling (none, single-layer, multi-layer)
- Irrigation representation (seasonal and daily protocols)
- Frozen-soil treatment

## 5.4.2 Land-surface models used in land data-assimilation systems

The use of LDASs has two motivations. One is to initialize the land-surface conditions (e.g., soil moisture and temperature) in atmospheric-model integrations. Even though these quantities are forecast by the LSMs that are run with the atmospheric model, they develop forecast errors just as do the atmospheric variables. Therefore, using the forecast values of land-surface variables as initial conditions for a subsequent forecast can result in the accumulation of error associated with model biases. Thus, LDASs are needed in order to provide realistic land-surface initial conditions (IC). The second motivation is to diagnose surface properties that are too difficult or expensive to measure directly. For example, a LDAS can be run for a forest, and the soil moisture and information about the vegetation can be used to estimate wildfire potential in remote areas. And, a LDAS can be run for an agricultural area, and the output used to diagnose regional variations in the soil-temperature and -moisture profiles that affect crop growth. Or, it can be run for a watershed, and the analyzed soil moisture used as input to a flash-flood forecasting system.

An example of a global LDAS is the Global Land Data Assimilation System (GLDAS, Rodell *et al.* 2004) developed by the US NOAA and NASA. It merges ground- and space-based measurements, which can be used as input to any of three LSMs: Mosaic (Koster and Suarez 1996), the Common Land Model (Dai *et al.* 2003), and Noah (Chen *et al.* 1996, Koren *et al.* 1999). Table 5.1 lists the input and output variables for GLDAS. The atmospheric-forcing variables in the left column are estimated based on observations, these data are used as input to an LSM, and the LSM diagnoses the output variables on the right. The GLDAS has basic grid-increment options of 0.25°, 0.5°, 1.0°, 2.0°, and 2.5°. Higher-resolution, mesoscale LDASs are used for regional applications, but their operation and purpose is the same as on larger scales. These are run using typical mesoscale grid increments of 1–10 km. Chen *et al.* (2007) report on the WRF-based High-Resolution Land Data Assimilation System (HRLDAS), which employs the Noah LSM. Reanalysis systems, described in Chapter 16, also often provide land-surface conditions as part of the archived output, but these are not the same as LDASs because the LSM input is from the model and not observations.

When an LDAS is used for model initialization, it is typical to use the exact same LSM, use the same input data (e.g., substrate properties), and employ identical computational grids for both the LSM in the model and the LSM in the LDAS. This avoids the problem that it is challenging to translate soil-moisture and temperature profiles from one LSM (in the LDAS) to another (in the forecast model) because of different basic assumptions and

| Table 5.1 Forcing and output fields for the GLDAS | |
|---|---|
| Required forcing fields | Output fields |
| Precipitation | Surface albedo |
| Downward shortwave radiation | Canopy transpiration |
| Downward longwave radiation | Soil moisture in each layer |
| Near-surface air temperature | Snow depth, fractional coverage, and water equivalent |
| Near-surface specific humidity | Plant canopy surface-water storage |
| Near-surface wind vector | Soil temperature in each layer |
| Surface pressure | Average surface temperature |
| | Surface and subsurface runoff |
| | Bare soil, snow, and canopy surface-water evaporation |
| | Latent, sensible, and ground heat flux |
| | Snow phase-change heat flux |
| | Snowmelt |
| | Net surface shortwave and longwave radiation |
| | Aerodynamic conductance |
| | Canopy conductance |
| | Snowfall and rainfall |

*Source:* From Rodell *et al.* (2004).

formulations, and different grid structures. For rapidly adjusting atmospheric variables, modest conversion errors would be reconciled quickly. However, LDASs can require many months to reach an equilibrium, after being initialized with erroneous soil-temperature and -moisture profiles.

## 5.4.3 Land-surface models coupled with atmospheric models

When used this way, the LSM is an integral component of the entire modeling system, with the atmospheric and land-surface components integrated together and communicating at every time step. The LSMs are employed in both weather-prediction and climate-prediction models, and with LAMs and global models. For climate prediction, more degrees of freedom are needed because a changing climate will lead to the evolution of the vegetation species and density. For real-data simulations or forecasts, the land-surface variables in the LAMs are generally initialized using an LDAS, a described in the previous section.

## 5.5  Modeling surface and subsurface processes over water

At a minimum, the lower boundary of the model atmosphere over water must have the roughness, temperature, ice-coverage, and salinity (which affects saturation vapor pressure) specified. This approach of specifying such lower-boundary quantities over water suffices for short-range forecasts or simulations. An exception exists when hurricanes are being modeled, because the high wind speed produces intense vertical mixing, rapidly leading to a negative anomaly in the SST that must be represented in the simulation. Thus, incorporation of ocean boundary-layer mixing and the resulting SST change into model simulations has been shown to improve hurricane-intensity forecasts (e.g., Bao *et al.* 2000).

In general, for forecasts of longer than a week or two, variables such as water temperature and ice cover should be calculated internal to the atmospheric-model simulation. This requires the use of ocean-circulation models and sea-ice models. Even though it is not the purpose of this section to provide details on ocean-circulation and wave modeling, it is worth mentioning the methods that are used. As described in Chapter 16, for simulation of the Intergovernmental Panel on Climate Change (IPCC) climate scenarios, and for initial-value simulations of years to decades, relatively physically complete ocean models are used. For interseasonal predictions, sometimes ocean models are run separately from the atmospheric model. For weather prediction and research, on smaller scales in maritime environments, coupled ocean–atmosphere LAMs are used. For example, the COAMPS is a LAM that has been used for a wide variety of applications for processes in the open sea and in littoral zones (e.g., Pullen *et al.* 2006). Similarly, Bao *et al.* (2000) have coupled the MM5 mesoscale atmospheric model, an ocean-wave model, and a version of the Princeton Ocean Model for regional-process studies.

Sometimes wave-height forecasts are needed, and this requires that atmospheric-model forecasts be used as input to wave-height models. Because the output from the wave-height model does not typically get fed back to the atmospheric model (i.e., the coupling is one way), wave models are discussed in Chapter 14, which is about special-application models.

At land–sea boundaries, it is especially important that the model have grid points defined correctly in terms of whether they are land or water points. That is, the land–sea mask, as it is called, which defines the coastline, must be accurate. This requirement sounds trivial, but the complex configuration of many coastlines means that sometimes grid points or observation points are defined to be on the wrong side of the coast. Thus, a land observation may be erroneously compared with a water grid point, and even though the points are close to each other, the model solution verifies poorly for obvious reasons.

## 5.6  Orographic forcing

The forcing of the atmosphere by orography is important on all scales, from the global to the mesoscale. Thus, except for models that are used for pedagogical applications, or those that employ less-than-complete physics to allow for simpler interpretation of the results in studies of physical processes or numerical methods, virtually all models allow for variable orography. Gridded data sets of terrain elevation are available with a variety of horizontal resolutions, with grid increments ranging from tens of meters to tens of kilometers. There are two considerations when deciding upon the best degree of smoothness to use to define the lower-boundary elevation in a model. If there is too much variability of the elevation in the $2$–$4\,\Delta x$ wavelengths, this will be reflected in energy in these wavelengths in the model solution, which can require heavier filtering to avoid the development of nonlinear instability. However, one of the motivations for using high horizontal resolution in a model is to permit the development of atmospheric features that result from small-scale forcing at the lower boundary. Thus, the terrain data set should be of as high a resolution as possible, while avoiding the wavelengths that will generate troublesomely short wavelengths in the model solution.

Any terrain-elevation data set is going to show some effects of smoothing relative to the actual terrain, with elevation maxima that are less than observed, and elevation minima that are higher than observed. That is, valleys will be less deep than they should be, and mountains will be less tall than they should be. The use of such smoothed data sets in models has implications for the accurate representation of processes that depend on the orographic extremes. For example, the amplitude of standing planetary waves, the drag of mountains on the atmosphere, and the effectiveness of the blocking of tropospheric synoptic-scale features are dependent upon the height of the lower-boundary obstacles. Indeed, it has been argued that the large-scale flow responds to an envelope that somewhat intersects the mountain peaks, rather than to the mean elevation of the mountains and valleys. That is, the height of terrain obstacles should be preserved in models, regardless of the smoothness of the terrain data set. This elevation-preserving topography is called *envelope topography*. Wallace *et al.* (1983) first tested this approach with the ECMWF model, because it was noted that model solutions showed persistent negative height biases over mountain ranges. They enhanced the orographic elevation by adding to the grid-box average value an increment that was proportional to the subgrid-scale variance in the true orography. Their tests showed forecast improvement for longer lead times, but a degradation for shorter times. Other approaches have been used to define the envelope orography. For example, Mesinger *et al.* (1988) defined the grid-box-average elevation as the tallest actual value for the area. Other applications of this approach have led to mixed results (e.g., Tibaldi 1986, Lott and Miller 1997, Georgelin *et al.* 2000). Even though this method has conceptual appeal, its implementation should be thoroughly evaluated in particular applications. For satisfying the particular need of improved mountain drag, Catry *et al.* (2008) suggest an alternative to the use of envelope orography in the French ARPEGE/ALADIN model.

An example is shown in Fig. 5.7 of how horizontal resolution can affect the ability of a model to correctly define the orography with which the atmosphere interacts. Shown is the terrain elevation for two different model horizontal resolutions, for the same region of complex orography in the southwestern USA. In one, the grid increment is 30 km, and in the other it is 3.3 km. No methods for defining envelope orography were employed. The coarser-resolution grid defines the significant regional topographic features with only a couple of grid points, and any atmospheric response to the orography on these scales is going to be filtered strongly by the model.

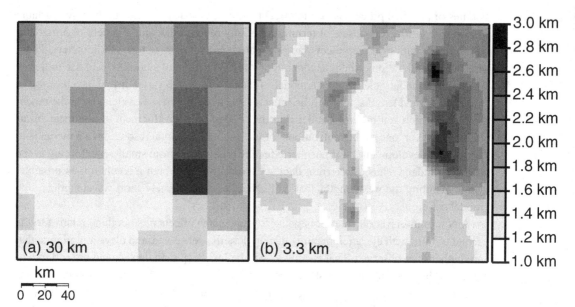

km
0  20  40

**Fig. 5.7**    The terrain elevation (see gray shades on the right) for two different model horizontal resolutions, for the same region of complex orography in the southwestern USA. In (a) the grid increment is 30 km and in (b) it is 3.3 km.

## 5.7 Urban-canopy modeling

There exists a number of modeling approaches for representing the dynamic and thermo-dynamic effects of urban areas on the atmosphere. One is to employ a Computational Fluid Dynamics (CFD) model that explicitly represents the effect of each building or other structure on the atmosphere. However, as described in Chapter 15, these models are very computationally expensive to use. In contrast to such fine-scale modeling, what is commonly needed is simply a way of representing the bulk effects of built-up areas on mesoscale processes. The simplest approach is to employ standard LSMs by defining the surface properties so that they approximate the artificial surface conditions in cities. For example, roughness length can be increased to represent the drag from buildings; albedo can be decreased to account for the existence of asphalt pavement, dark rooftops, and the trapping of shortwave radiation in street canyons; the heat capacity and thermal conductivity of the substrate can be elevated above standard values for asphalt and concrete to account for heat storage in building walls; the substrate water capacity can be decreased to reflect the prevalence of impermeable surfaces; and the green-vegetation fraction can be reduced. Liu *et al.* (2006) used this approach to successfully simulate urban–rural boundary-layer differences for Oklahoma City, USA. However, other methods need to be used to represent more-complex processes that are associated with the existence of buildings and non-natural substrates. Tools to accomplish this are called Urban Canopy Models (UCM, also known as urban canopy parameterization), which represent the model grid-cell-averaged effect of the building structures on the dynamics and thermodynamics. Such UCMs parameterize the aggregate effects of the urban morphology, but individual buildings and street canyons are not explicitly represented (Masson 2000, Kusaka *et al.* 2001, Martilli *et al.* 2002). Many UCMs consider the geometry of buildings and roads to represent the radiation trapping and wind shear in the urban canopy. Such an approach requires detailed three-dimensional, urban land-use data sets, and the input of a number of parameters that define the urban geometry, where these parameters need to be calibrated for each individual city. Because of the cost of mapping the three-dimensional geometries of tens of thousands of structures, these data sets are not available for many cities. Figure 5.8 illustrates one of the many factors that can be included in a UCM; in this case it is the shadowing that results from a particular configuration of structures. Two different sun angles are illustrated, where the smaller zenith angle ($\theta_z$) shades one side of a building and part of the street, while the larger one also shades part of the building on the opposite side of the street canyon.

Most UCMs to date have been single-layer parameterizations. That is, even though the vertical effects of buildings are represented, the fluxes of heat, moisture, and momentum are defined at the bottom boundary of the lowest atmospheric layer in the model. In contrast, multi-layer UCMs allow direct interaction between buildings and multiple layers in

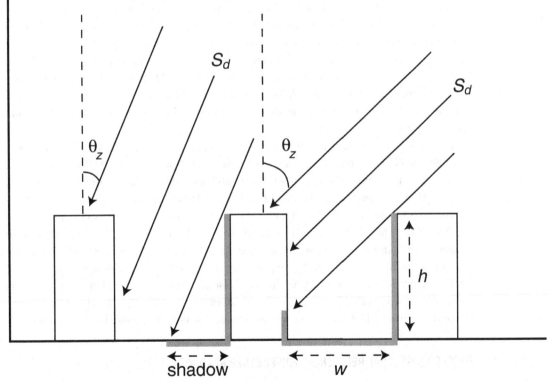

**Fig. 5.8**   Illustration of how UCMs must use the three-dimensional morphology of urban landscapes to calculate the illumination of the surface of the street canyon and the sides of the buildings. Similarly, this geometry is required to calculate the longwave radiation trapping in the street canyons. Shown are the direct solar radiation ($S_d$) and the zenith angle ($\theta_z$). Adapted from Kusaka *et al.* (2001).

the atmospheric model (even though obviously the buildings are not explicitly resolved). See Kusaka *et al.* (2001), Chin *et al.* (2005), Kondo *et al.* (2005), Holt and Pullen (2007), and Martilli (2007) for further discussion of multi-level UCMs.

## 5.8 Data sets for the specification of surface properties

As noted before, there are two general kinds of landscape variables. One defines the substrate type, and vegetation type and density, as a function of location. There are a number of national and regional data sets that represent surveys of these properties. For example, for the USA, the Geological Survey's Earth Resources Observing System (EROS) 1-km dataset (Loveland *et al.* 1995) defines vegetation type, and the State Soil Geographic (STATSGO) 1-km database defines soil type (Miller and White 1998). The EROS data set can need significant correction based on field reconnaissance.

The second kind of variable represents the time and space variability of the specific physical properties of the substrate and vegetation, such as temperature and soil-moisture content for the substrate, and the leaf-area index or the green vegetation fraction for the vegetation (which can vary as a function of season and antecedent rainfall). The variables of both kinds need to be defined for any model application over land. The LDASs described above can be used to define values for the substrate physical properties. For example, the GLDAS data set provides the variables listed on the right of Table 5.1. When the same land-surface parameterization is used in both GLDAS and the LSM that is coupled with the atmospheric model, no soil-moisture conversion is required. Numerous satellite-based methods are used to define the state of the vegetation (e.g., Gutman and Ignatov 1998).

For water, surface-temperature analyses are available from a variety of sources. For example, the NCEP Version 2.0 global SST data set (Reynolds *et al.* 2002) is defined on a $1° \times 1°$ grid and updated daily. Also, the Real-Time Global (RTG) analysis (Thiebaux *et al.* 2003) from the Marine Modeling and Analysis Branch of NCEP produces a two-dimensional variational analysis of data from buoys, ships, and satellites over the preceding 24 hours. The product is incorporated into the NCEP North American mesoscale Model (NAM) and the global forecast model at ECMWF. Since 2001, the RTG analysis has been available daily on a grid with pixel size of 0.5 deg latitude and longitude. In 2005, a 1/12-deg product became available. There is also an optimum interpolation SST analysis from NOAA. It is available weekly, and with a pixel size of 1/3 deg. The product uses *in-situ* and satellite SST observations, and incorporates a weekly median ice concentration.

### SUGGESTED GENERAL REFERENCES FOR FURTHER READING

Bonan, G. (2008). *Ecological Climatology*. Cambridge, UK: Cambridge University Press.
Hillel, D. (1998). *Environmental Soil Physics*. San Diego, USA: Academic Press.
Martilli, A. (2007). Current research and future challenges in urban mesoscale modelling. *Int. J. Climatol.*, **27**, 1909–1918.

Oke, T. R. (1987). *Boundary Layer Climates*. London, UK: Methuen.

Stensrud, D. J. (2007). *Parameterization Schemes: Keys to Understanding Numerical Weather Prediction Models*. Cambridge, UK: Cambridge University Press.

Stull, R. B. (1988). *An Introduction to Boundary Layer Meteorology*. Dordrecht, the Netherlands: Kluwer Academic.

## PROBLEMS AND EXERCISES

1. Why are microclimate conditions less extreme when high soil diffusivities prevail?

2. Miller (1981) describes the following situation. When a cold air mass ($-17$ °C) moved over an area near Leningrad, the surface flux was 45 W m$^{-2}$ over frozen bare soil with an admittance of 1000 J m$^{-2}$ K$^{-1}$ s$^{-1/2}$, but was only 15 W m$^{-2}$ over an adjacent snow-covered surface with an admittance of about 330 J m$^{-2}$ K$^{-1}$ s$^{-1/2}$. Explain these measurements in the context of the definition and meaning of admittance.

3. List the ways in which land-surface properties in urban areas differ from those elsewhere, and explain how these differences can be incorporated into LSMs in order to reasonably represent urban land-surface effects.

4. Perform a literature search on the use of coupled ocean–atmosphere LAMs, and describe why it can be important to represent regional ocean processes in forecasts having weather time scales.

5. How would the choice of the depth of the substrate layer that is modeled depend on whether the model is being used for weather, interseasonal, or long-term (multi-decadal) climate prediction?

6. Show mathematically that the depth to which the annual cycle's temperature wave penetrates into the substrate is about 14 times greater than the penetration depth of the diurnal cycle's temperature wave.

7. Based on physical arguments, describe what specific vegetation effects need to be modeled to account for the vegetation's influence on the surface-energy budget, and the atmosphere.

# Model initialization

## 6.1 Background

As we have seen in Chapter 3, solving the equations that govern the physical systems that we are modeling is an initial- and boundary-value problem. The lateral, upper, and lower boundary conditions are discussed in Chapters 3 and 5. In this chapter will be described the procedure by which observations are processed to define initial conditions for the model dependent variables, from which the model integration begins. This process is called model *initialization*.[1] There are essentially two requirements for the initialization. First, the dependent variables defined on the model grid must faithfully represent conditions in the real atmosphere (e.g., fronts should be in the correct location), and second, the gridded mass-field variables (temperature, pressure) and momentum-field variables (velocity components) should be dynamically consistent, as defined by the model equations. An example of the mass–momentum consistency requirement is that, on the synoptic scale, the gridded initial conditions should be in approximate hydrostatic and geostrophic balance. If they are not, the model will generate potentially large-amplitude inertia–gravity waves after the initialization shock, and these nonphysical waves will be overlaid on the meteorological part of the model solution until the adjustment process is complete. The final adjusted condition will prevail after the inertia–gravity waves have been damped, or have propagated off the grid of a LAM. However, the model solution will be typically unusable during this adjustment period, which is one reason for the common, historical recommendation that model output not be used for about the first 12 h of the integration. On the smaller mesoscale and convective scales, ageostrophic circulations, such as associated with horizontally differential surface forcing and convection, should ideally exist in the initial conditions. Otherwise, such features will need to *spin up* during the early period of the model integration.

Historically, there have been two approaches for accomplishing the initialization, although modern methods have blurred the distinctions between them. One is called *static initialization*, where observations are first interpolated to a model grid (*data analysis*), and then the resulting variables may be adjusted using diagnostic, dynamical constraints to

---

[1] The terminology here is not used consistently in the community. Some employ the term initialization to refer only to the process of defining a dynamic balance in the initial conditions.

make them more consistent with each other and with the model equations. In contrast to this diagnostic method, *dynamic initialization* involves the preforecast integration of the model to produce an initial state that is dynamically consistent with the equations used for the forecast.

The commonly used terms *data assimilation* and data analysis both refer to processes that employ observations to construct a gridded data set that defines the spatial variability of model dependent variables at the initial time of a forecast. However, the expression data assimilation typically means that a meteorological model is employed, where many approaches will be discussed throughout this chapter. The objective of data assimilation can be the production of initial conditions for operational forecasts, or the construction of long-term reanalyses of the state of the atmosphere (see Chapter 16 for an explanation of the latter).

Because the initialization of the land surface was treated in the last chapter in the context of LDASs, this subject will not be treated here. But, it should be remembered that these land-surface variables, such as substrate temperature and moisture, the state of the vegetation, etc., are time dependent, and their accurate specification in the model initial conditions is an important part of the initialization process.

## 6.2 Observations used for model initialization

### 6.2.1 Sources of observations used for model initialization

Meteorological observations can be classified as either *in situ* or remotely sensed. Obtaining the former involves the use of sensors that measure the local value of a variable. Remote sensing employs sensors that perform measurements from a distance, through the use of either active or passive methods. Passive methods employ the measurement of naturally emitted radiation. With active methods, the sensing system emits radiation and measures the response of the atmosphere to that radiation. Radiosondes are examples of *in-situ* sensors. Satellite-borne radiometers that measure radiances (the emissions spectrum) from the atmosphere are passive remote sensors, while radars that emit microwave energy and measure that fraction which is reflected by hydrometeors are active remote sensors. For either remote-sensing method, a *retrieval algorithm* is often needed in order to translate the information obtained by the sensor into meteorologically useful information (values of dependent variables). In the case of the radiometer, the algorithm translates the sensor data into temperature, and for the radar data the echo strength is converted to precipitation intensity. In contrast to the use of retrieval algorithms, variational-analysis methods, to be discussed later, allow the direct use of raw sensor information in the analysis process. Non-satellite-based measurement platforms that are commonly used to provide initial conditions for NWP models are listed below.

- *Radiosondes* – Measure temperature, relative humidity, and pressure; and tracking the balloon displacement provides wind speed and direction. This is still the primary method for defining the three-dimensional structure of the atmosphere on the synoptic and global scales, for model initialization. Even though the most-common frequency for radiosonde ascents is every 12 h, at 0000 UTC and 1200 UTC, in some countries it is every 24 h. These two standard radiosonde launch times define the most-common initialization times for models.

- *Near-surface weather stations* – Typically measure temperature, humidity, pressure, wind speed, wind direction, and precipitation. A challenge associated with using these observations is that it is difficult to estimate how far into the model boundary layer to spread their influence (i.e., over how many model levels) when defining the initial conditions on the model grid. This is important to know because vertical mixing in the model can quickly eliminate near-surface information that is incorporated in the initial conditions, if the atmosphere above is not analyzed with vertically consistent structures. Another challenge is that it is difficult to consider these observations in the context of any dynamic balance, simply because of the dominance of local forcing. Near-surface variables may be reported at intervals of 5 minutes, 15 minutes, 1 hour, 3 hours, or 6 hours. For near-surface winds, the averaging that is done to remove turbulence can be over periods of 5 minutes to 15 minutes. The height above ground at which near-surface measurements are made also varies. The standard is for winds to be measured at 10-m AGL and temperature and humidity to be measured at 2-m AGL; however, some observation networks do not adhere to this. The spatial distribution of observations varies considerably, on the scales of countries, and on smaller scales depending on population density. Contributing to the spatial-density variation is the fact that numerous special-purpose mesoscale networks exist, for example those that are established to meet air-quality and highway-maintenance needs. Buoy data are another type of near-surface measurement.

- *Commercial aircraft* – Onboard sensors measure wind speed and direction, temperature, pressure, and humidity. Some also measure turbulence intensity. Sloping profiles are provided at takeoff and landing, and a near-horizontal series of observations is available at cruising altitudes. Instrumented commuter aircraft, with shorter flight segments, generate a large number of vertical profiles in the lower troposphere. The reporting frequency varies, but observations are available at an interval of 60 seconds or less during ascent and descent, and approximately every 3 minutes at cruising altitude. Other aircraft sensor packages produce observations at specified pressure and horizontal-distance intervals. See Moninger *et al.* (2003) for additional information.

- *Doppler radar* – Measures the reflectivity from hydrometeors, and the radial wind speed relative to the radar. It scans a three-dimensional volume, and in modeling applications is used primarily for initialization of convective-scale models.

- *Doppler lidar* – Measures the radial wind speed relative to the lidar. It scans a three-dimensional volume on the convective scale, and is used primarily for initialization of mesogamma-scale and smaller-scale models of the boundary layer.

- *Wind profiler* – Upward-pointing radar that measures the horizontal wind vector in a column, with an hourly frequency. Often collocated with the wind profilers are Radio Acoustic Sounding Systems (RASS) for measuring temperature profiles.

Satellite-based measurement platforms include the following. There are many others that are described in the literature.

- QuikSCAT SeaWinds sea-surface winds from NASA are disseminated by the NOAA National Environmental Satellite, Data, and Information Service (NESDIS). The SeaWinds instrument on QuikSCAT is an active microwave radar that measures the backscatter from ocean-surface waves, and winds can be obtained in all conditions except for moderate to heavy rain. A function is used to relate the measured backscatter to the 10-m Above Sea Level (ASL) neutral-stability-equivalent winds. The QuikSCAT Level 3 gridded ocean wind vectors are available on an approximate $0.25° \times 0.25°$ global grid with separate maps for the ascending and descending passes. The data are available for the period 2000 to the present. See Bourassa *et al.* (2003) and Hoffman and Leidner (2005) for additional information about data properties.
- Radio-occultation soundings of temperature and water vapor are obtained by using satellite-borne receivers to measure the phase delay of radio waves emitted from Global Positioning System (GPS) satellites, as the waves are occulted by Earth's atmosphere. These soundings are available globally, and can provide data where there are voids in other observation networks. Additional information can be found in Anthes *et al.* (2008).
- The Tropical Rainfall Measurement Mission (TRMM) product (Huffman *et al.* 2007) combines precipitation estimates from multiple satellites (retrievals from measurements in the microwave and infrared regions of the spectrum) as well as gauge-based analyses on a $0.25° \times 0.25°$ grid that extends from 50° N to 50° S for the period from 1998 to the present. Latent-heating rates inferred from these rainfall analyses are used during the model-initialization process.
- The NASA Earth Observing System Terra and Aqua platforms have a MODerate-resolution Imaging Spectroradiometer (MODIS) sensor with visible, near-infrared, and infrared bands. The MODIS provides information on a suite of meteorological variables, including temperature and moisture. See Seemann *et al.* (2003) for an example of the retrieval of temperature and moisture.
- Special Sensor Microwave Imager (SSM/I) and Total Ozone Mapping Spectrometer (TOMS) provide data that have been used widely for model initialization. See Okamoto and Derber (2006), Goerss (2009), and Monobianco *et al.* (1994) for examples of the assimilation of SSM/I data.
- The Geostationary Operational Environmental Satellite (GOES) allows the calculation of hourly feature-track winds derived from infrared, visible, and water-vapor imagery (Gray *et al.* 1996, Nieman *et al.* 1997, Le Marshall *et al.* 1997). Also, estimates of precipitation based on the GOES Precipitation Index (GPI) can be used in diabatic initializations.
- The infrared Spinning Enhanced Visible and InfraRed Imager (SEVIRI) sensors on the geostationary Meteosat Second Generation satellites provide information about temperature and humidity (Di Giuseppe *et al.* 2009).

## 6.2.2 Observation-quality, -frequency, and -density variability

There is a great deal of space and time variation in the availability of observations used to initialize models. For example, *in-situ* measurements of the model-dependent variables are reported at time intervals (frequencies) that vary greatly, depending on the observation network. Figure 6.1 illustrates the spatial-density variability of near-surface observations, using North Africa as an example. The smaller number of population centers in the arid region is responsible for the paucity of observations there. In addition to the fact that the standard data-reporting interval varies among different observation networks, missing data are a frequent problem in some areas of the world. Such gaps in the data record can occur because meteorological observing stations are staffed only during the day, meteorological- and communications-equipment malfunctions occur, political unrests cause observing stations to be shut down for long periods, and late observations or communications-network delays cause data to arrive too late to be used to initialize a forecast. As an example of one of the more-continuous data records for West Africa, Fig. 6.2 shows the observed relative humidity for a surface station in Benin for a winter season. Clearly there are significant data gaps. Lastly, the suitability of instrument locations can vary considerably. Even though there are standards for the land-surface properties that should exist at an observation site, and for minimum distances between the observation site and physical obstructions, there is nevertheless considerable variability in the quality of the instrument

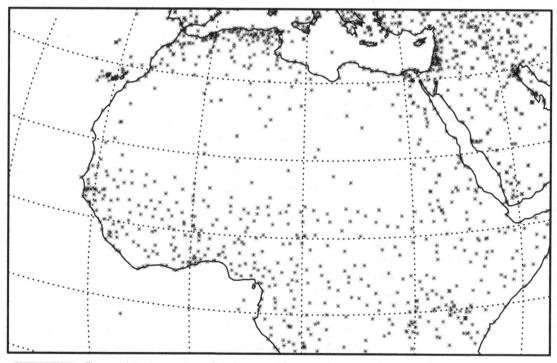

**Fig. 6.1**  The spatial distribution of near-surface, relative-humidity measurements in northern Africa.

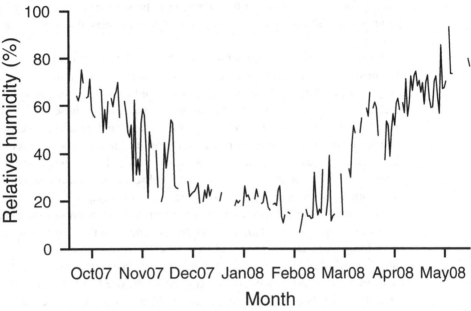

**Fig. 6.2**   A meteogram of the near-surface relative humidity at an observation location in Benin, illustrating the existence of gaps in the data reported during a winter season.

locations. For example, some measurement sites in cities are located on rooftops, where the thermal properties of the surface can be extreme, and where the winds are distorted by the structure. The next section describes specific approaches for ensuring the quality of observations used in a model.

### 6.2.3 Observation quality assurance and quality control

The term *Quality Assurance* (QA), in the context of meteorological observations, refers to the overall protocol that is employed to ensure the availability of quality observational data for use in NWP models and other applications. It is, in fact, a formal plan for accomplishing this goal, and typically might include the specifications for the instruments to be deployed, the instrument-siting requirements, the schedule for instrument calibration, the schedule for field inspection of instruments, and the routine numerical checks to be applied to the data (the *Quality-Control* (QC) process). See Shafer *et al.* (2000) for discussion of a complete QA procedure for a mesoscale network of sensors.

Historical (e.g., more than a day old) meteorological observations are publicly available from a variety of sources, some of which are mentioned in Section 10.10. In contrast, current observations for some nations are only available for a fee, which makes it difficult to establish real-time modeling systems for use in research or operations. Whether from historical observational-data archives, or from real-time observation networks, most observations have already undergone some quality checking. But, it is still important for those using the observations in a model initialization to perform checks of their own. A single grossly incorrect observation that erroneously passes the QC tests can have its negative influence spread over a large area in a model, and potentially damage the entire model solution.

There are a variety of causes for observations to be incorrect in some respect. A measurement may be of good quality, but the time, date, or geographic-location identifier can be wrong, resulting in the observation being applied at the wrong place or time. Or, the electronic transmission of the observation may have compromised it in a major or subtle way. The observation itself can have systematic and random errors. The systematic error is often related to incorrect instrument calibration. Another type of observation problem is referred to as representativeness error, which is discussed in more detail in Section 9.5.2. It results from the fact that an observation typically represents conditions at a point in space, and sometimes an average in time (e.g., winds), while the variables defined in model initial conditions represent grid-box-area averages, and apply at a specific time. Thus, the use of an observation to define conditions on a model grid can cause very-local properties to be spread over an erroneously large area. For example, if the observed wind and temperature associated with a convective outflow boundary are interpolated to a synoptic-scale-model's grid, and influence five to ten model grid boxes, an area of hundreds of square kilometers would be impacted by the small mesogamma-scale event.

Some simple, commonly employed QC tests include the following. More discussion about such checks can be found in Liljegren *et al.* (2009).

- *Limit tests* – Observations are compared with physical limits, sensor limits, and climatological limits. A physical limit, or constraint, for relative humidity would be that it cannot be less than 0% or much greater than 100%. And, wind speed cannot be less than zero. Similar absolute limits on the value of an observation can be defined in terms of the physical limits of a sensor or in terms of climatology (e.g., the minimum temperature ever observed at a station).
- *Temporal-consistency checks* – Successive observations of a variable define a rate of change, and this is compared with likely values. Because of rapid changes that can occur during convective weather, this check can be turned off when precipitation is occurring.
- *Spatial-consistency checks* – This is sometimes referred to as a buddy check, because observations are compared with horizontally or vertically adjacent data points. Or, an observation can be compared with an average calculated using a number of nearby observations. The resulting difference is compared with the historical maximum difference observed at that point, based on archived observations.

As will be seen later in this chapter, many modern data-assimilation systems merge observations with the most-recent gridded forecast that is valid at the observation time. Specifically, the gridded forecast is adjusted based on differences from the observations, and the result is used to initialize the next forecast cycle. But frequent large differences between the short forecasts and the observations at a particular location often result more from errors in the observations than from forecast errors. In effect, the volume of atmosphere that was initialized with accurate measurements is advected, during the short forecast, over the locations of new observations, and the statistical difference over a long period is used to judge observation quality. For example, Hollingsworth *et al.* (1986) describe how the operational ECMWF data-assimilation system can be used to monitor observation quality. This automated and economical approach to the QC process allows suspect instruments to be identified and corrective action taken, without routinely visiting and inspecting every instrument.

### 6.2.4 Other observation processing

Whether winds are observed and reported in terms of the individual components or as speed and direction, the measurements may need to be converted to the model wind components. This is because the model $u$ that is defined to be parallel to the grid-point rows, and the model $v$ that is parallel to the grid-point columns, generally differ from the geocentric $u$ and $v$ that are defined relative to latitude and longitude lines. For every vertical column of grid points (the same $i, j$ coordinate), the mathematical transformation will be slightly different. This necessity may be most easy to accidentally overlook when the model coordinates are Cartesian, and the grid-point rows and columns are approximately oriented east–west and north–south.

Software that interpolates (analyzes) observations to a model grid operates in the framework of the model's horizontal coordinate system. Thus, because observation locations are typically defined in terms of latitude and longitude coordinates, there needs to be a transformation to the horizontal coordinates of the model, if it is $x$–$y$ and not latitude–longitude based.

Lastly, the units of the observations may need to be transformed to those employed by the model. For example, wind speeds are often reported in knots, but models generally use the meter–kilogram–second (mks) system. And it is common for humidity observations to require conversion as well.

### 6.2.5 Metadata

Metadata (also called meta-knowledge) accompany the observations themselves, and provide information necessary for their use. Essential types of metadata include the file structure, data format (e.g., NetCDF), the variable (e.g., wind speed), the units (e.g., mks), and the time and three-dimensional-spatial coordinates of the observation. Optional, but useful, information includes the instrument type, the date of the most-recent calibration, and a photo of the instrument site and surroundings. The concept of metadata also applies to model-generated data as well, although the relevant information will obviously be different.

Conventions have been established for the format of metadata. For example, the NetCDF (Network Common Data Format) Climate and Forecast (CF) Metadata Convention is a well-documented standard for observational and forecast metadata, which is designed to promote the processing and sharing of files created with the NetCDF Application Programmer Interface [NetCDF API]. The CF conventions generalize and extend the convention of the Cooperative Ocean/Atmosphere Research Data Service, a NOAA/university cooperative group whose goal is the sharing and distribution of global atmospheric and oceanographic research data sets.

## 6.2.6 Targeted or adaptive observations

Economic and other constraints limit the number of observations that are made of the atmosphere, and thus it is reasonable to want to obtain observations from locations where they will have the largest positive impact on model-forecast accuracy, for a particular prevailing weather situation. Methods have been developed to satisfy this need, where the measurements are referred to as adaptive or targeted observations. However, it is clearly not economically feasible to deploy mobile observation platforms on a day-to-day basis. But, there are high-impact weather events, such as hurricanes or severe extratropical cyclones, for which special aircraft observations are made. If the aircraft can be routed so as to provide observations from locations for which the forecast skill is very sensitive to the accuracy of the initial conditions, the procedure can save lives. The routine use of targeted aircraft observations may become more common with the continued development of unmanned aerial vehicles.

Various strategies for observation targeting have been evaluated as part of the following field programs.

- Fronts and Atlantic Storm Tracks EXperiment (FASTEX; Emanuel and Langland 1998; Bergot 1999, 2001; Bishop and Toth 1999; Joly *et al.* 1999; and Bergot and Doerenbecher 2002)
- NORth Pacific EXperiment (NORPEX, Langland *et al.* 1999, Majumdar *et al.* 2002a)
- Atlantic THORPEX (The Hemispheric Observing-system Research and Predictability EXperiment) Observing System Test (Langland 2005)
- Annual US NWS Winter Storm Reconnaissance (WSR) programs (Szunyogh *et al.* 2000, 2002; Majumdar *et al.* 2002b)

The following notational framework for viewing the adaptive-observation problem is provided by Berliner *et al.* (1999), Majumdar *et al.* (2006), and others. Let $\mathbf{X}_i$, $\mathbf{X}_a$, and $\mathbf{X}_v$ represent $n$-dimensional vectors that define the state of the atmosphere at times $t_i$, $t_a$, and $t_v$, respectively, in terms of the grid-point values of variables or spectral coefficients. The initial time, $t_i$, is when the decision must be made, based on $\mathbf{X}_i$ information, about the types and locations of special observations to be collected at time $t_a$ (the targeted observation time, and the analysis (initial) time of the operational forecast), where the objective is to optimize the statistical properties of a forecast $\mathbf{X}_v$ at the verification time $t_v$. Within the interval $t_a - t_i$, the observing platforms need to travel to the target locations so that observations can be made at $t_a$ for use in initializing the forecast. The time interval $t_a - t_i$ is chosen based on logistical considerations associated with planning the surveillance mission, launching the aircraft, and getting the aircraft to the necessary locations to make the observations. The data set $\mathbf{X}_a$ is the result of assimilating standard observations and the special targeted observations, and is used as the initial conditions for the forecast.

A practical example of the above process is as follows. Assume that a 72-h forecast from a standard operational model run predicts very-heavy, flood-producing rainfall over New York City, associated with a coastal cyclone. At $t_i$ it is decided to deploy dropsondes at $t_a$, 24 h in the future, at locations where they will have the greatest impact on the 48-h precipitation forecast over New York City. The best location for making the measurements (the target area) will depend on the variable whose forecast must be improved (rainfall) and the verification region (New York City). Most adaptive-observation strategies allow the association of the observation target area with a specific verification region in the model. An exception is the ensemble-spread method discussed below.

There are a number of approaches for defining the locations and types of observations that will have the greatest positive impact on the quality of forecasts. A few of these are summarized below. Discussions of other methods can be found in Palmer *et al.* (1998), Bishop *et al.* (2001), Aberson (2003), and other references cited in this section. Berliner *et al.* (1999) focused on a statistical framework for the adaptive-observation problem.

- *Ensemble variance/spread* – This is a simple approach, described by Aberson (2003), that can improve tropical-cyclone-track forecasts by locating supplemental observations in areas where the variance is largest among members of an ensemble prediction that is valid at the analysis time. In regions of large ensemble variance, it is assumed that there is also a large uncertainty in the wind analysis, implying the need for additional observations. Unfortunately, there is no way to propagate the uncertainty or error at the analysis time, $t_a$, into another region at the forecast verification time, $t_v$. Nevertheless, Aberson (2003) showed that observations made in areas with large ensemble variance improved tropical-cyclone-track forecasts more than did uniformly distributed observations.

- *Adjoint methods* – In Chapter 3 it was noted that the adjoint operator, which is based on a linear version of a nonlinear forecast model, produces sensitivity fields that indicate the quantitative impact on a particular aspect of the forecast of any small, but arbitrary, perturbation in initial conditions, boundary conditions, or model parameters. Thus, given a specific characteristic of the forecast for which the sensitivity will be calculated, for example the minimum pressure in a cyclonic storm, the area in the initial conditions to which the characteristic is most sensitive can be defined. Thus, the implication is that this region should be better measured. This kind of analysis is also discussed in Chapter 10 in relation to the design of sensitivity studies. Palmer *et al.* (1998), Pu *et al.* (1998), Bergot (1999), Buizza and Montani (1999), and Bergot and Doerenbecher (2002) describe the use of the adjoint method for targeting observations. Using the terminology described above, at $t_i$ the forward linear version of the nonlinear forecast model is integrated from $t_i$ to $t_v$. Then, the adjoint of the linear model is used to define the sensitivity of conditions at $t_a$ to the forecast error at $t_v$. This sensitivity information defines the target area for observation platforms, which are deployed with sufficient time to reach the defined area and make the measurements at $t_a$. Shortly after time $t_a$, the operational model is initialized with the available data, and the forecast is performed. Figure 6.3 illustrates the process by which this method is applied. An issue with this approach is that a verification region must be defined, where error growth is to be

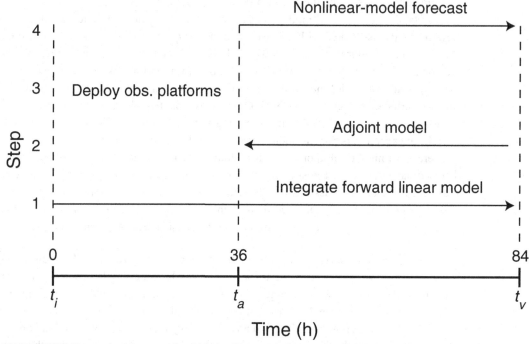

**Fig. 6.3** Schematic of the process by which the adjoint method is used to define the region where special observations may be targeted to improve the skill of an operational forecast. At time $t_i$ the forward linear model and the adjoint model are run (steps 1 and 2) to identify target areas for special observations, and the time when aircraft need to be deployed to a target area. The time $t_a$ is when the special observations apply, and when the forecast is initiated. The numbers along the dashed line to the left define steps in the process.

minimized, and this is problematic for phenomena spanning a large area or for situations where there are multiple regions of interest.

- *Ensemble transform Kalman filter* – The ensemble transform technique (Bishop and Toth 1999; Szunyogh *et al.* 1999, 2000) and the subsequent ensemble transform Kalman filter (ETKF, Bishop *et al.* 2001, Majumdar *et al.* 2002a,b) employ information from ensemble forecasting systems to identify regions where sampling would lead to forecast improvements. Advantages of the ETKF technique compared with the adjoint method include the lack of a requirement for an adjoint of the model, its low computational cost, the fact that it is based on nonlinear (ensemble) forecasts, and the fact that it provides quantitative estimates of the reduction in forecast error (not simply sensitivity metrics).

Because all of the above methods employ a model, the obtained target locations for observations will depend on both the method and the model. Different models can produce quite different estimates of locations.

A commonly noted practical limitation of adaptive-observation methods is that aircraft observations, whether they are made from the aircraft itself or with dropsondes, can only

measure a relatively small volume of atmosphere. Thus, even if the targeting region is calculated accurately, it is often not logistically possible to measure a sufficiently large area to
adjust the position or amplitude of large-scale features such as fronts or baroclinic waves
in the model initial conditions. This is especially problematic in a region that is otherwise
a data void. A related issue is that data-assimilation systems are sometimes more appropriate for observations made over larger areas than are observable with a modest number of
observing platforms. Therefore, the impact of targeted observations on forecast skill can
depend on the data-assimilation scheme. For example, Bergot (2001) shows that targeted
observations from 20 FASTEX cyclogenesis cases have a greater positive influence on
forecast skill when used in a four-dimensional rather than a three-dimensional variational
assimilation system.

Figure 6.4 provides an example of the impact of targeted observations on forecast
skill. It is a scatter plot of the RMS 500 and 1000 hPa height errors for 30, 36, 42, and
48 h forecast lead times from the ECMWF global model for five FASTEX case studies. The model was run with and without the use of the targeted observations. Each
point represents an average error for the verification region, and corresponds to a particular verification height, FASTEX case, and verification time. With a few exceptions,
the errors were less when dropsonde data were used. Of course adding observations
anywhere in the model domain might be expected to reduce forecast errors, so interpretation of these results in the context of the effectiveness of the targeting method
needs to be done cautiously. See Montani *et al.* (1999) for information about the targeting method used.

### 6.2.7 Optimal siting of permanently located observations

In contrast to the targeted observations just described, conventional, permanent observation platforms are distributed geographically to allow convenient access for their maintenance. However, there are approaches that could be used to locate such fixed platforms so
as to improve model initial conditions and therefore predictive skill. For example, if a
LAM is being run primarily to forecast a specific type of severe-weather event in a particular area, such as wind shear in the vicinity of an airport, one of the above-described
observation-targeting methods could be applied for a large number of historical cases,
and the results used to define the best overall permanent locations for observations.
Another approach that has been evaluated is called a field-coherence technique (Stauffer
*et al.* 2000, Tanrikulu *et al.* 2000), which is based on a statistical analysis of the model-
simulated atmospheric structure. The spatial and temporal coherence, as defined here, is a
measure of the distance scale over which there is temporal consistency in the spatial
structure within a variable field. Thus, the coherence indicates how well a measurement
made at one location is able to serve as an estimate of the value of that field at another
location at a given analysis time. The concept is that the larger the field coherence in a
geographic area, the fewer measurement sites are needed to adequately resolve the dominant features of that field. Observing-system simulation experiments, discussed in
Section 10.2, can also be employed to evaluate the relative benefits of different spatial
distributions of observations.

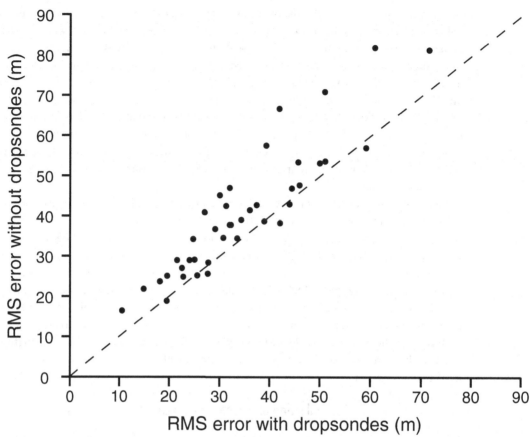

Fig. 6.4 Scatter plot of ECMWF global-model forecast errors of 500 hPa and 1000 hPa height, with and without the use of targeted observations. Each dot corresponds to RMS forecast errors in the verification region for one of five FASTEX cases, for four forecast lead times, and for two verification heights. From Montani *et al.* (1999).

## 6.3  Continuous versus intermittent data-assimilation methods

The processes of data assimilation and data analysis both have the objective of constructing a gridded data set that defines the state of a meteorological variable, and the terms are sometimes used interchangeably. That said, the use of the expression "data-assimilation system" generally means that a meteorological model is employed in the process. The overall purpose behind the use of the data-assimilation system can be the production of initial conditions for operational forecasts, or the construction of long-term reanalyses of the state of the atmosphere (see Chapter 16). The related expression "data-assimilation cycle" often encompasses the entire process of data quality control, the objective analysis, the initialization of the model (possible balancing), and the production of a short forecast to produce the next background field (Daley 1991). This section illustrates two general categories of data-assimilation systems, both of which involve the use of a model.

The qualities that are desirable in a computer-based objective-analysis process are well known by anyone who has constructed a manual, or subjective, analysis of observations. The following are traditional methods that have been used for decades in the manual analysis of observations.

- A first guess of the overall weather pattern is important. It provides the analyst with context for the observations, and can be based on the map constructed at the previous analysis time, a recent forecast, or personal knowledge of the typical regional weather patterns (the climatology).
- The variables should not be analyzed independently. For example, on large scales, areas with strong gradients in the height analysis are used to infer regions of high wind speeds when drawing isotachs.
- The overall weather patterns provide information that can be used in the interpolation between observation points. For example, when analyzing a jet maximum, isotachs are streaked out in the direction of the wind. And, at the analyzed position of fronts, isopleths of all variables reflect the transition in air-mass properties.
- The spatial density of observations is used in the analysis process. In areas where the observations are dense, the analysis is faithfully drawn to them, whereas in areas where the observations are sparse or nonexistent the analysis is based on knowledge of the background (climatology or the prior analysis). Also, an observation in a cluster of observations that is inconsistent with the rest is ignored, or given less weight, in the analysis.
- The smoothness of the analysis is made to be consistent with the density of the data and the known scales of the phenomena being analyzed.

### 6.3.1 Intermittent, or sequential, assimilation

Most operational data-assimilation systems use the intermittent, or sequential, approach. The general process is shown in Fig. 6.5. The cycle begins with an initial forecast. The next forecast in the cycle is initialized using a merger of observations (upper left) and a first-guess field (upper right). The latter is typically the output from the most-recent forecast, which is valid at the initial time of the current forecast. Observations that are made within a specific time window ($\pm n$ minutes in the figure) that spans the initialization time are aggregated and used in the analysis. The prior forecast in this process is called the first guess, or the prior estimate, or the background field. This use of the forecast in the analysis process allows the model solution to better fill spatial observation gaps than would interpolation over large distances between observations. In addition, the model solution can develop circulations in response to local surface forcing, and this allows those signatures to be included in the initial conditions (see Section 6.4). Because the merger of the forecast and the observations involves the use of information from different times, the process is referred to as Four-Dimensional Data Assimilation (FDDA). The initial conditions are then used to initialize the forecast, which will need LBCs if the model is a LAM. For global models, new forecasts are typically initiated every 6 hours, whereas for regional models this can occur as frequently as hourly. In either case, model fields are extracted at

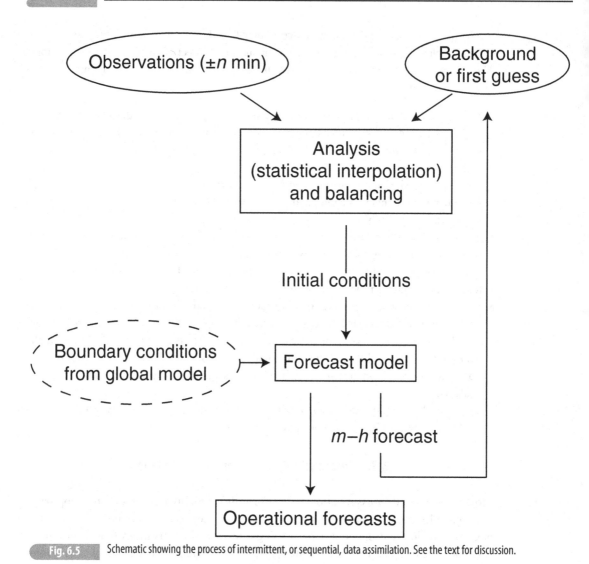

**Fig. 6.5**  Schematic showing the process of intermittent, or sequential, data assimilation. See the text for discussion.

forecast time $m$ to be used as the first guess for the next forecast cycle. This sequential approach to data assimilation serves as the basis for the optimal-interpolation, three-dimensional variational, ensemble Kalman filter, and other methods described later in this chapter. These represent particular approaches for accomplishing the process in the upper rectangular box of Fig. 6.5. See Fig. 6.6b for a different graphical depiction of this sequential-assimilation method.

### 6.3.2  Continuous assimilation

These continuous approaches involve the assimilation of observations at the times that they are made, rather than in batches, as with the sequential methods. Four-dimensional variational assimilation is a continuous-assimilation method, and is described later after

background material on statistically optimal methods is presented. The other major continuous-assimilation method, Newtonian relaxation, is summarized in this section. Data
assimilation by Newtonian relaxation (or nudging) is accomplished by adding nonphysical
nudging terms to the model predictive equations. These terms force the model solution at
each grid point to observations (observation, or station, nudging), or analyses of observations (analysis nudging), in proportion to the difference between the model solution and
the observation or analysis. The following equation illustrates the form of the relaxation
term in a prognostic equation, where $f$ is any dependent variable, $F$ represents all the
physical-process terms, $f_{obs}$ is the observed value of $f$ interpolated to the grid point,
and $\tau$ is a relaxation time scale. This relaxation-term weight can be separated into three
components: the factor that determines the magnitude of the term relative to the physical
terms in the equation ($G$), the function that defines the spatial and temporal influence of
observations ($W$), and the observation-quality factor ($\varepsilon$). In finite-difference space, this
equation applies at a particular grid point and at a particular time step:

$$\frac{\partial f}{\partial t} = F(f, \boldsymbol{x}, t) + \frac{f_{obs} - f}{\tau(f, \boldsymbol{x}, t)} = F(f, \boldsymbol{x}, t) + G(f)W(\boldsymbol{x}, t)\varepsilon(f, \boldsymbol{x})(f_{obs} - f).$$

If the relaxation time scale is too small, the model solution will converge to the observation too quickly, and the other variables will not have sufficient time to dynamically adjust.
If the time scale is too large, errors in the model solution will not be corrected by the
observations.

This approach has several advantages. It is efficient computationally, it is robust, it
allows the model to ingest data continuously rather than intermittently, the full model
dynamics are part of the assimilation system so that analyses contain all locally forced
mesoscale features, and it does not unduly complicate the structure of the model code.
Studies using Newtonian relaxation include Stauffer and Seaman (1990, 1994), Stauffer
et al. (1991), Fast (1995), Seaman et al. (1995), and Liu et al. (2006, 2008a). A finding of
these studies is that analysis nudging may work better than intermittent assimilation on
synoptic scales. Furthermore, Stauffer and Seaman (1994) and Seaman et al. (1995)
showed that nudging toward observations was more successful on the mesoscale than
nudging toward analyses. Leslie et al. (1998) found that the impact of observation-
nudging was similar to that of assimilating the same data in a four-dimensional variational
system (Section 6.11.1), with the former being practicable while the latter was too computationally expensive. Bao and Errico (1997) applied the adjoint method to illustrate the
impact of the nudging terms and some limitations of the method.

Figure 6.6 is a schematic that compares the intermittent and the continuous assimilation
processes. In both cases, the time increases from left to right. For the continuous assimilation (Fig. 6.6a), observations are ingested into the model at every time step, and forecasts
are launched at whatever frequency is desired (6 h in this example). The intermittent-
assimilation process (Fig. 6.6b) uses the same observations, except that they are aggregated temporally over some time interval to produce an objective analysis (Anal) that is
combined with a first-guess field from a short forecast. The resulting gridded fields may
undergo balancing (Initialization – Init), and are then used for the initial conditions of a

## (a) Continuous data-assimilation cycle

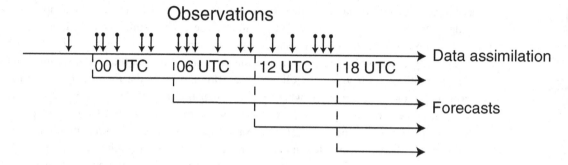

## (b) Intermittent data-assimilation cycle

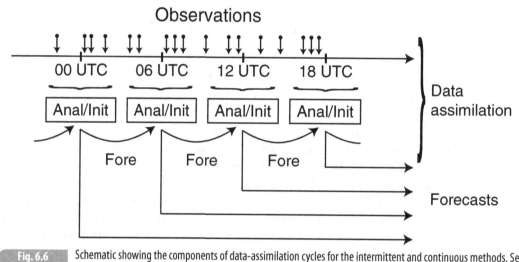

**Fig. 6.6** Schematic showing the components of data-assimilation cycles for the intermittent and continuous methods. See the text for details.

forecast. Figure 6.6b shows the same cycling process as does Fig. 6.5, but emphasizes the distinctions with the continuous-assimilation method.

A negative aspect of this type of continuous data assimilation is encountered when relaxing a mesoscale-model solution toward a synoptic-scale analysis of observations. Specifically, the model will develop fine-scale atmospheric features in response to differential surface forcing, but relaxation terms will damp these features if they are not properly represented in the analysis. Consider a situation where the model develops a sea-breeze circulation and a coastal front in response to differential surface thermal forcing at a coastline. Given the typical density of two-dimensional and three-dimensional observations, and the resulting lack of spatial detail in an objective analysis based on these observations, an analysis will only represent large-scale features and not the mesoscale detail. Thus, relaxing the model solution toward this analysis will damage the solution. This issue also exists when relaxing the model solution towards observations (rather than gridded analyses of observations) because isotropic functions for spatially spreading the influence of the observations do not respect linear mesoscale features and will also damage fine-scale features in the model solution. To avoid this problem, a method called spectral nudging has been developed. Here, the evolving model solution is filtered so that only the larger-scale features are differenced with the analysis, to define the correction term. The concept of spectral nudging is also discussed in Chapter 16, relative to climate modeling, because it is sometimes used when regional models are employed to downscale from global-climate simulations. In particular, the solution of the regional climate model is spectrally filtered, and the large-scale fields in the regional model are nudged toward the global-model solution, thus avoiding a drift of the large-scale solution in the regional model.

### 6.3.3  Hybrid intermittent–continuous methods

Even the assimilation methods that are referred to as continuous are, strictly speaking, intermittent because the data are inserted at the time-step intervals. Thus, it is perhaps understandable that there is not an especially clear definition of the terminology. For example, as the time period between analyses decreases (Fig. 6.6b), it is easy to see that the intermittent method approaches the so-called continuous one. This is not a hypothetical point, because the cycle, or update, frequency is now hourly in some operational data-assimilation systems. In addition, other methods combine aspects of the continuous and intermittent approaches. For example, in the analysis-correction system described in Lorenc *et al.* (1991), batches of data within 6-h time intervals are analyzed at each model time step, and the results are inserted into the model solution at each time step, with greater weight given to the observations whose valid times are closer to the analysis time. And, Bloom *et al.* (1996) describe an incremental analysis updating method wherein an analysis based on statistical interpolation is conducted every 6 h, and the *analysis increments* (the difference between the analyzed value and the first guess) are used as a continuous forcing during a 6-h integration. Even though these two methods retain some aspects of the intermittent approach, the data impact the model simulation at every time step.

## 6.4   Model spinup

Now that a couple of different types of data-assimilation methods have been discussed, it is appropriate to introduce the concept of model *spinup*. Because of the typical lack of spatial density in the observing network, especially in terms of observation platforms that provide information in three space dimensions, observations cannot generally define sharp cross-frontal gradients, the correct wind-speed amplitude of upper-level or low-level jet maxima, the structure of thermally forced boundary-layer circulations, the waves or channeling associated with orography, and the small-scale vertical motions and humidity gradients associated with clouds and precipitation. And, because the observations are not adequate to define these features, a simple analysis of them is not going to suffice.

However, the model itself can provide information about the atmosphere, to supplement what is in the observations. For example, land-surface properties (e.g., terrain elevation, land–water boundaries) are known with a horizontal resolution that is orders of magnitude greater than the resolution of our information about the three-dimensional structure of the atmosphere. Thus, after the model integration is begun, the lower troposphere will respond to the dynamic and thermodynamic forcing from the landscape at the lower boundary, producing thermally and dynamically forced wind circulations, contrasts in the boundary-layer temperature and humidity fields at coastlines, etc. The model dynamics have added this structural information to what was defined in the initial conditions based on observations. In addition, during the early period of a model integration the deformation at fronts will increase poorly resolved gradients, nonlinear interactions among larger waves will generate finer scales in the spectrum, and ageostrophic circulations will strengthen, creating vertical motions that can produce the saturation necessary for the development of cloud and precipitation in the model. This post-initialization development of realistic three-dimensional features during the model integration is called spinup.

Even though the spinup process allows the generation in the model solution of features that are not observed, it is problematic because it occurs during the model forecast. Thus, the early period of the forecast – perhaps 12 h in duration – does not contain properly rendered, potentially important, atmospheric processes. For example, precipitation during the first half-day of a forecast may not be realistic. Thus, there has been a great emphasis on developing initialization procedures that produce model initial conditions that are spun up, or largely so. This has led to subjective terminology such as *cold starts* for initializations that contain no spun-up processes, *hot starts* for the use of initial conditions that are completely spun up, and *warm starts* for the use of partially spun-up initial conditions.

When reading about the various data-assimilation strategies that are described throughout the rest of this chapter, the reader should keep in mind the desirability of having reasonably well spun-up initial conditions. For example, in the context of the so-called intermittent (or sequential) and continuous assimilation methods described in the previous section, the sequential method could produce less-well-spun-up initial conditions if the influence of the observations is distributed in a way that smooths out the model-produced background field. The historical motivation for all dynamic-initialization methods that employ a model during a preforecast integration period has been the desire for spun-up initial conditions.

## 6.5  The statistical framework for data assimilation

### 6.5.1  Introduction, and illustration with scalar relationships

This section describes mathematical concepts that form the basis for many approaches to data assimilation. Data assimilation is an analysis method wherein information from observations is accumulated, over a period of time, into a model state. The observational information is carried forward in time by the model, which imposes dynamic consistency among the variables and spreads the information both spatially and among the variables. There are three components to the data-assimilation process: observations; background information about the state of the atmosphere, perhaps based on a previous analysis or a model forecast; and dynamic constraints, perhaps based on a model.

In the following discussion, the term "vector" will be used to refer to a group of elements that defines a state of the model atmosphere, either in the form of gridded values or spectral coefficients. For example, the vector $x$ may be defined as $x = (x_1, x_2, \ldots, x_n)$. If this vector corresponds to the state of the atmosphere as defined in a grid-point model, the dimension $n$ will be the number of grid points multiplied by the number of dependent variables.

In the above example, the column matrix that is a collection of numbers that defines the state of a model atmosphere is referred to as the *state vector*, $x$. If this vector results from the use of an analysis system, it will disagree with observations because of errors in the analysis process, instrument error, and representativeness error that results from the finite spatial resolution of the analysis. The *true-state vector*, $x_t$, represents the best-possible state that can be defined on the model grid. This is not the same as a perfect-state vector, which corresponds exactly to the atmospheric state, because of the unavoidable representativeness error. The gridded background field, which is the first-guess estimate of $x_t$ before the analysis is conducted, is defined by the vector $x_b$. Lastly, the analysis is represented as $x_a$. The analysis problem is thus defined as finding a correction, $\delta x$, such that

$$x_a = x_b + \delta x$$

is as close as possible to $x_t$.

The observations used in an analysis are collected into an *observation vector*, $y$. In the analysis process, this observation vector needs to be compared with the state vector for the model-based first guess. Because each degree of freedom (the value of each variable defined at each grid point) in the state vector obviously does not have a corresponding observation (the observations being relatively few in number and irregularly located), for this comparison it is thus necessary to transform from model state space to observation space. This transformation is made by an *observation operator* (also called a *forward operator*) that is defined as $H(x)$. In the simplest sense, it corresponds to interpolating state variables from grid points to observation points. It also can involve the transformation of a state variable to an observed variable. In the data-analysis process, differences between the observations and state vectors are calculated. The difference

$$y - H(x_b)$$

is called the *innovation*, and the difference

$$y - H(x_a)$$

is the *analysis residual*.

These concepts can be used in a simple illustration of least-squares estimation, which will lead to a general framework for data assimilation. Suppose we have two estimates, $T_1$ and $T_2$, of the true value of a scalar, say the temperature, $T_t$, at a point. In order to combine them optimally, we need statistical information about the errors, $\varepsilon$, of these estimates. Let

$$T_1 = T_t + \varepsilon_1, \text{ and} \tag{6.1}$$

$$T_2 = T_t + \varepsilon_2, \tag{6.2}$$

where $\varepsilon_i$ are unknowns. Let $E(X)$ be the expected value of measurement $X$, or the value that would be obtained by averaging many measurements. It is assumed that the instruments that measure $T$ are unbiased. That is,

$$E(T_1 - T_t) - E(T_2 - T_t) = 0, \text{ or equivalently} \tag{6.3}$$

$$E(\varepsilon_1) = E(\varepsilon_2) = 0. \tag{6.4}$$

And, we assume that we know the variances of the observational errors:

$$E(\varepsilon_1^2) = \sigma_1^2 \text{ and } E(\varepsilon_2^2) = \sigma_2^2. \tag{6.5}$$

It is also assumed that the errors in the two observations are uncorrelated:

$$E(\varepsilon_1 \varepsilon_2) = 0. \tag{6.6}$$

Equations 6.4–6.6 define the statistical information that we need about the two observations. Our objective is to linearly combine the two estimates of $T$ in an optimal way, such that the result is the best least-squares estimate of $T_t$. Specifically, where $T_a$ is the best (optimal) estimate of $T_t$, let

$$T_a = a_1 T_1 + a_2 T_2, \text{ and} \tag{6.7}$$

$$a_1 + a_2 = 1. \tag{6.8}$$

The value of $T_a$ will be the best estimate of $T_t$ if the coefficients are chosen to minimize the mean-squared error of $T_a$. Specifically, using Eqs. 6.7 and 6.8,

$$\sigma_a^2 = E[(T_a - T_t)^2] = E[(a_1(T_1 - T_t) + a_2(T_2 - T_t))^2]. \tag{6.9}$$

Using the fact that $E(XY) = E(X)E(Y)$, for $X$ and $Y$ independent, and Eqs. 6.1, 6.2, 6.3, and 6.5, Eq. 6.9 becomes

$$\sigma_a^2 = a_1^2 \sigma_1^2 + a_2^2 \sigma_2^2. \tag{6.10}$$

Given Eq. 6.8, and defining $a_2 = k$, Eq. 6.10 becomes

$$\sigma_a^2 = (1-k)^2 \sigma_1^2 + k^2 \sigma_2^2. \tag{6.11}$$

To find the value of $k$ that corresponds to a minimum in the analysis variance, $\sigma_a^2$, differentiate Eq. 6.11 with respect to $k$, and set the expression to zero. This leads to

$$k = \frac{\sigma_1^2}{\sigma_1^2 + \sigma_2^2}. \tag{6.12}$$

Now, assume that our two sources of information, $T_1$ and $T_2$, are based on an observation and a background value. Eq. 6.7 becomes

$$T_a = kT_o + (1 - k)T_b \tag{6.13}$$

and Eq. 6.12 becomes

$$k = \frac{\sigma_b^2}{\sigma_b^2 + \sigma_o^2}, \tag{6.14}$$

leading to

$$T_a = \frac{\sigma_b^2}{\sigma_b^2 + \sigma_o^2}T_o + \frac{\sigma_o^2}{\sigma_b^2 + \sigma_o^2}T_b. \tag{6.15}$$

For example, if the observation is very poorly known, $\sigma_o$ is large and the analysis is weighted strongly toward $T_b$. Rearranging Eq. 6.13 yields

$$T_a = T_b + k(T_o - T_b). \tag{6.16}$$

Substitution of Eq. 6.14 into Eq. 6.11, and letting $\sigma_1 = \sigma_b$ and $\sigma_2 = \sigma_o$, yields

$$\sigma_a^2 = \frac{\sigma_b^2\sigma_o^2}{\sigma_b^2 + \sigma_o^2} = \sigma_b^2(1 - k). \tag{6.17}$$

This represents the uncertainty of the estimate $\sigma_a$ in terms of the uncertainties of the observation and the background. Note that $\sigma_a^2 \leq \sigma_o^2$ and $\sigma_a^2 \leq \sigma_b^2$, meaning that the analysis variance is smaller than the variance of both sources of contributing information. Stated differently, using even a large-variance source of information will reduce the uncertainty in the analysis. Equation 6.17 can be rewritten as

$$\frac{1}{\sigma_a^2} = \frac{1}{\sigma_o^2} + \frac{1}{\sigma_b^2}. \tag{6.18}$$

The inverse of a variance is called the precision (the larger the variance, the lower the precision). Thus, the precision of the analysis is the sum of the precisions of the observation and the background.

Alternatively, instead of minimizing $\sigma_a^2$ in Eq. 6.11 to find an expression for the best estimate of $T_a$, a different approach can be used. For any $T$, the distance between $T$ and $T_b$, and $T$ and $T_o$ can be measured by the following quadratic relationship:

$$J(T) = \frac{1}{2}(J_o(T) + J_b(T)) = \frac{1}{2}\left[\frac{(T - T_o)^2}{\sigma_o^2} + \frac{(T - T_b)^2}{\sigma_b^2}\right]. \qquad (6.19)$$

This function represents the square of the misfit of a variable ($T$) from each of the two sources of information, weighted by the precision of each of the estimators. It is often called a cost function or a penalty function. To define a value for $T$ that corresponds to a minimum in the cost function, $J$ is differentiated with respect to $T$, the resulting expression is set to zero, and it is confirmed that the extremum is, in fact, a minimum. The best estimate of $T$ defined by this expression is $T_a$. The result is the same as defined in Eq. 6.15. Figure 6.7 illustrates graphically how the two penalty terms, $J_o$ and $J_b$, in Eq. 6.19 are combined to produce the minimum in the analysis, $T_a$.

The above analysis involves the optimal combination of only two pieces of information, an observation and a background, or first-guess, value. That is, this has been posed as a simple scalar rather than a vector problem. It has also been assumed that these pieces of information are defined at the same location, and thus there has been no need for a forward operator to transform from model space to observation space. For application of these concepts in the framework of a model, the background state vector has a size in excess of $10^7$ (for a grid-point model, the number of grid points times the number of dependent variables). And the observation vector has perhaps a size of $10^6$. Fortunately, the above least-squares estimation methods have exactly the same form when applied to real multi-dimensional data-assimilation problems. The following summary of the most important points is based on Kalnay (2003).

- Equation 6.16 states that an analysis value is obtained by adding to the background (first guess), the innovation (the difference between the observation and first guess) multiplied by an optimal weight.
- The optimal weight, $k$, defined in Eq. 6.14, is the background error variance multiplied by the inverse of the total error variance (the sum of the background and observation error variances). The larger the background error variance, the larger is the correction to the background by the observation.
- Equation 6.18 states that the precision of the analysis is the sum of the precisions of the observation and the background.
- The rightmost part of Eq. 6.17 means that the error variance of the analysis is equal to the error variance of the background, reduced by a factor that is equal to one minus the optimal weight.

The application of the above least-squares methods to multi-dimensional and multi-variable problems is found in subsequent sections.

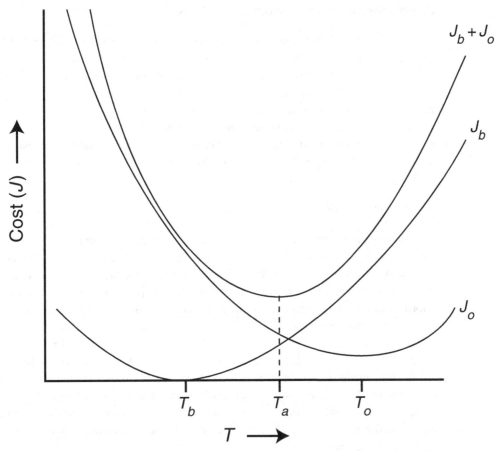

**Fig. 6.7** Schematic showing how the two penalty terms, $J_o$ and $J_b$, in Eq. 6.19, are combined to produce the minimum in the analysis error at $T_a$.

## 6.5.2 Statistical concepts for multi-dimensional problems

The following list summarizes the vectors and vector operators used in this and some of the following sections, as well as in the wider literature on this subject, and will serve as a reference for the discussion. The symbols are generally consistent with the unified notation proposed in Ide *et al.* (1997). State vectors $x$ may be defined on a model grid (or they can define spectral coefficients). Depending upon the setup of the analysis, the unknown analysis vector $x_a$ and the known background vector $x_b$ can define the values of a single variable in a two-dimensional space – e.g., $T_b(x, y)$. Or the vectors can represent the three-dimensional structure of a single variable – e.g., $T_b(x, y, z)$. Or they can define all the variables in the three-dimensional space – e.g., $x_b = (\mathrm{Psfc}_b(x, y), T_b(x, y, z), q_b(x, y, z), u_b(x, y, z), v_b(x, y, z)$, etc.). The model background and analysis fields are defined by column vectors ordered by grid point and by variable, where the vector length $n$ is the product of the number of variables and the number of grid points.

State and observation vectors are defined as follows.

- $x_t$   The true model state vector. As described in Section 6.5.1, it represents the best-possible state that can be defined on the model grid. It is not the same as a perfect-state vector, which corresponds exactly with the atmospheric state (perfect observations), because of the unavoidable representativeness error. The dimension is $n$.
- $x_a$   The analysis model state vector. The dimension is $n$.
- $x_f$   The forecast model state vector. The dimension is $n$.
- $x_b$   The background model state vector. The dimension is $n$. If a model forecast is used to define this, as in sequential initialization, $x_b = x_f$.
- $y$   Vector of observations. The dimension is $p$.

Error covariance matrices are defined as follows:

- **B**   The covariance matrix of the background (or forecast) errors. The background-error matrix has dimensions $n \times n$. It is very important to have reasonable estimates for this matrix because it controls the influence function for the analysis increment, in terms of its magnitude and shape. Regarding the shape, it defines the spreading of information from an observation to the analysis grid. And regarding magnitude, when background errors are large, observations are given greater weight. If this error covariance matrix is relatively accurate, a better adjustment of the gridded background to the observations will result, and observations will be used more effectively. That is, the information in the innovation vector that is defined at an observation point will be translated by **B** into a spatially variable analysis increment that is applied at surrounding grid points in such a way as to minimize the analysis error. In a scalar system, the background error covariance is simply the variance, or the average squared departure from the mean, where

$$\text{B} = \overline{(\varepsilon_b - \bar{\varepsilon}_b)^2}.$$

In a multi-dimensional system,

$$\mathbf{B} = \overline{(\varepsilon_b - \bar{\varepsilon}_b)(\varepsilon_b - \bar{\varepsilon}_b)^T},$$

which is a square, symmetric matrix with variances along the diagonal. For a very simple three-dimensional system,

$$\mathbf{B} = \begin{bmatrix} var(e_1) & cov(e_1,e_2) & cov(e_1,e_3) \\ cov(e_1,e_2) & var(e_2) & cov(e_2,e_3) \\ cov(e_1,e_3) & cov(e_2,e_3) & var(e_3) \end{bmatrix}.$$

The off-diagonal terms are cross-covariances between each pair of "variables" in the model, where, as noted earlier, the term variable here corresponds to the value of each physical dependent variable at each grid point. A variable pair can be the same model dependent variable at two different points, or it can be two different dependent variables. The number of variables, and the dimension of the matrix, is the product of the number of physical variables and the number of grid points. There are three approaches for estimating the covariance matrix.

1. *Precalculated error covariances* – Some data-assimilation methods use precalculated covariances that are based on (a) an average over many different observed states of the atmosphere, (b) theoretical considerations, or (c) model simulations. In any case, the statistics may be spatially and temporally homogeneous (i.e., not dependent on the specific meteorological regime or synoptic situation). Observation-based covariances are ideally calculated from a dense and homogeneous network of sensors with uncorrelated errors, where the innovation vector $[y - H(x_b)]$ (observation minus forecast) is calculated for varying separations between the locations of $y$ and $x_b$. In contrast, the so-called NMC (US National Meteorological Center) method is based entirely on model simulations, and is discussed briefly in Section 6.8 on three-dimensional variational (3DVAR) analysis. Examples of two different assumed decreases in correlation with increasing distance between $y$ and $x_b$ are seen in Fig. 6.8. See Schlatter (1975), Hollingsworth and Lönnberg (1986), Lönnberg and Hollingsworth (1986), Thiebaux *et al.* (1986), Bartello and Mitchell (1992), Xu and Wei (2001, 2002), and Xu *et al.* (2001) for additional discussion of the calculation of the non-regime-dependent error-covariance matrix.

2. *Nonoptimal, anisotropic spatial weighting* – One class of such methods employs information about the orographic elevation to control the spread of the innovation vector at lower elevations in the model atmosphere. The logic here is that covariances should be smaller between points on opposite sides of a mountain ridge, so an observation on one side has a weaker effect on grid points on the opposite side. So, the distribution of the analysis increment is anisotropic, with smaller increments (adjustments based on observations) on the opposite side of a barrier from an observation. Lanzinger and Steinacker (1990) employ this approach for an Optimal Interpolation (OI, see Section 6.7) analysis in the area of the Alps mountains. And, Miller and Benjamin (1994) made use of the fact that variables at two points will be better correlated if their potential temperatures and elevations are similar. So, the effective distance between an observation and a grid point was made proportional to the differences in elevation and potential temperature between the two points. Similarly, Dévényi and Schlatter (1994) spread their OI observation increments along isentropic surfaces.

3. *Fully regime-dependent error covariances* – The above methods do not account for the existence of "errors of the day" (Kalnay *et al.* 1997), which are weather-dependent errors in the background (forecast) field that should greatly influence the way that observations are analyzed. Ignoring these day-to-day variations in the covariance statistics can lead to large analysis errors. Some advanced data-assimilation methods, described in Section 6.11, calculate flow-dependent background-error covariances that evolve during the assimilation process. For example, Fig. 6.21 in Section 6.11.3 illustrates the spatial variability of these covariances.

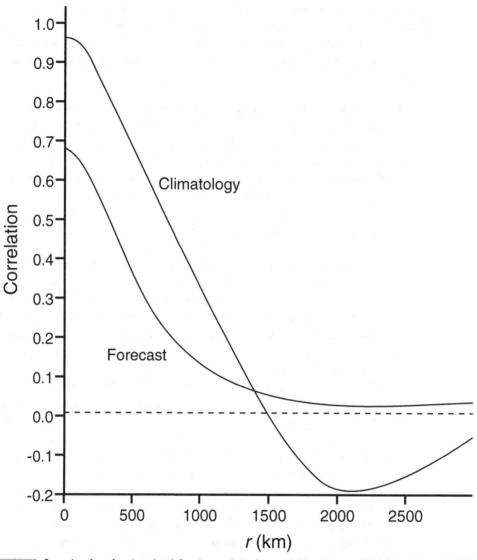

**Fig. 6.8** Examples of two functions that define the correlation between observations and background values for 500-hPa geopotential height over North America. One curve is based on the use of climatology as the background (Schlatter 1975) and the other is based on the use of model forecasts (Lönnberg and Hollingsworth 1986). Adapted from Daley (1991).

Figure 6.9 illustrates the difference between the use of a typical isotropic covariance matrix to spread the influence of an observation, and the use of one that is regime dependent, for a two-dimensional $(x,y)$ system. Shown are surfaces that define the value of the $u$ velocity component on a grid. The background value $(u_b)$ is spatially uniform (flat), and the observation $y$ produces a positive innovation. In Fig. 6.9a, the analysis increment is distributed isotropically on the grid, around $y$, producing a conical impact

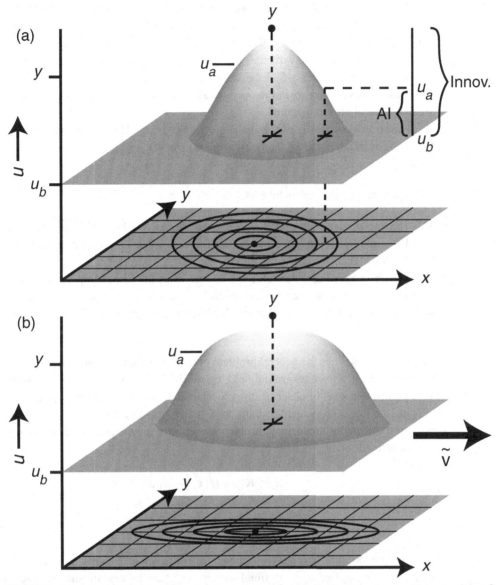

Fig. 6.9 Schematic illustrating the difference between the use of a typical isotropic covariance matrix to spread the influence of an observation (a), and the use of one that is regime dependent (b), for a two-dimensional $(x, y)$ system. See the text for discussion.

of the observation $(u_a)$. The shaded surface below shows isotachs for the resulting analysis on the grid. In Fig. 6.9b, the covariance matrix recognizes that, at this location, the wind speed pattern is streaked out in the direction of the total wind vector (shown), producing a nonisotropic, weather-regime-dependent distribution of the analysis increment and the isotachs.

- **R**  The covariance matrix of the observation errors ($\varepsilon_o = y - H(x_t)$). The dimensions are $p \times p$. Observation errors are often considered to be independent, especially when the observations are made by different instruments (e.g., in contrast to radiosonde profiles of observations). The variances are generally estimated based on knowledge of the instrument characteristics, which can be studied in the laboratory, even though representativeness errors and errors in the operator **H** can also be important. Most models of **R** are diagonal, or almost diagonal.
- **A**  The covariance matrix of the analysis errors ($x_a - x_t$). The dimensions are $n \times n$.
- **Q**  The covariance matrix of model forecast errors ($x_f - x_t$). The dimensions are $n \times n$.

Vector operators are defined as follows.

- **M**  The model dynamic operator. For example, $x_f(t + 1) = M[x_f(t)]$ refers to the fact that a model is used to advance the forecast value of vector $x$ from time ($t$) to time ($t + 1$). Dimensions are from $n$ to $n$.
- **H**  The observation operator. This is also known as the forward operator. Dimensions are from $n$ to $p$ because the transformation is from model state space ($n$) to observation space ($p$). Imagine interpolating a variable from model grid points to the location of an observation.

The following defines the general problem of finding an optimal analysis, $x_a$, of a set of model variables, given a background field, $x_b$, available at a two- or three-dimensional set of grid points, and a set of observations $y$ available at irregular locations $r$ (see Fig. 6.10). Analogous to Eq. 6.16, which pertains to a scalar problem, the following relationship applies to a full multi-dimensional problem, where the vectors and vector operators have just been defined:

$$x_a = x_b + \mathbf{K}(y - H[x_b]), \text{ where} \tag{6.20}$$

$$\mathbf{K} = \mathbf{B}\mathbf{H}^{\mathrm{T}}(\mathbf{H}\mathbf{B}\mathbf{H}^{\mathrm{T}} + \mathbf{R})^{-1}. \tag{6.21}$$

As before, the variable **K** is a weight matrix of the analysis. Exactly as in Eq. 6.16, the innovation is multiplied by an optimal weight, and this defines the analysis increment, $x_a - x_b$. The gain matrix is obtained by multiplying the background error covariance in the observation space and the inverse of the total error covariance (the sum of the background and the observation error covariances). The larger the magnitude of the elements of $\mathbf{B}\mathbf{H}^{\mathrm{T}}$, corresponding to an observation and an analysis variable at a grid point, the larger the weight with which the innovation vector is applied at that grid point. Regarding the inverse term $(\mathbf{H}\mathbf{B}\mathbf{H}^{\mathrm{T}} + \mathbf{R})^{-1}$, the larger the uncertainty in the observation, the smaller the observation increment will be weighted in the analysis. The vector $x_a$ is the optimal least-squares estimate. Most of the references listed at the end of the chapter, e.g., Kalnay (2003), can be consulted for a derivation of the gain matrix in Eq. 6.21.

## 6.6 Successive-correction methods

One of the first procedures to be used for interpolating observations to a grid is called the *Successive-Correction* (SC) method (Bergthorsson and Doos 1955, Cressman 1959). Variations of this approach are in use today because it is simple and robust. As mentioned above, a first-guess field represents a best estimate of the variable defined on the grid, and it is corrected using successive adjustments in which the observations influence surrounding grid-point values. The process is defined by the following expression:

$$x_i^{n+1} = x_i^n + \frac{\sum\limits_{k=1}^{K_i^n} w_{ik}^n (y_k - H(x_k^n))}{\sum\limits_{k=1}^{K_i^n} w_{ik}^n + \varepsilon^2}, \tag{6.22}$$

where $x_i^n$ is the $n$-th iteration estimation at grid point $i$, $y_k$ is the $k$-th observation of $x$ surrounding the grid point $i$, $H(x_k^n)$ is the value of the $n$-th estimate of $x$ interpolated from the surrounding grid points to the observation point $k$, and $\varepsilon^2$ is an estimate of the ratio of the observation-error variance to the first-guess error variance. The weights can be formulated in various ways. In Cressman (1959) they are defined as

$$w_{ik}^n = \frac{R_n^2 - r_{ik}^2}{R_n^2 + r_{ik}^2} \quad \text{for} \quad r_{ik}^2 \le R_n^2 \tag{6.23}$$

$$w_{ik}^n = 0 \quad \text{for} \quad r_{ik}^2 > R_n^2,$$

where $r_{ik}^2$ is the square of the distance between an observation point $k$ and a grid point $i$. Figure 6.10 illustrates an influence region defined on a field of regularly spaced grid points and irregularly distributed observations.

For the initial iteration, $x_i^0$ in Eq. 6.22 is the value of the first guess. For each grid point, $i$, in the first-guess field, each of the $K$ observations that is within the first radius of influence, $R_0$, of the grid point is used to adjust the first guess. The adjustment for each observation is weighted based on its distance from the grid point and on the difference between the observed value and the value of the first guess interpolated to the observation point. The difference between the first guess and the observation is, as before, called the innovation, and its weighted distribution around each observation point is isotropic. This process is repeated with successively smaller radii, given the constraint that there should be at least a few observations within each area of influence. Thus, the first guess is adjusted on the broader scales based on the effects of a large number of observations, and a progressively smaller number of nearby observations is reused in subsequent iterations to account for local effects.

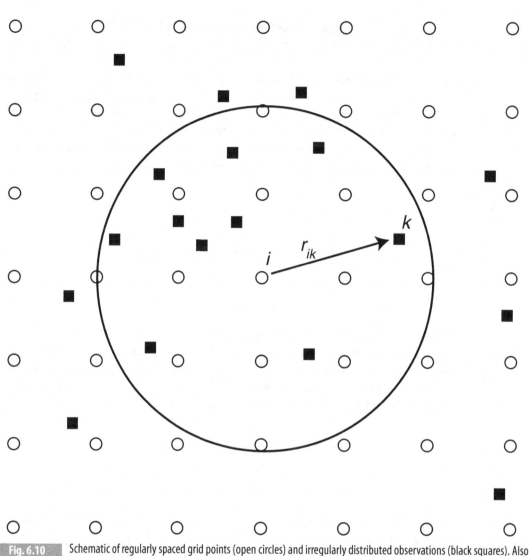

**Fig. 6.10** Schematic of regularly spaced grid points (open circles) and irregularly distributed observations (black squares). Also shown is a circular influence function around grid point $i$, with a displacement vector $r$ between that grid point and observation $k$.

If $\varepsilon^2$ is defined to be zero, the implication is that the observations are perfect. This means that the procedure will faithfully analyze to the observations, to the point of resulting in a bulls-eye pattern in the isopleths that encircle a bad grid-point value. By using a realistic value for $\varepsilon^2$, the adjustment to the observation in Eq. 6.22 is smaller and the first guess is given more weight. With $\varepsilon^2 = 0$, the impact of a single bad observation can be reduced by not using small radii of influence, so that multiple observations influence each grid point. Figure 6.11 shows a one-dimensional schematic of the generation of an analysis through the correction of a background field within an influence region of observations.

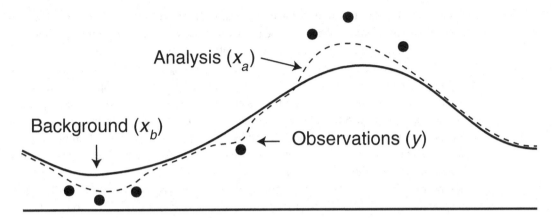

Distance ⟶

**Fig. 6.11**  Schematic showing the generation of an analysis (dashed line) through the correction of a background field (solid line) within an influence region of observations (black circles). The background and analysis are coincident outside the influence of the observations.

Another version of this method was developed by Barnes (1964, 1978), where one of the advantages is that no independent first guess is needed. In fact, the first guess is defined by a weighted sum of the observations within a radius of influence:

$$x_i^0 = \sum_{k=1}^{K_i} w_{ik}^0 y_k .$$

This has advantages on small scales where there may be no operational model to produce a background estimate. The Barnes formulation is similar to that in Eq. 6.22, except that $\varepsilon^2$ is assumed to be zero because we have no first guess (i.e., the error is large). The weights are given by

$$w_{ik}^n = e^{-r_{ik}^2 / 2R_n^2} ,$$

where the radii of influence

$$R_{n+1}^2 = \gamma R_n^2$$

are reduced by a constant fraction ($\gamma$) for each iteration. Additional discussion of the SC method, and examples of recent applications, can be found in Daley (1991), Barnes (1994a,b), and Garcia-Pintado *et al.* (2009).

The above use of an isotropic distribution of the analysis increment around each observation point obviously makes this method easy to implement, but it is an unnecessary approximation. Alternatives are based on the aforementioned idea that the spatial influence of observations can be related to prevailing meteorological structures. For example, observations should probably not be used to influence grid points on the opposite side of a front. And, in the atmosphere, scalar variables will be spread out more in the along-stream rather than the cross-stream direction. Regarding the latter point, an elliptical weighting function whose aspect ratio is proportional to the wind speed is an option (Benjamin and Seaman 1985). If the flow is curved, the semi-major axis of the ellipse can be curved accordingly. Such a weighting pattern has been called a banana function, for obvious reasons. And, Stauffer and Seaman (1994) adjusted the weight of an observation at a low-level grid point based on the surface-elevation difference between the two points. Obviously these weights are not optimal in any statistical sense. See Otte *et al.* (2001) for a description of other methods of weighting observations based on prevailing meteorological structures. And, see Bratseth (1986) for a discussion of how the SC method can be formulated so that it is equivalent with the statistical optimal-interpolation method (see next section).

## 6.7  Statistical interpolation (optimal interpolation)

Statistical interpolation is sometimes referred to as optimal interpolation. Equations 6.20 and 6.21 serve as the basis for OI. The unknown analysis and known background can be two-dimensional fields of a single variable, or three-dimensional fields for all the model dependent variables. The benefit of the OI method and three-dimensional variational (3DVAR) assimilation described below, relative to the SC method above, is that the spatial distribution of analysis increments is defined by the background-error covariance matrix that is based on archived model solutions or climatology (observations). In contrast, with the SC method the weight is often isotropic and somewhat arbitrary, only depending on distance from the observation. A fundamental computational-cost-saving concept in OI is that for each model variable (again, a model variable is one dependent variable at one grid point), only a few nearby observations are considered important in determining the analysis increment. These observations are selected based on empirical criteria, where it is assumed that distant observations would have small background error covariances $\mathbf{BH}^{\mathrm{T}}$.

The OI approach has been most often applied in intermittent-analysis schemes, such as depicted in Figure 6.6b. That is, the OI is used for the "Anal" at the beginning of each cycle. The model is integrated from the time of one analysis to the time of the next one. This provides the background vector $x_b$, while all the observations that are available in the analysis time window are used to build the vector $y$. Methods of defining $\mathbf{B}$ for OI approaches are described in Thiebaux and Pedder (1987) and Hollingsworth and Lönnberg (1986), and involve differencing the short forecast and the radiosonde observations (see the discussion of the background-error covariance matrix in Section 6.5.2). An advantage of OI is the simplicity with which it can be implemented, and the modest cost if the necessary assumptions can be made (e.g., observation selection).

## 6.8 Three-dimensional variational analysis

It was shown in Section 6.5.1 that there is a correspondence between two methods for the optimal analysis of a scalar: (1) minimizing the analysis error variance (by finding the optimal weights through a least-squares approach) and (2) using a variational approach (finding the analysis that minimizes a cost function that is a measure of the distance of the analysis to both the background and the observation). This correspondence also holds true for analyses of multi-dimensional fields, as described in the previous section for OI.

Lorenc (1986) showed the formal equivalence of the approach used in OI (where an optimal gain matrix $\mathbf{K}$ is found that minimizes the analysis-error covariance matrix) and a particular variational-assimilation problem. The latter is used in 3DVAR analysis, and corresponds to finding an optimal analysis field, $x_a$, that minimizes a cost function, such that the cost function is defined as the sum of (1) the distance between $x$ and $x_b$, weighted by the inverse of the background-error covariance and (2) the distance to the observation $y$ weighted by the inverse of the observation-error covariance. Mathematically, the cost function is

$$J(x) = J_b(x) + J_o(x) = \frac{1}{2}(x - x_b)^T B^{-1}(x - x_b) + \frac{1}{2}(H(x) - y)^T R^{-1}(H(x) - y),$$

(6.24)

and the gradient with respect to $x$ is

$$\nabla J(x) = 2B^{-1}(x - x_b) - 2H(x)^T R^{-1}(y - H(x)).$$

Note the parallel between Eqs. 6.24 and 6.19. The control variable, the variable with respect to which the cost function is minimized, is the state vector, $x$. The minimum of the cost function can be found analytically (e.g., Kalnay 2003), but in practice it is far less computationally demanding to estimate it iteratively by performing multiple evaluations of both equations. The minimum is approached by using a minimization, or descent, algorithm such as the conjugate-gradient or quasi-Newton methods. Only a small number of iterations are used, to produce an approximate minimum. Figure 6.12 illustrates the minimization process for a two-variable model space. The quadratic cost function has the shape of a paraboloid. The initial point in the iteration is generally taken to be the background value, $x_b$, and the final point is $x_a$, the approximate location of the minimum of $J$. Each step of the iteration moves the estimate down the gradient of the cost function.

Despite their formal equivalence, 3DVAR has a few advantages relative to OI. They are listed below, and are discussed more extensively in Kalnay (2003).

- In 3DVAR, there is no selection of only a limited number of observations that are within an influence region of a grid point. All observations are used simultaneously, which leads to a smoother analysis.
- The forecast- or background-error covariance, $B$, is defined using fewer assumptions in 3DVAR. In particular, the so-called NMC (now NCEP) method (Parrish and Derber 1992)

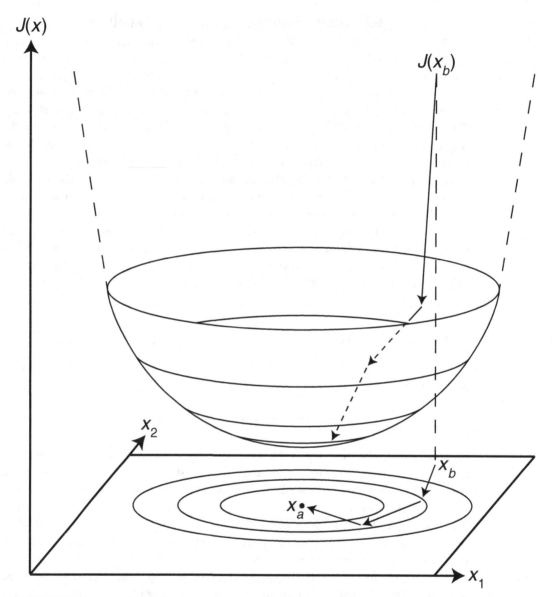

**Fig. 6.12** Schematic showing the variational cost-function minimization for a two-variable $(x_1, x_2)$ model space. See the text for details.

is generally employed. This does not depend on measurements, as with OI (Hollingsworth and Lönnberg 1986, Thiebaux and Pedder 1987) but rather the error covariance is based on the average (over perhaps 50 instances) difference between forecasts valid at the same time. Even though any time lag and lead time can be used, the following example is based on 24-h and 48-h forecasts.

$$\boldsymbol{B} \approx \alpha E\{[\boldsymbol{x}_f(48h) - \boldsymbol{x}_f(24h)][\boldsymbol{x}_f(48h) - \boldsymbol{x}_f(24h)]^T\}.$$

Even though this is the covariance of the forecast differences, which is only a surrogate for the background- or forecast-error covariance, it has been shown to produce better results than the methods used in OI, where forecasts and observations are employed. Parrish and Derber (1992) and Rabier *et al.* (1998) point out that the radiosonde network is not sufficiently dense to properly estimate structures. Nevertheless, the covariances in 3DVAR are typically isotropic and climatological (e.g., Fig. 6.8) – i.e., they are not situation (case, regime) dependent – which is a major disadvantage.

- Additional constraints can be added to the cost function, such as those related to dynamical-balance relationships. For example, Parrish and Derber (1992) employed an additional penalty term in Eq. 6.24, forcing the analysis increments to approximately satisfy the balance equation. In contrast, it was often found necessary to follow an OI analysis with a Nonlinear Normal-Mode Initialization (NNMI, see Section 6.10.3). Importantly, with the implementation of 3DVAR it became unnecessary to perform a separate balancing, or initialization, step in the analysis cycle (cf., Fig. 6.6b).

- Prior to the availability of 3DVAR, satellite radiances had to be processed through a retrieval algorithm that generated values of a model dependent variable that would be assimilated. But with 3DVAR, the radiances themselves can be assimilated directly.

## 6.9 Diabatic-initialization methods

*Diabatic initialization* involves the use of observations of precipitation and other variables to produce estimates of the four-dimensional distribution of latent-heating rate, and the associated humidity and divergence fields, in a model during the initialization process. This has two motivations. One is that it employs precipitation observations in a model initialization, and the other is that it helps produce reasonable vertical-motion and moisture fields at the initial time of a forecast. The latter goal is motivated by the fact that, typically, initial conditions do not have realistic vertical motions and humidities, and there is the resulting spinup period during which the model must internally develop such precipitation-scale circulations and humidity fields.

Some early diabatic-initialization methods simply inserted estimated latent-heating profiles into the model grid columns during a preforecast dynamic-initialization period. The column-total latent heat was based on satellite-, rain-gauge-, or radar-estimated rain rates, and the vertical distribution of the heating was typically defined to be consistent with the model's parameterizations (Fiorino and Warner 1981, Danard 1985, Ninomiya and Kurihara 1987, Wang and Warner 1988, Monobianco *et al.* 1994). Sometimes, additional observations are assimilated during the preforecast period using Newtonian relaxation. Figure 6.13 shows a schematic of this method (a), and of other methods to be described shortly. The black bars on the time axis of Fig. 6.13a indicate the insertion of latent-heating-rate information at appropriate grid points. Also shown is the simultaneous use of Newtonian relaxation to assimilate other observations during the dynamic-initialization period. A related approach, referred to as latent-heat nudging, is described in Jones and Macpherson (1997), and is applied in Leuenberger and Rossa (2007) to a

(a) Latent heat insertion

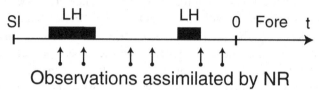

(b) Static initialization

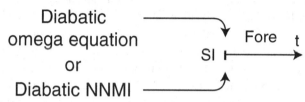

(c) Physical initialization

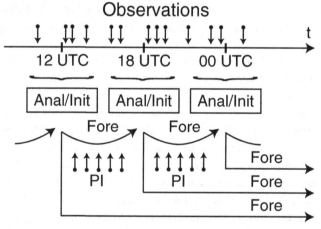

**Fig. 6.13** Schematics of three different types of diabatic initializations: (a) insertion of latent-heating (LH) profiles based on observed rain rates during a preforecast dynamic-initialization period, with possible simultaneous Newtonian relaxation (NR) using other observations; (b) the use of a diabatic nonlinear normal-mode initialization (NNMI) or a diabatic omega equation to incorporate precipitation processes in a static initialization (SI); and (c) a physical initialization where information about model dependent variables obtained from a physical initialization (PI, see text) is assimilated in an intermittent-assimilation cycle to improve the first-guess field.

mesogamma-scale rainfall simulation. Application of the method involves correcting the model's latent heating at each time step based on the ratio of the observed and model-simulated surface precipitation. Another approach is the use of a diabatic omega equation to define the vertical motion and divergent component of the wind in a static initialization

(Fig. 6.13b, Tarbell *et al.* 1981, Salmon and Warner 1986). Turpeinen *et al.* (1990) and Raymond *et al.* (1995) include a summary of these early studies.

Simultaneous with these efforts was the development of a diabatic version of the NNMI method (Wergen 1988) described in Section 6.10.3. This process incorporated estimated latent-heating rates into a static initialization (Fig. 6.13b), where the objective was also to provide initial conditions that did not require significant spinup of the precipitation processes during the early period of the forecast. Examples of the use of NNMI for diabatic initialization can be found in Puri (1987), Heckley *et al.* (1990), Turpeinen (1990), Turpeinen *et al.* (1990), and Kasahara *et al.* (1996).

The so-called physical-initialization method, described in Krishnamurti *et al.* (1991), also enables the use of rainfall-rate estimates and other observations during a preforecast integration to provide model initial conditions that contain spun-up vertical motion, horizontal divergence, and humidity fields. The details of the method can vary, but a common feature is that reverse algorithms are employed for relationships in the model that involve the moisture variables: the parameterizations for convection and outgoing longwave radiation (OLR), and similarity theory. For example, convective parameterizations provide the convective rain rate as a function of grid-resolved variables such as the vertical moisture profile. "Reversing" the convective-parameterization algorithm allows observations of the rain rate to be translated into an estimate of grid-resolved model dependent variables, which can be ingested into model initial conditions using Newtonian relaxation or they can replace simulated variables during a preforecast integration (Fig. 6.13c). Following the conceptual explanation in Treadon (1996), consider the equation $y = f(x)$ that represents a simple model parameterization relationship, where $x$ is a grid-resolved variable, and $y$ is an observed, parameterized variable. For example, let $y$ be OLR, and $x$ the variables that are used in its parameterization (e.g., temperature, specific humidity, etc.). Thus, based on measurements of OLR, the reverse algorithm provides estimates of the large-scale forcing, $x$. This estimate can be assimilated, or it can be fed back into the forward algorithm to provide an improved value after iteration. Physical initialization methods have been applied most often in the tropics, where the scarcity of conventional observations increases the dependence of the forecast on a reasonable first guess. Krishnamurti *et al.* (1994, 2007) and Shin and Krishnamurti (1999) show that this method leads to a significant improvement in the skill of forecasts of tropical rainfall, global cloudiness, land-surface hydrology, and tropical cyclones. The method has been tested with the Florida State University spectral model, the US Navy's NOGAPS model (Van Tuyl 1996), and the NCEP Global Data Assimilation System (Treadon 1996). A recent application of the method is described in Milan *et al.* (2008). Physical initialization has also been employed for ensemble prediction (Chaves *et al.* 2005, Ross and Krishnamurti 2005) as well as with LAMs (Li and Lai 2004, Nunes and Cocke 2004).

The above methods are not as likely to be effective for midlatitude precipitation events that are associated with large-scale forcing. For example, using observed precipitation-related fields to develop vertical motion and latent-heating rates in the initial conditions for a case of frontal precipitation is not going to be effective if the front, or the cyclone itself, is in the wrong location in the model solution. In this case, the lack of correct large-scale forcing for the specified precipitation in the model will cause the precipitation to dissipate during the early period of a forecast. Similarly, precipitation imposed in the solution through diabatic initialization will not persist in situations of air-mass convection if the large-scale environment is not sufficiently unstable. In contrast, for convection within air masses with realistic stability, and for tropical cyclones, the methods will be more likely to improve predictive skill.

Other methods, such as three- and four-dimensional variational data assimilation (discussed elsewhere), can also be used to assimilate observations that represent precipitation-scale processes in model initial conditions.

## 6.10  Dynamical balance in the initial conditions

This section focusses on the need to have a realistic dynamic balance in model initial conditions, the consequences of not having that balance, and methods that have been employed to ensure that a reasonable balance exists. The first section reviews the relevance to initialization of the concept of geostrophic adjustment, the second describes how integrating a model for a period prior to the initialization can contribute to an improved balance, and the third briefly summarizes the use of diagnostic relationships for achieving a balance.

### 6.10.1  Geostrophic-adjustment concepts, and relevance to initialization

The atmosphere on synoptic and planetary scales is in approximate geostrophic balance. Thus, if the mass and wind fields on these scales are significantly inconsistent with each other in the model initial conditions, inertia–gravity waves will cause those fields to adjust toward balance during the early period of the integration. Of course, the large-scale mass and wind fields in the real atmosphere are always in a state of continuous imbalance and adjustment. But, inertia–gravity waves that result from poorly defined initial conditions are not physically based, can be of large amplitude, and are potentially problematic in a model solution. This adjustment process is relevant to numerical modeling for a few reasons. First, if the inertia–gravity waves that effect the adjustment are of significant amplitude, they can mask the true meteorological features in the model solution until the waves are damped or propagate away from the area of interest. This could mean that the model solution will not be useful for the first 12–24 h after initialization. Also, the resulting waves can have troublesome interactions with the LBCs of LAMs. Lastly, the geostrophic-adjustment process can cause good-quality observations in one field to compensate for poor-quality observations in the other, or poor-quality observations in one field can negate the value of good-quality observations in the other.

To better understand the geostrophic-adjustment process, consider the simple geopotential-height pattern in Fig. 6.14. A parcel of air at location 1 is moving with a speed that is in geostrophic balance with the local pressure gradient. As the parcel moves toward location 2, the local pressure gradient becomes greater and the motion is subgeostrophic. The fluid system can respond in two ways to regain a balance: Either the parcel's speed can

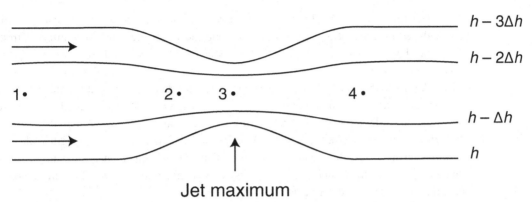

$$h - 3\Delta h$$

$$h - 2\Delta h$$

$$h - \Delta h$$

$$h$$

## Jet maximum

**Fig. 6.14**    Schematic of a jet maximum in the height field, where the flow is from left to right. The numbers represent locations of air parcels, where the discussion focusses on whether the wind or the mass (height) fields change as the parcels move to the right.

increase as it moves into the geostrophic jet maximum, or the pressure-gradient maximum can move downstream, causing the pressure-gradient at location 2 to weaken. Of course a similar question could be posed about the response of the system as parcels move farther downstream, away from the pressure-gradient maximum (from location 3 to 4). To resolve this, consider the ways in which the mass and wind fields can adjust in such situations. The shallow-fluid equations (see Chapter 2) represent a simple framework for addressing this, which nevertheless contains all the relevant dynamics. Two of the admissible wave solutions are gravity waves and inertia waves. Both mechanisms operate simultaneously to reconcile an imbalance. The inertia waves modify the winds, and the gravity waves modify the mass field (the fluid depth, in the shallow-fluid system). As an indicator of how much of the adjustment results from changes in each of the mass field and momentum field, consider the periods of these waves. For inertia waves $T_i = 2\pi/f$, and for gravity waves $T_g = L/\sqrt{gH}$, where $L$ is the length of the gravity wave (defined by the horizontal scale of the imbalance) and $H$ is the depth of the imbalance (the vertical scale). Given that both types of waves simultaneously act to adjust the atmosphere toward the geostrophic state, the wave mode with the shortest period accomplishes most of the adjustment. To define the condition where there is equal adjustment from both types of waves, the expressions for the two periods can be equated. Solving for the wavelength yields

$$L_R = \frac{2\pi\sqrt{gH}}{f},$$

where this length scale is the Rossby radius of deformation for the shallow-fluid system. For wavelengths shorter than this value, redistribution of the mass field through gravity waves is responsible for most of the adjustment, whereas for longer wavelengths, modification of the windfield by the inertia waves accomplishes most of the adjustment. The value of $L$ for deep (tropospheric) midlatitude adjustments is ~15 000–18 000 km. So, for planetary scales and very-long synoptic scales, the winds adjust to the mass field when there is an imbalance. That is, the mass field does not change much, so we say that it dominates the adjustment. This is consistent with the fact that large-scale weather maps are analyzed in terms of geopotential height, and the winds are sometimes inferred from that field. For much smaller scales, the winds change little when there is an imbalance because the gravity waves act quickly to adjust the mass. At these scales, we say that the winds dominate the adjustment process. A related important point is that the short periods of the gravity waves on small scales means that adjustments take place quickly, relative to the situation on larger scales where the time scale is that of the inertial period (~17 h at 40° latitude). In tropical latitudes (small $f$), for an imbalance of a given length scale, the windfield changes less than in higher latitudes. For shallow adjustments, for example those that are limited to the boundary layer or lower troposphere, the mass field is more dominant than with deeper adjustments.

As mentioned earlier, the geostrophic-adjustment process has many implications for model initialization and data assimilation. First, the model initial conditions need to be in a realistic state of balance, or the resulting excited gravity waves can mask the meteorological pressure field. As an example, Fig. 6.15a shows the evolution of the surface pressure at a point, after a well-balanced and a poorly balanced initialization of a LAM. For the poorly balanced initial conditions, the amplitude of the gravity waves created by the adjustment decreases with time, because the waves have propagated away from the area and possibly been damped by the model, but the sea-level pressure prediction for the first 6–12 h would have been relatively unusable by a forecaster. Also shown in the figure is the

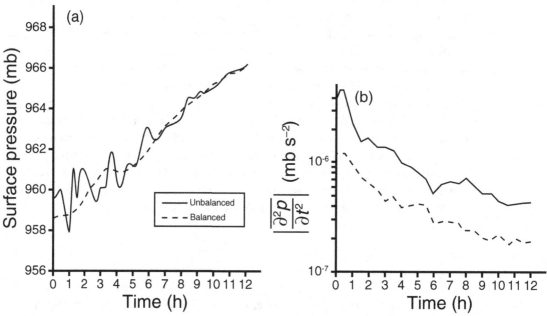

**Fig. 6.15**  Example of the model-simulated surface pressure at a grid point during the first 12 h of a LAM simulation, based on well-balanced and poorly balanced initial conditions (a). Also shown (b) is the computational-domain average of the absolute value of the second time derivative of the surface pressure, a measure of the intensity of inertia–gravity wave activity, for two LAM initializations with different degrees of initial imbalances. Part (b) is adapted from Tarbell *et al.* (1981).

computational-domain average of the absolute value of the second time derivative of the surface pressure, a measure of the intensity of inertia–gravity-wave noise, for two initializations with different degrees of initial imbalances. In both cases, the inertia–gravity-wave intensity decreases by a factor of 5–10 during the first 12 h of the integration, but there is still a residual benefit after 12 h of the better-balanced initial conditions. Ballish *et al.* (1992) show similar plots for a global model, where the unbalanced initial conditions led to high-frequency aphysical surface-pressure oscillations during the first 12 h, with a change of over 5 hPa in 2 h. However, in that study the use of one type of NNMI (see Section 6.10.3 below) to produce a balance filtered the gravity waves too effectively, such that the real semi-diurnal tidal oscillations were removed during the first 24 h of the simulation. An alternative NNMI approach removed the large-amplitude spurious waves, but retained the tidal oscillations.

Knowledge of the adjustment process can also inform decisions about the variables that need to be better observed. For example, because of the dominance of the windfield during the adjustment in the tropics (even though the atmosphere is less geostrophic in those latitudes), the model initialization should emphasize the use of wind observations. Information from pressure observations will be lost in the adjustment. Similarly, at all latitudes, on the scales where the winds dominate the adjustment, assimilated mass-field information can be viewed as redundant.

The concept of adjustment can be viewed in the context of the errors in the initial fields. If the observations and the analysis system are perfect, the generated model initial conditions will be in perfect balance relative to the model equations, and there will be no artificial adjustment. If only the winds have errors, and the winds dominate the adjustment, error will be induced in the mass field during the adjustment process, and vice versa. This means that there should be an interconsistency or compatibility of initial-condition mass-field and windfield errors on the scale of the motion being studied. This concept of initial-condition-error interconsistency for large-scale motions was discussed extensively as part of the Global Atmospheric Research Program (Jastrow and Halem 1970). In the context of data assimilation, the idea is that assimilated observations should have error consistency. The problem of error inconsistency exists because of the exchange of error-related energy through dynamic adjustment. A convenient way of demonstrating the transfer of errors among model variables is through the use of a stochastic-dynamic model (Fleming 1971a,b). Here, the statistical moments of the dependent variables are explicitly predicted with the model equations. Figure 6.16 illustrates a simulation with a one-dimensional $(x)$, spectral, stochastic-dynamic, shallow-fluid model, where the smallest resolved wave had a length of 1250 km. In this experiment, there was no initial error in the $u$ component, but a spatially uniform standard error of 23 m was defined in the initial height field. The plot shows the temporal evolution of the grid-average second moment of the fluid depth for different spatially uniform initial errors in $v$. When there was no initial error in $u$ or $v$, the $h$ error decreased with time because the perfectly known $v$ component caused an adjustment of $h$ toward a compatible value. For larger initial $v$ errors, there was a greater consistency in the initial mass and wind errors, and there was less net change in $\bar{\sigma}_h$. These results are independent of the first moments that are specified for the variables.

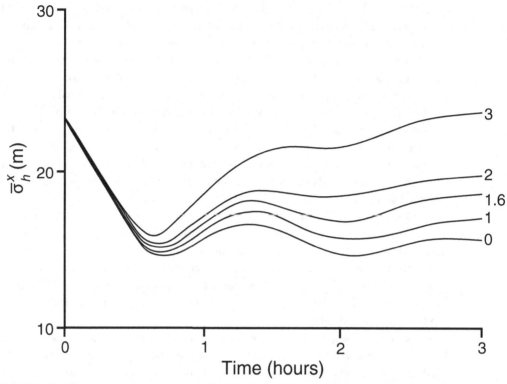

Adjustment of the uncertainty in the fluid depth in a one-dimensional ($x$), spectral, stochastic-dynamic, shallow-fluid model, where the smallest resolved wave had a length of 1250 km. A spatially uniform standard error of 23 m was defined in the initial height field. The different curves are the simulated height-field errors resulting from the use of different initial standard errors (m s$^{-1}$) in the $v$ wind component (curve labels on the right).

## 6.10.2  Preforecast integration of the model equations for achieving a dynamic balance

The objective of preforecast integrations has been mentioned before in the context of allowing ageostrophic circulations, related to precipitation and other processes, to spin up before the initial time of a forecast. Observations can also be assimilated during this period. In addition, initial imbalances can be reconciled during this preforecast integration, if there is some way to eliminate the inertia–gravity waves that are produced. Figure 6.13a is an example of a preforecast integration during which observations are assimilated, ageostrophic circulations spin up, and initial imbalances are reconciled. Another approach is to utilize a damping differencing scheme, such as the Euler-backward method described in Chapter 3, during a backward–forward integration of the model equations (e.g., Nitta and Hovermale 1969). A version of the forecast model that includes only the reversible processes is used; e.g., precipitation and explicit diffusion terms are removed. As the reversible model is integrated backward and forward by one or more time steps, inertia–gravity waves are generated by the adjustment process, and they are damped by the differencing scheme. At the end of each forward–backward cycle, the better observed of the wind or mass field can be recovered. This effectively forces the more-poorly observed field to adjust to the better-observed one. See Figure 6.17 for a schematic of the process. For additional discussion of this type of initialization method, see Fox-Rabinovitz and Gross (1993), Fox-Rabinovitz (1996), and Kalnay (2003).

### 6.10.3  The use of diagnostic relationships for achieving a balance

It is important to recognize that the use of a hypothetical method that provides a balance between the gridded initial wind and mass fields that is perfect with respect to the real atmosphere, will nevertheless result in the generation of gravity-wave noise in the model. This is because the model obviously represents a numerical approximation of the real atmosphere, so it has its own unique balance that is a function of the numerical methods and their associated truncation error. Thus, any initialization method, whether it is static or dynamic, should employ equations that use numerical approximations that are similar to those in the forecast model. Otherwise, the truncation error inconsistency between the forecast model and the initialization method will be a source of inertia–gravity wave noise in the model solution.

Early models used simple large-scale balance relationships to define somewhat compatible mass and wind fields in the initial conditions. The earliest and most primitive

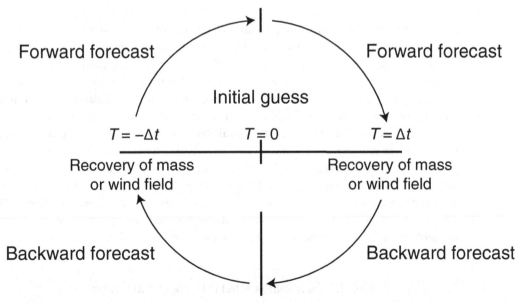

**Fig. 6.17**  Schematic of a model-initialization method that utilizes a backward–forward integration, with a reversible version of the model, to permit the geostrophic-adjustment process to be completed before the initial time of a forecast. The integration is performed with a damping differencing scheme. Adapted from Nitta and Hovermale (1969).

approach was to use the geostrophic relationship. More-complete balances were achieved through other diagnostic equations such as a combination of a divergent balance equation (see Holton 2004) and a vertical–velocity equation (e.g., Tarbell *et al.* 1981).

Another method of diagnosing a balanced set of initial conditions is the previously mentioned NNMI, introduced by Machenhauer (1977) and Baer and Tribbia (1977). It is applied after the analysis step (e.g., that uses OI). As the name implies, it requires the determination of a model's "normal modes" (i.e., solutions of a linearized version of the model equations) as a first step, and then the high-frequency (inertia–gravity waves) and low-frequency (quasi-geostrophic) components of the model input data are separated. The high-frequency modes are assumed to have no meteorological significance, and are removed. This method has been used in the initialization of many operational modeling systems.

Kalnay (2003) summarizes a few shortcomings of the standard NNMI. One is that physically meaningful fast modes are removed with the rest. And, the importance of diabatic processes in the tropics led to the need for a diabatic NNMI (Wergen 1988). Ballish *et al.* (1992) describe a so-called incremental NNMI procedure in which the process is applied to analysis increments rather than to the complete analysis fields. This procedure substantially reduces the aforementioned problems with NNMI, as well as others. See Daley (1991) for additional information about NNMI.

## 6.11 Advanced data-assimilation methods

With the OI, 3DVAR, and SC sequential methods of data assimilation, the observations employed are made at or near the time of the analysis, and the process is repeated at regular intervals that are defined by the update cycle (e.g., every 6, 12, 24 h). In each case, the model is used to propagate the observations and background information from one analysis to the next. Unfortunately, many types of observations are not available at regular intervals. These so-called asynoptic observations are abundant, and are provided by satellites, aircraft, radars, etc. But, unless they are available near one of the standard initialization times, they are not very useful with sequential methods. Of course it is possible to perform some sort of temporal adjustment for off-time observations, but this is not a very satisfying process. And, with the above methods the background-error covariance matrix remains the same throughout a simulation, as if the forecast errors were statistically stationary. The solution is thus to employ a data-assimilation method that can use observations that are available at any arbitrary time, and one for which the background-error covariance evolves during the forecast.

### 6.11.1 Four-dimensional variational initialization

The method of four-dimensional variational (4DVAR) assimilation is a generalization of 3DVAR, to allow inclusion of observations that are distributed in time within an interval $(t_0, t_n)$, where the subscript 0 denotes observations at the initial time of the assimilation period and the subscript $n$ corresponds to the time of the last ($n$-th) observation that is assimilated. The following cost function must be minimized, where the control variable is $x(t_0)$, the model state vector at the initial time of the forecast:

$$J[x(t_0)] = \frac{1}{2}[x(t_0) - x_b(t_0)]^T B_0^{-1}[x(t_0) - x_b(t_0)] + \frac{1}{2}\sum_{i=0}^{n}[H(x_i) - y_i]R_i^{-1}[H(x_i) - y_i].$$

(6.25)

The summation is over the number of observations, $n$. The zero-th observation applies at the beginning of the assimilation window. Note that, if observations are only available at a single time $t_0$, the cost function is the same as that used for 3DVAR (Eq. 6.24). That is, observations for $t > 0$ appear as additional penalty terms. This minimization problem is subject to the strong constraint that the sequence of model states, $x_i$, represents solutions to the model equations. That is

$$\forall i, \quad x_i = M_{0 \to i}(x),$$

where $M_{0 \to i}$ is a model forecast operator that is applied from $t = 0$ to the time of the last observation. The 4DVAR process is illustrated in Fig. 6.18, where the ordinate is the model state vector $x$ and the abscissa is time. With observations distributed throughout the assimilation period, the objective of the 4DVAR process is to estimate the state vector $x_a$ that produces a model solution $M$ that minimizes the cost function that has terms that (1) represent the distance to the background (the previous forecast) at the beginning of the

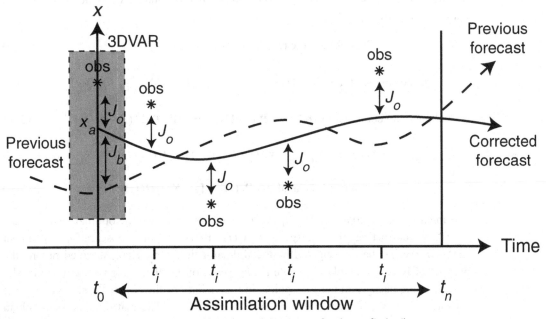

Fig. 6.18    Schematic illustrating the 4DVAR data-assimilation process. See the text for details.

interval and (2) are based on the observational increment computed with respect to the model solution at the time of the observations. That is, the process defines an initial condition that produces a forecast that best fits the observations within the assimilation window. In practice, if a forecast model is run with a 6-h update cycle, the assimilation window would extend from the time of the previous initialization to the time of the current one.

To solve the minimization problem, we must differentiate Eq. 6.25 with respect to the control variable, and solve for $x$ at the minimum. For $J = J_b + J_o$, the differentiation of $J_b$ is identical to the process for 3DVAR. However, the evaluation of $J_o$ and $\nabla J_o$ requires a model integration from $t_0 \rightarrow t_n$, as well as an integration of the adjoint model defined as the transpose of the model operator $\mathbf{M}_i$ ($\mathbf{M}_i^T$). Note that $\mathbf{M}$ is the tangent linear version of $M$. Constructing this adjoint model from the forecast model is a complex process, and, furthermore, the adjoint model must be maintained (bug fixes, improvements) in parallel with the forecast model. See Kalnay (2003) for mathematical details and additional references.

## 6.11.2  Extended Kalman filtering

A specific implementation of the least-squares analysis method is called the Extended Kalman Filter (EKF, Ghil and Malanotte-Rizzolli 1991, Bouttier 1994, Kalnay 2003, Hamill 2006). Equations 6.26a–e below summarize the process. As in OI, the EKF is based on the least-squares analysis method applied in the framework of sequential data assimilation, where each background is produced by a forecast that is initiated from the previous analysis. However, now the background error-covariance matrix is time dependent. The background (i.e., forecast) and analysis error-covariance matrices are now represented as $\mathbf{P}_f$ and $\mathbf{P}_a$, respectively. Compare the constant background error covariance $\mathbf{B}$ in the weight matrix used for OI (Eq. 6.21) with the time-dependent $\mathbf{P}_f(t)$ below in the gain matrix defined by Eq. 6.26c.

$$\text{State forecast } x_f(t+1) = M_{t \rightarrow t+1}(x_a(t)) \tag{6.26a}$$

$$\text{Error-covariance forecast } \mathbf{P}_f(t+1) = \mathbf{M}_{t \rightarrow t+1}\mathbf{P}_a(t)\mathbf{M}_{t \rightarrow t+1}^T + \mathbf{Q}(t) \tag{6.26b}$$

$$\text{Kalman-gain computation } \mathbf{K}(t) = \mathbf{P}_f(t)\mathbf{H}^T(t)[\mathbf{H}(t)\mathbf{P}_f(t)\mathbf{H}^T(t) + \mathbf{R}(t)]^{-1} \tag{6.26c}$$

$$\text{State analysis } x_a(t) = x_f(t) + \mathbf{K}(t)[y(t) - H(t)x_f(t)] \tag{6.26d}$$

$$\text{Error covariance of the analysis } \mathbf{P}_a(t) = [\mathbf{I} - \mathbf{K}(t)H(t)]\mathbf{P}_f(t) \tag{6.26e}$$

Similar to Eqs. 6.20 and 6.21, Eqs. 6.26c and 6.26d estimate the optimal analysis state $x_a(t)$ by correcting the background $x_f(t)$ based on the observation increment $y(t) - H(t)x_f(t)$ that is weighted by the Kalman-gain matrix $\mathbf{K}(t)$. As noted before, the purpose of $\mathbf{K}$ is to spatially distribute the influence of the observation increment in order to correct the background at grid points in the vicinity of the observation. And, $H$ is the forward operator that maps the state to the observations. The matrix $\mathbf{H}$ is the Jacobian matrix of $H$, such that $\mathbf{H} = \partial H / \partial \mathbf{x}$. Equation 6.26e updates the background error covariance to reflect the reduced uncertainty that results from the assimilation of the observations. Here, $\mathbf{I}$ is the identity matrix. Equations 6.26a and 6.26b propagate the state vector and error-covariance vector forward to the time when observations are next available. The matrix $M$ is the nonlinear model-forecast operator that integrates forward in time from the analysis vector $\boldsymbol{x}_a(t)$ (initial conditions) to the time of the next analysis, where the model forecast $\boldsymbol{x}_f(t+1)$ will become the background. The matrix $\mathbf{M}$ is the Jacobian matrix of $M$, where $\mathbf{M} = \partial M/\partial \mathbf{x}$ and $\mathbf{M}^{\mathrm{T}}$ is its adjoint. The matrix $\mathbf{Q}$ is the covariance of model errors that accumulate during the update interval. Figure 6.19 shows a schematic of how Eqs. 6.26 are solved. See Kalnay (2003) and Hamill (2006) for the assumptions that are involved in the use of this method, and LeDimet and Talagrand (1986) and Lacarra and Talagrand (1988) for additional information about the mathematics.

A major benefit of the EKF relative to 3DVAR is that the forecast, or background, error-covariance matrix is explicitly advanced using the model itself, evolving during the forecast. A few of the distinctions between the EKF and 4DVAR methods are summarized as follows.

- The EKF explicitly evolves the covariance matrix, whereas the covariance evolution in 4DVAR is implicit.
- Unlike the EKF, 4DVAR assimilation is based on the assumption that the model is perfect (i.e., $\mathbf{Q} = 0$).

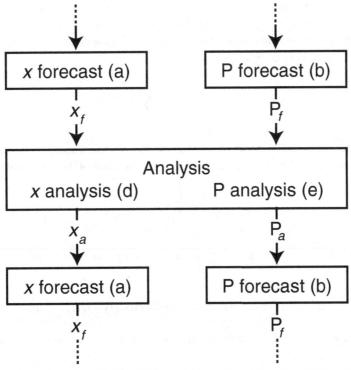

**Fig. 6.19** Schematic showing the organization of the computations for the solution of Eqs. 6.26 in an EKF assimilation. The letters refer to the specific equations in the Eqs. 6.26 series.

- The 4DVAR method can be used operationally in NWP because it is computationally much cheaper than the EKF. In contrast, with current computing resources the EKF is prohibitively expensive to use for all but the very smallest modeling systems.
- The 4DVAR method simultaneously uses all the observations within the update interval, whereas the EKF is a sequential method that assimilates observations that are grouped at the update times. The former situation is preferable.

An additional difficulty with the EKF is the fact that the accuracy of the assimilation depends greatly on the quality of the determination of $\mathbf{Q}$, but this is especially difficult to estimate.

### 6.11.3  Ensemble Kalman filtering

Ensemble-based data-assimilation methods are sequential in the sense that there is an ensemble of parallel short forecast and analysis steps. With the Ensemble Kalman Filter (EnKF) method, an ensemble of analyses is produced for a given time using (1) backgrounds that are produced by an ensemble of forecasts and (2) observations that have been perturbed by the addition of random noise that is drawn from a distribution of observation errors. The next ensemble of backgrounds is produced by running a short forecast from each of the members of the ensemble of analyses. The analyses are produced using the Kalman filter method described earlier, where each background is updated with a slightly different realization of the observations because of the addition of errors. For each cycle, the ensemble of backgrounds provides an estimate of the covariance matrix of the background error, and the ensemble of analyses allows the calculation of the covariance matrix of the analysis error. Specifically,

$$\overline{x_f} = \frac{1}{K} \sum_{i=1}^{K} x_{f,i} \tag{6.27}$$

defines the ensemble mean of the forecast state vector, where $K$ is the number of ensemble members, and the following represents the covariance of the sample of $x_f$,

$$\hat{\mathbf{P}}_f = \frac{1}{K-1} \sum_{i=1}^{K} (x_{f,i} - \overline{x_f})(x_{f,i} - \overline{x_f})^T, \tag{6.28}$$

where $\hat{\mathbf{P}}_f$ is an estimate of $\mathbf{P}_f$ from a finite ensemble. Similar to the state analysis in Eq. 6.26d,

$$x_a(t) = x_f(t) + \hat{\mathbf{K}}(t)[y_i(t) - H(t)x_f(t)],$$

where $y_i = y + y_i'$ are perturbed observations, and the Kalman gain matrix (Eq. 6.26c) is now

$$\hat{\mathbf{K}}(t) = \hat{\mathbf{P}}_f(t)\mathbf{H}^T(t)[\mathbf{H}(t)\hat{\mathbf{P}}_f(t)\mathbf{H}^T(t) + \mathbf{R}(t)]^{-1}. \tag{6.29}$$

Methods for simplifying and parallelizing the application of the EnKF are summarized in Hamill (2006), and include efficient ways of obtaining the Kalman gain without explicitly computing the background-error covariance matrix.

Figure 6.20 shows a schematic of the EnKF process, where time progresses downward from the top. An ensemble of analyses serves as the initial conditions for an ensemble of forecasts, where the forecast length is consistent with what is typically used for sequential initialization methods – e.g., 6 h. The ensemble of forecasts of $x$ at $t + 1$ is used to calculate the forecast, or background, error covariance using Eqs. 6.27 and 6.28. The resulting vector is then used in the Kalman gain calculation in Eq. 6.29, and this gain matrix is applied to the analysis increment to obtain the optimal correction to $x_a$ in order to obtain $x_f$. After another ensemble forecast, the Kalman filter is applied again, and the process continues.

The EnKF method unifies data assimilation and ensemble forecasting. Using a Monte Carlo approach, the ensemble of forecasts provides a sample of the relationship between observations and state variables, from which the forecast-error covariance can be calculated. The statistics derived from this sample serve as the basis for an ensemble of analysis from which the next ensemble of forecasts is initiated. One of the strengths of the EnKF is

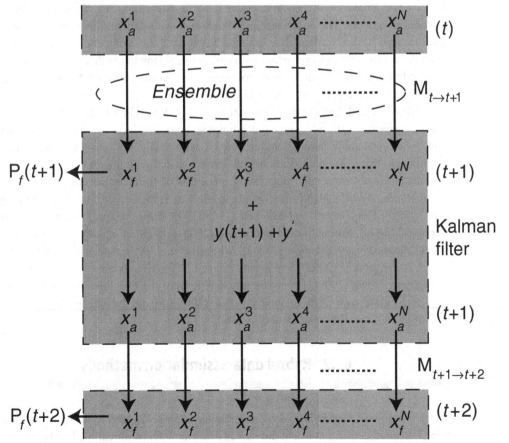

**Fig. 6.20**   Schematic showing the basis of the EnKF data-assimilation method. See the text for discussion.

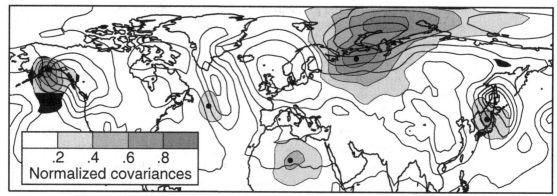

An example of the background-error covariance (gray shading) of sea-level pressure (solid lines) around the five indicated observation points (black dots). Adapted from Hamill (2006), based on experiments conducted by Whitaker *et al.* (2004).

the fact that the background-error covariance varies with location and time. Figure 6.21 shows an example of the covariance field for a particular case from a global-data-assimilation system. The covariance patterns are shown relative to five different observation locations in the Northern Hemisphere, based on a 100-member ensemble. The background-error magnitudes and patterns vary from one region to another, with greater values surrounding the grid points in northern Russia and to the south of Alaska. The pattern around the former location is especially complex.

The 4DVAR and EnKF methods are equivalently demanding computationally, but the EnKF approach has the advantage that adjoint and tangent-linear models are not required. And, where ensemble simulations are used operationally for reasons aside from data assimilation, the added computational burden associated with the EnKF approach to data assimilation is not especially great. See Lorenc (2003) for a further comparison of these two methods. Additional information about the EnKF method can be found in Burgers *et al.* (1998), Houtekamer and Mitchell (1998, 1999, 2001), Hamill and Snyder (2000), Keppenne (2000), Mitchell and Houtekamer (2000), Hamill *et al.* (2001), Heemink *et al.* (2001), Keppenne and Rienecker (2002), Mitchell *et al.* (2002), Anderson (2003), Evensen (2003, 2007), Lorenc (2003), Snyder and Zhang (2003), Houtekamer *et al.* (2005), Hamill (2006), and Zheng (2009). For recent practical tests of the EnKF method in realistic settings, see Fujita *et al.* (2007), Bonavita *et al.* (2008), Meng and Zhang (2008), Torn and Hakim (2008), and Houtekamer *et al.* (2009).

## 6.12  Hybrid data-assimilation methods

There are quite a few hybrid data-assimilation approaches, drawn from the aforementioned methods, that have had historically distinct development paths. For example, Hamill and Snyder (2000) suggest a hybrid of the EnKF and 3DVAR methods where the background-error covariance is a linear combination of the typically constant, isotropic, and homogeneous 3DVAR covariance (see Section 6.8) and the variable EnKF covariance. Specifically,

$$\mathbf{P}_f^{hybrid} = (1-\alpha)\mathbf{P}_f + \alpha\mathbf{B},$$

where $\alpha$ is a tunable parameter that varies from 0.0 to 1.0. The objective here is to compensate for the relatively small sample of ensemble members that are used to calculate $\mathbf{P}_f$ by also incorporating information that is represented in $\mathbf{B}$. Hamill and Snyder (2000) obtained the best results for $0.1 < \alpha < 0.4$.

Another hybrid method combines Newtonian relaxation and the adjoint of 4DVAR, the two approaches that are able to assimilate asynoptic observations at the actual measurement time. Recall that the adjoint equations compute the gradient of a cost function with respect to a control variable. In applications of the adjoint method for model initialization, such as in the 3DVAR and 4DVAR methods, the control variable is the model initial state. For model-parameter estimation, a vector of model parameters is the control variable. Examples of this hybrid approach include Zou *et al.* (1992) and Stauffer and Bao (1993), who employed the adjoint method for an optimization of analysis-nudging coefficients, which were the control variables.

Lastly, the 4DVAR and EnKF methods have been combined by Zhang *et al.* (2009). Using a somewhat idealized experimental setting, the EnKF-4DVAR system outperformed both the EnKF and 4DVAR methods.

## 6.13  Initialization with idealized conditions

For a variety of reasons, it is useful to be able to perform model simulations based on idealized (or synthetic) initial conditions. Even though this is discussed in Chapter 10, in the context of experimental designs of modeling studies, it will be mentioned here as well. The term idealized means that the initial conditions are not based on observations, but rather on a conceptual model of an atmospheric state. In general, this state is described by an analytic function that represents the dependent variables, which are defined at grid points. The general motivation for the use of synthetic initial conditions is that a single process or phenomenon can be isolated, in contrast to real-data initializations where there exist the inevitable complexities of the real atmosphere with many processes and scales. Specific purposes for using synthetic initial conditions include the following.

- *Instruction* – It is straightforward to demonstrate the effects of model numerics on a known solution. An example is a comparison of the correct phase speed with that produced by a given model solver. Or, simple experiments can be performed, for example of the geostrophic-adjustment process.
- *Testing a model for the existence of code errors* – There are some simple phenomena for which analytic solutions exist, and the model solutions can be compared with them to

assess model performance. Or a new model configuration can be tested against the solution that results from an older well-tested version of the model.

- *Dynamic-solver evaluation* – The same simple case can be run for many space and time scales.

Some modeling systems include software that allows the user to run a variety of preconstructed, idealized test cases. For example, the WRF system includes the following cases: flow over a bell-shaped mountain, a two-dimensional squall line, a three-dimensional supercell thunderstorm, a three-dimensional baroclinic wave, a two-dimensional gravity wave, a three-dimensional large-eddy-simulation case, and a two-dimensional sea breeze. See Chuang and Sousounis (2000) for an example of how idealized initial conditions can be implemented in a LAM.

### SUGGESTED GENERAL REFERENCES FOR FURTHER READING

Cohn, S. E. (1997). An introduction to estimation theory. *J. Meteor. Soc. Japan*, **75**, 257–288.

Daley, R. (1991). *Atmospheric Data Analysis*. Cambridge, UK: Cambridge University Press.

Daley, R. (1997). Atmospheric data assimilation. *J. Meteor. Soc. Japan*, **75**, 319–329.

Evensen, G. (2007). *Data Assimilation: The Ensemble Kalman Filter*. Berlin, Germany: Springer.

Hamill, T. M. (2006). Ensemble-based atmospheric data assimilation. In *Predictability of Weather and Climate*, T. Palmer and R. Hagedorn (eds.). Cambridge, UK: Cambridge University Press.

Kalnay, E. (2003). *Atmospheric Modeling, Data Assimilation and Predictability*. Cambridge, UK: Cambridge University Press.

Lewis, J. M., A. S. Lakshmivarahan, and S. K. Dhall (2006). *Dynamic Data Assimilation: A Least Squares Approach*. Cambridge, UK: Cambridge University Press.

Talagrand, O. (1997). Assimilation of observations, an introduction. *J. Meteor. Soc. Japan*, **75**, Special Issue 1B, 191–209.

### PROBLEMS AND EXERCISES

1. What processes can damp inertia–gravity waves that are generated by a model initialization?
2. What variables, in addition to the second time derivative of the surface pressure, might be used to diagnose the intensity of inertia–gravity waves associated with imbalances in the initial conditions?
3. Describe the relationship between the need for spectral nudging and the domain-size, in relaxation-based continuous data-assimilation systems.
4. Explain how a bad meteorological observation can have its negative influence spread over a large area of the computational domain.
5. Define what is meant by the expression Monte Carlo approach.

6. Conceptually and mathematically relate the SC analysis method (Section 6.6) to the statistical approaches to data assimilation (Section 6.5).

7. Explain possible ways in which near-surface observations can be used to infer model initial conditions for the boundary layer and lower troposphere. How could vertical mixing during the early part of the model forecast cause the value of the observations to be lost?

8. Derive Eqs. 6.9 and 6.10.

# 7 Ensemble methods

## 7.1 Background

As we have seen in previous chapters, there is a variety of generally unavoidable sources of model error, including

- initial conditions,
- lateral-boundary conditions for LAMs,
- land/water-surface conditions,
- numerical approximations used in the dynamical core, and
- parameterizations of physical processes.

Each of these input data sets or modeling approaches introduces some error in the modeling process, and ensemble prediction involves performing parallel forecasts or simulations using different arbitrary choices for the above imperfect data or methods. The objective of defining the different conditions for each model integration is to sample the uncertainty space associated with the modeling process in order to define how this uncertainty projects onto the uncertainty in the forecasts. As a preliminary example of the sensitivity of model forecasts to the above factors, Fig. 7.1 illustrates an ensemble of 5-day track predictions for hurricane Katrina in 2005. The forecasts are based on the ECMWF ensemble prediction system. The tracks are strongly dependent on the specific errors in the input observations as well as the model configurations employed.

An ensemble of forecasts is more useful than an individual, deterministic forecast for the following reasons.

- The mean of the ensemble of forecasts is generally more accurate than the forecast from an individual ensemble member, when the statistics are computed over a number of forecasts.
- The difference (spread, variance) among the ensemble members can be an indication of the flow-dependent quantitative uncertainty in the ensemble-mean forecast, given a proper calibration.
- The Probability Distribution (or Density) Function (PDF) of the frequency distribution of a variable can provide information about extreme events, which is especially useful information from a practical standpoint (e.g., issuing weather warnings).
- The quantitative probabilistic products can be more effectively employed in decision-support software systems.

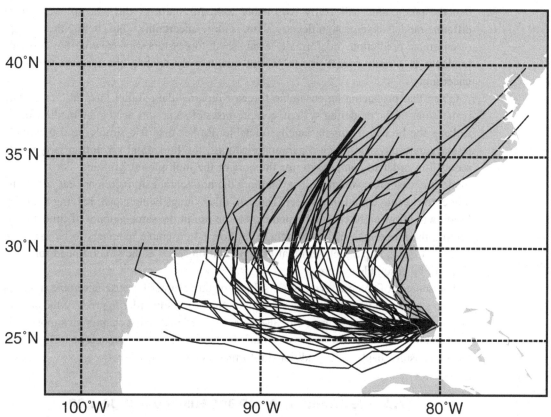

**Fig. 7.1**  An ensemble of track predictions for hurricane Katrina, initialized at 0000 UTC 26 August 2005, from the ECMWF ensemble-prediction system. The heavy line is the track forecasted by the ECMWF deterministic system. Adapted from Leutbecher and Palmer (2008).

The availability of stochastic forecast products is clearly a great advantage relative to the situation with a deterministic modeling system where a single realization of the future state of the atmosphere is produced, and the forecaster must guess at its veracity.

It should not be surprising that using ensembles of model simulations has led to improved forecast skill. Since the early 1960s it has been known that combining different forecasts from individual forecasters produces a group-mean probability forecast that is superior to the single probability forecast from the most skillful forecaster (Sanders 1963). These findings were confirmed through later studies by Sanders (1973), Bosart (1975), and Gyakum (1986). The recognition of the benefits of this statistical synthesis of human predictions has contributed to a similar process being applied to model predictions.

The potential of ensemble methods is also reflected in how forecasters have used model products for the last few decades. It has been well known by forecasters that when all available models are predicting a similar outcome, the probability is generally high that the predictions will verify reasonably well. In contrast, when the solutions from different models diverge significantly, there is more uncertainty. Thus, before the concept of ensemble prediction was formally established, forecasters were treating the available products as a multi-model ensemble and qualitatively relating the forecast spread to uncertainty.

Given that producing an ensemble forecast requires the parallel integration of many realizations of the modeling system, compromises of some sort need to be made in order to keep the process computationally tractable. Rather than use simpler and inevitably less-accurate physical-process parameterizations, the horizontal resolution is typically reduced in order to compensate for the cost of the multiple integrations. For the same model forecast area, we know that doubling the horizontal grid increment can allow the model to execute eight times faster. Thus, all other things being equal, halving the resolution will allow eight ensemble members to be run in the same amount of time. Quadrupling the horizontal grid increment will allow 64 ensemble members to be used. An issue that exists in parallel with that of computation speed is the general need for more memory by ensemble systems.

The sources of forecast error are a subject of this chapter because ensemble methods seek to sample the uncertainty associated with the sources in order to produce the ensemble. These same sources of error will also be considered in the next chapter on atmospheric predictability because that discussion must be based on the same concepts. The reader should consult Chapter 8 for additional information about that subject.

## 7.2  The ensemble mean and ensemble dispersion

### 7.2.1  The ensemble mean

One of the products of an ensemble-prediction system is the average of the members of the ensemble, which represents a forecast that can be interpreted in the same way as a deterministic (nonensemble) product. This ensemble mean, defined at the initial time or at any forecast lead time, is calculated simply by averaging together the gridded fields of the dependent variables from the ensemble members. The mean is typically more accurate than any arbitrarily chosen forecast from an individual member of the ensemble, when averaged over a number of forecasts. The averaging of the ensemble members appears to produce a nonlinear filtering that causes the unpredictable (random) aspects of the forecast to cancel each other, whereas the aspects of the forecasts on which the models agree are not removed in the averaging. Maps of the ensemble-average meteorological fields tend to be smoother than those from the individual members, especially for longer forecast lead times after the individual model solutions have diverged to a larger degree. Palmer (1993) has suggested that ensemble averaging will improve the forecast only up to the time when there is a change in meteorological regime – that is, when there is a bifurcation in the solutions of the members. For example, Fig. 7.2 shows a schematic of the model-state trajectories (dashed lines), for

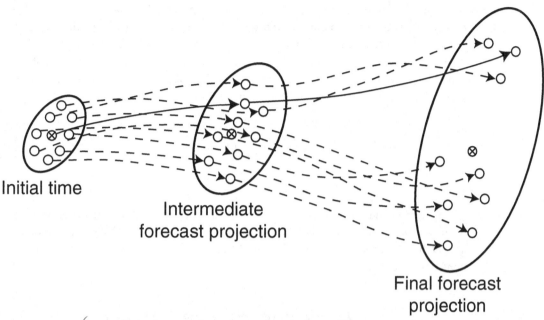

Initial time

Intermediate
forecast projection

Final forecast
projection

**Fig. 7.2** Schematic of model-state trajectories for simulations from an eight-member ensemble initialized from perturbed initial conditions. See the text for details. Adapted from Wilks (2006).

a two-dimensional phase space.[1] The initial conditions for the ensemble members are shown with open circles, as are the solutions at two times during the forecast. At each time, the ensemble mean is defined with an "x". In this example, the eight-member ensemble is constructed by perturbing a control set of initial conditions, and thus the initial states are defined by different phase-space coordinates. At the intermediate forecast time, the model solutions differ more than they did at the initial time, but the error growth has been somewhat linear. Later, at the final time, the trajectories of two of the forecasts have diverged from the trajectories of the other six forecasts; a regime change has taken place. Such a bifurcation in a real ensemble forecast might correspond to some forecasts defining rapid cyclogenesis while others are producing cyclolysis (see Mullen and Baumhefner 1989). In the illustration of Fig. 7.2, the ensemble mean after the bifurcation corresponds with neither branch of the solution, and thus might not represent an especially accurate forecast. An example of a bifurcation in a real forecast is seen later in Fig. 7.13, where the ensemble members tend to be grouped into two different patterns of midtropospheric troughs.

---

[1] Phase space has a dimension corresponding to each dependent variable in the system, and the coordinates in the phase space define the value of each variable. Thus, a trajectory in phase space represents the temporal change in the state of the system. Sometimes, spatial independent variables may also be dimensions of the phase space.

The above figure also illustrates that the time-dependent behavior of a forecast that is initiated from the ensemble mean (the solid arrow) is different from the time-dependent behavior of the ensemble mean itself. The reason for this can be understood by recognizing that an atmospheric model is a highly nonlinear function that transforms a set of initial conditions into a set of forecast conditions (Wilks 2006). For a nonlinear function $f(x)$, with $x$ the dependent variables and $n$ the number of ensemble members,

$$\frac{1}{n}\sum_{i=1}^{n}f(x_i) \neq f\left(\frac{1}{n}\sum_{i=1}^{n}x_i\right),$$     (7.1)

where the right side corresponds to the nonlinear function (the forecast model) applied to the mean and the left side is the average of the forecast fields. Stated another way, on average the best forecast does not result from the use of the best estimate of the initial conditions (the ensemble mean).

## 7.2.2  Ensemble dispersion, spread, or variance

Because of nonlinear interactions in the fluid, and interactions among the different sources of error in the modeling system, errors tend to grow during an integration. In fact, the errors will continue to grow until the forecast and the true state of the atmosphere (e.g., depicted by an objective analysis) will be as dissimilar as two randomly chosen observed states of the atmosphere. In a deterministic forecast setting we have no way of defining the future error growth. So, we employ an ensemble approach to estimate it by perturbing different aspects of the modeling system (the initial conditions or the model) and interpreting the degree to which the model solutions diverge. This divergence in the solutions, called the dispersion of the ensemble, can be related to the uncertainty in the ensemble mean and is an important component of the forecast. Figure 7.3 displays two arbitrarily chosen ECMWF ensemble predictions of temperature for London, UK and illustrates that the sensitivity of the model atmosphere to uncertainties in the inputs can be very flow (meteorological-situation or -regime) dependent. Clearly the ensemble members have much greater spread on one day than the other.

This relationship between the ensemble dispersion and the error of the ensemble mean is sometimes called the spread–skill relationship. It can be quantified by associating the variance in the ensemble members about their mean with the accuracy of the ensemble mean itself, for each of a large number of ensemble forecasts. Any of the standard metrics discussed in Chapter 9, such as the Root-Mean-Square Error (RMSE), can be used to quantify the accuracy of the ensemble mean. Calibration of the ensemble (see Section 7.5) is necessary in order to quantitatively relate the spread to the uncertainty. Discussions of the ensemble spread–error relationship can be found in Grimit and Mass (2007), and in references cited therein.

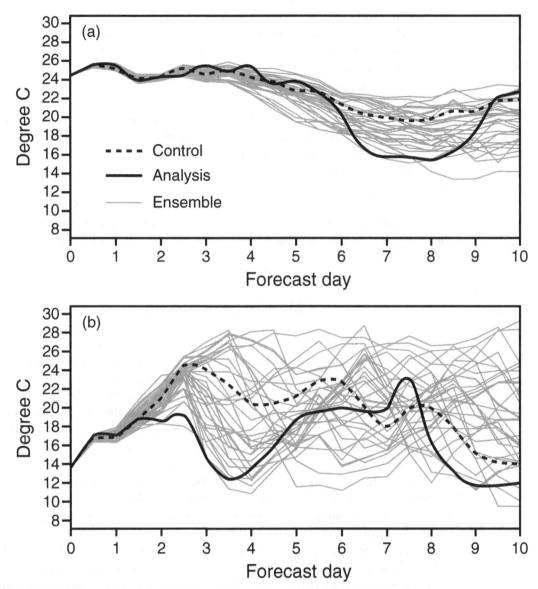

**Fig. 7.3** Two ensemble forecasts of 2-m AGL temperature for London, UK, based on ECMWF forecasts separated by 1 year. Note the large difference in ensemble spread. Based on Buizza (2001).

## 7.3 Sources of uncertainty, and the definition of ensemble members

The list in Section 7.1 provides a brief summary of the sources of forecast uncertainty. This section will offer additional discussion of this subject, as well as of how the uncertainties can be represented in an ensemble-prediction system. The accepted terminology is that forecast error can be divided into initial-condition error and model error. The latter refers to all aspects of the modeling system other than initial conditions.

There are two general approaches for defining ensemble members. In one, the normal configuration of a model is used for a control simulation, and then those conditions (initial conditions or model specifications) are perturbed to create the remainder of the ensemble members. It is reasonable to assume that the control simulation will be the most accurate because it has been optimized in terms of the tuning of the model physics parameterizations and numerics, and has resulted from the generation of an optimal initial state. The perturbations to this control may be somewhat less skillful. In the other approach, entirely different models are used for the different members of the ensemble. These are called multi-model ensembles.

A number of single-model ensembles, which each employ variations in initial conditions, model physics, etc., can be combined to produce a multi-model ensemble. This is referred to as a multi-model *superensemble* (Krishnamurti *et al.* 1999, Palmer *et al.* 2004).

### 7.3.1 Initial-condition uncertainty

We have seen in Chapter 6 that errors in the observations that are used to define model initial conditions can result from a number of sources. Possible sources include instrument-calibration errors, improper siting of the instrument, representativeness error, and data-transmission errors. In addition, the use of a dynamic-balancing method can introduce errors. Lastly, initial-condition uncertainty cannot be disentangled from the model error because, as we have seen in Chapter 6, modern data-assimilation systems employ the model in various ways. For example, sequential data-assimilation systems use a short model forecast from the previous cycle in order to define the background field for the analysis process. Thus, errors in any aspect of the model formulation will influence the quality of the first-guess field, which in turn will impact the initial-condition error, especially where observations are sparse. Improving the accuracy of the model can thus provide more-accurate initial conditions.

It is intuitive that the sensitivity of forecasts to the initial conditions is flow dependent. Imagine a synoptic-scale situation that is dominated by a large semi-permanent anticyclone, for example associated with a Mediterranean-type climate in summer. The forecast will be relatively insensitive to the details of the initial-condition error because the situation is dominated by the large-scale planetary forcing associated with the general circulation and the land surface. In contrast, there are situations where the atmosphere is close to an instability threshold and small initial-condition differences can cause the state of the model atmosphere to follow either one path or another at the bifurcation.

Initial-condition error could be created by adding to a mean state a random error that is consistent with the uncertainty in the observations. However, it has been found that simply adding random numbers to initial conditions will create ensemble members that are very similar to each other. This should not be surprising, given the discussion in Chapter 6 about geostrophic adjustment. That is, perturbations that are imposed independently on the mass and windfield variables on small scales will be dispersed as inertia–gravity waves. In contrast, the methods that are used in practice in NWP to generate initial-condition uncertainty in ensembles produce perturbations that have dynamically consistent structures.

Three different approaches are used to define initial-condition uncertainty in ensembles.

- Ensemble-based data assimilation is used to define a sample of initial states, as described in Section 6.11.3. The Kalman filter combines (1) background fields that have been created with an ensemble of forecasts and (2) observations, with their associated errors.
- An approach is based on so-called bred vectors, and samples the dynamically most sensitive modes in the initial conditions. It consists of the following steps.
  1. Random perturbations are added to the dependent variables that define an initial state.
  2. Both the perturbed and unperturbed states are used as initial conditions for model simulations with a duration of 6 h to 24 h.
  3. The two simulated states are subtracted, and the gridded difference field is scaled so that its magnitude is similar to the error in a typical analysis.
  4. The scaled perturbation is added to a new initial-state estimate, and the perturbed and unperturbed initial states are again used for a pair of parallel model simulations.
  5. This process of perturbation growth and rescaling is repeated, where the bred vector is the perturbation that results after a few iterations.

  The bred patterns are different from day to day, and reflect the features with respect to which the ensemble members are diverging most rapidly. See Ehrendorfer (1997), Toth and Kalnay (1993, 1997), and Kalnay (2003) for additional information about the breeding method.
- Singular vectors (Buizza 1997, Ehrendorfer 1997, Molteni *et al.* 1996, Ehrendorfer and Tribbia 1997, Kalnay 2003), also called optimal perturbations, are obtained by using tangent-linear and adjoint models, and define the fastest-growing patterns for the prevailing weather situation of the day. Linear combinations of these patterns, with the magnitudes scaled according to the expected analysis uncertainty, are added to a control analysis to define the ensemble members.

## 7.3.2 Lateral-boundary-condition uncertainty for LAM ensembles

For LAMs, the model solution depends strongly on the LBCs, especially for longer integration times, so errors in the LBCs can contribute significantly to the model error. In a forecast setting, the LBC-related error in the LAM depends on both the error in the forecast from the large-scale model as well as errors introduced by the algorithm used to couple the two grids. If the large-scale model is an ensemble, the individual ensemble members can be used to provide the LBCs for the LAM ensemble members. If the large-scale forecast is not an ensemble, the LBCs need to be perturbed in such a way that the process estimates typical errors from the large-scale model. Thus, the time scale, space scale, and amplitude of the errors need to be estimated, and used in the generation of LAM ensemble members.

### 7.3.3  Surface-boundary-condition uncertainty

Errors in the calculation of the land-surface properties during a forecast can result from initial-condition error or model error.

• The initial conditions for the time-varying land-surface variables are defined by a LDAS, as described in Chapter 5, and errors in the LDAS calculations can result from both errors in the LSM that is the basis for the LDAS and errors in the estimates of the non-time-varying physical properties of the substrate such as the thermal conductivity and heat capacity of the dry substrate, leaf-area index, etc. Thus, as with the atmospheric model, land-surface initial-condition error is intertwined with model error.
• Model error is associated with the LSM that is integrated in parallel with the atmospheric model, as described in Chapter 5. As with the atmospheric models, errors can result from the numerical approximations to the differential equations, as well as from the parameterizations of the physical processes.

This source of error can be represented by defining uncertainties in the parameterizations of the processes, in the estimates of the initial conditions of the time-varying physical properties of the substrate (moisture and temperature profiles), and in the time-invariant physical properties of the substrate (pore space, specific heat and thermal conductivity of the dry substrate). Many studies have evaluated the uncertainty in the atmospheric structure that results from uncertainties in different aspects of the land surface. See Pielke (2001), Sutton *et al.* (2006), and Hacker (2010) for references.

### 7.3.4  Errors in the numerical algorithms

We have seen in Chapter 3 that numerical approximations to space and time derivatives introduce errors, and they, along with the physical-process parameterizations, contribute to the model error. Even though these dynamical-core errors are initially largest on the small scales near the truncation limit of the model, through nonlinear interactions they can affect the scales of mid-latitude high- and low-pressure systems within a couple of days of model integration. The typical way to include in an ensemble the uncertainty associated with the particular properties of the dynamical core is to use entirely different models for different ensemble members. Ensembles constructed in this way are called multi-model ensembles. Because of the limited number of models available for this approach, the size of such an ensemble is limited to perhaps 10 members. A recent approach to including model error in ensembles is the use of stochastic kinetic-energy backscatter methods, which represent upscale propagating energy caused by unresolved subgrid-scale processes (Shutts 2005, Berner *et al.* 2009).

### 7.3.5  Errors in the physical-process parameterizations

As with the above method for representing model error associated with the dynamical core, errors related to physical-process parameterizations can also be represented with a multi-model ensemble. In addition, some modeling systems allow the user to choose from a list of options for each of the parameterizations. Even though some combinations of parameterizations are incompatible with each other, it is possible to create a significant number of ensemble members simply by varying the parameterizations used with a single dynamical core. And, as we have seen in Chapter 4 it is possible to vary uncertain parameters within a particular parameterization in order to create multiple ensemble members. Alternatively, Buizza *et al.* (1999) model random errors associated with physical-process parameterizations by multiplying a dependent-variable's time tendency from the parameterization by a random number that is sampled from a uniform distribution between 0.5 and 1.5. This method increases the ensemble spread and improves the skill of the probabilistic prediction. See Teixeira and Reynolds (2008) for an additional example of the use of a stochastic parameterization. Lastly, there are many studies that compare the use of initial-condition and model-physics uncertainties in ensemble systems (e.g., Stensrud *et al.* 2000, Clark *et al.* 2008a).

### 7.3.6 Multi-model ensembles

The use of multi-model ensembles is appealing when various models are already being routinely run operationally by different modeling centers for either weather or climate prediction. In these situations, the normally daunting challenge of creating a multi-model ensemble prediction is simply a matter of reconciling the outputs to a common grid for quantitative processing. Or, forecasters often view products from all the available models and qualitatively synthesize the information.

Regarding the latter approach, it was mentioned earlier that, for decades, forecasters have related the degree to which forecasts from different models agree to the overall uncertainty in the products. Fritsch *et al.* (2000), Woodcock and Engel (2005), and many others discuss the concept of consensus forecasting, which can involve the synthesis of forecasts made by humans as well as forecasts from different operational models.

Multi-model ensembles have been used especially extensively for seasonal prediction (Feddersen *et al.* 1999, Rajagopalan *et al.* 2002, Stefanova and Krishnamurti 2002, Barnston *et al.* 2003, Palmer *et al.* 2004, Robertson *et al.* 2004, Doblas-Reyes *et al.* 2005, Feddersen and Andersen 2005, Hagedorn *et al.* 2005, Hewitt 2005, Stephenson *et al.* 2005, Krishnamurti *et al.* 2006b), and climate prediction on longer time scales (Section 10.5.4 in Meehl *et al.* 2007, Section 16.1.6 in this text).

In its simplest form, a multi-model ensemble forecast can be produced by simply averaging the individual members using equal weights. However, more-complex methods for combining the model solutions are used, for example as described in Clemen (1989), Robertson *et al.* (2004), and Stephenson *et al.* (2005).

## 7.4 Interpretation and verification of ensemble forecasts

The ensemble mean and the individual members of the ensemble can be evaluated employing standard metrics that are also used for deterministic predictions, while other measures are used for the probabilistic aspects of the forecasts. This section summarizes some of these methods. Additional information on this subject can be found in Wilks (2006) and the Appendix of McCollor and Stull (2008a).

## 7.4.1  Ensemble – mean predictions

The predicted ensemble means can be evaluated using any of the conventional accuracy and skill metrics described in Chapter 9, which focusses on verification of nonprobabilistic model solutions. The accuracy measures include such familiar quantities as bias, RMSE, and Mean-Absolute Error (MAE). In contrast, the skill score is a way of comparing the accuracy of one forecast method with that of a reference forecast (Eq. 9.3). If the accuracy is no better than that of the reference forecast, the skill score is zero. This is relevant to ensemble prediction because a deterministic forecast can be used as the reference, and its accuracy can be compared with that of the ensemble mean. Lu *et al.* (2007) use the skill score in a comparison of the mean-absolute error of a time-lagged ensemble forecast (Section 7.6) with that of a deterministic counterpart.

The *Taylor diagram* (Taylor 2001) is used in atmospheric science to graphically summarize how well statistical properties of observed and forecast patterns match. Because it is easy to plot the statistical properties of multiple forecasts on the same diagram, it is used to display the performance of individual ensemble members as well as the ensemble mean. Figure 7.4 shows the form of the diagram, where the radial distance from the origin is proportional to the standard deviation of a pattern (of a forecast variable, in this application), and the azimuthal position is related to the correlation of the pattern with a reference field (the verification field). Plotted on these diagrams are points associated with the forecast fields, and a point corresponding to the analyzed field (open circle), where the latter has a correlation coefficient of 1.0 because it is perfectly correlated with itself.

In this example from Delle Monache *et al.* (2006a), the ensemble modeling system consists of coupled meteorological and air-quality models, and the forecast variable being plotted is ozone concentration. The numbers correspond to indices of forecasts of individual ensemble members, and the open square represents the ensemble mean. The objective is to graphically quantify the relationship between the forecast and verification ozone fields in terms of the two noted statistical metrics. Taylor (2001) shows how the forecast's Centered RMSE (CRMSE, the RMSE after the bias has been removed) can be plotted as well (the dashed line), because of its mathematical relationship to the original two statistical measures (the graph coordinates). The CRMSE is the distance between the two points that represent a forecast and the analysis, where the closeness of the two points is proportional to the accuracy of the ensemble member. In this case, the CRMSE coordinate associated with the ensemble mean ozone concentration is plotted. This CRMSE of the ensemble mean is smaller than the CRMSE of any of the ensemble members. Note that the bias of an ensemble must be represented separately, because the Taylor diagram shows the CRMSE, not the RMSE.

## 7.4.2  Probabilistic predictions

The following approaches are traditional ones used to evaluate the statistical properties of ensemble predictions.

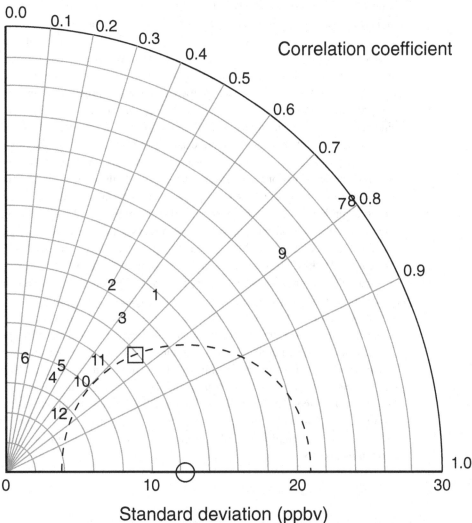

**Fig. 7.4** Taylor diagram, showing the statistical performance of ensemble members in a forecasting system that consisted of coupled meteorological and air-quality models. The forecast variable is ozone concentration, where the open circle corresponds to the analysis, the open square is the ensemble mean, and the numbers are indices of individual ensemble members. See the text for details. Adapted from Delle Monache *et al.* (2006a).

## Reliability diagrams

*Reliability* is an important attribute of ensemble forecasts of dichotomous events – ones that either occur or do not occur at a grid point or over an area – and reliability graphs are a device for easily visualizing the quality of probabilistic forecasts. Such discrete events include the existence of temperatures below freezing, or 3-h accumulated rainfall above a threshold value. Consider a set of ensemble forecasts of event E that are performed during a period of time, where, for each forecast, E is predicted to occur with probability $p$, for $0.0 < p < 1.0$. For the subset of the ensemble forecasts for which the forecast probability of occurrence is $p_f$, the observed frequency of occurrence is calculated to be $p_o$. For a perfect forecasting system, $p_f = p_o$. Figure 7.5 shows characteristic forms of the reliability graph (also called an *attributes diagram* or *calibration function*). Figures 7.5a and b illustrate unconditional biases, where the ensemble overpredicts (a) and underpredicts (b) the probability for all situations; that is, the sign of the bias is the same for all forecasts. Figure 7.5c shows a situation where the probabilities are predicted reasonably well in all situations. The plots in Figs. 7.5d and e show conditional biases in the model prediction of the event probabilities. In the former case, the model underpredicts the probability for low-probability situations and overpredicts it for high-probability situations. Here, it is said that the model forecasts have poor resolution (in a statistical sense). That is, the observed probability is similar over the full range of predicted probabilities. In contrast, in Fig. 7.5e there is good resolution because forecasts are able to identify situations with a variety of different probabilities, even though of course the forecasts are conditionally biased. See Wilks (2006) for additional discussion of this subject.

As an illustration of an actual reliability graph, Fig. 7.6 pertains to an ensemble of seasonal predictions, where the event is above-average 2-m temperature for the period February through April in the tropics. The panel on the left (a) is based on an ensemble that used a single model, where the ensemble was created by perturbing the atmosphere and ocean initial conditions. A number of other single-model-ensemble seasonal simulations were created, and all had reliability diagrams with a similar conditional bias. But, when the single models were combined into a multi-model superensemble, a considerably improved reliability resulted (b). This ensemble modeling was part of the Development of a European Multi-model Ensemble system for seasonal to inTERannual prediction (DEMETER), and employed models from seven institutions in Europe. See Chapter 16 for more information.

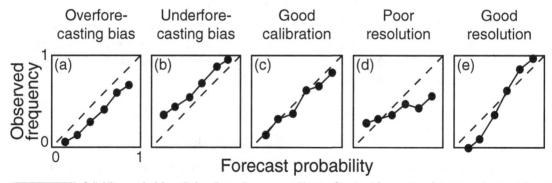

**Fig. 7.5**  Reliability graphs (also called attributes diagrams or calibration functions) for a variety of situations, showing different patterns of bias in predicted probabilities. The abscissa is probabilities predicted by an ensemble system over a large number of cases, and the ordinate is the corresponding conditional probabilities based on observations. See the text for details. Adapted from Wilks (2006).

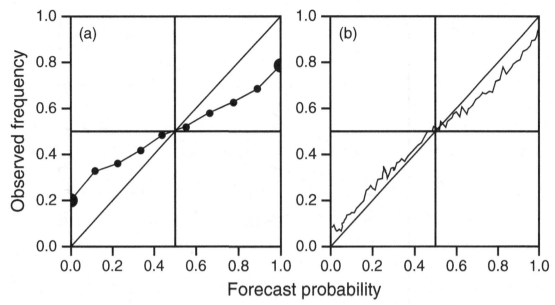

Reliability diagram, where the predicted event is above-average 2-m temperature for the period February through April in the tropics. The panel on the left (a) is based on an ensemble that used a single model, where the ensemble was created by perturbing the atmosphere and ocean initial conditions. A number of other single-model-ensemble seasonal simulations were created, and all had reliability diagrams with a similar conditional bias. But, when the single models were combined into a multi-model superensemble, a considerably improved reliability resulted (b). Adapted from Palmer *et al.* (2005).

## Rank histograms

*Rank histograms*, or verification rank histograms or Talagrand diagrams, are used to display the relationship between observations and forecasts from individual ensemble members. That is, they define the bias of probabilistic predictions. For a specific variable and the location of an observation, take an ensemble forecast of that variable at that location, and rank-order the forecasts from each of the members. Then define the $n + 1$ intervals bounded by the $n$ ordered forecast values. Figure 7.7 shows an example schematic with four ensemble members and five intervals, for a forecast variable $P$. For this location, and at time $t_{forecast}$, the observed $P$ (X obs) is lower than any of the forecast $P$s, and the observation is thus in interval $I_1$. If we follow a similar process for all other pairs of observations and forecasts at this time, we can calculate the total number of observations in each of the five intervals, or ranks, and plot a histogram of the frequency. This will provide a graphical view of how the ensemble of forecasts relates to the observations. A non-uniformity in the histogram's distribution will reveal systematic errors in the ensemble. Figure 7.8 shows four problems with an ensemble, which can be defined with rank histograms. In this hypothetical eight-member ensemble, the rank histogram has nine ranks, or intervals. In panels (a) and (b), many of the observations fall near the edge of the distribution of forecasts in the ensemble, or outside of the distribution entirely, corresponding to

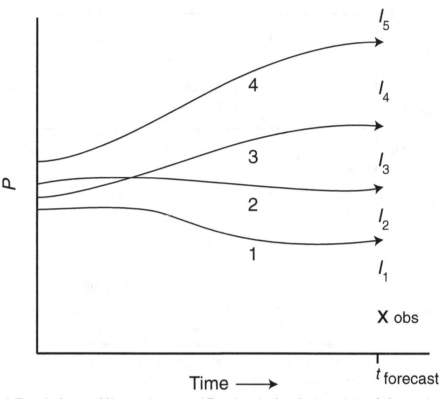

*P*

$I_5$

4

$I_4$

3

$I_3$

2

$I_2$

1

$I_1$

X obs

Time ⟶

*t* forecast

**Fig. 7.7** Schematic illustrating how a rank histogram is constructed. The trajectories show the time evolution of a forecasted variable *P* at the location of an observation, from a four-member ensemble. The values of *P* at the forecast time define the intervals (*I*) against which an observation of *P* (X obs) is compared. Each observation of *P* is assigned to one interval in the rank (even though the *P* values of the intervals will be a function of the observation location), and the resulting fractions of the total observations that fall into each rank are plotted in the histogram.

an overforecasting bias and an underforecasting bias, respectively. For example, in the situation depicted in panel (a), the most common situation was for all the ensemble members to forecast a value larger than the observation. Panel (c) shows a desirable rank-uniform distribution where observations are equally likely to fall anywhere within the distribution of ensemble members. In panel (d), many observations fall outside of, or near the edge of, the range of the forecasts from the ensemble members. Because the histogram is symmetric, there is no bias. In this case, the ensemble spread is too small, or in other words the ensemble is underdispersive. This situation is common, where the ensemble fails to always encompass the observations. Stated another way, the spread in the ensemble is less than the difference between forecasts and the validating analysis. The opposite situation is indicated in panel (e) where the ensemble is overdispersive. The ordinate can also be plotted as a probability, where the probability for each rank is defined by dividing the total number of times the verification occurred in the rank (the frequency) by the total number of forecast–observation pairs. These types of diagrams are further discussed in Anderson (1996), Talagrand *et al.* (1997), Hamill and Colucci (1997, 1998), Hamill (2001), and Wilks (2006).

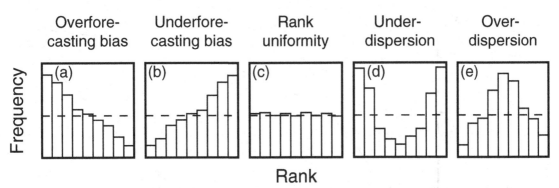

Fig. 7.8 Five rank histograms corresponding to different relationships between the ensemble members and the observations. The horizontal dashed line defines a perfectly rank-uniform distribution. See the text for details. Adapted from Wilks (2006).

### Relative Operating Characteristic (ROC) diagrams

The *ROC diagram*, used to evaluate probability forecasts of binary predictands, has the false-alarm rate ($F$) as the abscissa and the hit rate ($H$) as the ordinate, where $F$ and $H$ are defined in Section 9.2.2. Such a forecast might be whether the daily precipitation amount exceeds 1 cm, or whether the maximum daily temperature exceeds 30 °C. Wilks (2006) describes how to convert probabilistic forecasts, in this case from ensemble systems, into 2 × 2 contingency tables from which $F$ and $H$ can be calculated. The pairs of $F$ and $H$ are used to define points on the diagram, and along with the (0.0, 0.0) and (1.0, 1.0) points represent a curve. Better forecasts have a low $F$ and a high $H$, so more-accurate ones have points in the upper-left. The area under the curve has a maximum possible value of unity, corresponding to a perfect forecast. The diagonal corresponds to an unskilled forecast, and the associated area would be 0.5. Forecasts with ROC areas of ~0.75 or higher are considered to be good. Figure 7.9 shows an example ROC curve.

### Brier scores and Brier skill scores

The Brier Skill Score (BSS, Jollife and Stephenson 2003, Wilks 2006) is based on the Brier Score (BS), which assesses the accuracy of probabilistic predictions. The BS is defined as

$$BS = \frac{1}{n} \sum_{j=1}^{n} (p_j - o_j)^2,$$

and calculates the average squared difference between forecast probabilities ($p_j$) and observational outcomes ($o_j$), for $n$ forecast-event pairs, where $o$ is zero if the event does not occur and unity if it does occur. This expression is completely analogous to that presented in Section 9.2.1 for the Mean-Square Error (MSE) that is used for nonprobabilistic

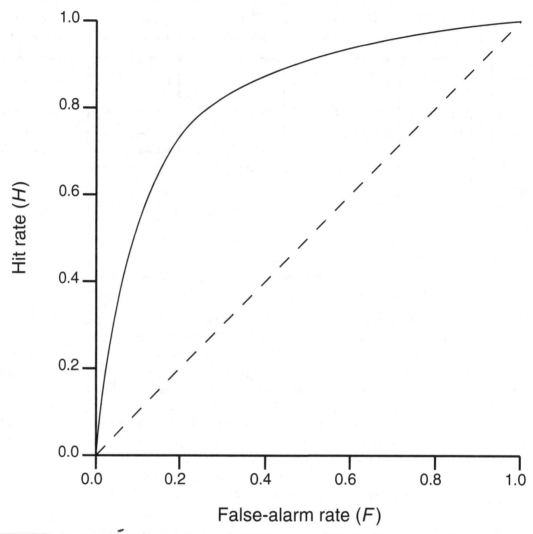

Fig. 7.9 An example Relative Operating Characteristic (ROC) diagram. The solid line is a ROC curve for a good forecast, the dashed line defines a random forecast, and the area under the ROC curve (fraction of 1.0) represents the overall performance in terms of this metric.

predictions. The BS ranges from zero to one, with lower values indicating better forecasts. The BSS is defined as

$$BSS = \frac{BS - BS_{ref}}{BS_{perf} - BS_{ref}} = 1 - \frac{BS}{BS_{ref}} \qquad (7.2)$$

where $BS_{perf} = 0$. The BSS is unity for a perfect forecast, and zero or negative for unskillful forecasts relative to the reference forecast.

### Rank probability skill score

The Rank Probability Score (RPS) describes the quality of categorical probabilistic forecasts for any number of event categories. The BS can be regarded as the special case of an RPS with two forecast categories. The Rank Probability Skill Score (RPSS) is defined analogous to Eq. 7.2, where the $RPS_{ref}$ is based on the climatological probabilities. See Wilks (2006) for the mathematical definition of the RPS and Weigel *et al.* (2007) for a comparison of the RPSS and BSS. An example of the use of the RPSS is found in the next section.

## 7.5  Calibration of ensembles

The *calibration* of ensemble forecasts of weather and climate is a post-processing step that removes the bias from the first moment (the ensemble mean) and possibly the higher moments. This is necessary in order to:

- provide greater accuracy in the ensemble mean,
- provide improved estimates of the probabilities of extreme events, and
- represent ensemble spread in terms of quantitative measures of the uncertainty in the forecast of the ensemble mean.

Figure 7.10 illustrates the calibration process in terms of its influence on the PDF. Both the mean and the spread of the distribution have been adjusted in the process.

Like many other statistical corrections that are applied to model output to remove systematic errors, a history of high-quality observations and ensemble forecasts is required to calibrate the operational forecasts. Historical, archived operational ensemble forecasts are not ideal for this purpose because models are continually being updated, and thus the required calibration changes as well. Ideally, reforecasts that use the current version of the ensemble system to recreate forecasts for a significant historical period should be used for the calibration. Discussion of ensemble reforecasts can be found in Hamill *et al.* (2004).

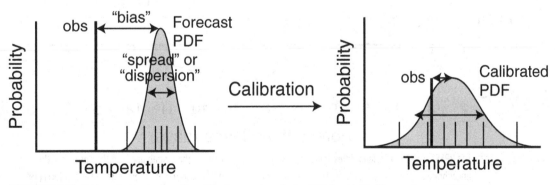

Fig. 7.10     Schematic showing how the calibration process adjusts a PDF. Provided by Thomas Hopson, NCAR.

There are several ensemble-calibration techniques. Hamill and Colucci (1997, 1998) employ the information in the verification rank histogram to interpret and correct the ensemble forecasts. It is important to note that a calibration performs better when it is done for as specific a set of conditions as possible. For example, calibrations can be dependent on weather-regime, forecast lead-time, geographic-area, season, etc. Also, separate calibrations are sometimes performed for different amounts of ensemble spread, for example as quantified by the standard deviation of the ensemble members. That is, the ensemble forecasts are divided into different bins based on their standard deviation, and separate rank histograms and calibrations are constructed for each group.

Eckel and Walters (1998) used a subset of a long history of Medium-Range Forecast model (MRF) ensemble predictions as a training data set with which they calibrated that model. They then used a complementary period for forecast verification. Both uncalibrated and calibrated forecasts were verified using the RPSS, which employed climatology as the reference forecast, to assess the benefit of the calibration of the MRF-ensemble Quantitative Precipitation Forecasts (QPF). Figure 7.11 shows the RPSS, based on calibrated and uncalibrated two-week forecasts. In one approach, the forecasts were calibrated using the "weighted ranks" method of Hamill and Colucci (1997, 1998). In another, an uncalibrated "democratic voting" method was used, where each ensemble member gets an equal vote regarding the occurrence of precipitation above some threshold. In the uncalibrated approach, the total number of ensemble members for which the precipitation exceeds the threshold, divided by the number of ensemble members, defines the probability. The skill score for a climatology-based forecast is zero, so positive scores beat climatology while negative ones do not. In this case, calibration of the forecasts extended the predictability by about one day. Other examples of the many calibration methods available are found in Bremnes (2004), Doblas-Reyes *et al.* (2005), Raftery *et al.* (2005), Roulston (2005), and Weigel *et al.* (2009). Weigel *et al.* (2009) address the issue of whether calibrated single-model ensembles are superior to multi-model ensembles.

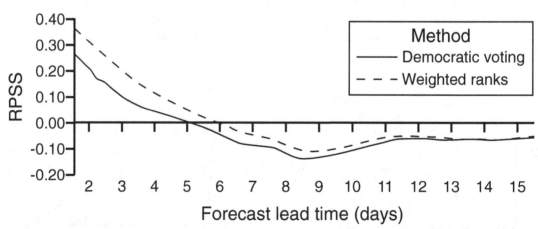

Fig. 7.11    The RPSS for all forecast lead times, based on two methods of defining the probability from the ensemble: the uncalibrated "democratic voting" method and the "weighted ranks" calibration method of Hamill and Colucci (1997, 1998). Adapted from Eckel and Walters (1998).

# 7.6  Time-lagged ensembles

Because contemporary forecasting systems typically involve the use of a production cycle in which new forecasts are initiated every 1 h to every 12 h, multiple forecasts are available for the same times when forecasts overlap. Figure 7.12 illustrates the concept. In the top example of a time-lagged ensemble, forecasts are initiated every 6 h as part of a normal deterministic operational system. These forecasts can be combined to form an ensemble, say for the time in the future corresponding to the dotted line. Only the initial conditions are different in this ensemble – different by an amount equal to the changes in the initial states between the 6-h initialization times. The bottom half of the diagram depicts the traditional approach for creating an ensemble, where multiple forecasts are initialized at the same time using different initial conditions or model configurations. The time-lagged ensemble can be created at no additional cost because the forecasts are part of a normal forecasting system, in contrast to the traditional ensemble approach where multiple forecasts are performed from the same initial time.

This approach was first proposed and evaluated by Hoffman and Kalnay (1983) and Dalcher *et al.* (1988). Lu *et al.* (2007) employ the RUC model with a 1-h update cycle in tests of very-short-range (1–3 h), time-lagged ensemble forecasts, where two approaches were used to create the ensemble mean. In one, the forecast values from each of the

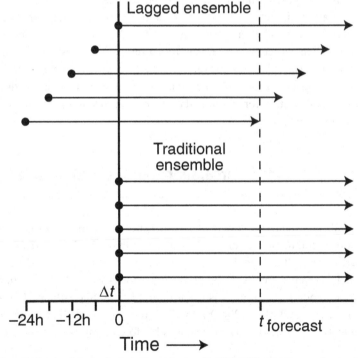

Fig. 7.12   Schematic showing forecasts in time-lagged (top) and traditional (bottom) approaches to ensemble prediction. See text for details.

ensemble members were equally weighted. In the other, different weights were used for the different time-lagged forecasts. Even though both methods provided improved forecasts relative to the deterministic-system products, the method with unequal weights was superior. Yuan *et al.* (2009) describe the verification of a multi-model time-lagged forecast system that employed the WRF and MM5 models running on a 1-h cycle. Because the cycle frequency was so high, a large number of forecasts were valid at the same time and thus it was possible to utilize a large ensemble. The ensemble-mean QPF had greater skill than did the forecasts from the deterministic NAM model running at the same horizontal grid increment (12 km), but other aspects of the verification were less positive. In Yuan *et al.* (2008) a larger time-lagged multi-model ensemble, based on MM5, WRF, and RAMS with a 3-km grid increment, was tested for hydrological applications. They concluded that such time-lagged ensemble systems can provide valuable ensemble-mean QPFs and probabilistic QPFs for water-management applications. For longer-range global-model forecasting, Buizza (2008) compared the performance of the 51-member low-resolution (T399L62) ECMWF traditional ensemble system with that of a higher-resolution (T799L91) 6-member lagged ensemble, for a 7-month period. The cycle frequency was 12 h, and the initialization times of the lagged forecasts spanned 60 h. The 51-member ensemble was superior to the lagged ensemble in terms of probabilistic measures, but the ensemble-member-weighted lagged ensemble had similar skill in predicting the ensemble mean out to forecast-day 4. Lastly, Delle Monache *et al.* (2006a) describe encouraging results from the use of an 18-member lagged-ensemble system for air-quality applications. Additional discussion of the lagged-ensemble method is found in Mittermaier (2007).

The relationship between forecast uncertainty and the spread of time-lagged ensemble members has a foundation in the way that forecasters have used operational models for decades. When consecutive forecasts from a series of forecasts in a cycle predict the same outcome at a particular verification time, the forecasters are confident in the model solution. On the other hand, models occasionally signal uncertainty by producing different outcomes in consecutive forecasts in the cycle. This results in the forecasters having less confidence in the products.

## 7.7  Limited-area, short-range ensemble forecasting

Mesoscale ensemble modeling with LAMs is becoming more prevalent in the research and operational communities. As noted earlier in Section 7.3.2, the existence of LBCs affects the error that must be sampled in the generation of the ensemble. Because the limited-area ensemble systems are used for producing shorter-range forecasts than are the global ensemble models, the process is often referred to as Short-Range Ensemble Forecasting (SREF). Eckel and Mass (2005) provide a summary of challenges posed by SREF, compared to medium- and longer-range ensemble forecasting with global models.

- Near-surface variables exhibiting fine-scale structures are important forecast quantities, but they are less predictable and their errors may saturate for short forecast lead times, thus limiting the use of an ensemble. Error saturation means that the forecast has no skill in the sense that it is completely uncorrelated with the verification field (see Fig. 8.1, and related discussion).

- Model error is poorly understood and difficult to quantify, and has a larger impact on near-surface variables for short-range forecasts (Stensrud *et al.* 2000).
- Methods for generating optimal initial-condition perturbations (e.g., the breeding method) were developed for the medium range, where nonlinear error growth generates a large spread of solutions. It is unclear how to generate initial-condition perturbations for SREFs, where error growth is initially linear (Gilmour *et al.* 2001).
- The use of LAMs may result in insufficient ensemble dispersion, even when the LBCs are perturbed (Nutter 2003).
- Very-high resolution may be needed in order to capture variability at small scales.

For additional examples of mesoscale ensemble prediction with LAMs, see Marsigli *et al.* (2005) and Holt *et al.* (2009).

# 7.8 Graphically displaying ensemble-model products

Probabilistic forecast information must be displayed in ways that are meaningful to both model developers as well as to the ultimate users of the forecast information. Even though the needs of those two groups, in terms of appropriate graphical products, are somewhat different, there is the commonality that more creativity is required than for the display of the state variables themselves. The following subsections review some of the common types of displays.

## 7.8.1 Spaghetti plots

One of the greatest challenges in interpreting the spread among ensemble members is graphically synthesizing the vast amount of information in an easily interpretable way. The use of small individual maps that show a particular variable field from each of the ensemble members is one approach (see Fritsch *et al.* 2000, Legg *et al.* 2002, Palmer 2002, and Buizza 2008 for examples). However, these "stamp maps" are small and can be difficult to interpret when details are important. An alternative is to define a meteorologically important and graphically simple aspect of the variable field, and display that for each of the ensemble members in the same image. Figure 7.13 shows an example of this approach that uses a single contour (5520 m) of the 500-hPa height field. Even though the entire field cannot be visualized, the shape and position of this contour reveal how the pattern evolves with time, and more importantly how it differs among the 17 ensemble members in this case. At the 12-h lead time (a), some of the ensemble members are beginning to produce a trough over central Canada, although they are all in good agreement elsewhere. Thus, except in the region of this trough the forecast would be interpreted with confidence. By 36 h (b), this trough development has continued in some of the members, leading to a

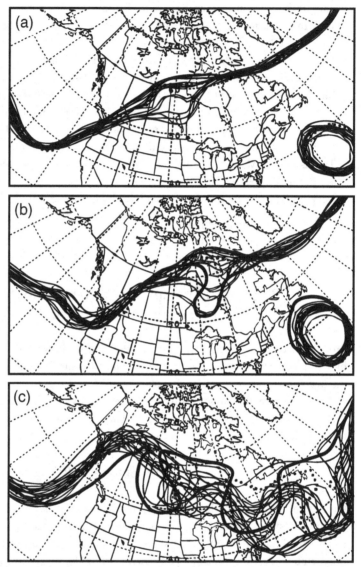

**Fig. 7.13**   Spaghetti plots of the 5520-m contour of the 500-hPa height field over North America, based on an ensemble forecast by NCEP, where 12 h (a), 36 h (b), and 84 h (c) lead times are shown. The light solid lines are the contours associated with each of the 17 ensemble members, the heavy solid lines in panels (b) and (c) are based on the verifying analyses, and the dotted lines show the control forecast. From Toth *et al.* (1997).

growing degree of uncertainty about the solution in this area. The control simulation (dotted line) shows no evidence of the trough, even though the verifying analysis (heavy solid line) indicates a large-amplitude feature. At the 84-h lead time (c), all members tend to agree with respect to the location and amplitude of the trough in the East Pacific, but over the continent and West Atlantic there is much scatter among all the members and much reason to be suspicious about the forecast accuracy.

### 7.8.2  Meteograms, or box plots

Ensemble forecasts of a single variable at a single location (or an average over an area) can be displayed, where the value of the variable is plotted as a function of time for each ensemble member. The spread of the members and the mean can easily be visualized as a function of forecast lead time. The ultimate user of the forecasts (not the forecaster) often has a particular variable of concern (e.g., precipitation rate) and a specific location (e.g., a city or watershed), so this type of plot makes more sense than a more-complicated mapping for a large area. An example of this type of display is shown in Fig. 7.14, in the form of plots of an ensemble of 6-month forecasts of El Niño SST anomalies for the NINO3 region in the eastern Pacific, based on an ECMWF coupled ocean–atmosphere model. The model solutions progressively diverge throughout the simulation period. Figure 7.3, shown previously, is another example of such a display, for near-surface temperature in London, UK.

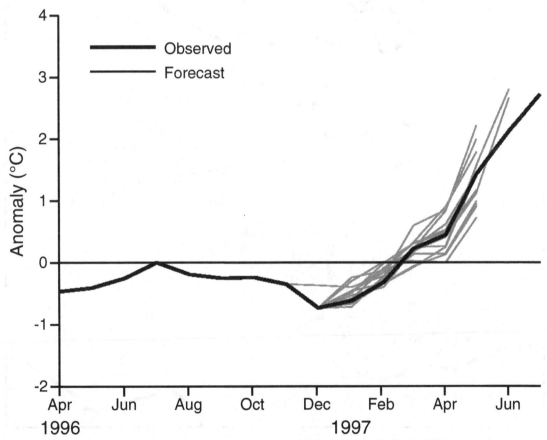

**Fig. 7.14**  An ensemble of 6-month forecasts of spatially averaged El Niño SST anomalies for the NINO3 region in the eastern Pacific, based on an ECMWF coupled ocean–atmosphere model. The simulation begins in December 1996. The solid line indicates observed average SSTs. Adapted from Palmer (2002).

### 7.8.3 Probability-of-exceedance plots

The PDF from properly calibrated ensemble forecasts can be interpreted in terms of the probability of an event occurring at a particular grid point. For example, 30 ensemble members will produce an equal number of estimates of the wind speed at a grid point at a particular forecast lead time. The number of members for which the speed exceeds a certain threshold can be used to define the probability that the speed will be exceeded at that point (this is the uncalibrated democratic voting method defined in Section 7.6). The resulting gridded field of probabilities can be contoured. An example of this type of plot is shown in Fig. 7.15. Based on a 50-member ECMWF ensemble, the map shows the probability that wind gusts will exceed 50 m s$^{-1}$ at the 42-h lead time of a forecast of a severe synoptic-scale storm that devastated parts of Europe on 26 December 1999. Similar exceedance plots

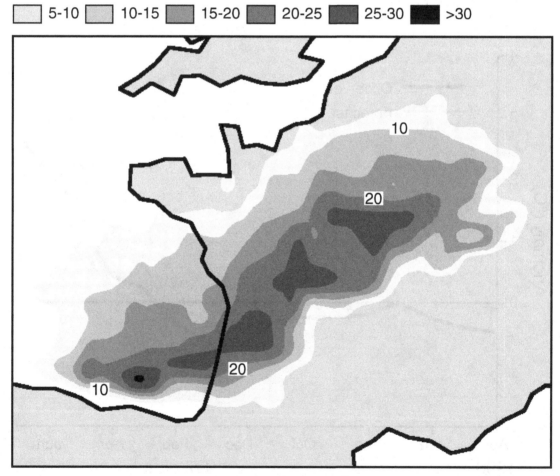

**Fig. 7.15** The probability (percentage) that wind gusts will exceed 50 m s$^{-1}$ at the 42-h lead time of a forecast of a severe synoptic-scale storm that devastated parts of Europe on 26 December 1999. This is based on a 50-member ECMWF ensemble. Adapted from Palmer (2002).

are used widely, where Delle Monache *et al.* (2006c) contains other examples. These types of products could, of course, be based on a PDF that has been adjusted through calibration.

### 7.8.4  Plots of some metric of ensemble variance

A number of measures are available for reflecting, in a single number, the spread of an ensemble, averaged over a model computational domain or defined at a point. For example, variance can be plotted as a function of forecast lead time in order to represent the spread of the ensemble as a function of time. Figure 7.16 is an example of this type of plot, and shows the variance in the 850-hPa specific humidity as a function of forecast lead time for a 19-member physics ensemble and a 19-member initial-condition ensemble that were run with MM5 for the same case of a long-lived mesoscale convective system.

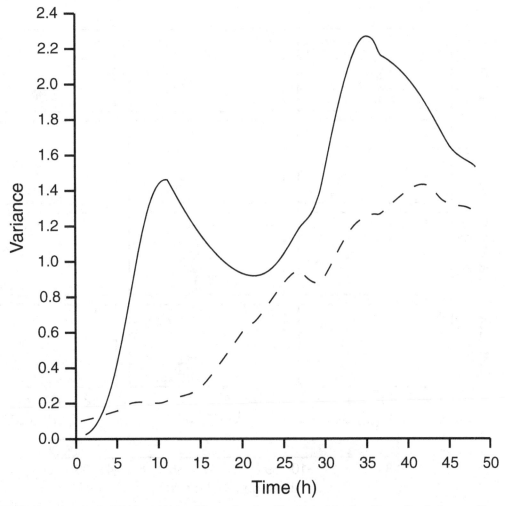

**Fig. 7.16**    The variance in the 850-hPa specific humidity as a function of forecast lead time for a 19-member physics ensemble (solid line) and a 19-member initial-condition ensemble (dashed line) that were run with MM5 for the same case of a long-lived mesoscale convective system. Adapted from Stensrud *et al.* (2000).

### 7.8.5  Ensemble plots from coupled special-applications models

Special-applications models that use atmospheric model output fields will be discussed in Chapter 14. This section shows example displays of variables produced by the secondary models that have used ensemble products from atmospheric models as input. For example, gridded ensemble forecasts or simulations may be used as input to air-quality models or plume-dispersion models that calculate the transport of gases or aerosols released into the atmosphere. Figure 7.17 illustrates stamp maps of dosages from plumes of gas whose transport and diffusion have been calculated using the Second-order Closure Integrated PUFF (SCIPUFF) plume model (Sykes *et al.* 1993), which has employed gridded meteorological products from a 12-member ensemble that is based on the MM5 regional model.

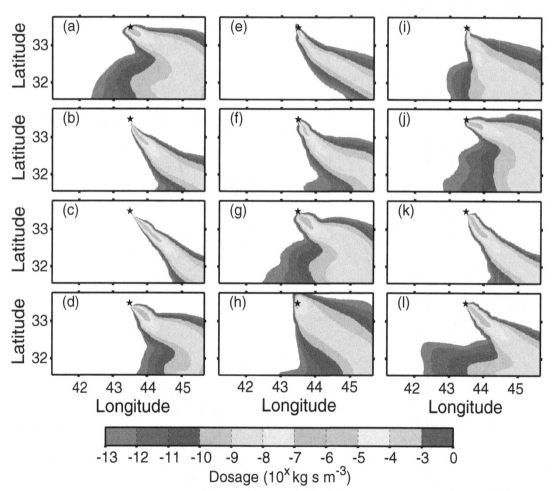

**Fig. 7.17**  Stamp maps of dosages from plumes of gas whose transport and diffusion have been calculated using the SCIPUFF plume model (Sykes *et al.* 1993), which has employed gridded meteorological products from a 12-member ensemble that is based on the MM5 regional model. From Warner *et al.* (2002).

The dosages show considerable sensitivity to the input from the meteorological ensemble. Also shown (Fig. 7.18) is a probability-of-exceedance plot that is based on the dosages in Fig. 7.17.

Atmospheric-model forecasts are also routinely used for estimating the future (e.g., next day) demand for electricity. The near-surface temperature, cloud cover, etc., are used as input to energy-demand models, and forecasts of the demand are plotted as a function of forecast lead time. When ensemble atmospheric models are employed, stochastic energy-demand products are produced, which can be plotted as lines on a meteogram or in the

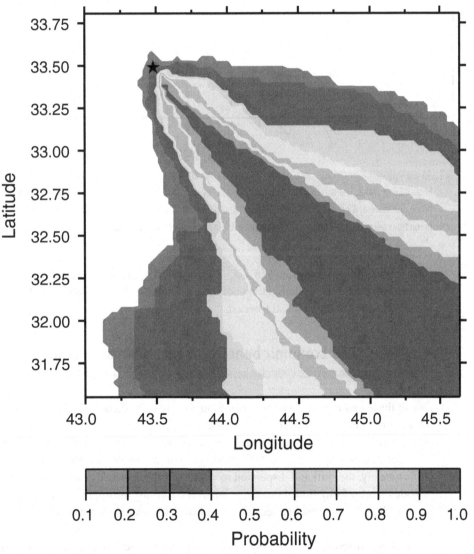

**Fig. 7.18** Probability that the dosage of a plume of gas will exceed a specified threshold, based on the ensemble of plume predictions shown in Fig. 7.17. The star in the upper left shows the location of the release. From Warner *et al.* (2002).

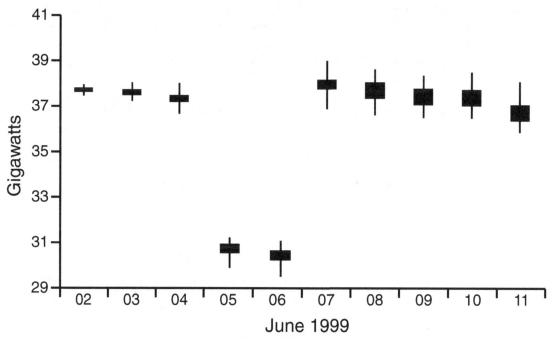

Fig. 7.19 An example of an "electricity-gram", which displays the uncertainty in a 10-day forecast of electric-energy demand for England and Wales. The meteorological input to an energy-demand model was provided by an ECMWF ensemble-prediction system. All of the predicted demand values lie within the vertical whiskers, and the middle 50% of the predictions fall within the box. Provided by James W. Taylor, University of Oxford.

form of the "electricity-gram" in Fig. 7.19. Here, forecasted average-daily demand for England and Wales is plotted in terms of the total range of the predictions and the range of the middle 50% of the demand forecasts.

## 7.9  Economic benefits of ensemble predictions

One of the stated motivations for performing ensemble predictions is that the probabilistic information can have greater value for making decisions than would the output from a deterministic prediction. Such value can be defined in the context of societal, environmental, or economic impacts, even though the economic benefits are more easy to quantify. Unfortunately, the skill and dispersion measures for the forecasts, discussed above, do not provide direct information about the value of the forecast information. This value, in fact, is dependent on the weather sensitivity and decision-making process of a forecast user group.

There are different frameworks for assessing value, where one of the most common is the *cost–loss model*. Given an uncertain prediction of whether an event will or will not occur, a decision maker has the option of choosing to either protect against the occurrence of the weather event or not protect against it. This is the simplest decision problem because there are only two possible actions (protect, not protect) and two possible outcomes (the event occurs, the event does not occur). Examples of the many potential events of economic consequence include sub-freezing temperatures that can damage agricultural crops, daily precipitation in excess of an amount that can produce flooding, heavy snowfall that can impact highway or air travel, or damaging wind speeds. A decision to protect against the event will incur a cost ($C$), whether or not the event actually occurs. A decision to not protect will result in a loss ($L$) if the event occurs. Figure 7.20 summarizes the cost–loss consequences of the different outcomes.

It is assumed that probabilistic forecasts for the dichotomous weather event are available, and, if their quality is sufficiently good, decisions with better economic outcomes will be possible. Assume that a calibrated ensemble forecast predicts that the probability of an event occurring is $p$. The optimal decision about whether to protect or not to protect will be the one yielding the smallest expected expense. If the decision is made to protect, the expense will be $C$ with a probability of 1.0. If no protective action is taken, the

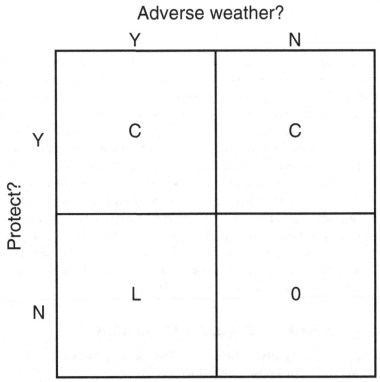

**Fig. 7.20**  The costs (C) and losses (L) associated with the four possible combinations of the occurrence of adverse weather and the decision to protect against it. A decision to protect against the event will incur a cost (C), whether or not the event actually occurs. A decision to not protect will result in a loss (L) if the event occurs. There is no economic effect (0) if the event does not occur and no protective action is taken.

probability-weighted expense will be $pL$. Therefore, protecting against the risk will result in the smallest expense whenever

$$1.0C < pL,$$

or

$$\frac{C}{L} < p.$$

Thus, protecting against an event is the optimal action when the predicted probability of its occurrence is more than the ratio of the cost to the loss. If the predicted probability is less than the cost/loss ratio, not protecting is the least-expensive action. Because the costs and losses are very strongly dependent on the particular situation, the threshold for protection will differ. The above discussion is only applicable if $C < L$, because otherwise the protective action could not result in a gain.

The economic value ($V$) of forecasts can be defined using an expression that is similar to that employed for skill scores for meteorological forecasts. Where $E$ is an expected expense,

$$V = \frac{E_{forecast} - E_{climate}}{E_{perfect} - E_{climate}}.$$

The quantity $E_{climate}$ is a default that represents the minimum of the expenses resulting from always protecting or never protecting. Always protecting incurs a constant cost, $C$, while never protecting results in losses $\bar{o}L$, where $\bar{o}$ is the climatological probability of the event occurring. With perfect forecasts, the protective action would only take place when the event was going to occur, so $E_{perfect} = \bar{o}C$. See Wilks (2006) and McCollor and Stull (2008b) for a complete discussion of the calculation of this value score.

The cost–loss decision model has been frequently applied to assess the economic benefits of ensemble prediction, for many specific applications that require decisions. These applications include hydroelectric reservoir operations (McCollor and Stull 2008a,b), medium-range flood prediction (Roulin 2007), temperature forecasts for the energy sector (Stensrud and Yussouf 2003), precipitation predictions (Mullen and Buizza 2002, Yuan *et al.* 2005), severe-weather forecasts (Legg and Mylne 2004), and air-quality prediction (Pagowski and Grell 2006). Additional discussion of the economic value of ensemble predictions is found in Richardson (2000, 2001).

## SUGGESTED GENERAL REFERENCES FOR FURTHER READING

Kalnay, E. (2003). *Atmospheric Modeling, Data Assimilation and Predictability*. Cambridge, UK: Cambridge University Press.

Leutbecher, M., and T. N. Palmer (2007). Ensemble forecasting. *J. Computational Phys.*, **227**, 3515–3539, doi:10.1016/j.jcp.2007.02.014.

Palmer, T. N. (2002). The economic value of ensemble forecasts as a tool for risk assessment: From days to decades. *Quart. J. Roy. Meteor. Soc.*, **128**, 747–774.

Palmer, T. N., G. J. Shutts, R. Hagedorn, *et al.* (2005). Representing model uncertainty in weather and climate prediction. *Annu. Rev. Earth Planet. Sci.*, **33**, 163–193, doi: 10.1146/annurev.earth.33.092203.122552.

Wilks, D. S. (2006). *Statistical Methods in the Atmospheric Sciences*. San Diego, USA: Academic Press.

## PROBLEMS AND EXERCISES

1. Choose a nonlinear function and demonstrate the correctness of the inequality in Eq. 7.1.
2. Given that we know from Chapter 9 that smoother forecasts verify better with conventional statistics (e.g., RMSE, MAE), and that the ensemble mean must be smoother than the solution from individual members, is it true that the smoothness of the ensemble mean contributes to its superior performance?
3. Access the web site of an operational ensemble prediction system and observe how the spread of the ensemble members varies from day to day, both within the same forecast and from one forecast to another in the cycle.
4. When ensemble modeling systems are coupled with special-application models, such as discussed in Section 7.8.5, describe how the system might be calibrated in terms of the variables predicted by the coupled model rather than the meteorological variables.

# Predictability

## 8.1 Background

The term atmospheric predictability may be defined as the time required for solutions from two models that are initialized with slightly different initial conditions to diverge to the point where the objective (e.g., RMS) difference is the same as that between two randomly chosen observed states of the atmosphere. In the practical context of a forecast, the no-skill limit that defines the predictability may be the forecast lead time when the model-simulated state has no greater resemblance to the observed state of the atmosphere than does a reference forecast based on persistence or climatology. Many of the other chapters in this text address the various components of the modeling process that limit predictability, from data-assimilation systems to numerical methods to physical-process parameterizations, as well as metrics for quantifying it. This chapter will review the general concept of theoretical and practical limits to forecasting skill.

## 8.2 Model error and initial-condition error

As shown in the previous chapter, error that limits predictability originates in both the model and the initial conditions. Refer to Section 7.3 for more information, especially about the various sources of error associated with the model. Often the concept of predictability is discussed in the context of the system's response to infinitesimally small perturbations in the model initial conditions. This predictability is an inherent property of the fluid system and not of the model. Indeed, it is sometimes assumed in this hypothetical discussion that the model is perfect. In contrast, the term predictability is most-often used in the literature in a very practical sense to refer to the average length of useful forecasts that are obtainable from a particular operational modeling system, where all the sources of uncertainty in the modeling process contribute to error growth. For example, the impact of a particular new data source, or data-assimilation system, or parameterization will be evaluated in terms of its effect on the predictability of a particular variable.

Lorenz (1963a,b) describes simple modeling experiments that served as the foundation for later studies on the inherent predictability of the atmosphere. Using a form of identical-twin experiment (see Section 10.2), he initialized the same model with initial conditions that differed only very slightly, in the digit a few places to the left of the decimal point. He found that, after a few weeks of simulated time the two model solutions differed by as much as two random solutions. Many modelers have unintentionally replicated this experiment with contemporary models by changing computer compilers, or compiler optimizations, in the middle of a series of controlled, long-running simulations. Different compilers often perform arithmetic operations in an equation in a different order, which leads to different roundoff or truncation errors. Even if modeling-system configurations (initial conditions and physics) are identical, the subtle compiler-introduced differences in the model solutions can amplify to define the same type of predictability time limit that Lorenz observed. This growth of small perturbations in the atmosphere (or the model atmosphere), regardless of the source, led Lorenz to refer to the possibility that the flutter of a butterfly's wings could, after passage of significant time, influence the large-scale weather.

A more-contemporary identical-twin experiment is shown in Fig. 8.1, and illustrates error growth from small initial-condition perturbations. Here, the same atmospheric GCM is used for a control simulation, and for a perturbation simulation in which the initial

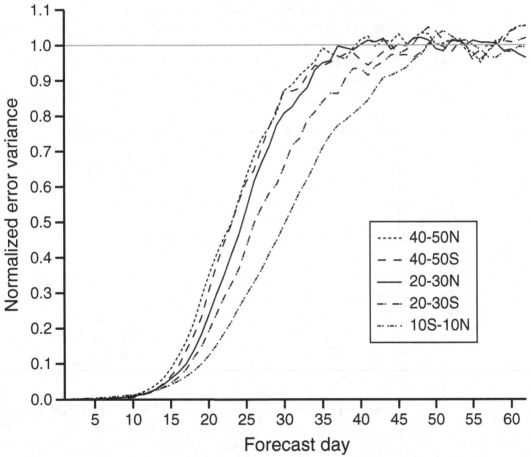

**Fig. 8.1** Error variance of the 850-hPa zonal wind component plotted as an average for different latitude bands. The variance is normalized in terms of the saturation value (maximum) for each curve. The error is the difference between a control and a perturbation simulation of an identical-twin experiment that employed an atmospheric GCM. Adapted from Straus and Paolino (2009).

conditions were slightly different. For the perturbation simulation, the initial conditions were perturbed according to the following method,

$$\delta A = 0.001 r A,$$

where $A$ is a model dependent variable, and $r$ is a random variable that ranges from $-1$ to $1$. This perturbation was applied independently at each grid point to temperature (K), both horizontal wind components, specific humidity, and surface pressure. The area-average difference in a variable between the control and perturbation experiments defines an *error*, and the square of this error is the error variance that is plotted in the figure. Specifically, the error variance in the $u$ velocity component is illustrated as a function of integration time. There is an induction period of 10–15 days during which the error grows very slowly, for the next 20 days it grows rapidly and approximately linearly, and then it reaches a saturation level where the two simulated fields are uncorrelated.

The relative contributions of model error and initial-condition error to the total forecast error, and therefore to the predictability, are not well understood nor are they easy to individually quantify. One approach is to define the growth in the total error through a comparison of the forecast with observations, or analyses of observations. Then, initial-condition-related error growth is estimated by calculating the difference between two forecasts that are initialized from slightly different initial times, for the same integration period. This is illustrated in Fig. 8.2 for the ECMWF ensemble-prediction system. The

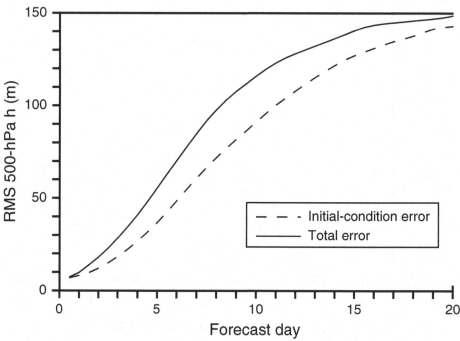

**Fig. 8.2** Growth of 500-hPa error associated with initial-condition error only (dashed line) and with both model and initial-condition error (solid line) in an ECMWF global model. The total error is based on a comparison of the model solution with observations and the initial-condition error results from the growth of differences between parallel simulations from the same model initialized 12 h apart. Adapted from Leutbecher and Palmer (2008).

upper curve corresponds to the total model error, and the lower one is based on the growth of differences in the initial conditions of forecasts whose initialization times differed by 12 h. In this example, the initial-condition error is a large fraction of the total error.

## 8.3 Land-surface forcing's impact on predictability

The existence of the diurnally and seasonally varying solar forcing of Earth's surface and atmosphere causes processes that are tied to this cycle. Thus, correctly defining this forcing in the model will produce circulations and structures that are seasonally and diurnally dominant, without the processes necessarily being well observed in model initial conditions. Examples of diurnally varying phenomena that require thermal forcing are some low-level jets, sea breezes, mountain-valley breezes, urban heat-island circulations, and a variety of moist-convective processes. Figure 8.3 illustrates the near-surface $u$ wind component observed over three diurnal periods near the slope of a north–south-oriented mountain range. If a model resolves the orography and represents the diurnal heating and cooling cycle reasonably well, these reoccurring winds, which are a dominant local feature, should be predictable, especially for weak synoptic-scale regimes. On seasonal time scales there are monsoons, the migration of the Hadley circulation and associated precipitation and trade winds, and the subtropical high-pressure systems whose seasonal migration with the Sun produces the Mediterranean-type climates described later. Getting the

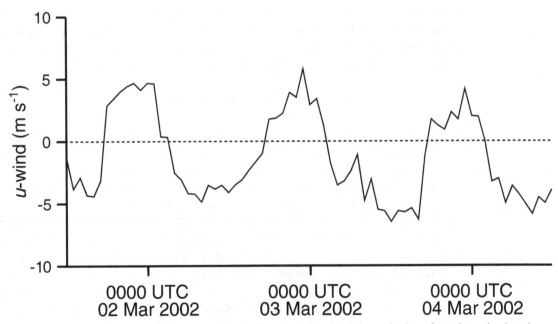

Fig. 8.3   The 10-m AGL $u$ wind component observed over three diurnal periods near the slope of a north–south-oriented mountain range. The weak 3-day-average wind for the period has been subtracted from each point in the series. Adapted from Rife et al. (2004).

solar forcing prescribed correctly in a model, and using a good LSM to translate this energy source into appropriate sensible-heating patterns at the lower boundary of the atmosphere, will positively impact predictability.

# 8.4  Causes of predictability variations

There are numerous dependencies of the predictability of models (and the skill of human forecasters) on the geographic region, the local climatology, the season, the weather-regime, the time scale of a phenomenon, and the dependent variable. The following sections discuss a few examples.

## 8.4.1  Regional and climatological variability

Because of the planetary-scale circulation of the atmosphere, there are some regions for which the weather patterns on both large scales and mesoscales vary little from day to day, so the weather is relatively easy to anticipate using even simple tools like diurnal persistence. For example, when the trade winds prevail in a tropical region the wind speed and direction tend to be very regular when they are not interrupted by convective events. Again for the tropics, if a location is near a coastline, the diurnal variation of the winds associated with a sea-breeze circulation, as they are superimposed on the trade winds, is very predictable. And, the development of cloud and precipitation associated with the inland penetration of the sea-breeze front are regular parts of the local climatology and easy to anticipate as well. Lastly, there are some climates that are so dominated by the planetary circulation that their weather is virtually identical every day of the year. For example, there is a large area of northeastern Africa that experiences cloudiness less than 2% of the time.

## 8.4.2  Seasonal variability

For some regions that are seasonally influenced by subtropical high-pressure centers, half the year is dominated by subsidence from the Hadley circulation. During those months, days are generally cloud free, with no precipitation or other disturbances. Again, simple forecasting methods such as diurnal persistence are difficult to improve upon. During the rest of the year, when the latitudinal positions of the subtropical high-pressure centers and the storm track are different, the region can be dominated by synoptic-scale storms that are challenging to predict. Regions having seasonal variability in the predictability of their weather, with cloud-free warm-season months and cyclones in the cold season, include the Mediterranean Sea and the west coast of North America.

## 8.4.3  Weather-regime dependence

Predictability varies by weather regime, and on longer time scales, for reasons that are sometimes understood, and sometimes not well understood. The existence of the longer time-scale trends suggests that the predictability variation is not random, and may be associated with low-frequency variability or oscillations. Of course, one of the benefits of ensemble prediction is that we are provided with information about predictability on a day-to-day, or regime-to-regime, basis. It is simply pointed out here that this predictability can vary significantly in an organized way over relatively long time-scale shifts in weather regimes. As an illustration, Fig. 8.4 is of the Northern Hemisphere anomaly correlation (see Chapter 9 for a discussion of this verification metric) between 108, 15-day global-model forecasts and corresponding analyses of observations, for approximately a 3-month period. Clearly the predictability varies on time scales of a few weeks to a month. The plots in Fig. 7.3, showing the great difference in the ensemble dispersion of two forecasts one year apart, could be illustrating a regime dependence, or some variation on shorter time scales. Various studies have formally documented the relationship between predictability and the existence of various types of flow patterns and low-frequency variability. For example, Tracton (1990) states that a strong association exists between atmospheric predictability and the existence of blocking events in midlatitudes.[1] The onset of a blocking pattern results in a dramatic drop in predictability, and the collapse of the blocking

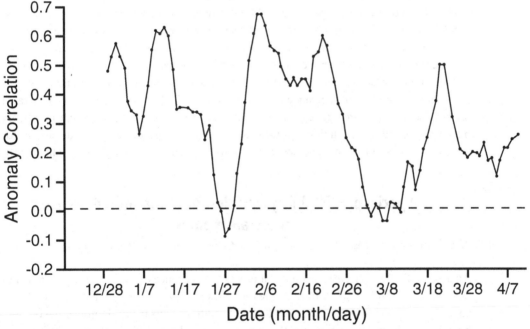

**Fig. 8.4**  The Northern Hemisphere anomaly correlation between 108, 15-day global-model forecasts and analyses of observations, for approximately a 3-month period. Adapted from Tracton (1990) and Tracton *et al.* (1989).

---

[1]  Blocking refers to a situation where there is an obstruction to the west-to-east progress of migratory cyclones and anticyclones in midlatitudes. This situation is normally associated with upper-tropospheric closed anticyclonic circulations at high latitudes and cyclonic circulations at low latitudes. It may be viewed as an extreme-amplitude pattern of the ridges and troughs that normally prevail in the westerlies. This anomalous circulation remains stationary or moves slightly to the west, and can persist for weeks.

situation leads to a recovery in the predictability. In general, when and where there exist states of the atmosphere that are close to an instability threshold, whether it be baroclinic instability or some type of convective instability, the predictability is less because small perturbations that are poorly observed or poorly resolved by the model can cause the system to take alternative trajectories in phase space.

### 8.4.4 Phenomenon time-scale dependence

As described in Chapter 9, it is illustrative to consider model predictability in the context of three time-scale regimes: periods longer than diurnal (super-diurnal), periods that are approximately diurnal, and periods that are shorter than diurnal (sub-diurnal). Meteorological features with longer-than-diurnal periods are generally associated with synoptic-scale or planetary-scale processes, and are therefore reasonably predictable by global or regional models. Diurnal time-scale motions of course are related in some way to the heating cycle. In the wind field, a diurnal signal could be related to stability-related vertical momentum mixing, mountain-valley circulations, coastal circulations, etc. Provided that the model reasonably represents the land-surface and boundary-layer processes, features with these time scales should be reasonably predictable. Motions with sub-diurnal time scales include mesoscale features or circulations that are not thermally forced by the diurnal heating cycle. They can result from orographic or other landscape forcing, perhaps far upstream, or from nonlinear interactions. Given the sparse nature of the radiosonde network, these mesoscale features are not represented well, or at all, in three dimensions by the observation network, and therefore are not in the model initial conditions. Unless they are locally generated through nondiurnal forcing, they are not deterministically predictable by any model, no matter how good the resolution and physics. Time-scale dependences of predictability for longer-period variations in the atmosphere are discussed by van den Dool and Saha (1990).

## 8.5  Special predictability considerations for limited-area and mesoscale models

An aspect of LAMs that affects predictability is the existence of LBCs. It was explained in Section 3.5 that information from upstream LBCs will sweep across a model grid during an integration, as information from the initial conditions exits the grid through the outflow boundary. This means that, unlike the situation with global models, the predictability of phenomena will depend less on initial-condition error for longer forecast lead times and more on LBC error. The importance of the LBCs to predictability with mesoscale LAMs is easily understood when one considers the number of important mesoscale phenomena that occur only when the large-scale atmospheric characteristics produce a conducive environment. Examples include frontal squall lines, mesoscale convective complexes, coastal fronts, and freezing rain. Also, land–atmosphere interaction has especially important controls on the solution of mesoscale LAMs (see Chapter 5). Thus, as noted in Section 8.3

above, the ability of the LAM to properly represent these processes will be an important factor in defining the predictability.

There is also perhaps a practical difference in the criteria used to define the predictive skill in forecasts by limited-area, mesoscale models, relative to synoptic- and global-scale models. For forecasts on the large-scale, predictive skill may arguably be defined with a continuous scale in terms of the phase error of waves or accumulated-precipitation amounts. On the mesoscale, in contrast, predictability may be in the context of whether individual high-impact events were forecast or not. The prediction of the existence of a severe-weather event, even in an incorrect location, may be viewed as a very successful forecast. In contrast, forecasters may consider the model solution to have zero utility beyond the point in the simulation when a major precipitation event was not forecast by the model.

Some of the smaller-space-scale processes that motivate the use of mesoscale models, such as moist convection, also have short time scales. And, it is sometimes stated that a practical limit to predictability is one life cycle of a physical process. This leads to a predictability limit of less than a few hours for individual convective events (not long-lived convective complexes).

Predictability based on mesoscale dynamic models has historically been considered to be somewhat complementary with the predictability based on algorithms that perform some sort of extrapolation from the current state. This is shown in the schematic of Fig. 8.5.

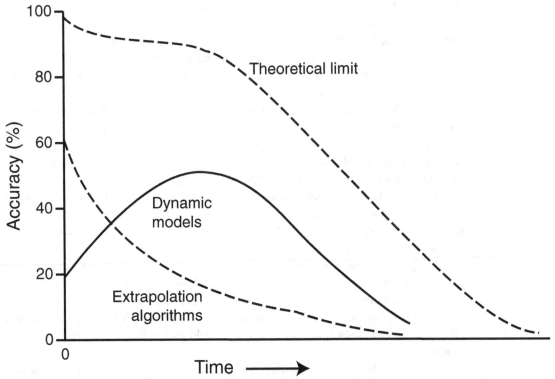

Fig. 8.5  Schematic of the accuracy of forecasts from dynamic models and extrapolation algorithms, relative to the theoretical limit to predictability.

Extrapolation methods, of which persistence is perhaps the simplest, begin with a best estimate of the current state of the atmosphere, and evolve it using statistical or other algorithmic approaches. Because there is no atmospheric dynamics involved, nonlinear processes
cause the accuracy to deteriorate quickly, and thus the predictability limit for this approach
is short. On the other hand, mesoscale models sometimes require a period of time early in
the forecast during which local, thermally driven circulations spin up. And, models can suf-
fer from a dynamic adjustment period after initialization, which causes the solution during
at least the first 6–12 h to be problematic. Both the algorithmic and dynamic approaches
begin from an imperfect state because of the coarseness of the observation network. Even
though the details of the curves will be model and data-assimilation-system and weather-
situation dependent, there is likely a period of time during which algorithmic methods have
greater predictability than do dynamic models.

## 8.6  Predictability and model improvements

New data-assimilation methods, numerical algorithms, and physical-process parameterizations are routinely being developed with the goal of increasing the predictability of
models for use in operational prediction or research. This section provides advice to consider when methods that are thought to represent improvement are implemented, but there
is no apparent benefit to the accuracy of the model products. One issue is that a new
parameterization, for example, may be very sophisticated and include more interactions
among components of the physical system than did the previous version. However, new
required inputs for the more-complex parameterization may be so poorly known that the
new method performs worse than the older and simpler approach – more complicated
methods do not necessarily lead to better performance. An example is that new multi-level
LSMs of one or two decades ago had difficulty producing more-accurate predictions than
were obtainable from single-substrate-layer slab models. Or, cloud-microphysics models
with more microphysical particle types will not perform better than simpler methods if the
conversion among types is not parameterized realistically.

Another consideration is related to the forecast verification metric that is used to assess
the impact of a model change on predictability. It will be shown in Chapter 9 that standard
metrics such as RMSE and MAE often produce better verification scores for smoother
model solutions (e.g., Fig. 9.4). Thus, if a change to the model increases its ability to represent small-scale features, the predictability may appear to decrease. Model changes that
could lead to this misleading response of the predictability metric are decreases in the horizontal grid increment; improvements in the filtering properties of the numerical algorithms such that finer-scale features are retained; or the use of higher-resolution estimates
of the variability of landscape properties, which could lead to finer-scale structures in the
boundary layer.

Lastly, there is the weak-link concept. The modeling system consists of many interacting components, and weaknesses in one of them can prevent improvements in another
from leading to a better model prediction. There are numerous possible examples of this,

but an obvious one is that a new observing system will not improve predictability if the data-assimilation scheme is not able to adequately use the observations.

## 8.7  The impact of post processing on predictability

The post processing of model output, described in Chapter 13, should be viewed as an integral part of the modeling system, at least for operational applications. This processing can take many forms, but one involves the use of methods for reducing the systematic error in the forecast products. The resulting better correspondence with observations will result in improved predictability for the entire system.

### SUGGESTED GENERAL REFERENCES FOR FURTHER READING

Holloway, G., and B. J. West (eds.) (1983). *Predictability of Fluid Motions*. New York, USA: American Institute of Physics.

Kalnay, E. (2003). *Atmospheric Modeling, Data Assimilation and Predictability*. Cambridge, UK: Cambridge University Press.

Leutbecher, M., and T. N. Palmer (2007). Ensemble forecasting. *J. Comp. Phys.*, **227**, 3515–3539.

### PROBLEMS AND EXERCISES

1. Read about Observing System Simulation Experiments (OSSEs) in Section 10.2, and explain how that type of experiment might be used to estimate the relative contribution of model and initial-condition error.
2. Refer to the weak-link concept in Section 8.6, which is related to how model changes impact predictability, and provide additional examples of why anticipated benefits may not be realized.
3. What types of mesoscale processes that can be predicted by LAMs have predictability limits of at least one day?
4. Referring to the accuracy curves for mesoscale models and algorithmic nowcasting methods, shown in Fig. 8.5, discuss what modeling-system, weather-regime, and scale factors will influence the shapes of the curves and their relationship with each other.
5. The predictability of clouds is quite low. Speculate on the reasons.

# Verification methods

## 9.1 Background

### 9.1.1 What is verification?

Forecast verification involves evaluating the quality of forecasts. Various methods exist to accomplish this. In all cases, the process entails comparing model-predicted variables with observations of those variables. The term validation is sometimes used instead of verification, but the intended meaning is the same. That said, the root word "valid" may imply to some that a forecast can either be valid, or invalid, whereas obviously there is a continuous scale that measures forecast quality. Thus, the term verification is preferable to many, and will be employed here. Special verification measures that are most applicable to ensemble predictions have been discussed in Chapter 7. There is an extensive body of literature on the subject of model verification, and students and researchers should read beyond the summary material in this chapter to ensure that they understand underlying statistical concepts and that they use the verification metrics that are most appropriate for their needs.

### 9.1.2 Reasons for verifying model simulations and forecasts

There are multiple motivations for evaluating the quality of model forecasts or simulations.

- Most models are under continuous development, and the only way modelers can know if routine system changes, upgrades, or bug fixes improve the forecast or simulation quality is to objectively and quantitatively calculate error statistics.
- For physical-process studies, where the model is used as a surrogate for the real atmosphere, the model solution must be objectively verified using observations, and if the observations and model solution correspond well where the observations are available, there is some confidence that one can believe the model where there are no observations. This is a necessary step in most physical-process studies.
- When a model is being set up for a research study or for operational forecasting, decisions must be made about choices for physical-process parameterizations, vertical and horizontal resolutions, LBC placement, etc. Objective verification statistics are employed for defining the best configuration.
- Forecasters learn, through using model products over a period of time, about the relative performance of the model for various seasons and meteorological situations. This

process can be made easier through the calculation of weather-regime-dependent and season-dependent verification statistics for the model.

- Objective decision-support systems, which utilize model forecasts as input, can benefit from information about the expected accuracy of the meteorological input data.
- Model-intercomparison projects, which compare model accuracy and skill in order to better understand the strengths and weaknesses of the participating models, are based on a foundation of objective model verification.

### 9.1.3  Some terminology related to forecast performance

It is useful to define some basic terminology that we will be employing. Further discussion of the following definitions can be found in Wilks (2006) and other general references on the subject.

- *Accuracy* – A measure of the average degree to which pairs of forecast values and observed values correspond. Scalar measures of accuracy summarize the overall quality of the forecasts in the form of a single number.
- *Bias* – A measure of the correspondence between the average of a forecast variable and the average of the observations.
- *Skill* – The accuracy of a forecast relative to a reference forecast.
- *Reference forecast* – This is an easily available, non-model-based data set that can be interpreted as a simple, minimal-skill forecast. See Section 9.3 for additional discussion of such reference forecasts.

## 9.2  Some standard metrics used for model verification

### 9.2.1  Accuracy measures for continuous variables

These measures apply to variables that are continuous in the sense that they can take on any value within a physically realistic range. For example, if temperature itself is the predictand, it represents a continuous variable. But, if the predictand is a binary "yes or no", regarding whether the temperature tomorrow will exceeds some threshold, it is a discrete variable (discussed in Section 9.2.2).

The MAE is the arithmetic average of the absolute difference between pairs of forecast and observed quantities. It is the average magnitude of the forecast error, and is defined as

$$MAE = \frac{1}{n} \sum_{k=1}^{n} |x_k - o_k|,$$

where $(x_k, o_k)$ is the $k$-th of $n$ pairs of forecasts and observations. In order for the MAE to be zero, the difference between each forecast and observation pair must equal zero. Another scalar accuracy measure for continuous variables is the Mean-Square Error

(MSE), which is the average squared difference between the forecast and observation pairs. It is defined as

$$MSE = \frac{1}{n} \sum_{k=1}^{n} (x_k - o_k)^2 .$$

Because the errors are squared, the MSE will be more sensitive to large errors than will the MAE. Sometimes, the square root of the MSE is used, such that $RMSE = \sqrt{MSE}$. This has the same physical dimensions as the forecast and observations. The above metrics represent both systematic and random components to the error.

An additional, commonly used measure of correspondence between observations and forecasts is the Anomaly Correlation (AC). As the name implies, it is designed to define similarities in the patterns of the departures (i.e., anomalies) of the observed and forecast variables from the climatological means. The AC can be calculated based on time series or spatial fields, and is designed to reward for good forecasts of the pattern (phase and amplitude) of the observed variable. See Wilks *et al.* (2006) for additional detail.

The bias is the same as the Mean Error (ME), such that

$$ME = Bias = \frac{1}{n} \sum_{k=1}^{n} (x_k - o_k) = \bar{x} - \bar{o} .$$

This is also known as the systematic error. Given that $\bar{o}$ is a simple way of defining the climatology of the variable (at least for the limited period of the verification), and $\bar{x}$ is the model climatology for the variable, the bias represents a comparison of the model and actual climatological values.

## 9.2.2 Accuracy measures for discrete variables

These measures apply when the verification question is defined in terms of a yes–no condition. For example, consider a precipitation forecast of whether the accumulated amount is above a specified threshold at a particular location. The observation at that point defines whether precipitation of that amount indeed occurred, a yes or no condition, and the forecast is also in the form of a yes or no. This problem can be illustrated with a 2 × 2 contingency table of the form shown in Fig. 9.1a. Of the $n$ forecast–observation pairs, $a$ represents the number of times that an observed event was correctly forecast (called hits), $b$ is the number of times that no event occurred but the forecast was for an occurrence (called false alarms), $c$ is the number of times that an observed event is forecast to not occur (called misses), and $d$ is the number of times that an event was correctly forecast to not occur (called a correct negative). An example is shown in Fig. 9.1b of areas where a condition (e.g., 24-h accumulated precipitation above a threshold) is observed and forecast. Each area is defined in terms of the elements of the contingency table.

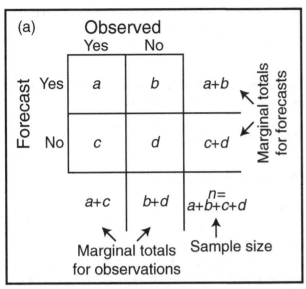
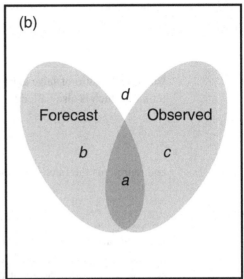

**Fig. 9.1** Contingency table showing the four possible outcomes of a forecast of a discrete variable (a). Also shown is a schematic example of the observed and predicted areas where a variable (e.g., accumulated precipitation) exceeds a specific threshold.

Many forecast-accuracy measures are based on the components of the contingency table. The Proportion Correct is defined as

$$PC = \frac{a + d}{n},\tag{9.1}$$

and represents the fraction of the forecasts that correctly anticipated the event or nonevent. A potential disadvantage of this score for some situations is that equal credit is given for correct positive or negative forecasts. If the forecast variable is the existence of sun in Cairo in summer, the correct forecast of sun is given equal credit as the more difficult correct prediction of the rare obscuring cloud. An alternative is the Threat Score (TS), which is useful when the yes-event to be forecast occurs much less frequently than the no-event. This is also termed the Critical Success Index, and is expressed as

$$TS = CSI = \frac{a}{a + b + c}.\tag{9.2}$$

The bias compares the average forecast and the average observation, and is defined as

$$B = \frac{a + b}{a + c}.$$

The False-Alarm Ratio,

$$FAR = \frac{b}{a + b},$$

is the fraction of yes forecasts that are wrong, and is different from the false-alarm rate

$$F = \frac{b}{b+d}$$

which is the ratio of false alarms to the total number of nonoccurrences of the event. The hit rate, which is also called the probability of detection, is defined as

$$H = POD = \frac{a}{a+c},$$

and represents the fraction of the event occurrences that were forecast.

### 9.2.3 Skill scores

As noted earlier, skill is defined as the accuracy of one forecast method relative to that of a reference forecast. The skill is usually represented as a Skill Score (SS), which is defined as a percentage improvement over the reference forecast. Mathematically, a SS can be defined as

$$SS_{ref} = \frac{A - A_{ref}}{A_{perf} - A_{ref}} \times 100\,\%, \tag{9.3}$$

where $A$ is the accuracy of a forecast, $A_{ref}$ is the accuracy of a reference forecast, and $A_{perf}$ is the accuracy of a perfect forecast. If $A = A_{perf}$, the skill score is 100%. If $A = A_{ref}$, the skill is zero, indicating no improvement relative to the reference forecast. If the forecast accuracy is less than that of the reference forecast, the skill score is negative.

A number of skill scores are based on the previously described 2 × 2 verification contingency table, and have the form of Eq. 9.3. One of the most-frequently used is called the Heidke Skill Score (HSS), and is based on the proportion correct (Eq. 9.1) as the accuracy measure ($A$, in Eq. 9.3). The reference accuracy measure, $A_{ref}$, is the proportion-correct value that would be obtained by random forecasts that are statistically independent of the observations. The expression for the HSS is

$$HSS = \frac{2(ad - bc)}{(a+c)(c+d) + (a+b)(b+d)},$$

where the derivation is described in Wilks (2006) and elsewhere. Analogously, the TS (Eq. 9.2) can be used as the basic accuracy measure, and the TS for random forecasts is used as the reference. This is called the Gilbert Skill Score (GSS) or the Equitable Threat Score (ETS), and is derived in Wilks (2006) as

$$GSS = ETS = \frac{a - a_{ref}}{a - a_{ref} + b + c}, \text{ where}$$

$$a_{ref} = \frac{(a+b)(a+c)}{n}.$$

Skill scores are also computed for continuous variables, using MAE, MSE, or RMSE, again based on Eq. 9.3. Climatology or persistence are generally used for the reference forecast. Using the MSE as an example, the accuracies for these references are

$$MSE_{Clim} = \frac{1}{n} \sum_{k=1}^{n} (\bar{o} - o_k)^2 \text{ and}$$

$$MSE_{Pers} = \frac{1}{n} \sum_{k=1}^{n} (o_{k-1} - o_k)^2,$$

where $o_k$ is the observation, $\bar{o}$ is the climatological mean of the observed variable, and $o_{k-1}$ is the previous value of the variable. Similar equations apply for MAE. For either metric, and for either reference forecast, the skill score, based on Eq. 9.3, can be written as

$$SS = \frac{MSE - MSE_{ref}}{0 - MSE_{ref}} = 1 - \frac{MSE}{MSE_{ref}}.$$

Many more skill scores, with various strengths and weaknesses, are described in Wilks (2006) and Gilleland *et al.* (2009).

## 9.3 More about reference forecasts and their use

The reference forecast defines a minimal-accuracy or zero-accuracy point on the scale – essentially a zero point on the accuracy scale. This is the forecast accuracy that can be achieved without running a model. Example reference forecasts include (1) a persistence forecast where it is assumed that present conditions prevail throughout the forecast period, (2) a diurnal-persistence forecast where it is assumed that the previous day's diurnal cycle is replicated, (3) a forecast based on seasonal climatological-average values of the forecast variables, and (4) the use of random forecasts. The first three approaches are self-explanatory. For the random forecast, the bootstrap technique of Efron and Tibshirani (1993) is an example of a method that can be used. Here, the available observations throughout the study period are repeatedly and randomly resampled (with replacement), to yield multiple synthetic samples (hundreds to thousands) of the same size as the set of observations that are used normally in the verification. These are the random forecasts, which are constrained by the climatological distribution of the observations over the study period. Note that randomly sampling the entire body of observations has the effect of removing the diurnal signal from the data set. Each of the random forecasts is compared with the observations at each verification time, and the average verification score at each time is then used to define the error.

The accuracy of the random no-skill forecast or one of the other reference forecasts, and the estimates described later in Section 9.5.2 of the maximum (or perfect-model) accuracy

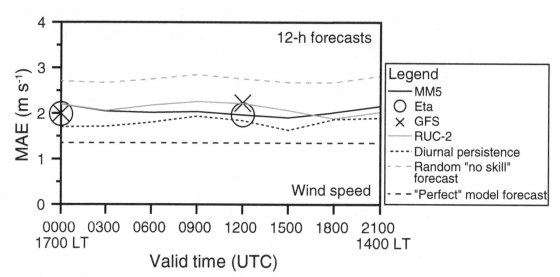

Fig. 9.2 The 12-h forecast MAE for 10-m AGL wind speed for the MM5, Eta, GFS, and RUC-2 models during the 3 February to 30 April 2002 period. Model forecasts from MM5 and RUC-2 were initiated every 3 h. Also displayed are the corresponding statistics from the diurnal persistence, random no-skill, and perfect-model forecasts. Adapted from Rife *et al.* (2004).

that is related to the existence of instrument and representativeness error, effectively produce a range within which falls the actual model-forecast accuracy. Figure 9.2 shows an example of the verification of 12-h wind-speed forecasts from four models, where the bounding perfect-model and no-skill curves are shown. Wind-speed MAEs from all models cluster around 2 m s$^{-1}$, within the range of about 1.5 m s$^{-1}$ between the two bounding curves. The verification statistics for the Eta, GFS, and RUC-2 models were calculated for the mesoscale area spanned by the MM5 LAM.

## 9.4 Truth data sets: observations versus analyses of observations

Verification of model solutions can be performed using either observations or analyses of observations. For the latter approach, operational analyses can be used for near-real-time verification of forecasts, or reanalyses can be used for verification of retrospective simulations. See Chapter 6 for a review of how operational analyses are produced, and Chapter 16 for a description of reanalyses. Using analyses for verification is advantageous in situations where conventional *in-situ* data are sparse, either temporally or spatially. In such situations, variational assimilation of satellite data in the analysis process can constrain the analysis and compensate to some degree for the paucity of *in-situ* observations. One problem with the use of analyses is that they are model generated, so one model is being verified with the products from another model. This is especially troublesome when verifying a variable such as precipitation, where the model that created the analysis often has not assimilated any precipitation observations – that is, the analyzed precipitation is entirely a creation of the model. Another issue is that only global-scale operational analyses or reanalyses are available for much of the world, so there is a clear scale mismatch if forecasts from high-resolution mesoscale models must be verified. The fine-scale details in the solution from the mesoscale model will not have counterparts in the analysis, and measures of accuracy will interpret the differences as errors. Thus, the mesoscale model will be penalized for successfully serving its intended purpose of providing information on the mesoscale. For verification using analyses, the forecast values can be interpolated to the grid points of the analysis, or vice versa. Because of the short distance between grid points, simple bilinear interpolation is often used.

In regions where observations are sufficiently dense, whether they be *in situ* or remotely sensed, it can be advisable to interpolate model forecasts or simulations to the observation points, and calculate the statistics there. *In-situ* observations include those from radio-sondes, near-surface mesonetwork stations, aircraft-borne sensors, and rain and snow gauges. Remotely sensed data that can be compared directly with model output may be from wind profilers (Doppler radars that point approximately in the vertical), radial winds from scanning Doppler radars and lidars, satellite cloud-track winds, satellite water-vapor-track winds, and satellite-estimated precipitation. Interpolation of model values from the grid to observation locations can be done through simple linear interpolation. Near-surface wind observations are often at 10 m AGL, although this height can vary, and corresponding temperature and humidity observations are generally at 2 m AGL. When the lowest model computational level is above the elevation of the observations, Monin–Obukhov similarity theory can be used to extrapolate the wind, temperature, and humidity predictions to the height of the observations (Stull 1988).

## 9.5  Special considerations

### 9.5.1  Orographic smoothing

Verification of a model solution is complicated by the fact that model orography is smooth compared to the actual orography. Unless envelope orography is used with a model, valleys are filled in and mountain ridges are lowered. Thus, a surface-based observation has a different elevation above sea level in the model compared to reality. As a practical example, assume that a model is used to forecast the winds at the 80-m AGL hub height of wind turbines that are located on a ridge near a coastline (see Fig. 9.3). And, assume that winds observed by anemometers at the hub height are available for model verification. The actual ridge is 100 m ASL, but the smoothness of the model terrain causes the ridge in the model to be only 20 m ASL. This raises the question about which model winds should be compared with the 180-m ASL observations. Based on the distance of the observation above sea level, 180-m ASL model winds (level 1) should be used. But, if distance above the local land surface is more physically relevant, the model winds at 100 m ASL (level 2) should be used. Note that this problem is not limited to wind prediction, and applies to temperature and humidity as well. In this example, the question takes on greater

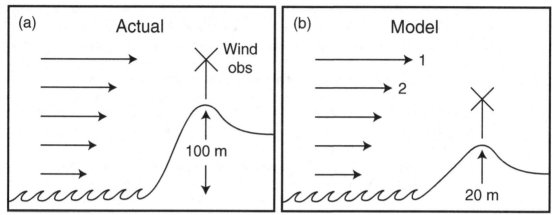

Fig. 9.3 Schematic showing a cross section of the orography near a coastline, based on the actual elevation (a) and that associated with the smoothed orography of a model (b). Wind observations are available at the top of an 80-m tall wind-turbine tower, and the issue is whether to compare the observation with model winds at the height ASL in the model (level 2) or the height ASL in nature (level 1).

importance when the prevailing large-scale wind is from the water. There is thus sometimes the question whether an observation should be assumed to apply at the correct location in the vertical (ASL) or at the correct distance above the model land surface. There is no easy answer to this question, but the modeler should be aware of the paradox because the choice will certainly influence verification statistics.

The above example might be interpreted as being somewhat isolated; however, in complex orography a systematic error will generally exist in forecasts of variables that have a significant climatological variation with height. For example, model forecasts will have a warm bias relative to observations located over mountains, where the surface elevation in the model is lower than in reality as a result of orographic smoothing. Similarly, wind speeds will be underforecast at higher elevations when the orography is smoothed.

### 9.5.2 The imperfect nature of verifying observations

It is well recognized that the errors produced by the model and the errors in the observations both contribute to the total error diagnosed in the verification process. The observations used for verification contain error associated with the accuracy of the instruments and the calibration error. In addition, there is another, somewhat less well-documented, cause of differences between the model solution and the observations that impacts the verification statistics. And, this error will always exist regardless of how much the model and instrument errors are reduced. This is the representativeness error.

Representativeness errors arise from the fact that there is a fundamental mismatch between the spatial and temporal scales represented by the models and the observations. Conventional ground-based instruments make instantaneous or time-averaged measurements at a point, whereas the model-predicted quantities represent spatial averages over each model grid-box volume. In addition, the numerics and explicit diffusion smooth the grid-box information so that it, in fact, represents an even-larger area (see Fig. 3.36). Representativeness error can be understood through the following idealized example. Suppose there exists a perfectly known near-surface wind field over a 1-km$^2$ area. The wind at the center of this area is used to create a "perfect" point observation of the wind. Next, the 1-km$^2$ spatial average is computed, which represents the corresponding grid-box-mean value of the wind predicted by a perfect model. Despite the fact that both the model-average and observation exactly characterize the wind field in their own way, the difference between the two will obviously not be zero. This difference is termed the representativeness error, and its magnitude is dependent on a number of factors including the prevailing weather regime, the amplitude of fine-scale atmospheric structures, and the geographic extent of the sampling area (or size of the model grid box).

It is worthwhile estimating this error because it contributes to the maximum model accuracy (see definition above) that is practically achievable, given the properties of the forecasting systems and the verifying observations. A tractable approach is to use an extremely high-resolution model to define the variability within a larger grid-box area. For example, Rife *et al.* (2004) used the model described by Clark and Farley (1984) and Clark and Hall (1991, 1996) to estimate the representativeness error in a verification of MM5 mesoscale-model wind simulations in complex terrain. The Clark–Hall model was run for a real-data case over the complex terrain near Pinewood Springs, Colorado, where the highest resolution grid in a nest had an increment of approximately 50 m and encompassed a nearly 36-km$^2$ area. To estimate the representativeness error, the spatially averaged wind speed and direction were computed from the Clark–Hall model output within a stencil having the dimensions of a 1.33-km grid box of the MM5 model. There were 676 Clark–Hall model grid points for each MM5 grid box. Next, the point values of the speed and direction were determined at the stencil center. This process was repeated until the entire Clark–Hall model domain had been sampled in a nonoverlapping fashion. The mean difference between the grid-box-average and point values of wind speed and direction from each unique sample (36 individual paired values) was computed to produce an estimate of the representativeness error.

Based on the above analysis, the representativeness errors for 10-m AGL wind speed and direction, under well-mixed boundary-layer conditions with this MM5 model resolution in complex terrain, are 1.15 m s$^{-1}$ and 14.6°, respectively. This estimate is conservative because the Clark–Hall model with a 50-m grid increment underestimated the true amount of spatial variability that would exist in the near-surface wind field. Conventional cup and vane anemometers are generally accurate to within ±0.3 m s$^{-1}$ and ±3° for wind speed and direction, respectively. This yields a practically realizable minimum error for a wind speed and direction forecast by a perfect model of 1.45 m s$^{-1}$ and 17.6°, respectively (assuming that the errors are additive).

The existence of representativeness error can extend beyond the influences of complex orography. For example, observations are made sufficiently far from natural obstructions, or those of human origin, such that the measured value of a variable is presumably not influenced by the obstacle. However, model grid-box-average values of surface properties, such as roughness length, are defined based on the average character of the surface over the grid-box area. Because grid boxes often contain obstructions, the average roughness length is defined accordingly. Thus, the observed wind experiences a different roughness than does the model wind, and this representativeness problem can lead to a difference between observed and modeled winds that has nothing to do with model accuracy. Strassberg *et al.* (2008) calculate that this effect can lead to small but nontrivial artificial errors in the wind verification. See the problem at the end of the chapter regarding how similar landscape representativeness problems can lead to errors in the verification of near-surface temperature or humidity.

The results of the above analyses of representativeness of course apply only to a specific configuration of the forecast model, the spatial structure of the meteorological field being sampled, and the error properties of the observing system. A separate analysis would need to be performed for other situations. Nevertheless, the example illustrates the potential importance of these factors to the verification process. Note that this sum of the representativeness error and the instrument error can be viewed as an upper bound on the forecast accuracy. That is, this is the accuracy of a perfect model.

### 9.5.3  Special issues related to the verification of winds

There are a few special issues that should be kept in mind with respect to the verification of winds. One is related to the fact that, unlike the other variables discussed here, wind is a vector quantity. We thus have the option of comparing the observed and forecast wind in terms of (1) separate statistics for the $u$ and $v$ components, (2) separate statistics for the speed and direction, and (3) vector differences. On the synoptic scale in midlatitudes, or with verification of upper-air winds, individual verification of $u$ and $v$ can make physical sense if the components align with the zonal and meridional direction. That is, the components represent the direction of the mean wind, and the perturbations to the mean wind. Wind speed and direction are intuitive metrics because they are geometric attributes of the vector, and the two types of error can be easily visualized. Alternatively, the vector difference between the observed and forecast winds is a way of representing the error.

If wind speed and direction are used for verification, account should be taken of the fact that low mean-wind speeds are associated with highly variable directions because turbulence will dominate the measurement. Given that we wish to verify the mean (nonturbulent) wind direction, it is common practice to not include in the verification, wind directions that are associated with speeds of less than some threshold, such as 0.5 m s$^{-1}$. Also, direction-error calculations are complicated by the fact that the direction scale is periodic, and this needs to be remembered when differencing the modeled and observed values.

A point that was made in Chapter 3 is that models with Cartesian grids, which are defined on map projections, have $u$ and $v$ wind components defined in terms of the rows and columns of grid points. That is, the $u$ component is parallel to the rows and the $v$ component is parallel to the columns. Thus, the model-defined wind components at a particular latitude–longitude are not the same as those reported in an observation at the same point. The latter components, of course, are parallel to the local latitude and longitude lines. Thus, just as with the model initialization process, the model and observed wind components need to be reconciled. In this case, if verification statistics are being computed at observation locations, the model wind components are transformed to the traditional geocentric components.

## 9.5.4  Response of the standard accuracy metrics to horizontal resolution

Conventional objective measures of forecast skill sometimes seem to show little improvement from increased horizontal resolution, in spite of the fact that a subjective assessment of model accuracy shows a definite positive impact. For example, Mass *et al.* (2002) describe the overall performance of a real-time mesoscale weather prediction system, and show that there were clear improvements in the objectively measured forecast skill as the horizontal grid spacing was decreased from 36 to 12 km. In contrast, there were only small improvements in the objective skill as the grid spacing was decreased from 12 to 4 km. However, in terms of subjective comparisons of observed and forecast structures, the coarser-resolution forecasts were often profoundly inferior to those from the highest-resolution grid. Similarly, Davis *et al.* (1999) showed that, in terms of conventional skill scores such as bias, MAE, and RMSE, a high-resolution (1.11-km grid increment) mesoscale model that was run operationally over the mountainous western USA provided only slightly better surface temperature forecasts than did the much coarser 80-km Eta model, with the two models exhibiting 10-m wind-field forecast errors of comparable magnitude. However, only the mesoscale model was able to accurately depict some important aspects of the observed locally forced circulations resulting from the regional orography and variations in other land-surface characteristics. Another study, for east-central Florida, compared objective skill scores from a mesoscale model, which employed a 1.25-km horizontal grid increment, to the scores from the 32-km grid increment Eta model (Case *et al.* 2002). The high-resolution model provided little objective improvement over the much coarser Eta model. However, a detailed subjective analysis indicated that the mesoscale model exhibited considerably more skill in predicting the observed Florida sea breeze, which strongly determines temperature and the timing and location of thunderstorm initiation.

One well-known characteristic of some standard verification metrics is that they can reward smooth solutions. That is, if output from a high-resolution model is progressively smoothed, the accuracy metrics calculated from the output may show progressively greater skill. Thus, the use of higher horizontal resolution, or the use of numerical methods that have small truncation error, can lead to poorer model verification. This is in spite of the fact that these model properties should lead to a more-realistic representation of fine-scale structures in the model solution. Figure 9.4 illustrates the common situation where model forecasts can have both phase and amplitude errors. The solid line represents the observed wind speed in a jet that is oriented perpendicular to the page. The dashed line shows a forecast from a high-horizontal-resolution model, where the correct amplitude of the jet is retained, but the maximum is displaced to one side. The dot-dash line and the dotted line show solutions from models that have less horizontal resolution, and therefore produce a smoother solution. The RMSE between the observed and forecast wind speed is greater for the model solution that better retains the correct amplitude of the feature.

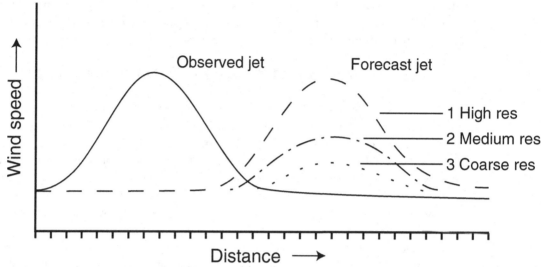

Fig. 9.4 Illustration of how a smooth forecast can lead to better verification statistics than a forecast with greater amplitudes in structural features. See text for details.

Another way to look at this influence of fine-scale spatial structures on the standard verification metrics is in the context of the decomposition of the definition for MSE. Murphy (1988) demonstrates that

$$MSE = (\bar{x} - \bar{o})^2 + \sigma_x^2 + \sigma_o^2 - 2\sigma_x \sigma_o r_{xo}.$$
$$\quad\quad (1) \quad\quad (2) \quad (3) \quad\quad (4)$$

Term (1) is the mean error, or bias; term (2) is the variance of the forecasts; term (3) is the variance of the observations; and term (4) contains $r_{xo}$, the coefficient of correlation between the forecasts and observations. Thus, all other things being equal, high-resolution forecasts or verification fields (observations) with larger variance, will lead to larger MSEs.

## 9.6 Verification in terms of probability distribution functions

Because the extremes in weather (temperature, precipitation) are often the most important situations that must be forecast with models, it is useful to use verification methods that provide specific information about how model accuracy varies for different values of the predictand. A simple approach would be to simply plot the frequency distributions of the observed and forecast variables for a point, based on a long series of forecasts. This will provide information about how well the model-forecast climatology verifies relative to the actual climatology for extreme values of the variable, which is important, but it does not quantify how accurately the extremes are forecast. To accomplish the latter, the joint distribution of the observed and forecast values of a variable can be plotted. For example, Fig. 9.5a

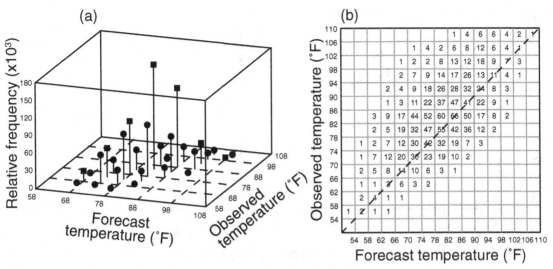

**Fig. 9.5** Examples of joint distributions of observed and forecast 2-m AGL temperatures. A bivariate histogram shows the joint distribution of observed temperatures and 24-h model forecasts of temperature, for one location in the warm season (a). Another type of display is also shown (b), based on a different data set, where the numbers plotted are the scaled frequencies of occurrence. Panel (a) is adapted from Murphy *et al.* (1989).

contains a bivariate histogram that shows the distribution of forecast temperatures for different values of the observed temperature. Similar information is provided in the simpler display in Fig. 9.5b, for a different set of forecasts. Alternatively, scatter plots showing observed and forecast values could be used to reveal errors in different parts of the distribution. No matter how the information is displayed, it can be used to better understand how well a model predicts the extreme values in the PDF, so that the modeling system can be improved as needed, and so that forecasters can develop a better knowledge of model strengths and weaknesses relative to forecasts of weather extremes.

## 9.7  Verification stratified by weather regime, time of day, and season

Model-verification statistics can be stratified by time of day, season, forecast duration, and weather regime. The resulting situation-dependent model-performance statistics can reveal information that is useful for isolating model shortcomings. For example, separating the precipitation-prediction skill by season provides insight into the model calculations of convective versus stable precipitation. Calculation of time-of-day-dependent statistics for near-surface variables can distinguish errors in the boundary-layer parameterization that manifest themselves differently during daytime, nocturnal, and day–night-transition regimes. Figure 9.6 shows examples of the verification of a mesoscale model by season and time of day. In Fig. 9.6a is the 2-m AGL temperature RMSE for forecasts of 24-h duration initialized at 3-h intervals, aggregated for all seasons. The location is the southwestern USA.

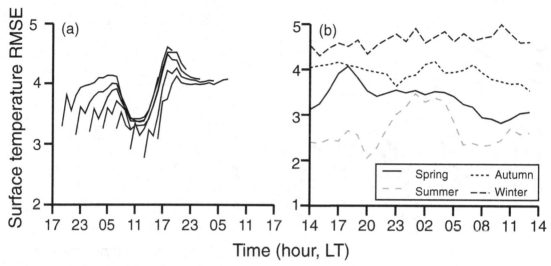

**Fig. 9.6** Examples of the verification of a mesoscale model by season and time-of-day (local). For a region in the southwestern USA, the 2-m AGL temperature RMSE is shown for forecasts of 24-h duration, initialized at 3-h intervals, and aggregated for all seasons (a). Also, for Alaska, the temperature RMSE is shown for the 10–12-h segment of mesoscale forecasts, calculated separately for each season and for different times of the day (b). Adapted from Liu *et al.* (2008b).

Regardless of when the forecast is initialized, the RMSE is a minimum in the late morning and early afternoon, and increases rapidly during the late afternoon and evening. This is potentially valuable evidence for identifying aspects of the boundary-layer parameterization that need improvement. In a second example (Fig. 9.6b), the temperature RMSE for the 10–12-h lead time of forecasts is shown separately for each season and for different times of the day. There is a clear trend for the errors to be greater in the winter. There is no diurnal variation in the cold-season RMSE because the forecasts are for central Alaska. But, the warm-season forecasts display an obvious error maximum during the nighttime hours.

Calculation of verification statistics for different weather regimes can also be revealing of model strengths and weaknesses. The process involves using some method to classify different weather regimes that prevailed during a large number of model forecasts. On global scales, the regimes might be extremes in a global cycle such as ENSO. In this case, forecast skill during the El Niño phase could be compared with the skill during the La Niña phase. On regional scales, cluster analysis methods (Wilks 2006) or the methods of self-organizing maps (Cassano *et al.* 2006, Seefeldt and Cassano 2008) can be used to aggregate weather patterns into different regimes, and the forecast error statistics can be calculated separately for each of the regimes. Or, more manual and subjective methods can be used to identify weather regimes, before the separate statistics are calculated. As an example of this concept, Fig. 9.7 shows the 2-m AGL temperature bias, computed separately for two large-scale weather regimes, based on mesoscale-model (MM5) summer-season forecasts for Greece, in the vicinity of Athens. There are two dominant wind regimes in this season. When the northerly Etesian winds are strong, they sweep across the peninsula and inhibit the development of a sea breeze. When the Etesian flow is weak, the

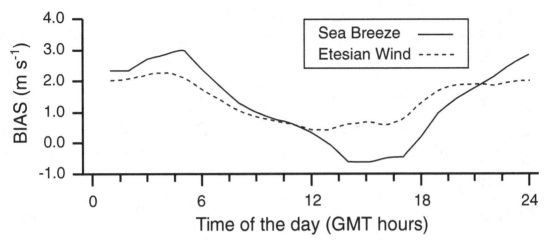

Fig. 9.7 The 2-m AGL temperature bias, computed separately for two large-scale weather regimes in the vicinity of Athens, Greece, based on three months of mesoscale-model summer-season simulations. One regime involves strong northerly Etesian winds and the other pertains to situations where the sea breeze dominates during weak Etesian forcing. Provided by Andrea Hahmann, Risø.

sea breezes from the north and south can dominate, penetrating to the center of the penin-sula. Even though the daily-mean temperature bias is not much different for the two regimes, the diurnal amplitude of the bias is over twice as large when the sea breeze dom-inates, likely indicating a significant contribution by model errors in some aspect of the simulated land–atmosphere interaction.

## 9.8 Feature-based, event-based, or object-based verification

The most useful information in weather forecasts is often related to changes or events, such as abrupt shifts in temperature or wind speed associated with frontal passages. Thus, model fore-cast verification can be especially meaningful if it is performed in terms of how well events are forecast. The terms objects, features, or events are used interchangeably in the literature.

An application of this approach to wind events is described in Rife and Davis (2005). Figure 9.8 illustrates an event in a time series of hourly wind observations. In this study, an event is defined as a 2-h change in the wind speed that exceeds one standard deviation from the 1-year average value at a given station and time of day. Two verification metrics are used. For one, a set of events is defined in a time series of observations, where $o_t - o_{t-2\Delta t}$ is defined as the event in the observations, and $\Delta t = 1\,\mathrm{h}$. For each observed event, the following quantity is calculated,

$$\frac{\sigma_o}{\sigma_x}\left(\frac{x_t - x_{t-2\Delta t}}{o_t - o_{t-2\Delta t}}\right),$$

where $x_t - x_{t-2\Delta t}$ is the change in the model solution for the location and time period of the observed event. The individual observed and forecast event magnitudes are normalized

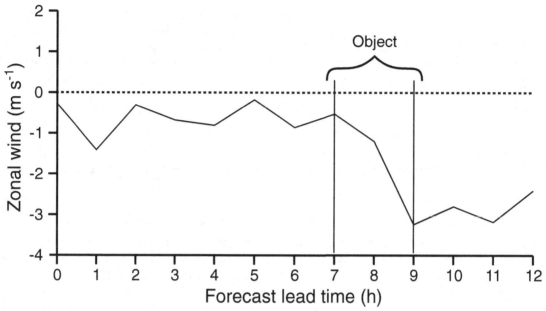

Fig. 9.8 Example of an object, or event, defined in terms of the speed of the zonal wind. Adapted from Rife and Davis (2005).

by the respective variances in the two time series for each location. This ratio is calculated for each station, where the value is +1 for a perfect forecast. In the second approach, again based on time series of observed and forecast values of a variable, both forecast and observed features are defined within every 12-h period. The resulting binary data set, of the existence of one or more features within each period, can be used to populate a contingency table of the sort shown in Fig. 9.3a, allowing the calculation of the many accuracy metrics that are based on discrete variables.

The verification of convective precipitation is well known to be especially problematic, and it lends itself to the use of feature-based methods. Other approaches in which analyses and forecasts of precipitation fields are overlaid, and the overlap regions used to compute scores (Section 9.2.2 above, Wilks 2006), sometimes do not adequately represent the accuracy or value of a forecast. Thus, alternative feature-based approaches have been developed that provide better metrics (Nachamkin *et al.* 2005, Ebert and McBride 2000, Davis *et al.* 2006a,b). Davis *et al.* (2006a,b) should be consulted for a summary of feature-based verification procedures applied to precipitation. In summary, the general approach involves (1) identifying features in the observed (e.g., radar-based analyses) and forecast precipitation fields using thresholds of precipitation amount; (2) describing the geometric properties of the features (e.g., number, location, shape, orientation, size, average precipitation intensity in the feature); (3) comparing the relative attributes of the observed and forecast features; and (4) associating features in the forecast and observed fields, where possible. Figure 9.9 illustrates the benefit of this method for verifying precipitation forecasts. Shown are different examples of observed and forecast areas of precipitation. Forecasts (a)–(d) all have identical basic verification statistics, with POD = 0, FAR = 1,

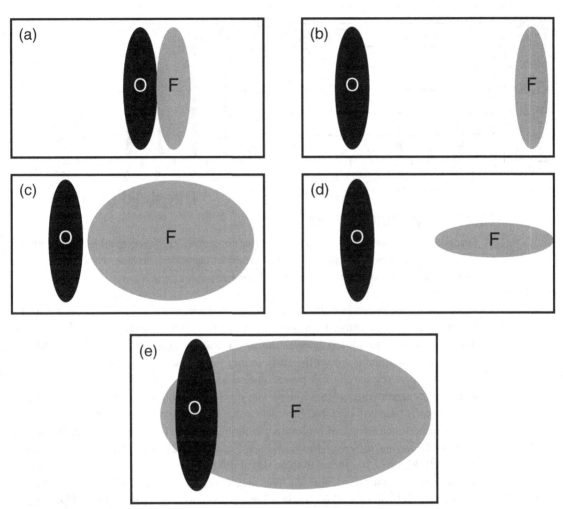

Fig. 9.9 Schematic example of different combinations of forecasts and observations.  From Davis *et al.* (2006a).

and CSI = 0 because the forecast and observed precipitation areas do not overlap (the area of "a" in Fig. 9.1b is zero). However, there clearly are differences in the forecast accuracy or value. Forecast (e) has some skill based on these metrics, but would likely not be judged as the best of the five. However, feature-based methods that compare the distances between features, their orientation, and area would provide a better ranking.

An example is shown in Fig. 9.10 of the use of feature-based methods to compare the skill of a few models at predicting summer-season precipitation in an area of complex orography in the southwestern USA. In particular, the method described in Davis *et al.* (2006a,b) will be employed here to compare the accuracy of short-range precipitation forecasts from the MM5, NAM, and RUC models. The horizontal grid increments are 10 km for MM5, 12 km for NAM, and 13 km for RUC. The precipitation forecasts are compared with the NCEP Stage-IV analyses, which are constructed by compositing

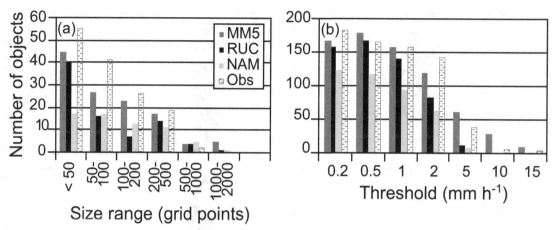

Fig. 9.10 Comparison of the number of observed and forecast precipitation features (based on closed isohyets) in each feature size range (number of grid boxes) (a) and for different precipitation thresholds (b), for three models, and observations based on radar and rain-gauge data. See text for details. Provided by Yubao Liu, NCAR.

WSR-88D radar and gage estimates (Fulton *et al.* 1998, Lin and Mitchell 2005). Precipitation features in the analyses and model forecasts are defined using thresholds of precipitation amount. Any closed isohyet defines a feature, whether or not it represents a maximum or minimum within the surrounding field. The forecasts were verified for the entire month of August 2005, a period during which the North American monsoon was active in the southwestern USA. Given that there were eight forecast cycles per day, this verification is based on over 200 forecasts, most of which have some rainfall.

The comparison here will be limited to the number of forecast and observed rainfall objects (1) having different area coverage (size) and (2) defined by different rainfall thresholds. Figure 9.10a shows the size distributions of the observed and forecast features for the 2-mm h$^{-1}$ intensity threshold. This measure reflects whether the models capture the degree to which the rainfall occurs in a scattered versus contiguous pattern. Figure 9.10b compares the number of observed features for each rain-rate threshold, with those in the model forecasts. For both measures, there are clear differences among the models.

## 9.9 Verification in terms of the scales of atmospheric features

A model solution should approximately preserve the observed spatial and temporal spectra of the dependent variables. Thus, the spectral power for the atmosphere and the model solution have been compared in numerous studies. Of course, unlike other verification criteria, model-simulated features do not need to be in phase with observed features in order for the model to verify well in this context. Rather, the model solution simply needs to contain the features on the correct scales. This type of verification can:

• help the modeler better understand the explicit and implicit spatial and temporal filters in a model;

- provide information about whether the lower-boundary forcing in the model is imparting the correct scales of motion in the lower troposphere and boundary layer;
- define the model's true resolution;
- illustrate the amount of fine-scale information that is contained in the initial conditions, provided by a data-assimilation system; and
- define the time required for the model to spin up scales of motion that are not in the initial conditions.

## 9.9.1 Temporal spectra

It is common to partition the temporal spectrum into three regions: periods longer than diurnal (super-diurnal), periods that are approximately diurnal, and periods that are shorter than diurnal (sub-diurnal). Features with longer-than-diurnal periods may be viewed as synoptic scale or planetary scale, and therefore reasonably representable by global or regional models that have typical horizontal resolution. Diurnal time scales of course are related in some way to the heating cycle. For example, in the wind field a diurnal signal could be related to stability-related momentum mixing, mountain-valley circulations, coastal circulations, etc. Provided that the model reasonably represents the land-surface and boundary-layer processes, features with these time scales should verify well. Motions with sub-diurnal time scales include mesoscale features or circulations that are not forced by the diurnal heating cycle. They can result from orographic or other landscape forcing, perhaps far upstream, or from nonlinear interactions. Comparison of the observed and model-simulated spectral power in each of these three regions is a way of verifying the ability of the model to simulate these types of features.

Figure 9.11a shows how the observed temporal spectrum, which must be represented by a model, depends on geographic location. Illustrated is the spectral power for time series of the observed zonal wind at three locations in Slovenia. The higher-elevation station on a mountain (M) is exposed to the synoptic-scale flow, more than are the other stations, so there is greater power in the longer time scales. The coastal station (C) has power maxima on approximately the diurnal time scale (12 h and 24 h peaks), as a result of thermally forced coastal circulations. And the valley station (V), which is protected by the orography from synoptic-scale features and has weak diurnal forcing, shows the flattest spectrum, the lowest overall power, and the largest fraction of the total power in the sub-diurnal range. A model that properly represents the various prevailing processes should replicate this spectrum, and have roughly the same percentage of the spectral power in each of the bands. This type of verification thus has the potential to reveal model strengths and weaknesses in an interesting way. Figure 9.11b shows how well zonal-wind power spectra from the ERA-40 reanalysis with a 125-km grid increment (see Chapter 16 for more information on this analysis) and the ALADIN model with a 10-km grid increment compare with the observed spectrum for the coastal station. Both the reanalysis and the model underestimate the power at all sub-diurnal scales, and overestimate the power on the synoptic scales. The relatively coarse-resolution ERA-40 analysis misses most of the diurnal component, but the ALADIN model captures its approximate amplitude.

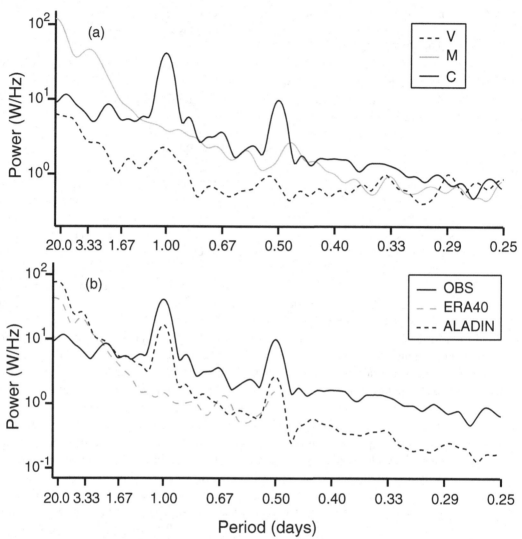

Observed spectral power in the time series of zonal wind speed at three locations: a mountain station (M), a valley station (V), and a coastal station (C) (a). Also shown, along with the observed spectrum for the coastal station, are spectra associated with the ERA40 reanalysis (125-km grid increment) and the ALADIN model (10-km grid increment) (b). Adapted from Žagar *et al.* (2006).

## 9.9.2  Spatial spectra

The above type of verification of model temporal spectra is made possible and convenient by the general availability of frequent measurements at surface stations. However, there is no equivalently good source of dense observations for use in comparisons of observed and model-based spatial spectra. Thus, the model solution is compared with field-program observations and theoretical solutions, for example of the shape of kinetic-energy spectra. As with the verification of the model temporal spectra, this spatial verification will also confirm the degree to which the model is faithful to the dynamics of the atmosphere.

See Skamarock (2004) for additional information about the characteristics of the kinetic-energy spectrum to be expected on different space scales. In summary, global-scale models should reproduce the large-scale, $k^{-3}$ slope of the spectrum. In mesoscale and cloud-scale model solutions, the slope should be $k^{-5/3}$. Examinations of model spectra have been used to verify possible negative impacts of explicit or implicit damping mechanisms (Laursen and Eliasen 1989). When model resolutions span the global scale and the mesoscale, as is the case with high-resolution global models, the verification of the existence of the slope transition in the kinetic-energy spectrum is a test that the model is faithful to the atmospheric dynamics (Koshyk and Hamilton 2001). And, analyses of kinetic-energy spectra have been used to verify the ability of a model to represent scales near the $2\Delta x$ limit of resolution (Bryan *et al.* 2003, Lean and Clark 2003, Skamarock 2004). This latter type of verification defines the *effective resolution* of the model. Figure 9.12 illustrates the concept of examining the kinetic-energy spectrum for this purpose. The sloping straight line is the anticipated spectrum, where the slope depends on the wavenumber range ($k$). Because of explicit and/or implicit dissipation in the model, there is some wavelength above the $2\Delta x$ limit where the model spectrum shows kinetic energy that is less than the expected value. This has been defined as the effective resolution because the model dissipation causes the kinetic energy to be unrealistic. The specific shape of the high-wavenumber part of the spectrum is dependent on the model. See Skamarock (2004) for illustrations of actual model spectra that demonstrate this concept.

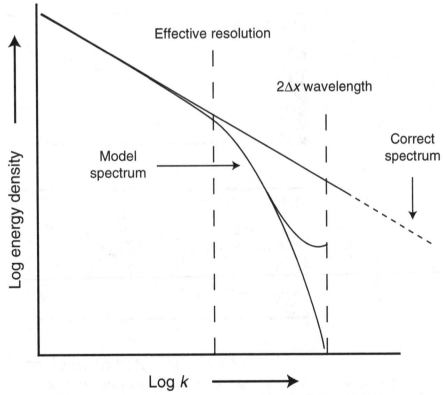

**Fig. 9.12**  Schematic of the theoretical and model kinetic-energy spectrum, showing the effective resolution. Two examples are shown of the decay in the kinetic energy at the high-wavenumber end of the model spectrum. Adapted from Skamarock (2004).

### 9.9.3  Variance

Comparing the observed spatial variance among observations with the corresponding variance based on the model output for the same locations is another way of estimating the realism of the simulated spatial structure. Figure 9.13 shows a comparison of the spatial variance for observed, and 12-h model forecast, 10-m AGL winds for a region of complex orography. The variances for a coarse-resolution global model (GFS) and for two resolutions of a mesoscale model (MM5) are shown. In each case, the model-forecast winds

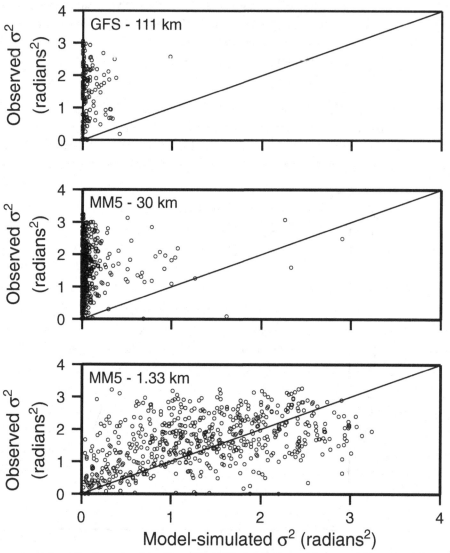

**Fig. 9.13**   Comparison between the observed spatial variance of 10-m-AGL wind direction and the corresponding variances from 12-h forecasts from the GFS model (111-km grid increment) and the MM5 model (30-km and 1.33-km grid increments) during a 3 February to 30 April 2002 study period. Each point corresponds to a single observation time. Adapted from Rife *et al.* (2004).

were interpolated to the locations of the observations before the variance was calculated. The analysis illustrates the benefit of the higher horizontal resolution toward replicating the observed variance.

## 9.10  The use of reforecasts for model verification

Chapter 10 defines reforecasts as retrospective forecasts that have been produced with a fixed version of a model, which is the same as what is currently used operationally. Verification statistics that are calculated from a long series of reforecasts provide a more meaningful description of model biases and other shortcomings than would statistics that are based on only a short recent history of operational forecasts. Additionally, rerunning forecasts with the current version of the model means that the verification statistics apply specifically to the current model, a situation that would not be the case if verification statistics were calculated from archived operational forecasts that were based on an evolving model. References on this subject are Hamill *et al.* (2004, 2006), Hamill and Whitaker (2006), and Glahn (2008).

## 9.11  Forecast-value-based verification

Assessment of the value of ensemble probabilistic forecasts was discussed in Chapter 7, but the importance of this process is worth repeating in the context of model verification. Forecast models are developed and employed because it is anticipated that there is some value to the products that they provide. This value associated with the numerical products can be defined in terms of money saved by business and government, lives saved, and public convenience. Thus, given that these are the ultimate goals of using the models, it is reasonable to verify the models in this value-centric context. That is, these values can serve as metrics that demonstrate, for example, the relative merits of different models. However, it is generally quite challenging to quantify the monetary value of forecasts. And, verification methods that are aimed at improving model performance must be based on accuracy that is expressed in terms of physical forecast variables. Nevertheless, for any operational model application, it is instructive for the modeler to at least qualitatively consider how the forecast value could be defined. There is a substantial body of literature on this subject, and Wilks (2006) and McCollor and Stull (2008b) are among many sources of information about the mathematical basis behind the assessment of forecast value.

## 9.12  Choosing appropriate verification metrics

There are obviously many choices of metrics available for assessing the accuracy or skill of model forecasts or simulations, and it is often difficult to decide which ones are the best for a particular situation. The reader should refer to the general references at the end of the chapter, to better understand the sometimes subtle differences in the properties of the different metrics. There are also choices with respect to the most appropriate variables on which to focus. In both cases, the answers depend on the ultimate use of the model output.

In terms of the relevant variables, if the most important use of the model output is to provide input to a flash-flood-prediction system, obviously hourly precipitation rate is most important, with an emphasis on verifying the higher-rate thresholds. Forecasting the precipitation in the correct watershed is important, so the geometry of the landscape can inform the selection of acceptable displacement criteria in feature-based verification methods. If the model is used to provide input to an air-quality model, errors in the low-level winds and boundary-layer depth contribute to errors in the boundary-layer ventilation. And, the depth and strength of surface-based inversions would be important aspects of the forecast. Ideally, if the atmospheric model is being used to provide input to specialized models, such as the above examples for flood and air-quality prediction, the verification should also be in the context of the ultimate variables – e.g., ozone concentration, water discharge in a river, etc.

## 9.13  Model-verification toolkits

Model verification can be made easier by taking advantage of toolkits available for this purpose. Some are model specific, and others are not. For example, the WRF model has the Model Evaluation Toolkit (MET, supported by NCAR). A free software environment for statistical computing and graphics, which can be used for verification of any model, is provided by the "R Project for Statistical Computing". Available on-line for virtually all the toolkits are the software itself, manuals, announcements of conferences for users to compare applications, frequently asked questions, newsletters, etc. Regardless of the model that is being used, it is worth inquiring about the availability of verification support services such as these.

## 9.14  Observations for model verification

Observations are, of course, required for both model initialization and verification. However, the availability of observations varies greatly for different parts of the world. Section 6.2 summarizes the observation platforms that can provide *in-situ* observations over land, as well as remotely sensed data that can be processed with retrieval algorithms to produce state variables or precipitation rates. Global data are archived by operational and research centers throughout the world, such as NCEP, ECMWF, the United Kingdom Meteorological Office (UKMO), NASA, the European Space Agency (ESA), and NCAR. The data are often available on-line, at no cost. As an example, Advanced Data Processing data sets are available from the NCAR Computational and Information Systems Laboratory's Data Support Section. The data represent a global synoptic set of hourly surface and

6-hourly data reports, operationally collected by NCEP. The surface data set includes mostly SYNOP[1] and METAR[2] land reports, but a few ship observations also exist. The upper-air data consist mainly of radiosonde soundings.

Because precipitation is a variable that often receives special attention for model verification, some special sources of such data are worth mentioning. Gauge data are, of course, useful, except they are only available over land. And their spatial distribution is far from uniform, and their locations tend to be biased toward more-populated lower elevations rather than the higher elevations where the precipitation is generally greater. Other data sets are based on a merger of satellite and gauge data. For example, the Tropical Rainfall Measurement Mission (TRMM) product (product 3B43; Huffman *et al.* 2007) combines precipitation estimates from multiple satellites (retrievals from measurements in the microwave and infrared regions of the spectrum) as well as gauge-based analyses on a $0.25° \times 0.25°$ grid that extends from $50°$ N to $50°$ S for the period from 1998 to the present. For verification of model climatologies, the Global Precipitation Climatology Centre (GPCC; Beck *et al.* 2005) data set provides gauge-based monthly precipitation totals from 1901 to the present on a $0.5° \times 0.5°$ global grid, but only over land. And, some national weather services produce analyses based on weather-radar data. For example, the US NCEP, 4-km, Stage-IV multi-sensor precipitation analysis is constructed using WSR-88D radar estimates, corrected by available gauge measurements (Lin and Mitchell 2005, Fulton *et al.* 1998).

In addition to national data networks, there are numerous regional mesonetworks whose data are often available in real time for no cost at a central repository. Examples in the USA are the Oklahoma (Brock *et al.* 1995) and MesoWest (Horel *et al.* 2002) mesonetworks.

## SUGGESTED GENERAL REFERENCES FOR FURTHER READING

Gilleland, E., D. Ahijevych, B. G. Brown, B. Casati, and E. E. Ebert (2009). Intercomparison of spatial forecast verification metrics. *Wea. Forecasting,* **24**, 1416–1430.

Jolliffe, I. T., and D. B. Stephenson (2003). *Forecast Verification: A Practitioner's Guide in Atmospheric Science*. Chichester, UK: Wiley and Sons Ltd.

Wilks, D. S. (2006). *Statistical Methods in the Atmospheric Sciences*. San Diego, USA: Academic Press.

## PROBLEMS AND EXERCISES

1. Explain sources of representativeness error in addition to the influence of orography on the low-level windfield. For example, how can differences between local and

---

[1] SYNOP reports are observations that are made at internationally agreed upon times, every 3, 6 or 12 hours, by meteorological observers. Specific practices are prescribed by the World Meteorological Organization, and adhered to by all national meteorological services.

[2] METAR reports are near-surface observations of the standard meteorological variables that are made hourly, or between hours when special observations are warranted, often at airports. METAR codes are regulated by the World Meteorological Organization.

grid-box-average landscape properties lead to artificial errors in the temperature and humidity fields?

2. On what factors do the spatial and temporal spectra of model solutions depend? What determines how well these properties of the model solution verify against the real atmosphere?

3. Why are there peaks in the power at both 12 h and 24 h for the coastal station winds in Fig. 9.11?

4. For the precipitation observation–forecast pairs in Fig. 9.9, how would you subjectively rank the forecasts in terms of accuracy? Explain your choices.

5. Is the so-called representativeness error really an error?

6. Describe different types of meteorological feature or events that could be used for forecast verification.

7. Using the general references above, summarize which accuracy and skill metrics are most appropriate for different purposes.

# 10 Experimental design in model-based research

The aim of this chapter is to provide a few examples of some common methods for using models in research studies. Other chapters also discuss experimental designs in the context of the specific subject being discussed. For example, there are many places in Chapter 16 describing experimental methods related to modeling studies of climate change. The summary here is far from complete because experimental methods are obviously closely tied to the objectives of a research project, which can vary widely. Nevertheless, the methods summarized are in wide use, and their strengths and limitations should be understood.

## 10.1 Case studies for physical-process analysis

Model simulations, generally for short time periods, are often used to study some aspect of a meteorological phenomenon. Sometimes the purpose is to better understand the predictability of a process, in terms of the necessary physical-process parameterizations or initial conditions. This is treated in Section 10.7 on predictive-skill studies. More often, the purpose is to use the model to help better understand the dynamics or kinematics of a physical process. The model is integrated from an initialization that is based on observations at the beginning of the study period. A next step in the process is to confirm that a good correspondence exists between the model simulation and the observations that are available during the simulation period. Good verification of the model skill at these observation locations is typically considered to be justification for believing the simulation in the space and time gaps between the observations. A benefit of using the model to fill the observation gaps is that the resulting fields are dynamically consistent, at least with respect to the dynamics embodied by the numerical approximation to the equations. Another benefit is that it is easier to analyze data on a quasi-regular matrix of grid points, compared to using the randomly spaced observations themselves. Additionally, the models respond to fine-scale local surface forcing that often adds information beyond what can be represented by the observations, and nonlinear wave interactions can add smaller scales than those represented in the observations.

The publicly available global or regional reanalyses described in Chapter 16 can be used for case studies. They are model-generated, and have all of the benefits mentioned above. And, these data sets are ready for analysis, requiring none of the investments associated with individual scientists running models. But, even though these data have been generated by trusted models from major forecasting centers, the resolutions are sometimes not sufficient to adequately represent some processes. Thus, it is often necessary to run a LAM to generate a fine-scale gridded data set for a case study. Generally, the LAM will use the large-scale reanalysis for LBCs. In predictive-skill studies, operational forecasts can be used instead. In addition to the benefits of added resolution, the use of a LAM for case studies allows the physical-process parameterizations to be chosen specifically for the phenomenon and geographic area being studied.

The following is a summary of the suggested components, and sequence of tasks, for a physical-process case study. It is especially important to understand the importance of first thoroughly analyzing the observations themselves, before running the model. Studying the observations will provide an understanding of the prevailing horizontal and vertical scales of motion, and the probable processes that are operating. This is essential information that is needed in order to properly set up the model.

- Clearly define the scientific objectives of the case study.
- Identify a candidate case to study, based on reviewing reanalyses or operational analyses, personal observation of a case, or the availability of special field-program observations.
- Obtain, quality check, and study all observations for the proposed study period. Analyze the vertical and horizontal structures in the atmosphere. Perform the best possible overall analysis of the process being studied – this could require months. Avoid the tendency to run the model before this phase is complete; running the model prematurely is a very common mistake (modelers like to model)!
- Determine how you would like the model to improve upon the above analysis of observations.
- Develop an experimental design for the modeling study. For example, will there be sensitivity studies? How will the model simulations be analyzed to satisfy the study objectives – cross sections, trajectory analyses, budget calculations?
- Based on the identified vertical and horizontal scales of motion associated with the processes being studied, choose appropriate horizontal and vertical grid increments. This should be based on the effective resolution of the model (Fig. 3.36), and not simply the grid increment. Evaluate the sensitivity of the model solution to the use of different horizontal and vertical resolutions.
- Based on a review of the literature, estimate the most appropriate physical-process parameterizations for the geographic area, the horizontal and vertical grid resolutions, and the process being simulated. Evaluate the sensitivity of the model solution to the use of alternative physical-process parameterizations.
- Define sources for the best possible model initial conditions, LBCs (if a LAM is being used), and land-surface conditions.
- If a LAM is being used, run test simulations to evaluate the sensitivity of the model solution to the domain size (LBC location).
- Perform a control model simulation, for use in verification.
- Compare the model solution with the observations available during the simulation period. If there are significant errors, adjust the model configuration accordingly (resolution, parameterizations).
- Based on the fact that the model compares favorably with available observations, use the gridded model output as a surrogate for the atmosphere in the physical-process

analysis. Chapter 11 reviews some methods that might be useful for analyzing model output.

The reader should have received the clear message above that there is much work that needs to be done in a physical-process case study before a researcher even thinks about using the model. In fact, the author's experience is that the sooner that the model is used in the process, the longer the study will take.

It is tempting to consider using continuous data assimilation, for example through Newtonian relaxation, to generate the gridded data sets for cases studies. After all, it can be argued that that process is better at integrating observations and model dynamics than one in which a model simulation only uses observations at the initial time. However, the relaxation, or nudging, terms are not physical, so the resulting model-generated data set does not exactly represent the thermodynamic or dynamical balance of the finite-difference equations. However, if this is less important than the correspondence of the model solution and the observations, the observations can be assimilated in this way. Nevertheless, it would still be advisable to first perform the simulation without assimilating observations throughout the study period, to allow those observations to be used as an independent check on the ability of the model equations to reproduce the processes.

When a single extreme weather event is to be analyzed using a case study, or if special field-program observations are available for a short period, it makes sense to use only one example of the process (i.e., one case) in the analysis. And, an in-depth analysis of even a single case can be very time consuming. However, it may be reasonable for some purposes to study a series of cases, to evaluate case-to-case variations in a process, or to make the conclusions more convincing. In such situations, it may be appropriate to focus on one or two aspects of each case, to make the analysis more tractable, rather than perform the analysis with the same level of detail that would be appropriate for a single case.

## 10.2 Observing-system simulation experiments

An Observing-System Simulation Experiment (OSSE) is a procedure for identifying the potential benefit to operational NWP of a yet-to-be-developed and -deployed observing system or observational strategy. Its use is motivated by the fact that observing systems are often extremely expensive to develop and deploy, so they must be first justified by a quantitative evaluation of the degree to which the possible new observations will improve the forecasts of operational models. This is accomplished by simulating the entire process, beginning with observing the atmosphere and ending with the verification of the forecast. Figure 10.1 shows the components of the OSSE process. The process begins in the upper left with the so-called nature run, where the best possible surrogate for the real atmosphere is generated by a model. Then, the measurement process is simulated by sampling the surrogate atmosphere in a way that is consistent with the existing and proposed

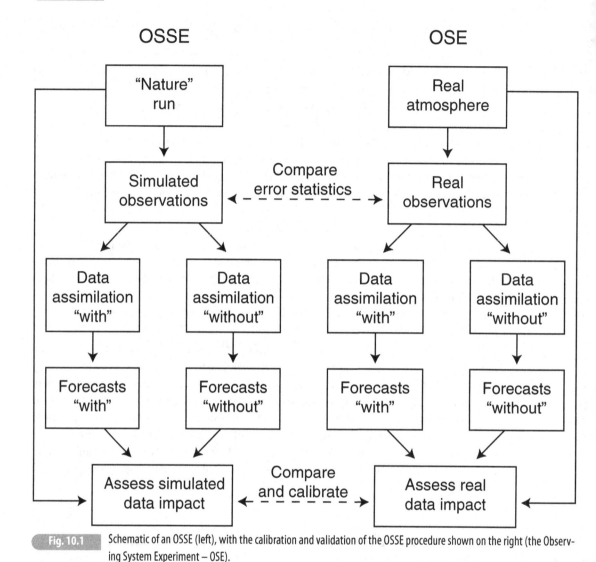

**OSSE**

**OSE**

Schematic of an OSSE (left), with the calibration and validation of the OSSE procedure shown on the right (the Observing System Experiment – OSE).

new observing systems. The simulated existing measurements are assimilated by the operational data-assimilation system, and this process is repeated by assimilating both the simulated existing and proposed new observations. Forecasts based on both sets of initial conditions are produced by an operational-class model (i.e., it must run in real time), and they are compared with the nature run in order to assess the impact of the new observations. The process is the same whether the impact is being evaluated of proposed new observing systems, or of new configurations (e.g., locations, numbers) of existing instruments. The OSSEs can be performed with global models alone, or with LAMs nested within global models. More discussion about each of these steps is provided below.

As suggested above, in addition to assessing the impact of potential new observing systems, OSSEs can be used to evaluate new observing strategies. For example, the influence of adding additional observations of a type that is already used can be estimated, as can the effect of moving present observations from one location to another.

## 10.2.1 The nature run

This simulation of an historical study period employs the highest model resolution and the best representation of physical processes that can be afforded with the available time and computational resources. It should therefore be expected that the nature run will more faithfully represent the real atmosphere than will forecasts from operational models that must run faster, on operational time scales. This is because the higher resolution in the nature run will result in (1) smaller truncation-related errors in approximating derivatives, (2) an ability to explicitly represent some processes rather than parameterize them, and (3) a better rendering of landscape forcing. The results of the nature run are archived at the native grid-point resolution and with high temporal frequency. This nature-run output has also been referred to as the "truth", "history", or "reference" atmosphere.

A goal is for the nature-run model to be as different from the operational model used later in the process as the real atmosphere is from the solution from the operational model. If the same model is used for the nature run and for the forecast, the surrogate atmosphere will have the same biases as does the forecast model. Thus, even if the resolution of the nature run is much greater than used in the forecast run, and the representations of some of the physical processes are much better in the nature run, the use of the same dynamical core will lead to common biases. The use of the exact same model configuration for both purposes is referred to as an *identical-twin experiment*. When the exact same model is not used for both purposes, but the models are not as different from each other as the forecast-model solution is from the real atmosphere, it is called a *fraternal-twin experiment*. Given that the nature-run simulation will be used to verify the forecasts, any biases that are common between the nature and forecast runs will make the forecasts appear better than they really are. An alternative is to use completely different models, and therefore different dynamical cores, for the nature run and forecasts, and this would normally qualify as a reasonable approach for an OSSE. But, even then it is well known that model solutions are sometimes more similar to each other than they are to the atmosphere.

## 10.2.2 Simulating the observations

There are two basic approaches for simulating an observation. The simplest is to interpolate from the nature run's model grid to the observation location, and add an error that is consistent with the known systematic and random error of the measurement system. However, the most thorough approach is to use what is called an *instrument-forward model* that, as explicitly as possible, represents the interaction of the sensor with its environment, to produce a measurement that may, in fact, not be one of the model dependent variables. The atmosphere from the nature run is used as input to the instrument-forward model, and the output of the model is the simulation of the sensor output. For example, an emulator for a satellite-based sensor would use the nature-run variables as input to software (a model of the sensor) that would emulate the functioning of the sensor's optics and electronics. The output from the sensor, for example radiances, would then be assimilated by the data-assimilation software.

It is worth being reminded that the truth atmosphere was generated by Reynolds-averaged equations, and thus the model solution does not represent the turbulence that exists in the actual atmosphere. Thus, a current area of research is the development of methods for representing the effects of the turbulence on simulated observations in OSSEs.

## 10.2.3  Data assimilation

If an OSSE is being employed to assess the impact of a completely new sensor, on a satellite that has yet to be launched, it could be 5–10 years before the new data become available. The operational assimilation systems and models in use at that time in the future will inevitably be different than the present-day systems used in the OSSE. Thus, because the impact of an observation on forecast skill depends greatly on the assimilation system and the model, the impact of the new observations as assessed by the OSSE will not reflect the future impact. Even though there is no obvious way to address this problem, it should be recognized as a source of error in the process. A separate point is that, just as the error characteristics of real observations are used in operational data-assimilation systems, the error characteristics associated with the instrument-forward models mentioned above should be used in the data-assimilation software employed in the OSSE.

## 10.2.4  Forecasting

For the reasons mentioned above in the context of the data-assimilation system, the forecast model used in the OSSE should, as closely as possible, approximate the operational systems that will be in use at the time that the potential new observations will become available. This is because the characteristics of the forecast model will affect the impact of observations. To understand this, remember that the entire modeling process has many components that contribute to the final errors in the forecast, where these include the model initial conditions, the dynamical core, the physical-process parameterizations, and the quality of the lower-, upper-, and lateral-boundary conditions. Large errors in any one of these components can limit forecast skill regardless of the sophistication of the other components. For example, the benefit of accurate, high-resolution data from a new sensor is not going to be realized if the forecast model has coarse resolution or large errors in the representation of the physics. Thus, the value of a new observing system can be limited by the properties of the model used to produce the forecasts. The implication is that the use in the OSSE of a forecast model that is inferior to the one that will be in use when the future observing system is implemented, will probably underestimate the positive impact of the new observations.

## 10.2.5 Assessing the impact of observations

The forecast-skill measures used here can be conventional ones, or they can be specific to a particular application of the forecast. For example, if the objective of the OSSE is to evaluate the impact of a new type of global satellite-based observing system on 72-h synoptic-scale weather forecasts, measures such as anomaly correlations would be reasonable. However, consider an OSSE that is conducted to estimate the effectiveness of different scan strategies for Doppler radars that are used to initialize forecasts of moist convection. In this case, some of the measures described in Chapter 9 for verification of precipitation forecasts would be more appropriate.

## 10.2.6 Calibrating the OSSE

The right side of Fig. 10.1 illustrates a way of verifying the ability of the OSSE procedure to properly represent all components of the process, based on the use of current, real observing systems. The process is called an Observing System Experiment, or an Observation Sensitivity Experiment (OSE), where the purpose is to evaluate the impact on forecast skill of the use of an existing observation network. Here, a nature run isn't needed because real observations are used to initialize the forecast. And, the data-assimilation and forecast steps are run both with and without the observations from the measurement system that is being evaluated. The comparison of the forecast variables and real observations during the forecast period defines the contribution of the withheld observations to forecast skill. Then, the OSSE procedure is applied to the same case, where a nature run is generated, and the observations are simulated using the method of choice. Again, the data-assimilation and forecast systems are run with and without the simulated observations. A comparison of the two forecasts with data from the nature run defines the impact of the observation type. If the impact is similar from the OSSE and the OSE, it provides some confidence that the OSSE will reasonably estimate the impact of a hypothetical new measurement system. If there is a difference, OSSE results can be calibrated such that the estimate of the impact of the new measurements will be more realistic. Comparison of the documented, known error statistics from real observations and those from the simulated observations is another way of evaluating the OSSE process.

## 10.2.7 Examples of OSSEs

Table 10.1 lists examples of the use of OSSEs to assess the impact on model-forecast accuracy of future observing systems. Note that this list represents only a small subset of the hundreds of OSSEs that have been conducted. The observing system whose impact was evaluated, and the associated references, are provided.

| Table 10.1  Example applications of OSSEs, arranged approximately chronologically | | |
|---|---|---|
| Purpose of experiment | Observing system, variable | References |
| Define the sensitivity of the accuracy of heat and moisture budget calculations, for convective situations, to the spatial and temporal density of soundings. | Radiosonde | Kuo and Anthes (1984) |
| Define the spatial and temporal frequency of soundings needed for calculation of accurate kinematic trajectories. | Radiosonde | Kuo *et al.* (1985) |
| Estimate the impact of a new surface-based observing system on mesoscale weather prediction. | Profiler network, winds and temperature | Kuo *et al.* (1987), Kuo and Guo (1989) |
| Assess the impact of potential new satellite observing systems on a global data-assimilation system. | Satellite Doppler lidar wind, microwave temperature and moisture | Hoffman *et al.* (1990), Zagar *et al.* (2008) |
| Assess the impact of potential new satellite observing systems on analyses and forecasts. | Microwave sensors; rainfall, water vapor, temperature | Nehrkorn *et al.* (1993) |
| Estimate the impact of a potential new satellite observing system on predictability. | GPS refractivity | Kuo *et al.* (1998), Ha *et al.* (2003) |
| Evaluate the impact of satellite winds on forecasts. | Satellite scatterometer winds | Atlas *et al.* (2001) |
| Evaluate the impact on forecast skill of a higher density observation network. | *In-situ* observations and satellite radiances | Liu and Rabier (2003) |
| Generate improved 13-month nature run with the ECMWF T511 General Circulation Model (GCM) | NA | Reale *et al.* (2007) |
| Assess the impact of potential new super-pressure balloon data on regional weather analyses and forecasts. | Balloon-borne pressure, temperature, humidity, and wind | Monobianco *et al.* (2008) |

## 10.3  Observing-system experiments

The OSE procedure described above was used to verify the validity of OSSEs. However, OSEs may also be used to quantify the contribution to model forecast skill of existing observations. This may be motivated by the need to eliminate individual observations, or perhaps entire types of observations, as a budget-cutting measure. The process is described in the schematic on the right side of Fig. 10.1. A pair of data-assimilation and forecast cycles is performed, with and without the use of the observations being evaluated. As with OSSEs, the results should be based on an analysis of the sensitivity for each season of the year, for as long a period as possible. The availability of special observing systems during a field program also provides an opportunity for using OSEs to evaluate the impact on forecast skill of adding new types of observations.

## 10.4  Big-Brother–Little-Brother experiments

These Big-Brother – Little-Brother (BB-LB) experiments have traditionally been used to evaluate the impact of LBCs on the model solution in dynamic-downscaling experiments. The procedure is to first generate a high-resolution, large-grid-area reference simulation, called the BB
simulation. This solution is then spatially filtered, so as to retain the scales typical of atmosphere–ocean general-circulation-model simulations. The identical model is then run for a
smaller grid that is within the area of the larger grid, using the filtered large-grid simulation for
ICs and LBCs. This is the LB simulation. The difference between the BB solution, and the LB
solution after it spins up, is entirely attributable to the numerical impacts of the nesting procedure (e.g., the size of the smaller grid, the LBC update frequency, the blending strategy at the
boundary) used in the downscaling process. Figure 10.2 shows a schematic of the procedure.

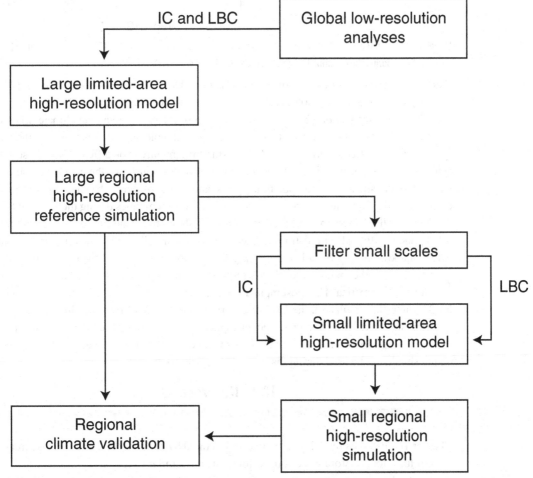

**Fig. 10.2**   Schematic of BB-LB experiments, as they can be used to test the LBC process for downscaling with regional climate
models. Adapted from Denis *et al.* (2002).

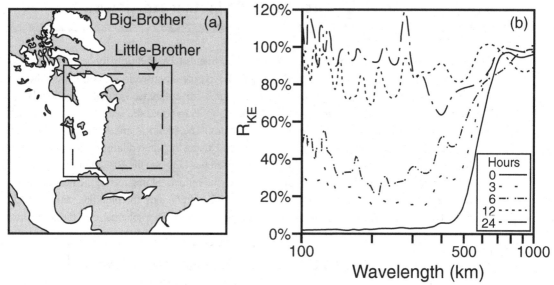

Fig. 10.3 Model grids used in the BB and LB experiments (a) and the spectrum of the ratio of the LB and BB low-level KE for different times during the LB simulation (b). Adapted from Denis *et al.* (2002).

Examples of the use of this type of experiment are in Denis *et al.* (2002), Castro *et al.* (2005), Dimitrijevic and Laprise (2005), Antic *et al.* (2006), Herceg *et al.* (2006), Diaconescu *et al.* (2007), and Køltzow *et al.* (2008). Even though these references focus on climate downscaling, this method can be used for other purposes. For example, Fig. 10.3 is based on the Denis *et al.* (2002) BB-LB experiment, and shows the time required for different spatial scales to spin up after initialization in the LB experiment. Panel (a) illustrates the area coverage of the large (BB) and small (LB) computational grids, and panel (b) shows the spectrum of the ratio of the LB and BB low-level Kinetic Energy (KE) for different times during the LB simulation. The ratio for the ICs (0 hour) of the LB simulation shows the result of the low-pass filter that was applied to the BB simulation. In fact, there was virtually no KE at this time in scales below 500 km on the LB grid. The ensuing KE growth in this part of the spectrum results at least partly from the model atmosphere's response to the Appalachian Mountains near the east coast of North America. The most rapid adjustment in the KE occurs between 6 and 12 hours. By 24 h into the simulation, the LB KE is very similar to that of the BB. This use of BB-LB experiments could define, for a new model application, the time required after initialization for model spin-up. The user of the simulations or forecasts would thus be aware that the model output should not be used during that period after initialization.

## 10.5 Reforecasts

Chapter 16 describes reanalyses, which are obtained by running the same data assimilation system for a long historical period. Reforecasts are similar to reanalyses, except that forecasts with the same numerical model are produced at regular intervals (e.g., daily) over an historical period using the reanalysis data set for initial conditions. The reforecasts can be deterministic forecasts or ensemble forecasts. These retrospective forecasts, which have been produced with a fixed version of the model, can be used for a variety of purposes. Biases can be calculated to help identify and correct weaknesses in the model. Or, they can be used for calculation of model-output statistics or, similarly, for training algorithms for statistically downscaling the forecasts. Lastly, predictability studies can be conducted. The obvious benefit of reforecasts over archived operational forecasts is the fact that operational models are routinely changing, with implementations of improved physics and numerics, and with code-bugs fixed.

Unfortunately, the computational requirements for generating decades of reforecasts are usually prohibitive for operational centers, given that operational-forecasting demands consume virtually all available resources. Alternatively, it has been possible for individual researchers to conduct a limited number of reforecasts to satisfy some of the above objectives. However, long periods of reforecasts are needed in order quantify model skill at forecasting infrequent, extreme events. References on this subject are Hamill *et al.* (2004, 2006), Hamill and Whitaker (2006), and Glahn (2008).

# 10.6 Sensitivity studies

A common motivation for performing a modeling study is to define the sensitivity of a model simulation to initial conditions, lateral or lower boundary conditions, or physical-process parameterizations. The following sections summarize some of the methods used for analyzing this sensitivity.

## 10.6.1 Simple sensitivity studies

A simple and historically common method of performing a sensitivity analysis is to produce a simulation with a control version of the model, and then change some aspect of the modeling process and perform a second simulation. By directly comparing the two simulations, or subtracting them to produce a difference field, an assessment can be made of the sensitivity to the modified process. For example, Fig. 10.4 shows the difference between two simulations, one with the Great Salt Lake and Utah Lake in western North America (the control experiment), and the other with the lakes replaced by the surrounding natural landscape. The purpose was to define the influence of the lakes on the regional wind field. A drawback to this approach to sensitivity analysis is that it only can answer simple questions. For example, in this case there are mountains near the lake shore, so the lake breeze at the time in the figure is influenced by the terrain, so what is seen is the interaction of the lake breeze with the orography, and not the result of the lake breeze alone. Nevertheless, if the objective of the analysis is to answer a practical question – in this case, how would the low-level wind field change if the lakes dried up – rather than to separate the different physical effects, this experimental design serves the purpose.

Table 10.2 lists a few of the many hundreds of model-based sensitivity studies that have been conducted. In addition to physical-process sensitivity studies, others isolate the impact of some aspect of the model configuration, such as resolution, physical-process parameterizations, LBCs, etc. Some of the listed studies, many of which had the aim of improving our understanding of the physical processes that prevailed in a particular meteorological situation, could have used the factor-separation method described in the next section. The last section of Chapter 16 discusses additional sensitivity studies that evaluated the impact of landscape changes on regional and global climate.

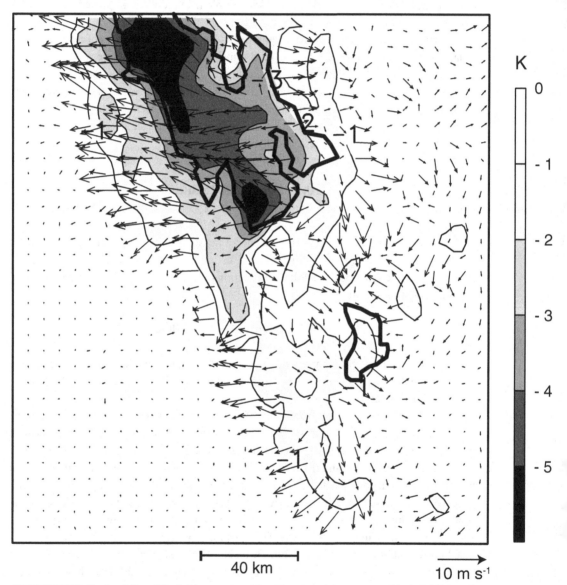

**Fig. 10.4**  Simulated 10-m-AGL wind (see vector scale) and 2-m potential temperature (shaded) difference fields (control minus no-lake experiment) for 1900 LT 14 July 1998. The potential temperature difference field is analyzed and shaded with a 1 degree interval, and the wind vector difference field is plotted at every second grid point. The heavy solid lines outline the Great Salt Lake and Utah Lake. Adapted from Rife *et al.* (2002).

**Table 10.2** Examples of simple sensitivity studies. They include evaluations of the sensitivity of simulations of synoptic- and convective-scale processes to various factors (impact variables), including surface conditions, surface fluxes, latent heating, static stability, resolution, and LBCs.

| Processes for which impacts are evaluated | Impact variables | References |
|---|---|---|
| Ice storms in the southeastern USA | Atlantic Ocean SSTs | Ramos da Silva *et al.* (2006) |
| Explosive maritime cyclogenesis in the Atlantic Ocean | Ocean sensible- and latent-heat fluxes, latent-heat release, initial conditions, horizontal resolution | Anthes *et al.* (1983) |
| Idealized maritime cyclogenesis | Ocean sensible- and latent-heat fluxes, latent-heat release, static stability, baroclinity | Nuss and Anthes (1987) |
| Convective initiation | Amount of assimilated data, lateral-boundary location, data-analysis procedure | Liu and Xue (2008) |
| Southern Hemisphere extratropical climate | Sea ice | Menendez *et al.* (1999) |
| Island convection | Wind speed and direction, surface fluxes, low-level moisture | Crook (2001) |

## 10.6.2 Factor-separation method

This method is similar to the previous approach to performing sensitivity analyses, except that, here, more modeling runs span all the different combinations of the chosen factors, and mathematical manipulation of the model results allows isolation of the contribution of each factor to a specific output variable, such as precipitation. As an illustration, consider the experiment in Stein and Alpert (1993), where the purpose was to evaluate the relative contributions of the surface heat fluxes and irregular orography to modifying the large-scale dynamical production of precipitation in the region of the eastern Mediterranean Sea. During the day, the boundary layer is warmed as a result of the existence of surface heat fluxes, this establishes horizontal pressure gradients where there are variations in surface elevation, and the resulting thermally direct circulations have ascent over the higher elevations. This can generate precipitation. Here, it is not possible to isolate the effect of either the surface heat flux or the orography using just two simulations. For example, comparing the results of a control simulation having all the factors represented, with the results of a simulation that does not have variable orography, will eliminate the thermally forced precipitation over the higher elevations. But, the difference between the two precipitation simulations does not represent the contribution of only the orography because the heat fluxes were necessary as well. Similarly, precipitation from a no-heat-flux simulation subtracted from the control precipitation might produce a similar difference field, but the difference would be associated with the orographic variation as well as the fluxes. The factor-separation method addresses this problem by conducting four experiments: a control simulation with all the factors, a simulation with variable orography eliminated, a simulation with no surface heat fluxes, and an experiment with no variable orography and no surface heat fluxes. From these experiments, the effects on precipitation of the interaction of the heat fluxes and orography can be isolated.

Using notation similar to that in Stein and Alpert (1993), let $f$ represent a variable field that results from a model simulation. For the above study, there are three factors that contribute to precipitation formation: large-scale dynamics ($d$), surface sensible-heat fluxes ($f$), and variable orography ($o$). Subscripts indicate the factors that are represented in a simulation; e.g., $f_{dfo}$ is the precipitation field that results from the inclusion of all three factors in a control simulation. The quantities $f_{df}$, $f_{do}$, and $f_d$ are similarly defined, and represent simulations where one or two of the local-forcing factors are not included. Now, let $\hat{f}$ be the variable field after factors have been separated from the solution, where subscripts refer to processes that have been isolated. For example, for the above situation, $\hat{f}_{df}$ is the precipitation that results from two factors, the large-scale dynamics and the heat fluxes; that is, the variable-orography factor has been eliminated. Based on Stein and Alpert (1993), the factor-separated fields can be calculated by the following equations:

$$\hat{f}_d = f_d, \tag{10.1}$$

$$\hat{f}_f = f_{df} - f_d, \tag{10.2}$$

$$\hat{f}_o = f_{do} - f_d, \text{ and} \tag{10.3}$$

$$\hat{f}_{fo} = f_{dfo} - (f_{do} + f_{df}) + f_d. \tag{10.4}$$

The last equation defines the precipitation field that results from the interaction of the surface heat flux and the irregular terrain elevation. Stein and Alpert (1993) describe the general form of the above equations, for an arbitrary number of factors. Note that multiple factors can be lumped together, if it is acceptable for their effects on the model solution to be aggregated in the sensitivity analysis. And, simulations do not necessarily need to be performed for every combination of factors. Even though there are clear benefits to the factor-separation method in terms of enabling the isolation of specific factors and combinations of factors, there are a few drawbacks.

- The method is time consuming. If $n$ factors must be completely separated, $2^n$ model simulations are required.
- It is often not possible to identify a priori the most important physical factors that contribute to a particular aspect of a model solution. The unidentified factors, as important as they might be, have their effects collectively represented in the simulation that has all the other factors removed.
- Knowing the quantitative effect, on a simulated variable, of interactions among factors provides no insight about the physical processes represented in the interaction. This issue becomes more challenging for larger numbers of factors.

Table 10.3 lists some examples of applications of the factor-separation method used in sensitivity studies, with information about the variable in terms of which the sensitivity is tested, the geographic area, the types of factors whose impact on the model solution is assessed, the numbers of factors, and references. Figure 10.5 illustrates one way of

**Table 10.3** Examples of applications of the factor-separation method

| Process or variable on which impacts are evaluated | Geographic area | Factors | Number of factors | References |
|---|---|---|---|---|
| MCS-precipitation forecast skill | High Plains of North America | Physics parameterizations, initial conditions | 8 | Jankov et al. (2005, 2007) |
| Lee cyclone Sea-Level Pressure (SLP) | Western Mediterranean Sea | Orography, upper-level Potential Vorticity (PV) anomaly, surface sensible-heat flux | 3 | Horvath et al. (2006) |
| Snowfall | North America | Different Great Lakes | 3 | Mann et al. (2002) |
| Extreme convective precipitation | Spain | Orography and latent heating | 2 | Romero et al. (2000) |
| Mesocyclone vorticity | Eastern Mediterranean Sea | Orography, sea-surface fluxes (latent and sensible) | 2 | Alpert et al. (1999) |
| Cool-season heavy precipitation | Western Mediterranean Sea | Orography and surface latent-heat flux | 2 | Romero et al. (1998) |
| Quasi-tropical cyclone SLP and precipitation | Western Mediterranean Sea | Orography, surface sensible- and latent-heat fluxes, latent-heat release, PV anomaly | 5 (not all are separated) | Homar et al. (2003) |
| Sea-breeze wind | Monterey Bay, California, USA | Coastline, coastal mountain, inland mountain | 3 | Darby et al. (2002) |
| Lee cyclone SLP | Alps Mountains | Lateral-boundary location, initial conditions, orography | 3 | Alpert et al. (1996) |
| Cyclone geopotential height, convective instability, wind | Coastal South Africa | Orography, surface sensible-heat fluxes | 2 | Singleton and Reason (2007) |
| Extratropical cyclone precipitation | Connecticut and Long Island, New York, USA | Orography, coastal differential friction | 2 | Colle and Yuter (2007) |

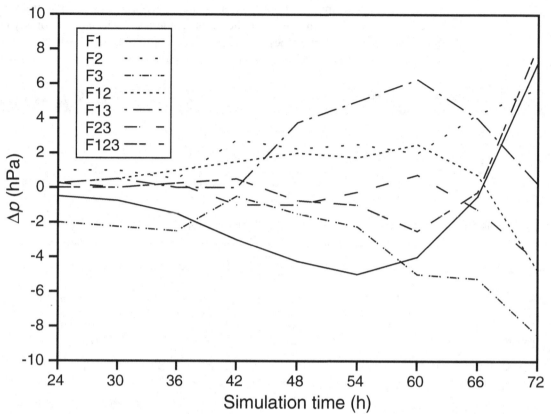

**Fig. 10.5** The contributions of three factors to the evolution of the central pressure of a deep cyclone in the Mediterranean Sea. The factors evaluated are the existence of Atlas Mountain orography (factor 1), the surface sensible-heat flux (factor 2), and an upper-level PV anomaly (factor 3). The first 24-h period of the simulations is not shown on the time axis. Factors and factor interactions associated with each line are defined in the legend. Adapted from Horvath *et al.* (2006).

viewing results from a factor-separation experiment, in this case from the Horvath *et al.* (2006) study listed in the table. The factors evaluated here are the existence of Atlas Mountain orography (factor 1), the surface sensible-heat flux (factor 2), and an upper-level Potential Vorticity (PV) anomaly (factor 3). The influence of these factors on the central pressure of a deep cyclone in the Mediterranean Sea is assessed. In the figure, the F1, F2, and F3 curves are calculated using an equation analogous to Eq. 10.2. They thus isolate the individual influences of the three factors on the pressure depth of the storm. The next three curves identified in the legend show the contributions of the synergies between pairs of factors to the evolution of the central pressure (Eq. 10.4). The last curve shows the results of the triple interaction among the factors. Another way to interpret factor-separation results is to compare plan views of a simulated variable, for the same time in each experiment. Or, the factor-separated fields defined by Eqs. 10.1–10.4 can be plotted. For another synoptic-scale cyclogenesis case in the eastern Mediterranean, Fig. 10.6 shows the 36-h

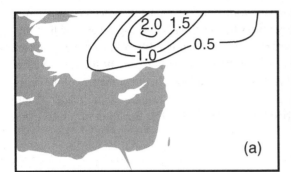
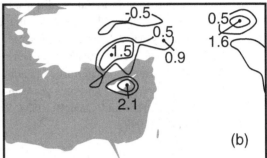
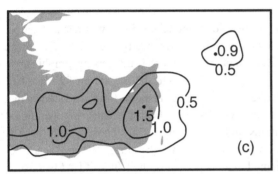
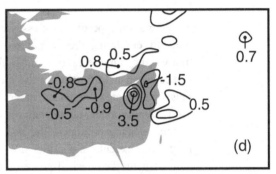

**Fig. 10.6** For a synoptic-scale cyclogenesis case in the eastern Mediterranean, the 36-h simulated precipitation totals (cm) associated with large-scale dynamics alone (a), orography alone (b), surface fluxes alone (c), and the interaction between orography and surface fluxes (d). The zero isohyet is not shown. Adapted from Stein and Alpert (1993).

precipitation totals associated with large-scale dynamics alone (a), orography alone (b), surface fluxes alone (c), and the interaction between orography and surface fluxes (d), based on Stein and Alpert (1993). The results are intuitively reasonable. The large-scale dynamics produces a relatively large, smooth area of precipitation along the storm track, the orographic effects are smaller in scale and near mountains, the surface fluxes have their greatest impact over the waters of the eastern Mediterranean and immediately downwind, and the synergistic effects of fluxes and orography are near the orographic forcing in the eastern Mediterranean.

### 10.6.3  Adjoint methods

Adjoint methods are discussed in Chapter 6 in the context of their use in variational model-initialization procedures. And, Chapter 3 describes variational techniques, employing an adjoint model, that were used to investigate the sensitivity of LAM forecasts to initial conditions and boundary conditions. The adjoint operator produces fields that indicate the quantitative impact, on a particular aspect of the forecast, of any small, but arbitrary, perturbation in initial conditions, boundary conditions, or model parameters. In the types of sensitivity studies described in the previous two sections, modeling systems with different physics, data inputs, LBCs, etc., are run, and the resulting differences in the simulated variables are the measures of sensitivity. The difference fields represent the sensitivity of the variable to the perturbed factors. With the adjoint approach, direct metrics of the sensitivity are provided. For a more in-depth discussion of this technique, the reader should consult Hall and Cacuci (1983), Errico and Vukicevic (1992), Errico *et al.* (1993), and Errico (1997).

## 10.7  Predictive-skill studies

This common type of study assesses an operational model's skill at weather prediction, typically emulating the operational environment in a retrospective setting. This can be motivated by an interest in testing a forecast model for perhaps some new geographic area, or because a modification to the model has been made and the impact on forecast skill must be evaluated. Such studies are different in a couple of respects from case studies that use research models rather than operational systems. Specifically, because an operational system is being emulated, the resolutions and model physics are chosen such that the model will execute in a sufficiently short period of time for the output to serve as a forecast. Also, unlike physical-process-oriented modeling studies, the LBCs of LAMs must be specified using archived operational global-model forecasts rather than reanalyses of observations. That is, operationally realistic errors should exist in the LBCs. An individual case can be used, or more-meaningful results can be obtained using a long series of forecast cycles.

Modeling system components that are often evaluated in the context of operational prediction are the data-assimilation system, the dynamical core (numerics), the formulation of the LBCs for LAMs, the physical-process parameterizations, and the land-surface specification. Also, the impact of new data types, in OSE or OSSE frameworks, is evaluated in tests that emulate operational models.

Criteria for evaluating a forecast's success in an operational setting should be related to the variables of greatest importance for the ultimate users of the forecast information. For example, if warm-season convective rainfall is an important quantity because forecasts of it are used for agricultural applications, that variable should be included in the verification process.

Every operational forecast center conducts studies such as these using off-line (nonoperational) versions of the operational models. Thus, the fact that many hundreds of such studies have been conducted makes it impractical to summarize the literature. Let a couple of examples suffice. Powers *et al.* (2003) describe the initial testing of a new operational regional modeling system for use over Antarctica, to support aviation, maritime, and land-based activities. And, Liu *et al.* (2008b) summarize the performance of an operational regional modeling system used for five locations in North America.

# 10.8 Simulations with synthetic initial conditions

The use of real meteorological situations for modeling studies introduces inevitable complexity, sometimes making it difficult to interpret results when many processes are interacting. A solution to this problem can sometimes be achieved by the use of synthetic, or idealized, initial conditions. For example, much can be learned about coastal circulations, urban-heat-island circulations, and mountain-valley circulations through model simulations that use simple, idealized large-scale flow conditions. Initial large-scale winds may be defined as horizontally uniform or calm, and in geostrophic or gradient balance with the mass field. The resulting thermally forced circulations are more easy to interpret when they are superimposed upon the smooth large-scale flow, compared to the situation with real-data simulations where other features exist. Or, if the model's ability to properly simulate Rossby waves is a subject of study, an appropriate large-scale wave in the initial conditions can be prescribed analytically, superimposed on a zonal flow in a channel model. This general approach has been used for simulation of many processes, including tropical cyclones (Frank and Ritchie 1999, Riemer *et al.* 2008), boundary-layer flow over a forest canopy (Inclan *et al.* 1996), conditional symmetric instability (Persson and Warner 1995), and mesocyclones (Klein and Heinemann 2001).

# 10.9 The use of reduced-dimension and reduced-physics models

The use of these reduced-dimension and reduced-physics models is motivated by the same reasons as the use of synthetic initial conditions described above – simplifying the experimental situation to allow for a more-clear interpretation of results. In addition, the simplification of the modeling framework results in the use of less wall-clock time and computational expense to perform simulations.

## 10.9.1 Reduced-dimension models

These models include the single-layer, shallow-fluid models ($x$–$y$) described in Chapter 2, cross-section models ($x$–$z$ or $y$–$z$), and column models ($z$). The shallow-fluid models are useful because the computer codes are simple, and there are generally no moist processes, radiation, or turbulence. They are typically used for evaluation of dynamical cores. Two-dimensional, vertical cross-section models often include a fairly complete representation of physical processes (to the extent possible with two dimensions), but the lack of the second horizontal dimension makes the models perhaps two-orders of magnitude less computationally intensive to use. Thus, higher vertical and horizontal resolutions can be used efficiently, and more computationally intensive numerical procedures and process parameterizations can be evaluated. Lastly, one-dimensional, column models are convenient for testing parameterizations of boundary-layer fluxes and growth, radiation, and moist convection – all being somewhat one-dimensional processes in terms of their representation in a model.

Some LAM systems allow a user to select an option that collapses the model to a cross-section configuration. If that option is not available, the user may be able to sufficiently reduce the grid dimension in one direction to achieve the same goal. For example, if only one row or column of grid points is needed to define the lateral boundary conditions at each edge of the grid, it may be possible to specify a grid dimension of three in the collapsed direction – one computational row or column, and boundary values replaced by interior values (zero-order extrapolation).

### 10.9.2  Reduced-physics models

Obviously these models that employ less than the full suite of physics need to be employed appropriately for situations where the lack of complete physics still allows the experimental objectives to be mct. One type of reduced-physics model that we have already seen is the shallow-fluid model. The shallow-fluid system has no moist processes, no radiation, and no turbulence parameterization because it is used primarily for studies that focus on numerical solutions to the equations, and on simple dynamical processes. As noted in the previous section, this is also a reduced-dimension model, with typically no variability in the vertical.

The above-mentioned column models are, out of necessity, reduced-physics models. For example, if studies of the radiative impact of dust on the vertical temperature profile are conducted with a column model, all processes except radiation can be excluded. Similarly, in boundary-layer studies, all processes other than those associated with the land surface and the vertical fluxes of heat, moisture, and momentum can be ignored.

In the context of operational weather prediction and climate-system modeling, there are many examples of some physical processes not being included in the modeling system. For example, for weather prediction on time scales of weeks or less, coupled ocean processes are not represented. On climate time scales, a spectrum of models with different complexities is available for answering specific questions (Randall *et al.* 2007). In addition to quite simple climate models, there are Earth-system Models of Intermediate Complexity (EMICs) that have somewhat simplified physics representations compared with full physics Atmosphere–Ocean General Circulation Models (AOGCMs). Because of their greater simplicity and computational speed, the EMICs can address climate processes and interactions that evolve on time scales too long for AOGCMs. The use of simplified EMICs also allows large ensembles to be employed.

## 10.10  Sources of meteorological observational data

Unless model initial conditions are idealized, or the verification of the model uses analytic solutions, observations will be needed for initialization and verification in research or operational applications. Observational data are available at no cost from a number of sources. The US NCAR archives operational observations in their Mass-Storage System, but late observations are not added to the data set, nor are data-transmission-related gaps in the record filled. In contrast, complete archives are maintained by the US NOAA National Climatic Data Center (NCDC). In fact, NCDC maintains the world's largest archive of climate data. Satellite products that define atmospheric and land-surface properties are available from the ESA and NASA. Also, there are many regional mesonets that make, primarily, near-surface data available on servers in real time. However, it is always the user's responsibility to ensure that the data, regardless of the source, have been adequately QCed. Reanalyses, and archivals of operational forecasts, can also be obtained from many sources including NCAR, NASA, NOAA, and ECMWF. The best way to investigate how to obtain data from these organizations is to see their websites.

## SUGGESTED GENERAL REFERENCES FOR FURTHER READING

### Observing-system simulation experiments

Atlas, R. (1997). Atmospheric observations and experiments to assess their usefulness in data assimilation. *J. Meteor. Soc. Japan*, **75**, 111–130.

Atlas, R., R. N. Hoffman, S. M. Leidner, *et al.* (2001). The effects of marine winds from scatterometer data on weather analysis and forecasting. *Bull. Amer. Meteor. Soc.*, **82**, 1965–1990.

### Big-Brother–Little-Brother experiments

Denis, B., R. Laprise, D. Caya, and J. Côté (2002). Downscaling ability of one-way nested regional climate models: the Big-Brother experiment. *Climate Dyn.*, **18**, 627–646.

### Reforecasts

Hamill, T. M., J. S. Whitaker, and S. L. Mullen (2006). Reforecasts: An important data set for improving weather predictions. *Bull. Amer. Meteor. Soc.*, **87**, 33–46.

### Factor-separation methods

Stein, U., and P. Alpert (1993). Factor separation in numerical simulations. *J. Atmos. Sci.*, **50**, 2107–2115.

### Adjoint methods

Errico, R. M., and T. Vukicevic (1992). Sensitivity analysis using an adjoint of the PSU– NCAR mesoscale model. *Mon. Wea. Rev.*, **120**, 1644–1660.

## PROBLEMS AND EXERCISES

1. Explain why the impact on model-forecast skill of a new type of observation will depend on the characteristics of the model and data-assimilation system.
2. It is claimed that, if the instrument characteristics are properly specified and utilized in the assimilation process of an OSSE, then the instrument should have a positive or

neutral impact on the forecast quality. In contrast, a negative impact indicates a problem in the OSSE system. Explain why this should be true.

3. An example is provided in this chapter about how a BB-LB experiment can be used to estimate the time required during a simulation for the atmosphere to spin up in response to local forcing. Describe alternative approaches for this, with relative advantages and disadvantages.

4. Why are reduced-dimension models also commonly reduced-physics models?

5. The first section discusses the use of case studies for physical-process analysis. Traditionally, the cases analyzed were only a few days in duration. Speculate about how longer-period model simulations can be analyzed in a practical way, to provide more robust analyses of physical processes.

# Techniques for analyzing model output

## 11.1 Background

This chapter describes methods for (1) the graphical display and interpretation of model output, and observations; (2) the calculation of derived variables from model output, which can help in the analysis of processes; and (3) the mathematical processing of model output, which can reveal properties and patterns that are not apparent from the dependent variables themselves. The comparison of the model output with observations is a type of analysis of course, but Chapter 9 on model verification is devoted to this subject. Also, the application of post-processing algorithms, for example to remove systematic error, is a special type of mathematical processing of the output, and this subject is treated in Chapter 13.

## 11.2 Graphical methods for displaying and interpreting model output and observations

Much of the material in this section is covered in courses on meteorological analysis; however, it is provided here because many students of NWP have not had such a course available to them. More in-depth material can be found in texts such as Saucier (1955) and Bluestein (1992a,b).

There have been so many creative ways of displaying model output, and comparing it with observations, that it is impossible to present a thorough treatment here. Nevertheless, some examples will be provided and the student is encouraged to review the literature and become familiar with typical techniques (see chapter Problems 1 and 3). This subject is important because successfully publishing research, whether it is model-based or not, depends on displaying the results in an easily and quickly understood format.

### 11.2.1 Eulerian analysis frameworks

In the *Eulerian framework*, the values of dependent variables are defined at grid points that are fixed in space. All standard software packages that are used for viewing model output include the option of plotting Eulerian plan views (quasi-horizontal), or maps, of the variables, where the analysis is performed on some reference surface such as pressure and is applicable at a particular time defined in terms of Greenwich time, local time, or forecast lead time. Maps of conditions near the ground may apply at a certain distance above ground level. For the atmosphere above the boundary layer, where diabatic heating associated with surface fluxes is negligible, performing the analysis of the meteorological conditions on isentropic surfaces can allow for a revealing interpretation of processes because, in the absence of diabatic effects, the flow remains on the analysis surface. Note that model output can be plotted and interpreted on isentropic surfaces, even if the model itself uses a different vertical coordinate.

An alternative Eulerian plotting option is the use of vertical cross sections. Here, a specific vertical plane is chosen on which model-output variables or analyses of observations are plotted. The orientation of the plane should ideally by chosen so that it best reveals particular processes or phenomena of interest. For example, Fig. 11.1 shows an east–west-oriented vertical

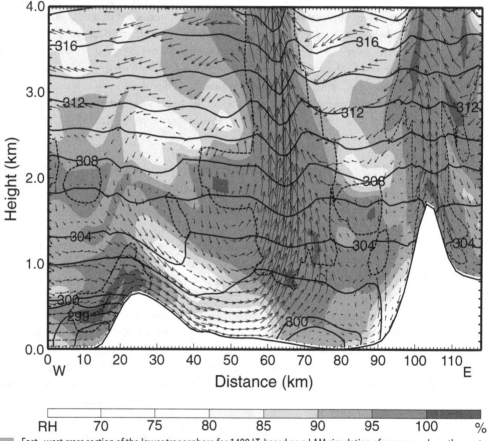

**Fig. 11.1** East–west cross section of the lower troposphere for 1400 LT, based on a LAM simulation of processes along the west coast of Colombia. The Andes Mountains are seen at the lower boundary. The model has a 2-km grid increment. Relative humidity is shaded (see the legend at the bottom), with cloud boundaries indicated by dashed curves. Arrows indicate wind components in the plane of the section (zonal and vertical). Potential temperature is contoured at an irregular interval: 2° above 300 °C and 0.5° below 300 °C. The simulation contains what appears to be a hydraulic jump, as the leading edge of the cool maritime air of the sea breeze surmounts the low coastal mountain range and flows eastward into the Atrato Valley, to the foot of the Western Cordillera of the Andes. From Warner *et al.* (2003).

cross section of model-simulated conditions over the Andes Mountains of Colombia. For special applications such as providing weather guidance to aircraft pilots, displays of forecast weather can be produced on vertical surfaces that follow an irregular flight path.

There are numerous other types of Eulerian analyses that show the time evolution of forecast variables at a point, or along a line of points. For example, Fig. 11.2 illustrates a type of display in which the diurnal evolution and seasonal evolution of a meteorological

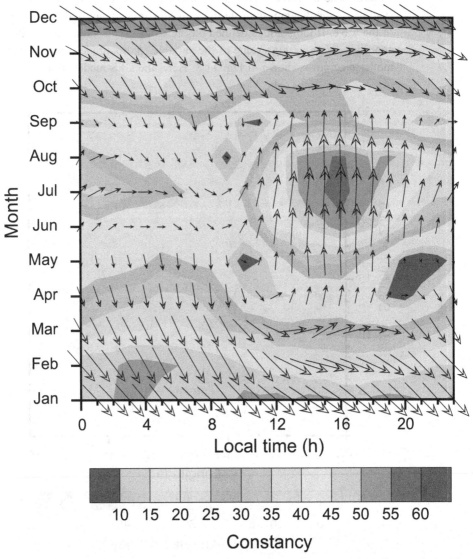

**Fig. 11.2** Winds at John F. Kennedy Airport, New York, USA, as a function of time of day and time of year. The gray shades define the constancy of the wind, which is the resultant of vectors summed for a period of time divided by the average wind speed. For this near-coastal station, the onshore winds are strong and constant during summer afternoons. Provided by Ming Ge, NCAR.

variable can be revealed for a single location. In this case, the winds (vectors) are plotted for John F. Kennedy Airport, New York. The vectors illustrate the strong sea breeze during spring and summer afternoons. The gray shades define the constancy of the wind, which is the ratio of the magnitude of the resultant wind vector and the average wind speed.

Two other types of plots are known as Hovmöller diagrams and time–height sections. The *Hovmöller diagram* is a commonly used approach for plotting meteorological data, either generated by a model or based on observations, to highlight the motion of waves or features. The abscissa is latitude or longitude, and the ordinate is time/date. Colors or shading on the diagram indicate the values of some quantity that varies with the position of the wave. If longitude is defined on the abscissa, the value plotted at a longitude–time coordinate is an average value of the variable for a latitude band. An example of a Hovmöller diagram is shown in Fig. 11.3. Here, the GFS-model-simulated precipitation rate

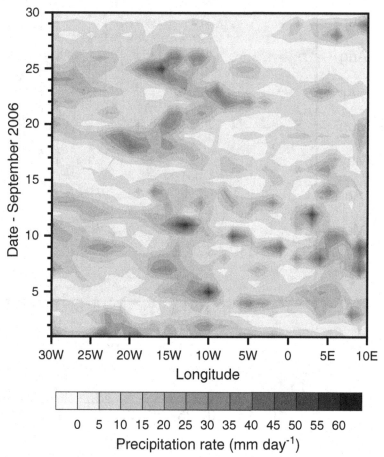

**Fig. 11.3** A Hovmöller diagram showing the precipitation rate (mm day$^{-1}$) from a series of 24-h NCEP GFS-model forecasts. The abscissa spans West Africa, and the time interval plotted is one month. The values plotted apply to the latitude band 5–15° N. The slope of the pattern shows that the precipitation features were moving from east to west. Provided by Erik Noble, NASA GISS.

averaged for 5–15° N latitude in West Africa is shown as a function of longitude and time. The slope of the pattern shows that the precipitation features were moving from east to west, where the period is about 4–5 days between the occurrence of heavier precipitation at a particular longitude.

An example of one type of time–height section is shown in Fig. 4.12a, which depicts the vertical structure of the boundary layer as a function of time of day. A more typical version of this diagram would show isopleths, or colors, or gray shades that define the variation of the vertical profile of a variable at a point in the horizontal, as a function of time.

## 11.2.2 Tracking the movement of parcels of air, or physical features: the Lagrangian framework

### Trajectory analysis

*Trajectories* are paths followed by parcels of air, and are thus often called parcel trajectories. A graphical display of trajectories applies to the period of time over which the parcel translation takes place. There are two common methods for calculating the parcel movement that defines the trajectory, where the difference is in terms of how the vertical velocity is computed. In what are called kinematic trajectories, the most-common type, the three velocity components are provided by a model, and these are used to define the parcel's three-dimensional motion. For $v$ the velocity vector and $x$ the position vector,

$$v = \frac{dx}{dt}$$

is integrated in time, where displacement in each coordinate direction can be calculated independently. Thus, in the east–west direction,

$$\int_{t_1}^{t_2} u\, dt = \int_{t_1}^{t_2} \frac{dx}{dt} dt.$$

Solving this numerically, using a time step $\Delta t$, we have

$$x_2 = x_1 + \left( \frac{u(x(t_1), t_1) + u(x(t_2), t_2)}{2} \right) \Delta t \approx x_1 + u(x(t_1), t_1)\Delta t$$

for small $\Delta t$. Thus, using high-frequency output of the three wind components from a model forecast or research simulation, the path of a parcel can be incrementally calculated. Typically a variety of initial points is chosen if the purpose is to reveal the pattern of fluid motion. For example, Fig. 11.4 shows a large number of trajectories calculated using model-simulated winds within the circulation of a hurricane. The trajectories were initiated in the low-level convergence zone, rise in the eye wall, diverge at upper levels, and provide a visual perspective on the circulation that would not have been possible using

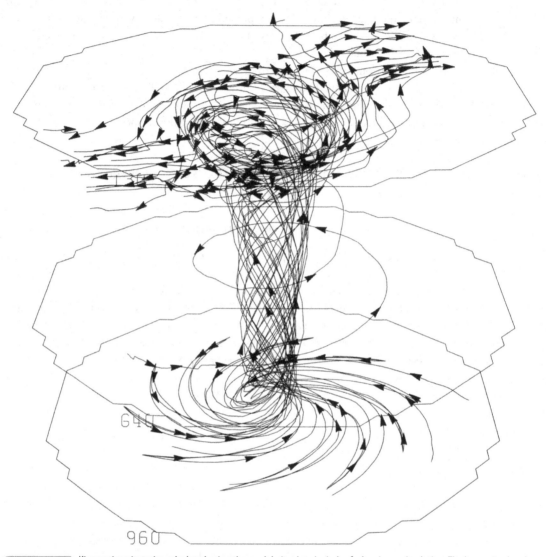

**Fig. 11.4** Kinematic trajectories calculated using the model-simulated winds of a hurricane circulation. The lower circular plane is at 960 hPa, the middle one is at 640 hPa, and the upper one is at 130 hPa. Arrowheads are shown every 9 h along the path of each trajectory. From Anthes and Trout (1971).

vectors or other Eulerian displays. The other approach for calculation of trajectories is to assume that the parcels remain on surfaces of constant potential temperature, where these are called isentropic trajectories. The vertical velocity is thus implicitly defined by the horizontal component of the motion and the slope of the surface. As with kinematic trajectories, the model-defined winds are used to compute the horizontal displacement of parcels. Sometimes it is desirable to calculate the origin of a parcel of air that arrives at a particular location, in which case the mathematical process can be reversed and *back trajectories* calculated. Figure 11.5 shows another example of how trajectory analysis can be used to visualize complex spatial patterns in the motion of a model-simulated fluid. A grid was placed over a simple flow pattern in the troposphere, associated with an approximately symmetric trough in the heights. Trajectories were used to define the paths of different regions in the fluid, as defined by the grid, and the resulting distortions in the grid were mapped.

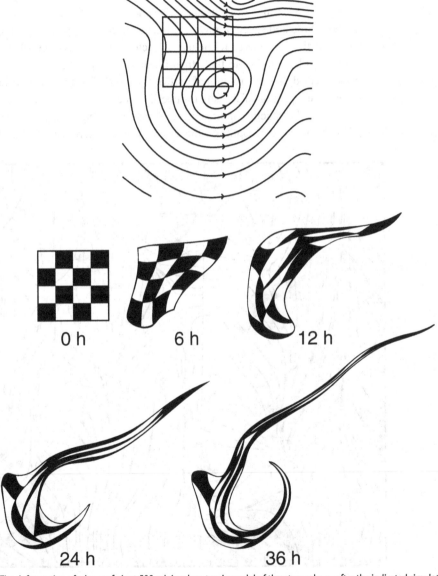

**Fig. 11.5** The deformation of a layer of air at 500 mb in a barotropic model of the atmosphere, after the indicated simulation times. The initial streamline pattern is seen at the top, along with the overlaid pattern whose subsequent evolution is traced with trajectories. From Welander (1955).

Physical features can also be tracked, where an example is the image of the tracks of the hurricane centers shown in Fig. 7.1. All that is required to produce this type of analysis is the ability to define the location of a feature in an automated way. This is fairly straightforward for a hurricane, but can be more problematic for extratropical cyclones or convective systems.

## Streamline analysis

*Streamlines* are lines that are drawn parallel to wind vectors at a specific time. They are thus different from trajectories because the lines do not follow the movement of parcels through both time and space. Rather, the lines simply make it easier to view the pattern in the wind direction, compared to the use of vectors or other symbols that only define wind direction at grid points. Note that streamlines are not the same as streamfunction lines, where the latter define the rotational part of the wind. The two may visually appear to be similar, but the spacing of the streamfunction lines is quantitatively related to the wind speed, whereas the spacing of streamlines is arbitrary and is determined by the analyst or analysis software so as to optimize the ease of visual interpretation. Figure 11.6 illustrates

**Fig. 11.6** Streamlines for ~15-m AGL based on a mesoscale-model simulation of a region in the western USA. There is much structure to the wind pattern because of the existence of complex orography. The shading shows wind speeds, where white is less than 5 m s$^{-1}$, and the gray shades have bandwidths of 5 m s$^{-1}$. Provided by Yubao Liu, NCAR.

~ 15-m AGL streamlines based on a mesoscale-model simulation for a region in the western USA. There is much structure to the pattern because the area is dominated by complex orography. In this display, the speed is also shown in the form of gray-shade bands.

## Isochrone analysis

Isochrones are lines (literally, lines that apply at a particular time) that define the location of a geometrically simple feature in the atmosphere, based on model output or observations. An example with which we are nearly all familiar is the synoptic-scale midlatitude front. But, in an isochrone analysis the frontal location would be shown for multiple times in order to characterize changes in its shape and position. Other features that could be similarly analyzed include convection-related gust fronts or outflow boundaries, sea-breeze fronts, dry lines, the edge of elevated mixed layers, the leading edge of a precipitation shield, etc. A necessary characteristic of the feature is that its geographic position must be definable in a simple way, so that when it is drawn for multiple times the image does not become too complex to interpret. An example is shown in Fig. 11.7 of isochrones of frontal position.

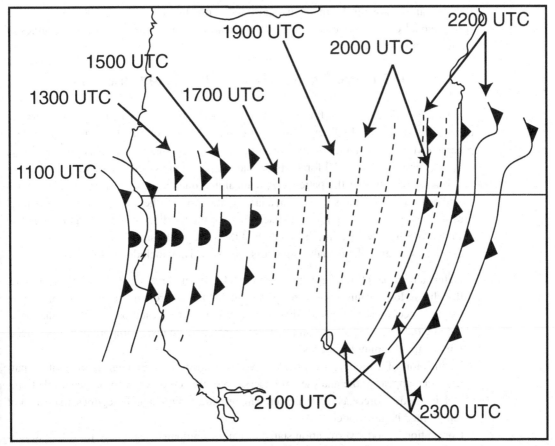

**Fig. 11.7**  Isochrones of a sequence of frontal positions over the northwest USA. Dashed occluded fronts are decaying, and dashed lines without frontal symbols represent trough positions. Adapted from Steenburgh *et al.* (2009).

### 11.2.3 Miscellaneous special plotting diagrams

A variety of special data-plotting diagrams can be used to help the modeler interpret output. Many of these have been described already in the context of the verification of forecasts, but a few are worth mentioning again. For example, there are many types of thermodynamic diagrams, such as the skew $T$–log $P$ plot, that can be used to display the vertical distribution of variables based on model output and observations. From these diagrams can be inferred many important properties of the atmosphere, such as static stability and associated variables such as CAPE and CIN, and the structure of the boundary layer. Another type of special display approach is the Taylor diagram in Fig. 7.4.

## 11.3 Mathematical methods for analysis of the structure of model variable fields

Rather than discussing the mathematical foundations for the following methods for analyzing model output, basic concepts and applications will be stressed, as will sources of additional information.

### 11.3.1 Grouping atmospheric structures by pattern analysis

A variety of techniques exist for grouping atmospheric structures, whether the structures be defined based on model output, analyses of observations, or observations themselves. This process involves the automatic identification of recurring weather patterns in a large data set (e.g., composed of model forecasts or analyses) and the association of variable fields in the data set with one of the patterns. A manual and qualitative equivalent of this process would be for an analyst to sort through a large number of weather maps, say of the sea-level pressure, and put each map into a pile according to the locations of troughs and ridges, the amplitude of the wave pattern, the strength of the average pressure gradient, etc.

Applications of such an analysis process are many. They include the following.

- Define a regional climatology based on model-generated reanalyses – For each variable, the climatology would consist of gridded fields that represent the different prevailing patterns, accompanied by the frequency of occurrence of each pattern. Note that cataloguing the extreme patterns and their frequency may be just as important as documenting the more-common patterns.
- Verification of a model's treatment of regime transitions – Sequences of weather patterns that appear in the analyses of observations are compared with sequences that prevail in model forecasts. Differences can provide insight about regimes that are not represented by the model.
- Conventional model verification statistics can be computed separately for different prevailing weather regimes. This can offer insight into the different components of the model that contribute to the errors. For example, Fig. 9.7 shows model bias calculated

for two of the most-common warm-season weather regimes in the area of Athens, Greece: strong Etesian flow from the north, and weak large-scale flow with a prevailing sea-breeze circulation.

Disadvantages of these automated pattern-sorting techniques are as follows.

- The patterns are not sorted using any dynamical constraints, so there will likely be different underlying processes within a category.
- The patterns represent a composite of the individual analyses, and thus they may not be internally consistent in a kinematic or dynamic sense. Thus, a "typical day" is sometimes chosen from the archive, where the pattern for that day closely matches the composite. This analysis for the selected day will be internally consistent.
- Because there is no predefined number of groups, this is an arbitrary choice that must be made by the analyst without any knowledge of the number of natural clusterings. Thus, some trial and error will be involved.

Two of the most common approaches for such pattern analysis are referred to as *Self-Organizing Maps* (SOMs) and *cluster analysis*. See Wilks (2006) for a summary of meteorological applications of cluster analysis, and Marzban and Sandgathe (2006, 2008) for examples of applications related to model verification. And, Kohonen (2000) describes the general method of SOMs, and an example of one of its many applications for analyzing model simulations of weather and climate output is provided by Seefeldt and Cassano (2008).

Figure 11.8 shows an example of a SOMs analysis for a small number of categories. Six patterns have been arbitrarily chosen for this analysis of 0000 UTC 700-hPa winds, based on one year of model-generated reanalyses for an area of the Middle East. The patterns are distinct in terms of wind speed and/or direction, and the frequency of occurrence of each classification ranges from 10.1% to 26.1%. The frequency refers to the percentage of the analyses that fall into each category. Because the number of categories was arbitrarily chosen, further analysis could involve repeating the process with additional degrees of freedom to determine if there exists considerable variance within any of the original categories.

## 11.3.2  Finding coupled patterns in model or observational data

A different group of methods is aimed at finding coupled patterns in data from models or observations. Bretherton *et al.* (1992) and Wilks (2006) describe and compare three of the most commonly used methods: Principal Component Analysis (PCA), Canonical Correlation Analysis (CCA), and Singular Value Decomposition (SVD). Additional comparisons are found in articles by Hannachi *et al.* (2007) and Tippett *et al.* (2008).

### Principal component analysis

This is also referred to as Empirical Orthogonal Function (EOF) analysis, where the objective of the mathematical procedure is to transform a data set containing a large number of

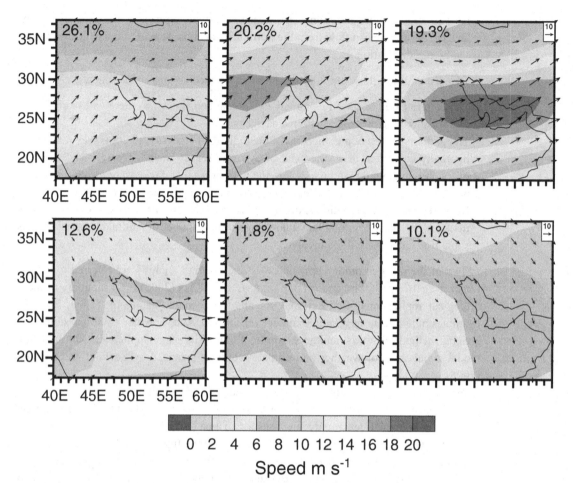

Fig. 11.8 Example of the use of SOMs for an analysis of 0000 UTC 700-hPa winds, based on one year of model-based analyses for an area of the Middle East. The percentage of the analyses that fall into each category is shown. Provided by Ming Ge, NCAR.

correlated variables into one containing many fewer uncorrelated variables, called principal components. The new variables are linear combinations of the original ones, with the first component being the linear combination that contains the largest variance, the second being the combination that contains the second largest variance, etc. Two or more variables can be combined in a PCA to reveal relationships between the fields.

There have been many applications of PCA in the atmospheric sciences. For example, Teng *et al.* (2007) use a form of PCA to show that an AOGCM supports three distinct circulation regimes, having a persistence period of about 7 days, and that analyses of observations have very similar regimes and persistence periods. The impact of greenhouse warming is interpreted in the context of changes in these regimes. Other studies include one by Smith *et al.* (2008) who use PCA to compare the diurnal cycles in a climate-model simulation and observations.

An early description of the PCA method in the context of applications in the atmospheric sciences was by Kutzbach (1967). Exhaustive treatments of PCA can be found in Preisendorfer (1988), which is aimed at geophysical applications, and in Jolliffe (2002), which is a more-general treatment.

## Canonical correlation analysis

Canonical correlation analysis is applied to two multivariate data sets and identifies coupled variability between them. The two data sets can apply to the same time period, or there can be a lag between them. In the latter case, the relationships between the lagged variable fields can be used for statistical weather prediction. Indeed, this was the first meteorological application of CCA, where in most of the subsequent efforts the forecast time scales have tended to be interseasonal. For example Barnett and Preisendorfer (1987) relate seasonal-mean SST anomalies over the Pacific Ocean to surface air-temperature anomalies over the USA during the following season. See Bretherton *et al.* (1992) and Wilks (2006) for other examples.

## Singular value decomposition

Singular value decomposition is similar to CCA in that it isolates combinations of variables in two fields that tend to be related to one another. See Bretherton *et al.* (1992) for references to example applications.

## 11.3.3  Spectral analysis

We have seen in Fig. 3.36 that the spectrum of model output can be computed, and interpreted to define the effective-resolution of the model. And, Section 9.9.2 discusses how a model solution can be verified in terms of its spatial spectrum. Here it will be shown that the same type of spectral decomposition can be used to interpret model output in creative ways. There is a variety of different wavelength bands that the modeler might desire to isolate using spectral analysis, but a common one is that which is associated with diurnal forcing. For example, Fig. 11.9 illustrates the variability (each dot is a location) in the amount of spectral power in this diurnal band. For time series of observations of 10-m AGL winds at each of 28 locations in a mountain valley, a spectral analysis separated the spectral energy into three bands: periods longer than the diurnal, a period of about 24 h, and periods of less than the diurnal. The diurnal power for each station is plotted against the average diurnal amplitude of the oscillation of the $u$ component of the wind at that station. This spectral decomposition has allowed us to learn that there is a large station-to-station difference in the diurnal power, and that the greatest diurnal power is located at those stations near a canyon or near a mountain slope (open circles) where the thermally forced circulation should be the strongest.

A recent, related method is called wavelet analysis. If there is a long time series of values of model-simulated or observed variables, performing a Fourier transform will convert the series to frequency space. But, frequently the time series is not stationary in the sense that frequencies change with time. An option for defining the frequency spectrum for short

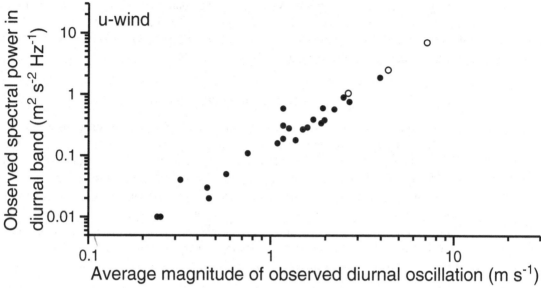

Fig. 11.9 Example of the amount of spectral power in the diurnal band in observations of 10-m AGL winds at each of 28 locations in a mountain valley. The diurnal power for each station is plotted against the average diurnal amplitude of the oscillation of the $u$ component of the wind at that station. Stations close to mountain canyons have open circles. Adapted from Rife *et al.* (2004).

periods, in order to characterize this variability, is a short-period Fourier transform. However, wavelet analyses are more suitable for this, and are capable of providing the time and frequency information simultaneously, thus giving a time–frequency representation of the signal. Torrence and Compo (1998) offer a practical guide to wavelet analysis in the atmospheric sciences, and Wilks (2006) provides additional references.

## 11.4  Calculation of derived variables

The model dependent variables have been used to calculate many derived variables that are useful in understanding atmospheric processes. For example, based on the model winds, the vorticity and divergence can be computed. Or, frontogenesis terms can be calculated. Showing the geostrophic and ageostrophic wind vectors can reveal interesting circulations. See one of the previously noted standard references on meteorological analysis for further examples.

## 11.5  Analysis of energetics

The analysis of model energetics was discussed in Chapter 3 in the context of ensuring that there are no erroneous sources or sinks of energy in the model. However, the calculation of energy terms and conversions based on the gridded output from model simulations, forecasts, or reanalyses can reveal physical processes as well as differences among models. Energy can be partitioned into kinetic, internal, and potential components, and into components that are associated with the mean flow and eddies that are superimposed on the mean flow. The potential energy can be separated into both available and unavailable components; the available part is convertible to kinetic energy while the unavailable part is related to the equilibrium base state of the system and cannot be converted.

The time-dependent equations for the different energy components are derived from the same fundamental equations that are the basis for the model. The rates of change in the energy components are then obtained by applying the model-defined values of the dependent variables in the terms on the right side of the energy equations.

Two general approaches to the calculation and display of energy-conversion terms involve the use of grid-averaged values as well as local, instantaneous formulations. As an example of the latter method, referred to as the "local energetics" (Orlanski and Katzfey 1991) approach, the eddy kinetic energy tendency would be represented on a grid, and displays of that quantity would be provided.

Energetics analyses have been performed in many types of studies. A few of the classes of studies, with references, are listed below.

- Midlatitude cyclogenesis – Orlanski and Katzfey (1995), Lackmann *et al.* (1999), Lapeyre and Held (2004), Moore and Montgomery (2004, 2005)
- Storm tracks in IPCC AOGCM simulations – Laîné *et al.* (2009)
- Madden–Julian oscillation in a climate model – Mu and Zhang (2006)
- Hydrostatic and geostrophic adjustment to thermal forcing – Fanelli and Bannon (2005)

## SUGGESTED GENERAL REFERENCES FOR FURTHER READING

Bluestein, H. (1992). *Synoptic-dynamic Meteorology in Midlatitudes. Vol. 1: Principles of Dynamics and Kinematics.* New York, USA: Oxford University Press.

Bluestein, H. (1992). *Synoptic-dynamic Meteorology in Midlatitudes. Vol. 2: Observations and Theory of Weather Systems.* New York, USA: Oxford University Press.

Saucier, W. J. (1955). *Principles of Meteorological Analysis.* Chicago, USA: University of Chicago Press.

## PROBLEMS AND EXERCISES

1. Survey a variety of journal articles, and list and describe the various types of plots that are used to display model output and compare it with observations.
2. Using the shallow-fluid model employed in the problems of Chapter 3, calculate the trajectory of a parcel of air at the surface of the fluid, as a gravity wave propagates past.
3. Access the websites of national and international modeling centers, both those doing operational forecasting and research, and describe the types of plots that are used to display model products.
4. What fundamental differences should exist between analysis methods that are designed for use by operational forecasters and those designed for researchers?
5. Using a hypothetical wind-field pattern of your own creation, illustrate Lagrangian and Eulerian approaches to its characterization.

# 12 Operational numerical weather prediction

## 12.1 Background

The application of models for operational NWP has much in common with their use for answering physical-process questions, and for satisfying practical needs related to the assessment of air quality, evaluating the potential utility of new observing systems with OSSEs, and testing new numerical methods and physical-process parameterizations. Nevertheless, there are some issues that are unique to operational modeling. These will be addressed in this chapter.

It could be argued that the student of NWP should not need this kind of operations-oriented information because only large national modeling centers with experienced staff and large, fast computers are involved in operational prediction. However, there is a rapid growth in the use of operational regional models by consulting companies, universities, and regional governments to satisfy specialized needs. Thus, the student should become familiar with some of the concepts associated with the operational use of models.

Figure 12.1 illustrates the various components of a very simple operational modeling system. It should be kept in mind that the modeling systems that are operated by national weather services have very large infrastructures, and that the one summarized here is more consistent with the many modest-sized, specialized, operational-modeling systems that exist throughout the world. Some of these system components have been discussed before in earlier chapters, for example related to model initialization. To begin with, the system must have real-time connectivity to operational observational-data networks (top box in the figure), where this generally involves separate access to a number of different data providers. The input data types include current land-surface conditions, meteorological observations from *in-situ* and remote sensors, and gridded analyses and forcecasts from operational centers. After observations are received, they must undergo a quality-control process (Chapter 6). If the observations (e.g., satellite radiances) are not in the form of the standard model dependent variables, some data-assimilation processes may require that a retrieval algorithm be applied to obtain these variables. In the figure, the analysis and the data-assimilation processes are listed separately; however, as we have seen, they are often closely coupled (Chapter 6).

If a LAM is being used, LBCs must be provided to the forecast model, and possibly the data-assimilation system, from a global-model forecast (or a forecast from a larger-area LAM). The gridded forecast fields from the global model must be acquired

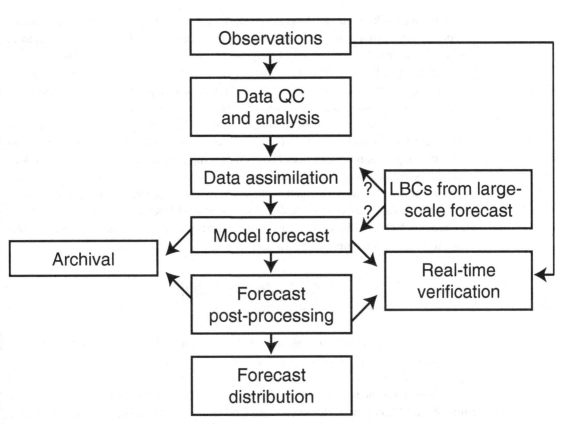

Fig. 12.1 Schematic showing the various components of a simple operational NWP system. See the text for discussion.

from a data service, unless the global model is run at the same facility as the LAM. Obviously, the global-model forecast must be completed prior to the integration of the LAM.

For each forecast of the model, observations that are assimilated, gridded initial conditions, and forecast products are typically written to a storage facility for archival. This archive allows retrospective reruns of the forecasts, to allow modelers to perform experiments to evaluate and correct the causes of especially poor model performance. Forecasts are also often verified in real time on a separate computing platform. The real-time verification statistics are sometimes made available to forecasters so that they can assess the objective accuracy of forecasts from recent cycles of the model.

The post-processing step will be discussed in Chapters 13 and 14, and can include the statistical correction of systematic errors. In addition, special post-processing codes (other models, in some cases) can be applied that derive unforecast variables – such as ocean wave height, river discharge, and air-pollution and dust concentration – from the forecast variables.

The forecasts are disseminated in graphical (analog) and digital form. Forecasters can use a web-based graphical interface to visualize the model output, or the gridded model output can be downloaded to a workstation on which special graphics software is installed.

When the atmospheric-model forecast products are to be used as input to the above-mentioned specialized post-processing models, users can access the digital, gridded forecast products from a data server. Lastly, sometimes specialized model products are produced for specific geographic locations such as major cities. A number of websites provide publicly available software for use in displaying forecast products. One of them is the Unidata site of the University Corporation for Atmospheric Research.

It is worth commenting on the evolving relationship between operational NWP and climate prediction. As will be seen in Chapter 16 on climate modeling, deterministic forecasts with coupled atmosphere–ocean (and other components) models are being produced on interseasonal and interannual time scales. These forecasts are now generated on a regular cycle like weather forecasts, and as the cycle frequency increases, the distinction will become even more blurred between them and operational NWP (see Toth *et al.* 2007). For example, the NOAA Climate Forecast System (CFS) is now running out to ~9 months, with a daily cycle and output every 12 h.

## 12.2  Model reliability

For research applications of models, the ability of the model to complete the integration with a very high reliability is not an especially great concern because the modeler has the opportunity to rerun the model with the problem corrected – perhaps eliminating a bad observation that made it through the QC process or reducing the time step to correct a violation of the CFL criterion. However, when fatal errors occur in an operational setting, the consequences are more severe, especially if the problem occurs in a model-based, sequential, data-assimilation system where each set of model initial conditions is based on the successful execution of a prior forecast. The resulting unavailability of a forecast in extreme-weather situations can result in the loss of lives. Interruptions to a forecast cycle can occur for the following reasons, among others.

- Violations of the CFL criterion can occur because of the existence of especially strong winds associated with the prevailing meteorological conditions. This could be prevented with an extremely small time step, but it is not especially sensible to use unnecessary computing resources during the 99.9% of the forecasts when the short time step is not needed in order to ensure trouble-free operations during the other 0.1% of the forecasts.
- Models rely on the timely availability of initial conditions. Problems with the data-assimilation system can cause cycles to be missed.
- Limited-area forecast models obtain their LBCs from previously run models, such as operational global models, that span larger areas. If this model does not complete its forecast cycle on time, or if there is a failure in the communication network over which data from the larger model are transmitted, the LAM forecast cannot run.
- The hardware on which the model computations are performed can have a catastrophic failure that will cause the entire system to become unavailable for executing the model, or a sufficient number of processors can become unavailable so that a forecast cannot be produced.

- When using LAMs operationally, for some meteorological situations and grid place-ments, the LBCs can generate sufficient noise to terminate a forecast. This is especially problematic when LAM grids are relocated operationally to focus the high resolution on specific prevailing meteorological features.
- There are many components of model code, especially related to physical-process parameterizations, that can fail when confronted with an unusual combination of meteorological inputs. Clearly one of the requirements for selecting a parameterization for operational use is its stability over a wide range of inputs.

## 12.3  Considerations for operational limited-area models

The above discussion highlights a couple of possible reliability problems that are unique to operational LAMs, related to generation of LBC noise and the availability of larger-scale data for defining the LBCs. There are additional issues that have to do with the efficiency of the operational LAM. One is the fact that the large-scale forecast that provides the LBCs must complete before the LAM forecast begins. Thus, there can be a substantial delay, of at least a few hours, in the start time of the LAM forecast, after the initial condi-tions have been defined.

Another point related to forecast-product timing applies when using a nested system of grids in an operational LAM – a common practice. If the grids are not two-way interact-ing, i.e. the information passage is only from the coarser grid to the finer grid, the coarse-grid forecast that is run first can be disseminated while the forecasts on the finer grids are still being generated. This sequential output of forecast products, as they are being pro-duced, puts model guidance in the hands of the forecasters faster than if a two-way-inter-acting grid nest is used. Whether this benefit is worth sacrificing the possible advantages of two-way interacting grids is a situation-dependent decision.

## 12.4  Computational speed

For operational prediction, it is of obvious importance that model (simulated) time advance much more rapidly than actual (wall-clock) time. This is one of the motivations behind the adaptation of model codes to allow them to run on massively parallel computer architectures. And it is why such a great effort is invested in the development of efficient algorithms for solving the model equations, and allowing the use of longer time steps. The following factors influence the wall-clock time required by a model to produce a forecast of specific duration.

- Before an analysis or data-assimilation process can begin, the system must wait suf-ficiently long for most of the observations to be available from the network. Different operational systems have different cutoff times for observation availability, after which the data processing is begun and later-arriving observations are not used. This cutoff time is often 60–90 minutes.

- For LAMs, the forecast must wait for the availability of LBCs from a larger model that has to finish executing first.
- The time step defines the rate at which the integration moves forward, given specific computing resources. And, this time step depends on numerical-stability considerations as well as on the amount of temporal truncation error that is acceptable. If the CFL ratio is a constraint, obviously the grid increment is a strong controller of the time step, as is the speed of the fastest wave on the grid.
- The number of points in the computational mesh is a strong constraint. If the computing cost per grid point does not change with the number of grid points, this is a linear relationship. Combining this factor with a requirement for the CFL criterion to be satisfied leads to the often-stated rule that halving the horizontal grid increment increases the computational burden by a factor of eight, given the same area coverage for the grid.
- Operational models are run on computing platforms that range from desktops to massively parallel systems with thousands of processors. In addition to the effect on model-forecast speed of the number and speed of the processors, the efficiency depends on how the total-model speed scales with the number of processors.
- Some atmospheric models that are used for specialized applications do not include the full suite of physical processes. For example, for short-range predictions of convective outflow boundaries, many processes, such as long- and short-wave radiation, do not need to be included.
- Model output is sometimes made available to the forecaster as it is generated. For example, forecast products might be made available to the forecaster at 12-h intervals (forecast time), while the model is still running. Thus, the forecaster does not have to wait for the entire integration to finish before benefiting from information at the shorter lead times.
- There is sometimes a trade-off that must be made, when deciding on the numerical constructs to use in a model, between execution speed (and accuracy) and the "friendliness" of the code. That is, fast numerical methods (e.g., implicit differencing) will allow forecasts to finish more quickly, but the codes can sometimes be cumbersome to work with. This is an issue when an effort is made to unify the operational and university-research modeling communities by using a common model that must be accessible by both graduate students and experienced modelers. How this compromise is made will affect the speed of the forecast.
- Large data input–output loads can slow down the model-execution speed.

## 12.5  Post processing

There are a few types of post-processing algorithms that are commonly used.

- Correction of systematic error – When models are used in research for the study of physical processes, it is important that the gridded output be faithful to the governing equations. However, for operational prediction the imperative is to have the best guidance for the forecaster. Thus, it is perfectly appropriate to apply statistical-correction algorithms to the raw model output, for example to remove systematic error, even though this will upset the dynamic compatibility of the gridded output. See Chapter 13 for additional discussion.

- Calculation of additional variables with simple statistical or physically based algorithms – As noted above, the post processing also includes the use of the forecast dependent variables as input to statistical algorithms to calculate quantities that are not well forecast by the model, or not forecast at all. The latter have historically included quantities such as freezing rain, fog, turbulence intensity, and visibility.

- Use of secondary models that are coupled to the atmospheric model, for simulating complex processes – These models are discussed in Chapter 14, and include air-quality models, models that predict the elevation of dust from the surface and its transport in the atmosphere, models for the prediction of wildfire behavior, models for predicting the development of agricultural and human infectious diseases, etc.

## 12.6 Real-time verification

Chapter 9 describes the basic concepts of model verification. Real-time verification is distinct from what is done for research applications because it is performed immediately after observations become available for use in verification. The resulting statistics inform the forecaster about recent errors in model performance, in terms of the error dependence on time of day, lead time, location, etc. A challenge is summarizing and displaying the error statistics in ways that are intuitive, and that can be understood quickly by forecasters who are operating under severe time constraints.

## 12.7 Managing model upgrades and developments

Many organizations that employ operational modeling systems also perform research that is aimed at carefully verifying the forecasts, and improving the predictive skill of the model in the context of the specific mission of the organization. The computing needs associated with this objective are two-fold. First, proposed model improvements must be thoroughly tested in the setting of the operational system, and this can be most effectively done by having two independent modeling systems running in parallel. One is the operational system, and the other is identical except that the system improvements are included. In this controlled setting, model performance, with and without a system change, can be compared for a long series of forecasts. Such real-time system testing obviously requires a second computer platform that has at least the capability of the primary platform. In addition, while these two systems are running in parallel, allowing a comparison of the performance of the existing and prototype system, researchers need to be able to conduct case studies and other tests in the process of developing future upgrades to the system. This requires a more-modest sized third computing platform, or additional processors on the secondary platform. An important message is that it is not advisable to perform research during idle time on an operational system, because there is too much risk of accidentally corrupting it.

## 12.8  The relative role of models and forecasters in the forecasting process

Decades ago, when models were not nearly as skillful as they are now, the experience of forecasters was especially crucial to the generation and accurate interpretation of the operational products provided to the public. In the intervening years, models have contributed a growing amount to the "human–machine mix". In the extreme, model products are sometimes translated directly into images and computer-worded forecasts for the public, without expert interpretation. This has led to an ongoing discussion of how forecasters can best add value to the final products, now and in the future. As an illustration of the directions of such conversations, the following points have been made about the relative roles of the models and forecasters.

- Forecasts of "routine" weather should be automated to allow forecasters to focus their efforts on high-impact weather (Sills 2009).
- There should be a greater emphasis on the use of science in operational forecasting, which would be based on improved forecaster knowledge and the use of a more scientific approach to forecasting (Roebber *et al.* 2004).
- Forecasters should be at the "heart of weather prediction", playing a vital role in forecasting high-impact weather.
- Product generation should be automated, with forecasters focussing on analyzing the prevailing meteorology.
- Forecasters can make important contributions to forecast quality by manually modifying model gridded output, using a software system that maintains dynamic consistency among variables (Carroll and Hewson 2005).
- The use of higher-resolution models provides the forecaster with products having ever-increasing complexity. Tools thus need to be available to the forecaster to allow for the easy exploration and analysis of the model output (Roebber *et al.* 2004).
- As the role of forecasters evolves, benefit could be derived from entraining other disciplines that are involved in the cognitive psychology of decision making (Doswell 2004).

See the references in Sills (2009) for additional information.

### SUGGESTED GENERAL REFERENCES FOR FURTHER READING

The reader should access the websites of organizations that run operational modeling systems and become familiar with the models, the organizational missions, and the weather products that are provided.

## PROBLEMS AND EXERCISES

1. Compile a list of operational modeling systems that are employed by organizations other than national weather services.
2. Why have LAM systems been developed to predict weather for limited geographic areas? Do they compete in any way with national weather services? Are they complementary?
3. Speculate on the role of forecasters as models continue to improve in accuracy.

# 13 Statistical post processing of model output

## 13.1 Background

The statistical post processing, or calibration, of operational NWP-model output is common because it can result in skill metrics that are equivalent to many years of improvement to the basic model. And, the greater skill is achieved at relatively little day-to-day expense, compared to other traditional approaches of trying to improve skill, such as through increasing the model resolution.

Historically, statistical post-processing methods were used to diagnose variables that could not be predicted directly by the low-resolution, early-generation NWP models. Standard model dependent variables associated with the large-scale conditions were statistically related to other poorly predicted or unpredicted weather variables such as freezing rain, fog, and cloud cover. However, many current-generation, high-resolution models can explicitly forecast such variables, and statistical correction methods are primarily employed to reduce systematic errors.

There is a variety of ways of classifying statistical post-processing methods. They may be categorized in terms of the statistical techniques used, as well as by the types of predictor data that are used for development of the statistical relationships. And, distinctions are made between static and dynamic methods. With static methods, statistical algorithms are developed for removing systematic error using a long training period that is based on the same version of the model, and the algorithms are applied without change for a significant period of time. Because of the computational expense associated with the calculation of the statistical relationships, models cannot be upgraded frequently because doing so requires recalculation of the relationships. Even when significant code errors are revealed, they cannot be corrected until new relationships are created. In contrast, with dynamic methods the calibration equations are recalculated on a regular basis.

Statistical post-processing methods are not appropriate for use in research applications of models. For physical-process studies, it is important that the model output be consistent with the dynamic equations, so artificial adjustments in the output would not be appropriate. Also, in such studies it is straightforward to optimize the model for a particular case to reduce systematic error (e.g., by testing different physical-process parameterizations, model resolutions, etc.), so there is less need for statistical adjustment than with operational NWP. And, research is often aimed at improving the model in order to reduce the systematic error.

The NWP-model forecast variables are sometimes used as input to specialized models that provide information about other quantities. For example, air-quality models contain continuity equations for various gaseous and aerosol species, they calculate the transport and diffusion of these contaminants, and they represent their chemical transformations. This use of appended and specialized models may be viewed as one type of post processing of the NWP-model output, but it is sufficiently specialized that it will be treated separately in Chapter 14.

The following section reviews some different approaches for statistical correction of model output. Another type of model post processing employs what are called *weather generators*, which are summarized in Section 13.3. Weather generators, also discussed in Chapter 16 in the context of climate models, take model output that is typically smoother than reality in terms of space and time variability, and define a more-realistic statistical structure. This kind of post processing is important for some model applications, for example in predicting flooding where the local short-time-scale variations in rainfall intensity are highly relevant for estimating the partitioning of rainfall between runoff and infiltration. In the last section is a brief discussion of how some types of downscalings of model output – that is, processing that can define the modulation of the large scales by local forcing, such as from orography – represent a form of statistical post processing.

## 13.2 Systematic-error removal

The following subsections review various methods for statistically correcting NWP-model forecasts in order to reduce the systematic error. The static methods require the use of a lengthy period of model reforecasts in order to define the statistical corrections based on relationships between past model output and past observations. These statistical relationships are not updated frequently. In contrast, the dynamic methods perform corrections to forecasts based on much shorter periods of training. In both cases, the goal is to reduce the error in current forecasts using estimates of past error.

Note that only systematic error is reduced by these methods. The random errors that result from numerically induced phase errors in the propagation of features, the smoothing of small-scale propagating features associated with insufficient model resolution, and other causes, will remain in the solution and cannot be statistically removed. However, the systematic error can represent a significant fraction of the total error, especially near the ground, so removing it through the use of post-processing methods can be very beneficial. For example, Fig. 13.1 shows the systematic and random forecast errors in the near-surface temperature (2 m AGL) and wind speed (10 m AGL), based on regional mesoscale-model simulations for the southwestern USA. The systematic temperature errors are clearly larger at some of the observation sites, presumably because local forcing is not represented well in the model.

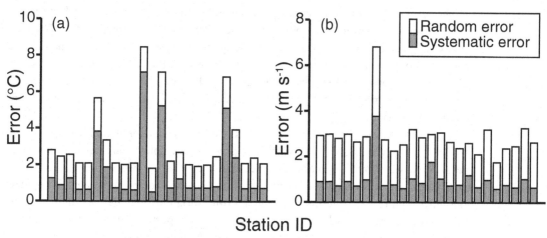

Fig. 13.1 Systematic and random errors in the near-surface temperature (a, 2 m AGL) and wind speed (b, 10 m AGL) calculated based on regional mesoscale-model simulations for southwestern USA. Each bar corresponds to a different observation location. Adapted from Hacker and Rife (2007).

### 13.2.1 The "perfect-prog" method

The earliest approach to statistical post processing is known as the Perfect-Prog (PP, perfect-prognosis) method (Klein *et al.* 1959). Here, observations of quantities that are predicted by the model (the predictors) are statistically related to observations of a predictand that may or may not be predicted by the model. The regression relationships are then applied to NWP-model forecasts of the predictors to produce forecasts of the predictands. Because the statistical relationships are not generated using model forecasts, they do not correct for model error. They simply statistically translate predicted variables into unpredicted or poorly predicted variables. In effect, it is assumed that the model prognosis is perfect. Because, as noted earlier, current models can explicitly predict many of the quantities that previously had to be statistically inferred through the PP method, this approach is less used operationally. A benefit of the PP method is that it is not dependent on the model to which it is applied, and thus the statistical relationships do not need to be recalculated when the model is modified. Figure 13.2 schematically compares the PP approach with the method of Model Output Statistics (MOS) described in the next section.

### 13.2.2 Model output statistics

The calculation of MOS involves statistically relating previous forecasts of a variable and the corresponding observations of the variable, in order to quantify the systematic forecast errors (bias) for each observation point. This bias results from many factors, including shortcomings in the physical-process parameterizations, and the inability of the model with a particular resolution to represent small-scale processes. The bias for each observation location is then used to correct future forecasts at the respective points. There are a number of MOS-type approaches, that differ in terms of the length of time over which the previous

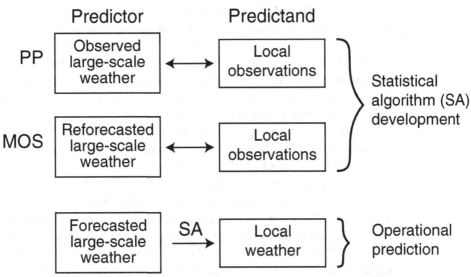

**Predictor**                          **Predictand**

PP | Observed large-scale weather | ⟷ | Local observations

MOS | Reforecasted large-scale weather | ⟷ | Local observations

⎫ Statistical algorithm (SA) development

Forecasted large-scale weather | SA → | Local weather | ⎫ Operational prediction

**Fig. 13.2** Schematic of the PP and MOS approach to statistically relating variables or features that are reasonably well forecast by a model (predictors, left column) with those that are not (predictands, right column). With the PP method, historical archives of observations or analyses are used in the definition of the statistical relationship, whereas with the MOS approach, model reforecasts of historical cases are employed. The sections on MOS describe methods that use training periods of different length.

forecasts are used to define the bias. The classical, historical approach, referred to above as the static method, has been to calculate the statistical relationships over a multi-year historical period. The use of this long training period leads to stable statistics, but it is so computer-resource intensive that operational models cannot be frequently updated with improvements because that would invalidate the statistics (that would then need to be recalculated). As a result, other MOS-type methods with shorter training periods have been developed. An advantage of very-short training periods is that systematic errors can be weather-regime dependent, so adjustments based on recent model performance can be beneficial. Thus, a balance must be reached between (1) a short learning period that is vulnerable to missing data that are needed for training, and to the occurrence of extreme weather events with unrepresentative errors and (2) a long learning period that produces stable statistics but that is arguably too computationally expensive to be practical. The following sections review a few different MOS-based approaches to systematic-error reduction.

### Conventional MOS

This approach, requiring statistics that are generated by forecasts from the same model over a period of at least two years, is summarized in Glahn and Lowry (1972). Because MOS requires the separate calculation of statistics based on forecast–observation pairs for each forecast lead time, for each observation location, and for each variable, a large number of equations are involved. Even though there has been a clear trend toward the use of shorter training periods with MOS-based methods, Hamill *et al.* (2004, 2006) present results that show that, for challenging situations such as long-lead-time forecasts, forecasts of rare events, or forecasts of surface variables with significant bias, long training periods can be beneficial. And Clark and Hay (2004) illustrate the great potential benefit of conventional MOS for producing improved forecasts for hydrological applications.

Jacks *et al.* (1990) provide a summary of an NCEP MOS system, where predictors included forecasts of temperature, temperature advection, thickness, precipitation amount, precipitable water, relative humidity, vertical velocity, horizontal wind components, wind speed, relative vorticity, vorticity advection, stability, and moisture convergence. These predictors are often defined at different levels in the model. The resulting system was very computationally demanding, involving the use of many thousands of statistical equations. It is interesting to note that in the late 1980s the MSC replaced its operational MOS system with PP products, which were used throughout the 1990s. See Brunet *et al.* (1988) for a discussion of the relative statistical characteristics of the PP and MOS methods.

Figure 13.3 shows an example of the benefit of the application of the conventional MOS approach, in this case in the context of mesoscale LAM simulations of 10-m AGL winds. The MM5 model was used for operational prediction during the 2002 Winter Olympics, for the Salt Lake City area, which is dominated by the complex orography of the surrounding area. The MOS equations were derived using three winter seasons of forecasts and observations for 18 mountain and valley locations. The grid increment of the model used for the generation of the statistics and for the operational prediction was 12 km, even though a 4-km nested grid was also employed in order to assess the benefit of higher horizontal resolution. The 4-km grid did not feed back to the 12-km grid, so it did not affect the MOS correction. However, the model version did change during the period of the MOS-equation development. The figure shows the wind speed MAE for both the 0000 UTC and 1200 UTC forecast cycles, based on the Direct Model Output (DMO) from the

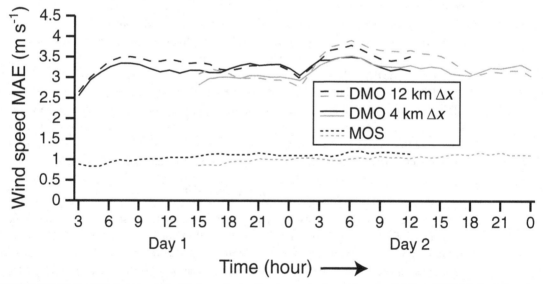

**Fig. 13.3**    The MAE of wind speed for 18 observation locations in complex terrain, based on DMO from MM5 LAM simulations that used horizontal grid increments of 12 km and 4 km, and MOS simulations that were based on the 12-km grid increment version of the model. The statistics for 36-hour forecasts from both the 1200 UTC (gray) and 0000 UTC (black) cycles are shown. Adapted from Hart *et al.* (2004).

12-km and 4-km models, as well as based on the MOS from the 12-km model. Based on the four curves for the DMO, the higher horizontal resolution produced no significant benefit because much of the orographic variability was still subgrid-scale. However, the use of MOS reduced the average MAE from about 3.5 m s$^{-1}$ to about 1 m s$^{-1}$.

### Updatable MOS

Updatable MOS (UMOS, Wilson and Vallée 2002, 2003) allows frequent and automatic updating of statistical forecast equations soon after changes are made to the NWP model. This is accomplished through user-controlled weights, such that, after a model change is implemented, the statistical properties of the new model forecasts and those for the old model can be weighted and blended in the operational statistical relationship. That is, instead of training a new algorithm using a frozen version of the new model for a long period of time, independent of the operational system, with UMOS the statistical properties of the new model are gradually given more weight as a longer history is accumulated. In the MSC UMOS implementation of this system, the blending of the old and new systems begins after 30 cases from the new model have accumulated. After 300–350 cases with the new model have been included, the influence of the old model is neglected. Figure 13.4 compares the

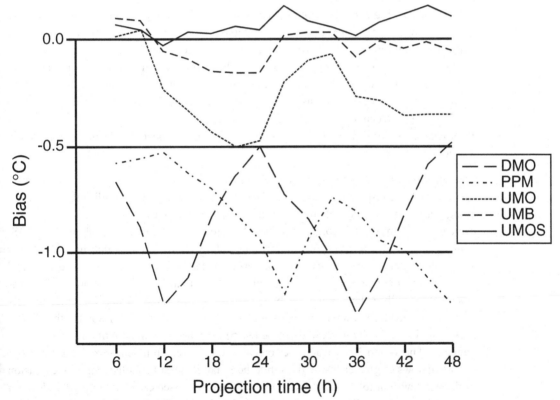

Fig. 13.4  Bias of winter-season temperature forecasts for about 250 Canadian stations, based on three statistical-correction methods related to Updatable MOS (UMOS, UMO, UMB) and the Perfect Prog Method (PPM). Also shown is the bias associated with uncorrected Direct Model Output (DMO). Adapted from Wilson and Vallée (2003).

bias associated with a hierarchy of three applications of UMOS (UMO, UMB, complete UMOS) with that of the PP method and the statistically uncorrected DMO for 250 Canadian stations for a winter season. The UMO statistical equations contain no new-model data, so this method ignores the effect of the change in the model, using the old statistical equations with the new model. With UMB, some new-model data are used, but none that are close in time to the test period. The DMO has a negative bias that varies between $-0.5\,°C$ and $-1.3\,°C$, depending on forecast lead time. Because the PP method does not correct for model biases, the PP method curve has a similar average bias. In spite of the fact that the UMO correction was based on old-model-version statistics, there is still significant bias correction. The complete UMOS approach has the smallest bias, averaged over all forecast lead times. See Wilson and Vallée (2002, 2003) for additional information about this method.

## Very-short-update-period dynamic MOS

A review and evaluation of implementations of MOS with different training periods is provided in McCollor and Stull (2008c), where the model employed was the CMC GEM model (Côté *et al.* 1998a,b). Four bias-calculation approaches tested are summarized below.

- *Seasonal-mean error* – For cold-season forecasts, the average mean forecast error was calculated for the six-month period encompassing the previous cold season. Similarly, warm-season forecasts were corrected using errors from the previous warm season.
- *Moving average with uniform weighting* – The average mean forecast error was calculated using an unweighted average of the bias error from the previous *n* days.
- *Moving average with linear weighting* – Same as above, but using a linearly weighted average, with recent errors weighted more heavily.
- *Moving average with nonlinear weighting* – Same as above, but using a nonlinearly weighted average.

The objective of the weighting of course was to provide greater weight to the recent forecast errors, to be responsive to regime changes, while employing a significantly long averaging period to enhance statistical stability. Averaging windows from 1 to 24 days were evaluated in terms of their ability to reduce the forecast error. Figure 13.5 shows the MOS-adjusted errors in forecasted maximum temperature for the method that used the linear weighting. Each curve corresponds to a particular lead time within 8-day forecasts, and shows the error at that lead time as a function of the length of the different averaging periods involved in the calculation of the bias. For all lead times, the greatest incremental forecast improvement associated with adding days to the averaging period was for the shorter averaging times. The longer lead times benefited the most from the use of longer averaging windows. The 1- and 2-day lead time forecasts did not benefit much from the use of averaging windows of greater than 5 days, but the 8-day forecast benefited from the extension of the windows out to 15 or 20 days. Other studies have used error-weighting windows of 7 days (Stensrud and Skindlov 1996, Stensrud and Yussouf 2003), 12 days (Stensrud and Yussouf 2005), 14 days (Eckel and Mass 2005, Jones *et al.* 2007), 21 days (Mao *et al.* 1999), and 15–30 days (Woodcock and Engel 2005).

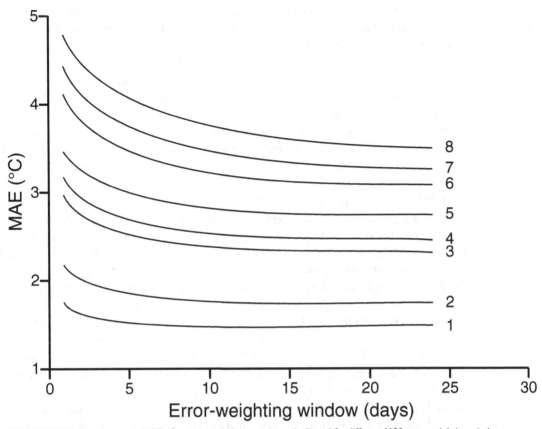

**Fig. 13.5** Forecast errors (MAE) of maximum daily temperature (ordinate) for different MOS error-weighting windows (abscissa). Each curve corresponds to a forecast lead time (days). In the calculation of the bias correction used in a forecast, the biases from the previous *n* days (the window length) were averaged, using a linear weighting where the recent errors were weighted more heavily. Adapted from McCollor and Stull (2008c).

### 13.2.3 Kalman-filter methods

The use of Kalman filters (KF) is another automatic post-processing method for employing past observations and forecasts to estimate model bias in future forecasts. Delle Monache *et al.* (2006b) review the mathematical basis for the method. Analogous with Eq. 6.16, which describes the application of least-squares estimation with Kalman filtering to data assimilation, the following equation pertains to the bias-estimation problem:

$$B_{t+\Delta t} = B_t + \beta_t(y_t - B_t).$$

The variable $B$ is the estimate of the bias in some forecast variable. The quantity $B_{t+\Delta t}$ is the estimate of the bias in the variable at a forecast lead time $\Delta t$, $B_t$ is the estimate of the bias at the end of the previous forecast, $y_t$ is the observed forecast error (both systematic and random) at the end of the previous cycle (difference between the forecast and observations), and $\beta$ is the Kalman gain. Say the forecasts are of 24 h duration ($\Delta t$), and we desire to estimate the bias in the forecast of $B$ at this time ($B_{t+\Delta t}$) so that we can correct for it. For the forecast valid at the present time, the bias had previously been estimated to be $B_t$, and this is used as the first guess. Because the forecast valid at time $t$ has completed, the total error $y_t$ has been calculated. This is differenced with the previous estimate of the bias, and multiplied by the weighting factor $\beta_t$. Thus, the future bias is estimated to be the most-recent estimate of the bias that is adjusted by a weighted difference between this bias estimate and the observed total error. See Delle Monache *et al.* (2006b), and Appendix A therein, for a discussion of the calculation of the Kalman gain.

Delle Monache *et al.* (2008) illustrate the error reduction associated with the application of a KF procedure for each member of a multimodel ensemble of 24-h forecasts of ozone concentration (Fig. 13.6). Each of the first eight names in the legend corresponds to a particular photochemical model and meteorological model combination that was used in the construction of the ensemble. Position on the coordinate axes corresponds to the RMSEs associated with the systematic error (abscissa) and the random error (ordinate) of model forecasts of ozone concentration. The distance between the origin and any coordinate represents the total RMSE. The coordinate of the tail of each vector represents the RMSE values for the DMO, and the coordinate of the head of the vector represents the RMSE for the KF-corrected forecast. The direction and length of the vector show the amount of change in the systematic and random errors that results from the application of the KF correction. The "E" refers to the ensemble average of the forecasts, where the vector tail defines the RMSEs of the ensemble average of the DMO, and the head defines the RMSEs after the KF correction is applied to the ensemble average. Here, the ensemble averaging is done before the filtering. The tail of the "EK" vector applies to the ensemble average of the individual KF-corrected forecasts, and the position of the head results from the application of the KF correction a second time. Here, the filtering is done before the averaging. There was clearly a large decrease in the systematic error that resulted from the application of the KF to each of the ensemble members and to the ensemble mean. Note that this method does not require an extensive statistical database for training.

### 13.2.4   Gridded bias-correction

Statistical corrections based on standard MOS methods apply at observation points only, and thus it is not possible to straightforwardly infer the model bias in a more general way at any arbitrary location for which a forecast is desired. This ability to provide spatially distributed information about systematic error is important for many applications, such as when using forecast precipitation in a gridded hydrologic model, or when forecast temperatures interact with the land surface at every grid point to control evaporation and sensible-heat fluxes. Hacker and Rife (2007) show how computation of error covariance matrices can allow the definition of systematic error on a grid, and describe the implementation of the method in an operational LAM.

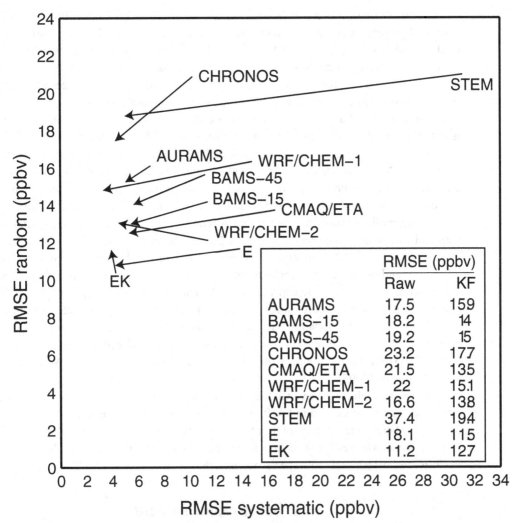

Fig. 13.6 The RMSE of DMO and KF-adjusted forecasts of ozone concentration for members of an ensemble. Each of the first eight names in the legend corresponds to a particular photochemical–meteorological model combination that was used in the construction of the ensemble, and "E" and "EK" refer to the application of the filtering and averaging in different orders (see the text). Position on the coordinate axes corresponds to the RMSEs associated with the systematic error and the random error of model forecasts of ozone concentration. The tail of each vector defines the RMSE values for the DMO, and the head represents the RMSE for the KF-corrected forecast. The direction and length of the vector show the amount of change in the systematic and random errors that results from the application of the KF correction. Adapted from Delle Monache *et al.* (2008).

## 13.3  Weather generators

Even though model time steps may be relatively short - perhaps tens of minutes – much of the short-time-scale variability associated with some phenomena is not represented in the model solution. For example, precipitation rates in nature can be highly variable, as rain bands, or other small convective features that are in various stages of their life cycle, pass across a location. Variability on these time scales is not represented in most operational models. This is especially true for large AOGCM grid boxes for which the time series of variables are smoother than those that apply to single points, simply because of the averaging that is implied over the large area. Unfortunately, high-frequency rain rates are needed for many hydrological applications, where the rate determines the partitioning of the rainwater between runoff and infiltration. Another example of high-frequency variability that is not represented in NWP models is wind gustiness, which is needed in models of dust elevation and transport, ocean waves, etc. To address such needs for high-frequency information from NWP and climate models, synthetic high-resolution time series can be generated with what are called *stochastic weather generators*. These methods essentially post-process the model-generated time series, adding realistic higher-frequency variability.

For NWP model simulations, the weather generators can add high-frequency spatial and temporal variations in the precipitation rate. For climate projections, which perhaps only have output at a monthly frequency, these generators can simulate the temporal distribution of wet and dry spells, the typical number of days with and without precipitation, etc. The generators can be tuned to apply to particular current and recent weather types. Information about the application of stochastic weather generators for climate-change studies, especially related to precipitation rate, can be found in Katz (1996), Semenov and Barrow (1997), Goddard *et al.* (2001), Huth *et al.* (2001), Palutikof *et al.* (2002), Busuioc and von Storch (2003), Katz *et al.* (2003), Wilby *et al.* (2003), Elshamy *et al.* (2006), Wilks (2006), and Kilsby *et al.* (2007).

Analogously, high-frequency wind-speed variability, sometimes known as gustiness or turbulence, is not represented in NWP or climate models, but it is important for predicting ocean-wave height, the elevation of dust from the surface in dust models, and dangers to aircraft. Application of weather generators for predicting gusts and turbulence will be discussed in Chapter 14, which deals with specialized models that are coupled to NWP models.

## 13.4  Downscaling methods

The concept of downscaling large-scale analyses and forecasts of weather and climate, such that small-scale features are estimated based on input about the larger-scale structure of the atmosphere, is discussed in Chapter 3 regarding the use of nested grids, and in Chapter 16 related to defining regional climates based on large-scale analyses or projections. The statistical downscaling of climate simulations, from interseasonal to century time scales, is described in Section 16.3.1, and has much in common with the MOS-based statistical methods described above.

### SUGGESTED GENERAL REFERENCES FOR FURTHER READING

Hamill, T. M., J. S. Whitaker, and S. L. Mullen (2006). Reforecasts: An important data set for improving weather predictions. *Bull. Amer. Meteor. Soc.*, **87**, 33–46.

McCollor, D., and R. Stull (2008c). Hydrometeorological accuracy enhancement via post-processing of numerical weather forecasts in complex terrain. *Wea. Forecasting*, **23**, 131–144.

Wilks, D. S. (2006). *Statistical Methods in the Atmospheric Sciences*. San Diego, USA: Academic Press.

## PROBLEMS AND EXERCISES

1. Speculate on the possible sources of systematic and random errors in NWP-model forecasts, in addition to those listed in this chapter. Distinguish between the two sources, and if necessary explain why the error is in one category or the other.

2. In reference to Fig. 13.1, why might the wind have a larger percentage of the error associated with the random component than does the temperature?

3. Again in reference to Fig. 13.1, describe situations that could be responsible for the considerably larger error at some locations compared to others.

4. Why might there be a greater need for statistical correction of model error for levels near the ground?

# Coupled special-applications models

## 14.1 Background

Sometimes the standard dependent variables of NWP and climate models are all that are required for making decisions. But, frequently these meteorological variables influence some other physical process that also must be simulated before a weather-dependent decision can be made. As we will see, there are myriad examples of such situations. These models that are coupled with the atmospheric model may be referred to as special-applications models or secondary models. Examples include the following.

- Air-quality models
- Infectious-disease models
- Wave-height models
- Agricultural models
- River-discharge, or flood, models
- Wave-propagation models – sound and electromagnetic
- Wildfire-behavior and -prediction models
- Electricity-demand models
- Dust-elevation and -transport models
- Ocean-circulation models
- Ocean-drift models
- Aviation-hazard models – turbulence, icing, visibility

Sometimes the secondary model is embedded within the code of the atmospheric model, and the coupled system is run simultaneously. And, sometimes there are two distinct model codes that are run sequentially. When the code that represents the secondary process is run within the atmospheric model, the secondary process may interact with the atmospheric simulation. Or, the flow of data may be in one direction only, where the atmospheric variables are used in the secondary model without feedback. There are some secondary-model processes that have strong feedbacks to the atmosphere, and for their prediction there is of course a greater need to have a two-way exchange of information between the atmospheric and secondary models. Examples include dust models wherein the dust influences the atmospheric radiation budget, wildfire behavior models where the fire modifies the atmospheric circulation, atmospheric-chemistry models where gases and particles that are involved in reactions influence the radiation budget, and wave-height models where the waves influence the evaporation rate and roughness length. These coupled models are applied on time scales of daily weather prediction, seasonal prediction, and multi-decadal climate prediction.

Even though it is implied above that the coupling is between two models, the atmospheric model and the application model, there are some situations in which more models are involved in the process. For example, assume that a disease pathogen is released into the atmosphere.

- *Atmospheric model* – The atmospheric model will define the transport winds, boundary-layer turbulence, humidity, etc.
- *Plume model* – A plume model will calculate the transport and diffusion of the aerosol, and the dosage (time-integrated concentration) or human exposure footprint at ground level.
- *Disease models* – A disease model can simulate the spread of a disease in an organism (humans or agricultural livestock) or among organisms (e.g., human beings) in a population.
- *Treatment models* – These define an optimal course of treatment based on many factors including the time since exposure, the size of the exposed organism, etc. This will most likely be a simple protocol that is based on previously run pharmacokinetics (how the drug moves around the organism) models and pharmacodynamics (how the drug acts on the organism) models.

The secondary and tertiary (etc.) processes or variables may be defined using physically based equations, such as chemical reactions in air-quality models or ocean currents in ocean-circulation models. Or, the secondary model may simply be a set of statistical or empirical algorithmic relationships that relate the predicted atmospheric state to some other variable. The simpler relationships are often call translation algorithms because they translate atmospheric conditions to the state of some other quantity. Thus, a hierarchical ranking of coupled models, from the simplest to the most sophisticated approaches, is as follows.

- *Type 1: Decision-Support Systems (DSSs)* – These can be simple or complex systems that formalize a decision-making process, using meteorological and other input data. Even though it can be argued that these are not models at all, DSSs nevertheless do post process model output, interpreting the large amount of data to allow decisions to be made in an intelligent and repeatable way. As shown below, these DSSs can be used to interpret atmospheric-model or coupled-model output.
- *Type 2: Translation algorithms* – These are simple physical equations or statistical relationships that use, as input, the variables predicted by the atmospheric model to define ancillary, sometimes nonmeteorological, variables that are required. An example would be algorithms that calculate atmospheric visibility or radio refractive index based on model output.
- *Type 3: One-way coupled models* – Even though the above translation algorithms do not feed information back to the atmospheric model, and are therefore one-way coupled, they are sufficiently simple that it can be argued that they should be distinguished from codes that are larger in size and might actually be considered a model in a traditional

sense. Examples of one-way coupled models are codes that model the spread of infectious diseases, some dust elevation and transport models that do not feed back, and some flood models.

- *Type 4: Two-way coupled models* – These are generally large pieces of code that may be embedded in the atmospheric model, for example as a subroutine. An example may be an ocean-wave model or an ocean-circulation model.
- *Type 5: Specialized atmospheric models* – Sometimes an atmospheric model is merged with specialized-applications codes in such an extensive way that the entire modeling system becomes specialized. For example, some atmospheric-chemistry models involve a thorough merger of the atmospheric and chemistry codes to produce an integrated, specialized model (e.g., WRFChem).

Because the coupled, secondary models are typically employed in order to provide information that can be used to make practical decisions, the secondary-model output is often used as input to a formal DSS. This DSS translates the data provided by the secondary model, and perhaps the driving atmospheric model, into a decision about whether to take an action – protect an agricultural crop against freezing, apply an anti-icing agent to a highway or an aircraft, vaccinate a population against an infectious disease, or evacuate a town that is threatened by flooding. The DSS may include an analysis of the relative benefit versus cost of taking alternative actions. This sequence of software components is summarized in Fig. 14.1.

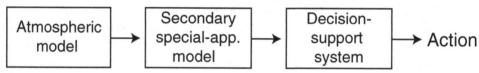

**Fig. 14.1**  Sequence of software components that are involved in providing the basis for a decision (action) that is weather dependent.

There is a hierarchy of methods for verification of coupled modeling systems, and this is illustrated in Fig. 14.2. First, it is reasonable to want to verify the accuracy of the atmospheric model alone, in the context of the geographic area and atmospheric variables of interest. In the figure, this is referred to as a Type-1 verification, where archived cases are used and the model retrospective forecasts are compared with meteorological observations or reanalyses. A Type-2 verification would again involve the use of historical cases, but the full coupled model (the atmospheric model and the end-user, or secondary, model) would be employed to produce a forecast of the secondary variable. This would be compared with a forecast from the secondary model that used meteorological input from observations or analyses. This tests the coupled system, but no observations of the secondary variable are used for verification, so the veracity of the secondary model is not evaluated. For the Type-3 verification, the coupled model is used for a retrospective forecast, but the forecast secondary variable is compared with observations.

This chapter will not provide a detailed discussion of the coupled models themselves. Rather, the focus will be on how the coupled modeling systems or algorithms are used to address practical problems. References will be cited for additional reading.

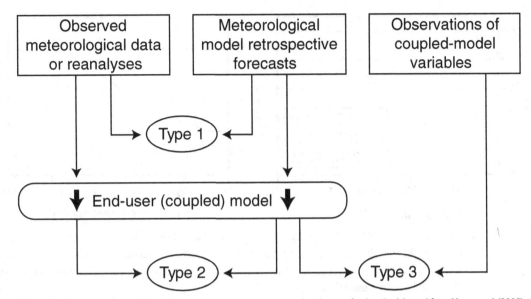

Three types of verification for coupled modeling systems. See the text for details. Adapted from Morse *et al.* (2005).

## 14.2  Wave height

There are various practical and modeling-related reasons for wanting to model the properties of wind-driven waves that exist on oceans and other water bodies.

- Heights of waves and swells impact the safety of recreational and commercial maritime activities, and therefore must be forecast. In the extreme, it would also be desirable if the probability of "freak" or "rogue" wave occurrences might be predictable.
- Wave action in littoral zones can be used to generate electricity, and thus wave forecasts are related to power forecasts.
- Evaporation of spray from waves releases aerosols into the atmosphere, which can influence cloud microphysical processes that are parameterized in a model.
- The albedo of the water surface is a function of wave activity, and is needed in the calculation of the ocean's energy budget.
- The evaporation rate is a function of the amount of sea spray, and this affects atmospheric temperatures.
- The roughness of the ocean surface that is experienced by the atmospheric-model's surface layer is a function of wave properties.
- Wave activity is associated with vertical mixing in the upper layer of water, which influences the water temperature at the lower boundary of the model atmosphere.

One or more of the above effects of waves on the atmosphere can be individually parameterized directly in the atmospheric model, or the predicted atmospheric variables can be used as input to a separate model that diagnoses wave properties. Wave-height predictions can be verified against buoy observations, as can the near-surface wind predictions that are

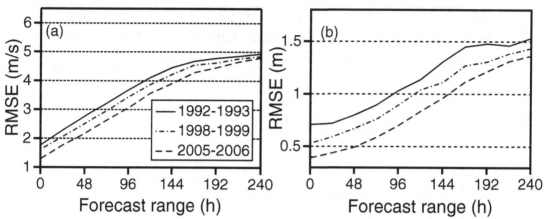

Accuracy of wind-speed (a) and wave-height (b) forecasts, based on ECMWF global-atmospheric and wave-height models. Statistics are shown for three years during a 13-year evolution of the models. The wave model employed wind predictions from the atmospheric model. Verification of both variables used buoy observations during October through March. Adapted from Janssen (2008).

used in the wave models. Indeed, wind-prediction accuracy is critical to wave-height prediction. For example, Fig. 14.3 shows the improvement in the accuracy of ECMWF wave predictions over a 13-year period, as well as the associated improvement in the prediction of the near-surface wind speed. Janssen (2008) calculated that 25% of the improvement in the wave-height forecasts, whose accuracy is shown in this figure, resulted from improvements in the wave model itself. The rest of the improvement was a consequence of more-accurate wind predictions. Growth in the wave-height RMSE was approximately linear between about 48 h and 168 h, with the error beginning to saturate after that. For additional discussion of wave-model verification, see Bidlot *et al.* (2002) who compare buoy observations with ocean-wave forecasts from a number of operational centers.

Note that wave heights are a function of wind gustiness as well as of the mean wind that is predicted by Reynolds-averaged model equations. The previously mentioned concept of a weather generator can be used here to infer gustiness based on air–sea temperature differences, where the gustiness metric can be used in the wave model (Abdalla and Cavaleri 2002).

## 14.3 Infectious diseases

The atmosphere can influence the spread of human and agricultural infectious diseases through a few different mechanisms, which are summarized in Fig. 14.4.

- *Health of the pathogen* – The health of the disease organism may be related to atmospheric variables such as temperature, relative humidity, the intensity of ultraviolet radiation, and precipitation.

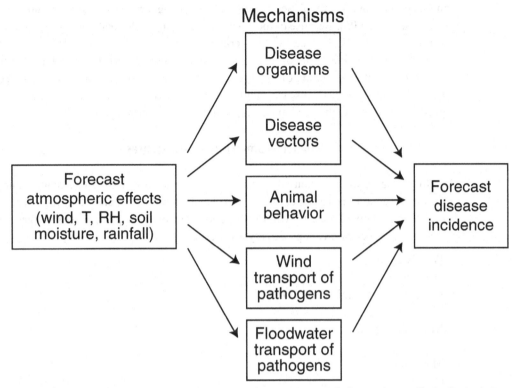

Fig. 14.4  Mechanisms by which atmospheric processes influence the spread of infectious diseases. The implication is that predictions for the atmosphere can be translated into predictions of disease emergence and spread.

- *Disease vectors* – The disease may spread through vectors such as fleas, mosquitoes, or rodents, and the number and health of the vectors can depend on temperature, relative humidity, vegetation greenness, and soil moisture.
- *Animal behavior* – Animal behavior is related to atmospheric conditions (e.g., for humans, the amount of time spent indoors in proximity to other people), and this behavior can influence the spread of disease.
- *Wind transport* – The wind can transport disease organisms and expose new populations.
- *Flooding* – This can increase the incidence of many diseases as a results of the compromise of fresh-water supplies, forced migration, and the production of a favorable environment for disease vectors.

Through knowledge of the statistical or physical relationships between disease incidence, for example outbreaks, and weather or climate conditions, it is possible to translate predictions of the atmosphere into predictions of disease spread or incidence. Medium-range forecasts of 7–10 days can allow redistribution of vaccines and medical personnel to locations that will be in greatest need. And interseasonal forecasts, e.g., of the ENSO cycle, can provide long-lead-time information for disease early-warning systems, which can guide the manufacture of vaccines and inform aid agencies about future requirements (Thomson *et al.* 2006). Because of the existence of complex physical, biological, and societal aspects to the links between atmospheric conditions and disease, correlations used for prediction are sometimes employed without a good knowledge of the underlying mechanisms. Given that some period of time exists between the occurrence of atmospheric conditions that are related to disease incidence, and the response of the incidence itself, lagged correlations can be used to develop statistical relationships for prediction.

## 14.3.1  Human infectious diseases

The following major infectious diseases have been shown to have some relationship to atmospheric conditions (weather or climate). The meteorological and other factors can be temperature, relative humidity, wind speed and direction, precipitation, sea-surface temperature, and vegetation-canopy density and health.

- West Nile virus
- Dengue Fever (DF)
- Dengue Hemorrhagic Fever (DHF)
- Valley fever
- Rift Valley fever
- Malaria
- Meningitis
- Cholera
- Typhoid fever
- Leptospirosis
- Hepatitis A

For human infectious diseases, there is an especially strong potential connection between weather or climate and human activities that can spread pathogens. For example, drought can cause migrations. Windy, dusty, rainy, or cold conditions can cause humans to congregate inside and spread diseases through contact. Thus, disease-spread prediction models must incorporate societal/behavioral factors as well as physical and biological processes.

An example of a success at establishing correlations between weather factors and human disease is described in Fuller *et al.* (2009), who explain 83% of the variance in weekly DF/DHF cases in Costa Rica from 2003 to 2007 using a simple regression model that incorporates lagged ENSO-related SST and MODIS vegetation indices. Another example involves the production of retrospective forecasts of malaria. The atmospheric model simulations were produced by seven institutions in Europe that participated in the DEMETER project (Palmer *et al.* 2004). The output from these models was used in a Malaria Transmission Simulation Model (MTSM, Hoshen and Morse 2004). The DEMETER-MTSM system was verified using the Type-2 method described earlier (Fig. 14.2), where ERA-40 gridded analyses (see Chapter 16) served as the atmospheric verification data set. The malaria predictions were shown to be skillful for the 1-month lead seasonal predictions, and for the 4–6-month lead for the seasonal malaria peak. Other discussions of the use of seasonal weather predictions for anticipating malaria outbreaks are found in Thomson *et al.* (2000) and Thomson and Connor (2001). Reviews of the overall potential for predicting human infectious diseases using weather and climate forecasts can be found in Kuhn *et al.* (2005) and NRC (2001).

As another illustration of how atmospheric-model reanalyses and predictions can be used to make decisions about infectious-disease management, consider the problem of meningitis in the Sahel of Africa. Even though the mechanisms are unclear, meningitis outbreaks tend to occur as the relative humidity decreases in the winter season, when harmattan winds bring dry, dusty air from the Sahara. When the relative humidity increases with the beginning of the Guinea monsoon, the number of cases decreases. Forecasting the spatial pattern of the seasonal rise in relative humidity is important to allow the appropriate distribution of the remaining vaccine. We begin by defining a somewhat arbitrary threshold for cessation of meningitis susceptibility as the first occurrence of five continuous days of relative humidity of at least 40% for any point on a grid. Those areas of the Sahel for which this threshold is reached at times that vary considerably from year to year could benefit from forecasts. If there is little year-to-year departure from the climatological date when this condition is met, the climatology can be used. Figure 14.5 shows a map of the standard deviation of the date on which this criterion is first met, based on the NCEP-NCAR Reanalysis Project archive (NNRP, Section 16.2, Kalnay *et al.* 1996) for

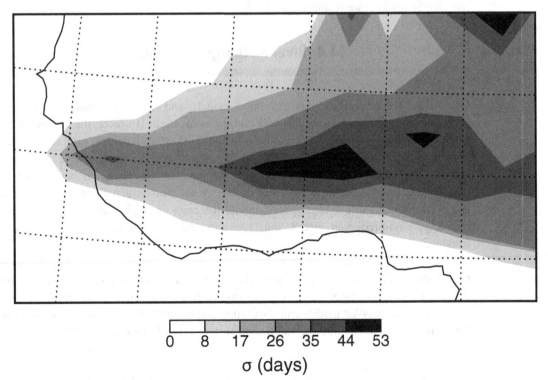

**Fig. 14.5** The standard deviation in the date on which five continuous days of relative humidity of at least 40% first occur in the Sahel, based on 50 years of the NNRP reanalysis. Provided by Thomas Hopson, NCAR.

1949 through 2009. The standard deviation approaches or exceeds a month over a significant area of the central Sahel, indicating where forecasts of this threshold, or a similar one, would be useful.

### 14.3.2 Agricultural diseases

Many agricultural diseases are related to atmospheric conditions, so models for disease probability or spread use NWP forecasts or analyses as input. Examples of plant-disease forecasting systems used by the US Department of Agriculture for diagnosis and prediction of insect pests, fungal diseases, mildew, etc., are NAPPFAST (Magarey *et al.* 2007) and ipmPIPE (Isard *et al.* 2006). These are interactive web-based systems that generate a number of weather-dependent crop and disease maps and other graphical products that allow farmers to save money by targeting mitigation strategies (e.g., pesticide applications) only where and when they are needed. Also, some plant and livestock diseases are spread by aerosols (e.g., spores, bacteria, viruses) that are carried by the wind, so the plume models discussed in Section 14.5.1 can be used with input from NWP-model analyses or forecasts to track the movement of these pathogens. Similarly, insect pests are carried by wind from regions where they begin their life cycle to regions where they can impact agriculture, so plume-type models can predict these processes as well. Many different systems that employ atmospheric-model products have been developed to address the specific needs of agriculture, worldwide.

## 14.4  River discharge, and floods

Operational river-discharge[1] models, flood models, and flash-flood models often use precipitation estimates from radars and rain gages as input. However, the forecast lead time for flooding at a particular location on a river is thus limited to the time it takes the rainwater (and possibly resulting snowmelt) to travel along the water course. This may allow insufficient time to respond, whether the forecast information is being used to warn or evacuate people near the water course, or to release water from a downstream dam. A solution is to use atmospheric-model forecasts of precipitation as input to the discharge/flood-prediction model, thus providing much additional lead time for the response.

Discharge models have a wide range of complexities, and the partitioning between those aspects of the surface hydrologic cycle that are treated in the atmospheric model and those that are represented in a separate discharge model varies considerably. Many current-generation NWP models employ a simple representation of the hydrologic cycle in their land-surface model; they partition some water as runoff in each grid cell, but they do

---

[1]  Discharge is the volume of water flowing in a river or stream channel, and is generally defined in cubic meters per second.

not route or track the water from grid cell to grid cell through channels across the land surface, and thus they cannot explicitly predict discharge. This would need to be accomplished by a coupled runoff/discharge model. Obviously, the processes represented in coupled discharge and atmospheric models need to be compatible, and collectively define the entire surface-hydrologic system. Alternatively, some atmospheric models represent the entire process, including surface routing and calculation of discharge.

Flash floods are defined as events wherein the water goes from base-flow to flood level in less than 6 hours, they are most common in complex orography, they are typically caused by convective–precipitation events, and they take a significant number of lives because of the short warning time available. Unfortunately, for a variety of reasons, radar estimates of precipitation are often unreliable or nonexistent in areas of complex terrain. And, rain gages are sparse in mountains, and their estimates are not representative of a larger area. Thus, a useful degree of predictability of flash floods in mountainous terrain, if that is attainable, may have to rely on precipitation forecasts with convection-resolving atmospheric models that accurately represent local orographic and other landscape forcing. There have been occasional successful simulations in research settings that give us hope for possible eventual operational predictability in such situations. For example, Nair *et al.* (1997) describe a successful model simulation of the convective storm that produced the severe 1972 Black Hills flash flood in a mountainous area of South Dakota, USA. Also for an area in the USA with complex orography, Fig. 14.6 shows a discharge simulation from a coupled mesoscale LAM (MM5) and a discharge model (Precipitation-Runoff Modeling System, PRMS, Leavesley *et al.* 1983) for a convection-related flash-flood event (Yates *et al.* 2000, Chen *et al.* 2001). The simulations were part of a research study to estimate the effects of a recent wildfire on the severity of the flood. Shown are discharge calculations from PRMS based on the use of radar-estimated and model-simulated precipitation, and for land-surface parameters in MM5 and PRMS that were defined with the burn area (fire) and with the pre-burn natural vegetation (no fire). Also shown is the discharge calculated by simply totaling the amount of model-simulated rainwater that is partitioned to runoff, where no routing through the stream channels is calculated. The discharge estimated from the high-water marks along the water course is indicated. Discharge was calculated accurately by PRMS using the radar-estimated rainfall (with the fire, solid line), but the use of MM5-simulated precipitation in PRMS underestimated the peak discharge by a factor of three (dotted line). However, the base flow for the stream was normally only a few meters per second, so a significant event was still predicted. The direct-MM5 discharge peak was earlier than the observed peak because the time required for the rainwater to flow to the location of the verification was ignored. And, no water was lost to infiltration during overland flow, so the peak discharge was greater than that produced by PRMS. Thus, the reasonably accurate value of this peak was probably obtained for the wrong reason.

As an example of the use of this type of coupled system for larger scales (continental USA) and longer lead times (to 8 days), Clark and Hay (2004) evaluated discharge forecasts for four study basins, based on the PRMS model coupled with the NCEP MRF model. The MRF precipitation forecasts showed considerable error in many regions, so the temperature and precipitation forecasts were corrected using MOS (Chapter 13). The

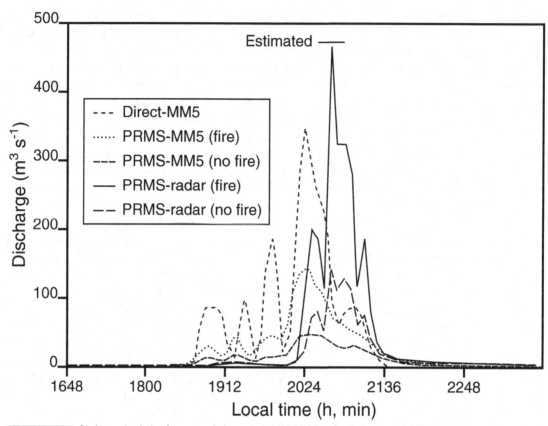

**Fig. 14.6** Discharge simulation from a coupled mesoscale LAM (MM5) and a discharge model (PRMS) for a convection-related flash-flood event. Shown are discharge calculations from PRMS based on the use of radar-estimated and model-simulated precipitation, and for land-surface parameters in MM5 and PRMS that were defined with the burn area and with the pre-burn natural vegetation. Also shown is the "discharge" calculated by simply totaling the amount of model-simulated rainwater that is partitioned to runoff, where no routing through the stream channels is calculated. See text for details. Adapted from Yates *et al.* 2000.

MOS correction of systematic error provided improved discharge forecasts in only the snowmelt-dominated river basins because the MOS was only able to improve temperature and not precipitation forecasts.

    The predictive skill of discharge forecasts from coupled atmospheric–discharge models is obviously no better than that of the precipitation forecasts. And, of course we know that precipitation forecast skill has shown the slowest benefit, relative to other dependent variables, as a result of improvement in all aspects of atmospheric-modeling systems. On the convective scale, operational models have virtually no skill at deterministically predicting precipitation events – i.e., correctly locating individual convective cells. Thus, referring to the convective events that commonly lead to flash flooding, even though a mesoscale model may be able to routinely predict the general area and severity of the convection, there is relatively little hope of using the forecasts as a basis for evacuating residents when the watersheds are of small to modest size.

Discharge models are often used in conjunction with global or regional climate models in order to couple the hydrologic cycles over land and over the ocean. In addition, there is a need to assess changes to the hydrologic system associated with future-climate scenarios (e.g., Bell *et al.* 2007, Bronstert *et al.* 2007, Charles *et al.* 2007, Fowler *et al.* 2007).

# 14.5 Transport, diffusion, and chemical transformations of gases and particles

There are a few types of models that are used operationally and for research to track the transport, turbulent diffusion, and chemical transformation of particulates and gases in the atmosphere. These types of models are summarized in the sections below. Note that there can be considerable overlap in the purpose, numerical approaches, and processes represented in the different types of models. Dust elevation and transport models are used to simulate or predict the elevation of mineral dust from the surface, and the consequent dust storms that result. Volcanic-ash models also track dust, but specifically the material that is ejected forcefully into the troposphere and stratosphere from a volcano. Air-quality models are specialized or general models that simulate gases or particles whose concentrations are often regulated for environmental or human-health reasons, where the sources can be numerous and distributed over a large area (e.g., an entire city). Plume models tend to be used for single sources of contaminants, or perhaps a few sources.

## 14.5.1 Plume models

A plume is a volume of air, containing particles or gases that have been released into the atmosphere, that spreads horizontally and vertically from its source. Models that simulate plumes are generally Type-3 models, which do not feed back to the atmospheric model that provides them with time-dependent winds, thermal properties, humidity, precipitation, and possibly turbulence intensity. All of these meteorological variables can influence plumes in different ways. Plume models can have either an Eulerian or Lagrangian framework. With an Eulerian approach, the gaseous or particulate contaminant is released in a grid box within a three-dimensional array of points, and the model calculates the transport of the material from the source to downwind grid boxes through transport by the mean wind and turbulence. In contrast, with Lagrangian methods, puffs or particles of material released from the source are tracked, where their size or concentration, and location, are again controlled by atmospheric processes. Thus, Eulerian methods are grid-centric and Lagrangian methods are plume-centric. In either case, the plume model essentially solves a continuity equation for the gas or particles released.

Plume models have a variety of practical applications, including predicting

- the impact of smoke from wildfires;
- the movement of hazardous chemical, biological, and radiological material that has been accidentally released (an industrial or transportation accident) or intentionally released (a terrorist attack) into the atmosphere; and

- the airborne movement of insects or naturally occurring pathogens that are related to human or agricultural infectious diseases.

If the atmospheric model produces a forecast, the plume evolution is a forecast. If model-based reanalyses are used as input to the plume model, the simulated plume can be viewed as a reconstruction of a real or hypothetical historical event. The plume-modeling process involves a few steps. First, the atmospheric variables must be predicted by a model or diagnosed by a data-assimilation system. Then the source of the material being tracked needs to be estimated in terms of the amount of material released (instantaneously or continuously) and the location of the source (moving or stationary). The plume-model equations, in Eulerian or Lagrangian form, are then integrated in order to calculate the downwind advection and diffusion of the plume.

## 14.5.2  Air-quality models

Air-quality models typically represent (1) multiple sources and species of contaminants, (2) chemical reactions among contaminants, and between them and naturally occurring gases and particles, (3) transport and diffusion, and (4) interactions of contaminants with cloud, precipitation, and radiation. They are used for research, regulatory purposes, and forensic analysis. Research applications attempt to improve knowledge of physical and chemical processes associated with the existence of pollutants in the atmosphere. Regulatory applications take a number of forms, where an example is the use of the model to assess the impact on air quality of a proposed new source of pollution, where the results of the study would serve as the basis for a decision about whether to permit the source to operate. A forensic analysis could involve the use of a model to establish a source–receptor relationship between a region that produces pollutants and regions that are impacted by them. Air-quality models can be of Type 3, 4, or 5. Examples and brief discussions of each type of air-quality model application are provided in the following sections. Summaries of these models, and extensive reference lists are provided in the review papers by Russell and Dennis (2000) and Carmichael *et al.* (2008), and the text by Jacobson (1999).

## Research applications

Research applications are likely to use model Types 4 and 5, because complete interaction among the various physical and chemical processes is desirable. An example of a Type-5 model is the WRFChem LAM system (Grell *et al.* 2005) that is based on the integration of chemistry into the framework of the community WRF atmospheric model. When used with some type of urban-canopy parameterization, it is applicable for research studies of atmospheric, physical, and chemical processes in urban areas. For example, Jiang *et al.* (2008) used WRFChem to estimate the impact on surface ozone of climate change related to greenhouse gases and urban growth in Houston, USA for the 2050s. And, Zhang *et al.* (2009) applied the model for Mexico City for the period of the MILAGRO field campaign, and verified chemical-species concentrations against special observations.

An example of a global air-quality system is the Model for Integrated Research on Atmospheric Global Exchanges (MIRAGE, Easter *et al.* 2004), which is designed to study the impacts of anthropogenic aerosols on the global environment. The MIRAGE system consists of a chemical transport model coupled with the Community Climate Model, Version 2 (CCM2). Zhang (2008) summarizes the history, current status, and outlook for coupled atmospheric and chemistry models, and includes the MIRAGE and WRFChem models in the discussion. The Community Multiscale Air Quality (CMAC) model is an example of a Type-3 coupled system that is used for research.

## Operational applications

Air-quality models are used to provide operational next-day predictions of various measures of air quality, such as ozone concentration, in order to inform the public about the need for possible avoidance of exposure. Such models are used worldwide, and address the specific local air-quality issues. An example is the US National Air Quality Forecast Capability (NAQFC), which utilizes the NCEP Eta meteorological model coupled with the CMAQ modeling system (Byun and Schere 2006). Others are the MM5-CMAQ-based system, which is employed for Europe (San José *et al.* 2006) and the Australian Air Quality Forecasting System, which is applied to the regions of Melbourne and Sydney (Cope *et al.* 2004).

## Forensic analysis

Forensic studies can take a number of forms, and can be viewed as research investigations that have a specific practical objective (in contrast to improving our knowledge of processes). For example, if a particular geographic region experiences poor air quality in terms of some chemical species or aerosol, perhaps not attaining the minimum standards prescribed by government, air-quality models can be used to help estimate whether the contaminant is being transported into the area from external sources, or how different local mitigation strategies will improve the air quality. The above noted CMAQ model is often used for such studies, where models or reanalyses provide the input meteorological variables.

### 14.5.3 Dust elevation and transport models

For both research and operational-prediction applications, it is important to be able to model processes that involve mineral dust that has been elevated into the atmosphere by high winds. Because dust has strong influences on atmospheric short- and longwave radiation, and it affects cloud microphysical processes that in turn influence precipitation, its effects should be represented in weather- and climate-prediction models. An example of this importance is that aerosols of Saharan origin have been shown to affect the development of tropical cyclones and hurricanes in the Atlantic Ocean (Karyampudi and Pierce 2002). In addition, dust elevated into the atmosphere has numerous environmental consequences that are important to represent in coupled-modeling systems. These include contributing to climate change; modifying local weather conditions; producing chemical and biological changes in the oceans that can lead to blooms of toxic algae and coral-reef mortality; transporting bacteria and other pathogens over long distances; and affecting soil formation, air quality, surface water and groundwater quality, and crop growth and survival. Societal impacts include disruptions to air, land and rail traffic; interruption of radio services; the effects of static-electricity generation; property damage; and health effects on humans and animals.

Physical processes that must be simulated by a dust model include the lifting of the dust from the surface, which requires accurate representation of the near-surface wind speeds, including the parameterization of gustiness. Correct calculation of this source term also relies on the use of a good land-surface model for predicting soil moisture, and the correct estimation of the density and size of vegetation that can shield the surface from high winds. Winds aloft must also be simulated well by the meteorological model in order to accurately estimate the distance and direction of the horizontal transport. Lastly, the size distribution of aerosol particles needs to be estimated in order for settling velocities and surface deposition to be accurately calculated.

There are numerous dust models used for research and operational prediction world-wide. Examples of Type-4/5 models include (1) the US Navy Aerosol Analysis and Prediction System (NAAPS), which is a global operational aerosol model that involves a coupling of the Navy Operational Global Atmospheric Prediction System (NOGAPS) meteorological model and a dust-transport model (Westphal et al. 2009) and (2) a companion LAM, the Coupled Ocean–Atmosphere Mesoscale Prediction System (COAMPS) with an embedded dust-modeling capability (Liu et al. 2007). Type-3 models are the Community Aerosol and Radiation Model for Atmospheres (CARMA, Toon et al. 1988, Barnum et al. 2004, Su and Toon 2009) and the DUst-emission MOdule (DuMo, Darmenova and Sokolik 2007). Lastly, the Barcelona Supercomputing Center operates the Type-3 Dust REgional Atmospheric Model (DREAM, Nickovic et al. 2001), which uses meteorological input from the NCEP Eta model.

### 14.5.4 Volcanic-ash models

Contemporary volcanic-ash models are based on atmospheric models coupled to special transport and diffusion models. Their use is motivated by the need to evacuate populations because of the negative health effects of the dust (called tephra) immediately downwind. And, the dust can seriously damage aircraft engines, so commercial airline flights must be rerouted to avoid the dust plumes. The need for ensuring aircraft safety has motivated most of the historical model applications. One of the earliest examples of the operational use of such a coupled modeling system employed NCEP (then NMC) regional and global models to provide input to the Volcanic Ash Forecast Transport And Dispersion (VAFTAD) model (Heffter and Stunder 1993). Operational products included relative ash concentrations in three aircraft flight layers. Other operational plume-tracking systems are used by the Canadian Meteorological Centre (CMC). The simplest is a three-dimensional trajectory model (see Section 11.2.2) that uses input from the CMC global data-assimilation and forecast systems. The second capability is the CANadian Emergency Response Model (CANERM), which is a three-dimensional Eulerian model for calculating the medium- to long-range transport of pollutants (volcanic ash, radioactive plumes, etc.) in the atmosphere (Pudykiewicz 1988). It uses the same operational CMC global modeling system for meteorological input. More recent coupled models have focussed on predicting tephra impacts and accumulations at the surface. For example, the

RAMS atmospheric model coupled with the HYbrid Particle And Concentration Transport (HYPACT) model was used to simulate the ash dispersal from the 1995 and 1996 eruptions of Mount Ruapehu, New Zealand (Turner and Hurst 2001). More recently, Byrne *et al.* (2007) verified simulations from the MM5 atmospheric model used with trajectory and particle-fall models against observed tephra accumulations for the Cerro Negro, Nicaragua, volcano.

# 14.6 Transportation safety and efficiency

## 14.6.1 Aviation

Airport ground operations, the routing of aircraft by traffic controllers, and real-time pilot decisions are all affected by weather, and in many cases DSSs are employed to translate weather observations and forecasts into decisions. The example topics below focus on weather impacts on in-flight safety, and discuss associated coupled models.

### Turbulence

Turbulence that impacts aviation safety results from a variety of meteorological situations, such as convection and wind shear. Models that diagnose the probability of turbulence use forecasts from NWP models as well as pilot reports of turbulence as input. The resulting fields are used by dispatchers and pilots for turbulence avoidance. Graphical products displaying turbulence potential are available on the web for different flight levels. See Sharman *et al.* (2006) for a description of one such turbulence diagnostic model. This is a Type-3 coupling.

### In-flight icing

Aircraft icing, which results from flight through supercooled liquid water, is a significant cause of aircraft accidents. An example of an operational system for predicting the likelihood of aircraft icing is the Current Icing Product (CIP) algorithm (Bernstein *et al.* 2005), which combines analyses and very-short-range forecasts from the RUC model (Benjamin 2004a) with real-time satellite, radar, surface, lightning, and pilot-report observations to create an hourly three-dimensional diagnosis of the potential for the presence of supercooled large droplets and icing. First, the volume of atmosphere that is occupied by clouds and precipitation is estimated. Then, fuzzy-logic methods use temperature, relative humidity, vertical velocity, pilot reports of icing, and explicit model fields of supercooled liquid water to estimate the presence of icing. A Future Icing Product (FIP) system produces forecasts of icing based on longer-lead-time RUC products. Resulting icing-advisory and icing-severity maps are available on the web, as are flight-path tools that define conditions along a prescribed flight track. This is a Type-3 coupling.

## Cloud ceiling and visibility

Cloud ceiling and visibility must be predicted for a couple of reasons. Airport capacity, in terms of the minimum spacing of aircraft on approach and departure, is generally a function of the prevailing visibility. And, noncommercial pilots with only "visual-flight-rules" certification must avoid flight within clouds and in low-visibility situations. To obtain ceiling and visibility from model-simulated, state-of-the-atmosphere variables, Stoelinga and Warner (1999) developed a translation algorithm based on empirical and theoretical relationships between model hydrometeor characteristics and light extinction. This is a Type-2 coupling.

### 14.6.2 Surface transportation

Weather affects highway and rail traffic in almost as many ways as it does air travel. Forecasts from NWP-models have been used as input to algorithms and DSSs that are employed for

- deployment and prepositioning of snow-removal equipment in advance of a winter storm;
- evacuation of the public in advance of a hurricane;
- estimating regions where rail lines and highways will flood as a result of heavy precipitation;
- defining the amount and type of chemicals to be applied to highways to melt ice and snow; and
- deployment and prepositioning of electrical and communications workers in advance of a natural disaster such as a hurricane or midlatitude winter storm.

Visibility is a hazard that surface and air transportation have in common. Some models now have sufficient skill to enable them to predict visibility, based on impairments from aerosols and fog. For example, Clark *et al.* (2008b) and Haywood *et al.* (2008) report on the operational version of the UKMO Unified Model in terms of its ability to predict visibility. This would be a Type-4 modeling system.

## 14.7 Electromagnetic-wave and sound-wave propagation

Electromagnetic (EM) energy is refracted through, primarily, vertical gradients of temperature and moisture. This is a practical issue because it is important, for various applications of radars, to know from where in space a reflection originates. This requires knowledge of the refraction, obtained through the use of a propagation model. An example of an EM propagation model that simulates this refraction, based on atmospheric variables provided by a model, is the Advanced Refractive Effects Prediction System (AREPS). The AREPS is a complete suite of software modules that can be applied for a wide range of propagation applications over sea and land.

Sound propagation through the atmosphere is sensitive to vertical profiles of meteorological variables as well as the nature of Earth's surface, so predictions of the sound intensity at various distances and azimuths can be produced using input from an NWP model. Practical applications of such coupled modeling systems include the timing of explosions associated with military testing or commercial excavation, such that the resulting sound will have minimal impact on surrounding structures and personnel. Another application could involve mitigation of the impact of aircraft-engine noise on developed areas around airports. An example of a prediction from a coupled sound-propagation model and a mesoscale NWP model is shown in Fig. 14.7. The sound propagation model is the Noise Assessment Prediction System (NAPS), where the physical basis for the model and references can be found in Sharman *et al.* (2008).

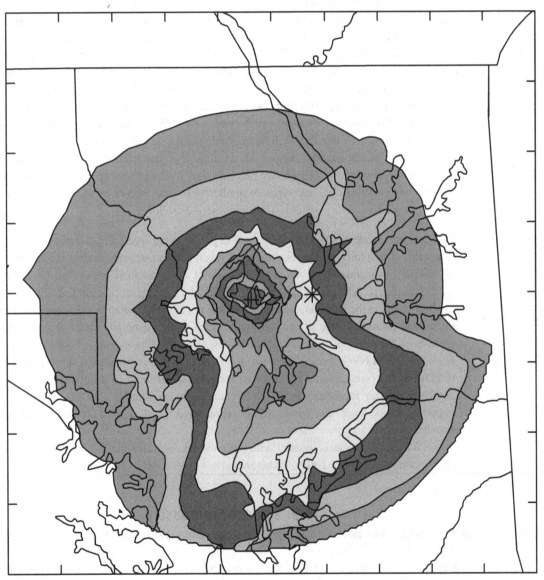

**Fig. 14.7**  Predicted sound intensity from an explosion in a coastal area in the eastern USA. Output from the MM5 LAM was used as input to the Noise Assessment Prediction System. The sound intensity in the outer shaded band is 100–105 dB and for the innermost band it is 140-145 dB. Adapted from Sharman *et al.* (2008).

## 14.8  Wildland-fire probability and behavior

The probability that a wildland fire will occur in a particular region depends on many factors, but one of the most important is the amount of moisture in the natural fuel (live and dead vegetation). This fuel-moisture level is a function of antecedent temperature, relative humidity, wind speed, and rainfall. In order to supplement *in-situ* and remotely sensed estimates, specially adapted high-resolution land data-assimilation systems (see Section 5.4.2) can be used to provide continuous gridded fields of fuel-moisture estimates. Or atmospheric-model forecasts can provide the input variables. An example of the latter approach is described in Fiorucci *et al.* (2007), wherein the atmospheric variables predicted by the LAM described in Doms and Schättler (1999) are used as input to an operational Italian dynamic wildfire-danger assessment system. Analogous wildfire-danger assessment systems in the USA and Canada employ observed meteorological conditions. The resulting fire-threat assessment can be used in a DSS that optimizes the location of fire-fighting assets such that they will be available quickly in the event of a fire.

Models that directly simulate, or provide qualitative guidance about, the behavior (e.g., direction and rate of growth of the fire perimeter) of wildland fires span a wide range of spatial scales, complexities, and types of applications. The categories of operational and research coupled-modeling systems are summarized as follows.

- Standard mesoscale NWP models are used to predict wind speed and direction, temperature, relative humidity, and precipitation, all quantities that strongly impact the evolution of existing or potential fires. Accurate forecasts of these variables are critical for use by wildfire managers who must (1) define the strategy for fighting an existing fire, and deploy the firefighters in the most safe and effective way or (2) decide whether to proceed with an intentional controlled burn of dead fuelwood. Such models generally have horizontal resolutions on the mesogamma or mesobeta scales. These are Type-1 coupled systems.
- High-resolution NWP models, or even-higher resolution computational fluid-dynamics (CFD) models (see Chapter 15), provide meteorological input to a fire-growth model, but there are no feedbacks from the fire to the atmosphere – i.e., fire impacts on humidity, temperature, and winds are not represented (Fujioka 2002). These are Type-3 coupled systems.
- The NWP or CFD models interact with a fire model, such that the fire feeds back to the atmospheric state (Coen 2005). These are Type-4 or -5 coupled systems.

## 14.9  The energy industry

There are a number of sectors of the energy industry that use special models that are driven by atmospheric predictions. For example, hydropower-generation facilities use decision models to determine how much water to release in anticipation of a heavy rainfall event over their supply watersheds (Section 14.4). And, energy companies that have nuclear facilities use air-quality and plume models to assess the potential impact of accidental releases from their facilities on public health (Section 14.5). Thus, the following sections only illustrate a few examples of the use of coupled models by this large industrial sector.

### 14.9.1  Electricity-demand models

Because electric-power providers benefit from being able to anticipate the demand, they use models to estimate this quantity from the many governing meteorological and nonmeteorological factors. Meteorological variables that are relevant are cloud cover, wind speed, temperature, and humidity. Taylor and Buizza (2003) summarize the use of atmospheric ensemble models for providing 10-day lead time probabilistic demand forecasts. Electricity-demand models are also used with input from climate projections to provide insight into long-term trends in requirements (e.g., Miller *et al.* 2008). These would typically be Type-2 or -3 coupled systems.

### 14.9.2  Wind-power prediction

Because available wind power is a function of the wind speed at 80–100 m AGL at the farms where the turbines are located, estimates of future power require wind-speed forecasts for these heights. The forecast wind speeds are translated to power production using an algorithm that is based on the number of turbines operating, the mix of turbine types, the efficiency of each turbine, etc. These also would typically be Type-2 or -3 coupled systems.

Forecasts of wind-power production are required in order to plan how to balance the load among the various available sources, such as gas, coal, nuclear, and wind. Especially problematic for energy companies are wind "ramp events" that are not correctly forecast. In these cases, the speed increases or decreases precipitously because of a frontal passage, the variation in the height of the shear zone below a low-level jet, orographically forced lee waves, or the passage of a convective outflow boundary. All of these can be challenging for a model to predict well.

As wind power becomes a greater percentage of the total power supply, regionally at least, coupled-model forecasts of power from this source must become increasingly accurate in order to avoid (1) brownouts if the wind speed decreases unexpectedly and (2) wasting fossil fuels and releasing greenhouse gases unnecessarily if the wind speed increases unexpectedly. This will be especially challenging because of the mesoscale character of many of the above processes, and because advantageous locations for wind farms are in complex terrain and in complex littoral zones.

### 14.9.3  Wind-power resource assessment

Wind-power resource assessment, or prospecting, involves the generation of short- to long-term reanalyses (see Section 16.2) of the near-surface climate. The resulting statistics of the windfield and air-density field, when used as input to power-production algorithms or models, enable wind-farm developers to determine where installation of turbines would be economically successful. Current climatological maps that have been produced show the most favorable regions to be over water in littoral zones, over some elevated terrain, and where low-level jets prevail. Because low-level wind-speeds vary greatly in space because of landscape forcing, especially high-resolution regional reanalyses are desirable, but it is very computationally demanding to produce them for long periods. Note that the PDF of the wind speed is needed for such wind-farm-siting decisions because it is important for the winds to not be excessively intermittent. Figure 14.8 shows an example of a wind-speed climatology for North America based on an MM5, 40-km grid increment, global reanalysis. This map defines the 120-m AGL wind speed at 0600 UTC for July 1997, and is especially relevant to wind-resource assessment because it shows the importance of the warm-season, nocturnal, low-level jet in the Southern Plains. See Landberg *et al.* (2003) and Petersen *et al.* (1998a,b) for an overview of wind-resource estimation, and Section 16.3 for a discussion of dynamical and statistical methods for downscaling coarse-resolution analyses to represent finer scales.

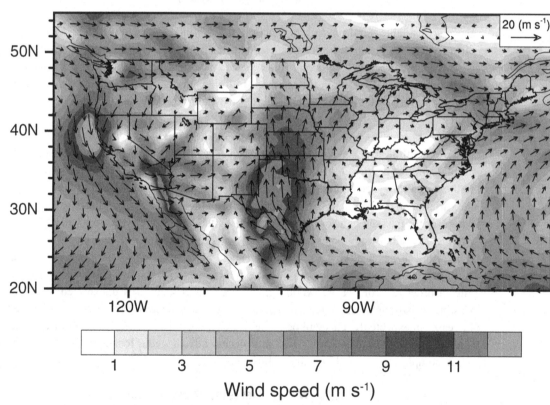

**Fig. 14.8**  The 120-m AGL wind-speed climatology of the US for 0600 UTC for July 1997, based on a 40-km grid increment, 21-year, MM5 global reanalysis. Every sixth wind vector is shown. This month and time are especially relevant to wind-resource assessment because they show the importance of the warm-season, nocturnal, low-level jet in the Southern Plains. Provided by Daran Rife, NCAR.

# 14.10 Agriculture

There are many agricultural applications of atmospheric models.

- *Planting and harvesting* – This often requires dry conditions, and necessary planning is based on weather predictions of at least 12–72 h lead time. The use of appropriate soil models coupled with atmospheric-prediction models can allow the diagnosis of soil trafficability by farm machinery.
- *Application of pesticides* – Integrated pest-management systems require that chemicals be applied a specific number of hours before rainfall, and at appropriate temperatures and relative humidities. Low-wind conditions also allow for more-accurate application.
- *Application of herbicides* – Herbicides must not be allowed to drift with the wind into areas where unintended damage to vegetation will take place.
- *Application of fertilizers* – Chemical or natural fertilizers should not be applied too soon before rainfall because the fertilizer will run off into waterways, not serving its intended purpose and contaminating waterways with nitrogen and other chemicals.
- *Insect movement* – Insects that can damage crops are carried by winds from their breeding grounds to agricultural areas, and predictions of weather patterns can allow for assessment of this risk.
- *Crop selection* – For agricultural areas in which irrigation water is not available, crops can be selected for planting that will be appropriate for the weather conditions that are expected during the growing season. For example, if dry conditions are forecast to accompany the onset of a particular phase of the ENSO cycle, crops can be selected accordingly.
- *Development and spread of plant and animal disease* – This is discussed in Section 14.3.2.
- *Crop yield estimation* – This is an especially important agricultural application of ensemble NWP systems (Cantelaube and Terres 2005, Challinor *et al.* 2005, Marletto *et al.* 2005).

Two well-established crop-yield modeling systems are summarized in Mera *et al.* (2006), who studied the impact of climate-related changes in radiation, temperature, and precipitation on crops, specifically soybeans and corn. The two models are CROPGRO (soybean) and CERES-Maize (corn), which are part of a Decision Support System for Agrotechnology Transfer (DSSAT). Both are predictive, deterministic models that simulate physical, chemical, and biological processes in the plant as a function of weather, soil, and crop-management conditions.

# 14.11 Military applications

Many military requirements for coupled models are very similar to those discussed above, for example related to forecasting quantities that are important for aviation safety and efficiency, and calculating the transport and diffusion of hazardous material in the atmosphere. There are, however, additional types of coupled models that specifically address the needs of military activities. A few are noted below.

- *Soil trafficability* – Heavy vehicles have difficulty operating on soils that are wet or too loose. Special types of land-surface models, called soil-trafficability models, employ analyses and forecasts of meteorological variables that affect substrate wetness (precipitation, temperature, wind speed, and humidity), and are used to estimate the ability of the substrate to support different vehicle types.
- *Guided and unguided missile trajectories* – Winds, turbulence, and air density affect the trajectory of missiles. The aerodynamic impact of observed and modeled meteorological conditions on the trajectories is calculated using a trajectory model.
- *Electro-optical visibility* – Weapons targeting systems are sometimes optical, so atmospheric turbulence and aerosol influences on feature detection by existing and proposed systems are anticipated using a model that employs atmospheric-model input.

See Sharman *et al.* (2008) for additional discussion of some of these applications.

## SUGGESTED GENERAL REFERENCES FOR FURTHER READING

Kuhn, K, D. Campbell-Lendrum, A. Haines, and J. Cox (2005). Using climate to predict infectious disease epidemics. Geneva, Switzerland: World Health Organization.

NRC (2001). *Under the Weather: Climate, Ecosystems, and Infectious Disease.* Washington, DC, USA: National Research Council, National Academy Press.

Palmer, T. N. (2002). The economic value of ensemble forecasts as a tool for risk assessment: From days to decades. *Quart. J. Roy. Meteor. Soc.*, **128**, 747–774.

Sharman, R., Y. Liu, R.-S. Sheu, *et al.* (2008). The operational mesogamma-scale analysis and forecast system of the U.S. Army Test and Evaluation Command. Part 3: Coupling of special applications models with the meteorological model. *J. Appl. Meteor. Climatol.*, **47**, 1105–1122.

## PROBLEMS AND EXERCISES

1. Predictability of atmospheric processes is an important topic in NWP. Speculate on the likelihood that the coupling between an atmospheric model and a secondary model will be such that there is a nonlinear sensitivity of the error in the solution of the secondary model to the error in the solution of the atmospheric model. Provide an example of a type of coupled-model application that might have a high degree of sensitivity to the accuracy of the atmospheric forecast.

2. Perform a literature search to determine the ways in which atmospheric information (observations, analyses, forecasts) is used in disease surveillance, early-warning, and response systems, and summarize them.

3. The use of coupled atmospheric models and decision models by many businesses and industries is not discussed in this chapter. Speculate on such applications of coupled modeling systems.

# 15 Computational fluid-dynamics models

## 15.1 Background

The expression Computational Fluid Dynamics (CFD) modeling comes from engineering, and refers to methods that can be used for the simulation of very-fine scales of motion. The terminology is confusing in the context that weather and climate modeling also involves the use of computational methods to solve the dynamic equations for a fluid. When the term CFD modeling is used in its conventional way in the atmospheric sciences, it refers to the simulation of motions that can synonymously be referred to as occurring on the sub-mesogamma scale, the microscale, or the turbulence scale.

Because we are revisiting the concept of the scales of motion that are represented by a model solution, a reminder of the pertinent discussions in Chapter 3 is appropriate. There is a tendency to think of the $2\Delta x$ length scale as the resolution limit of a model, although it has been shown by Skamarock (2004) (e.g., Fig. 3.36) and others that spatial filters associated with the finite-differencing scheme and the explicit diffusion in a model can cause the effective resolution to be quite different from this limit. Motions unresolved by the model can generally be referred to as the subfilter-scale (SFS).

## 15.2 Types of CFD models

There are three general categories of CFD models, although there are myriad methods for solving the equations, just as with larger-scale models.

- Reynolds'-Averaged Navier–Stokes (RANS) equations serve as the basis for one type, where, as described in Chapter 2, averaging operations relegate the turbulence effects on the mean motion to Reynolds-stress terms that must be parameterized, and the dependent variables in the equations pertain to the nonturbulent part of the motion. These RANS-type CFD models resolve small-scale flows around obstacles such as complex terrain and buildings, but the solution represents an average over the turbulent eddies that can dominate the motions in these situations. Thus model solutions remain steady, as long as the large-scale conditions defined by the LBCs do not change. An example of a RANS CFD model is described in Coirier *et al.* (2005).
- Large-Eddy Simulation (LES) models do not use averaging to eliminate the turbulence, but explicitly simulate the larger energy-containing eddies. A SFS parameterization

(also called a model) is used to represent the effects of the smallest-scale turbulence on the resolved scales.

- Direct Numerical Simulation (DNS) models capture all of the relevant scales of turbulent motion, so no parameterization is needed of the effects of unresolved scales. This is by far the most computationally demanding type of CFD modeling, and has limited use for complex processes.

The LES-type models are the ones most commonly used for research and practical applications in the atmospheric sciences. As an example of LES-model applications, an intercomparison of simulations of the stable boundary layer by eleven LES models was undertaken as part of the Global Energy and Water-cycle EXperiment (GEWEX) Atmospheric Boundary-Layer Study (GABLS). See Holtslag (2006) for a description of GABLS and Beare *et al.* (2006) for a summary of the LES models used in the study.

## 15.3  Scale distinctions between mesoscale models and LES models

Using the terminology of Wyngaard (2004), let $\Delta$ represent the scale of the spatial filter associated with the solution of the equations of motion and $l$ be the scale of the energy-containing turbulence. Figure 15.1 shows a schematic of a turbulent-energy spectrum, as

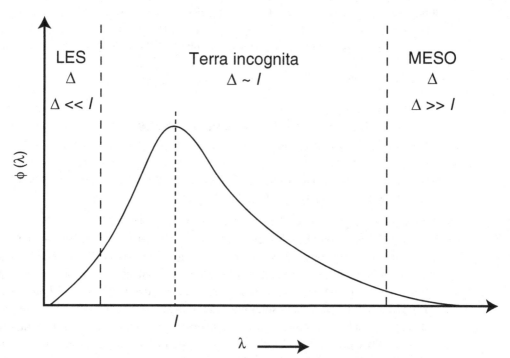

**Fig. 15.1** Schematic of a turbulent-energy spectrum, as well as spatial-filter length scales ($\Delta$) for LES and mesoscale (MESO) models. The variable $l$ is the scale of the energy-containing turbulence, $\Phi$ is turbulent energy, and $\lambda$ is wavelength. See the text for details. Adapted from Wyngaard (2004).

well as spatial-filter length scales for LES and mesoscale (or larger scale) models. For model spatial-filter scales in the "MESO" region on the right, long-wavelength, side of the graph, the turbulent energy is clearly in the unresolvable SFSs ($\Delta \gg l$). This is appropriate because for meso- and larger-scale models, the turbulence should be parameterized and not resolved. But, for LES models that resolve the energy-containing turbulence, the scale of the spatial filter must be sufficiently small relative to the turbulence scales ($\Delta \ll l$). Thus LES spatial filters should be in the short-wavelength region on the left side of the graph. The point of Wyngaard (2004) is that it is not clear how to apply models with spatial-filter scales within the part of the spectrum containing the turbulent energy (the terra incognita).

## 15.4 Coupling CFD models and mesoscale models

Because CFD model domains only span a limited area, they must obtain their LBCs from observations, analyses of observation, or larger-model grids with perhaps resolutions on the mesoscale. The CFD model may be run with temporally constant LBCs, or the LBCs may vary as the large-scale flow evolves. Initial conditions are typically defined from a relatively smooth, or even horizontally uniform, variable field. The local forcing, e.g. from orography or structures, will then allow microscale features to develop. For example, when CFD models are used to simulate the impact of a building on the winds in an urban area, the initial conditions will represent the "skimming flow", well above the rooftops, and the forcing from the building will generate channeling in the street canyons, and vortices on all sides of the building, during the simulation.

In some cases, the same dynamical core can be run as a traditional mesoscale model and as an LES model, with inner grids using LES closures and the outer grids run with standard mesoscale-model closures. In this case, there is generally two-way interaction between the mesoscale and the LES scale. In contrast, when distinct models simulate the two scales, it is more typical to use one-way coupling. Note that the use of the same model dynamical core to span the LES scale and the mesoscale with a series of nested grids can lead to the scale-separation issues described in the last section, if a standard ratio of 3–5 is used for the resolution of adjacent grids.

A significant issue with LES-model LBCs is that the inflow boundary will generally be defined by an atmosphere in which the turbulence effects are parameterized. Thus, there will be no turbulence structures entering the grid, and because of the short residence time of the air flowing over such small computational grids there may not be sufficient time for the turbulence to develop before the air exits at the outflow boundary. This situation is similar, in principle, to that discussed in Section 3.5, where a significant buffer zone is needed between the upwind boundary and the area of meteorological interest on the grid. This allows small-scale processes to spin up as the air enters the central region of the grid. Unfortunately, the advective time scales are the same for mesoscale and LES-scale models, even though the sizes of the computational grids are much smaller in the latter case. For example, an LES model may have a grid increment of 5 m and a computational domain with a length scale of 1 km. If the inflow wind speed is 5 m s$^{-1}$, the air will reach the center of the computational grid in 100 s, quite possibly an insufficient amount of time to develop the turbulence. This would render the LES inadequate for the intended purpose. One approach to this problem is to try to specify turbulence structures in the inflowing air, but this is challenging.

As with any system of coupled models, a situation-dependent aspect of the coupling between LES and larger-scale models is the sensitivity of the LES model solution to errors in the LBCs. For example, wind-direction errors in a mesoscale model may be consistent with the state of the science, and these errors may not have a significant negative impact on the value of forecasts of sensible weather. But, there may be particular applications of a CFD model such that this error in the initial conditions and LBCs produces a profound error on the CFD-model scales. For example, Fig. 15.2 shows the simulated concentration of a plume of hazardous material that has been released at street level into the atmosphere on the south side of Oklahoma City, USA. A large-scale wind was used as the input to a RANS-type CFD model (Coirier *et al.* 2005), and the resolved wind flow within the street canyons was used as input to a transport and diffusion model. In one case, the large-scale wind was from the south-southwest, in which case the plume covered the east side of the urban area (a). When the large-scale wind direction was changed by 22.5°, to southerly, the plume's impact changed from the eastern half to the western half of the city (b). Here, the existence of the street canyons causes a large response in the low-level flow of the plume to a small change in the large-scale wind direction. Note that this wind-direction difference is consistent with the expected errors in forecasts of low-level winds from mesoscale models.

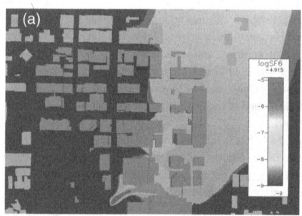
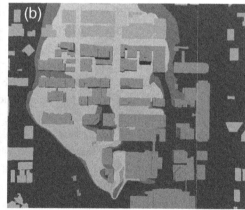

**Fig. 15.2**  A large-scale wind was used as the input to a RANS-type CFD model (Coirier *et al.* 2005), and the resolved wind flow within the street canyons (black area between rectangular buildings) was used as input to a transport and diffusion model. In one case, the large-scale wind was from the south-southwest, in which case the plume (irregular gray shapes) covered the east side of the urban area (a). When the large-scale wind direction was changed by 22.5°, to southerly, the plume's impact changed from the eastern half to the western half of the city (b). Provided by William Coirier, Kratos/Digital Fusion, Inc.

## 15.5 Examples of CFD-model applications

Applications of CFD models fall into two categories, as do the uses of NWP models – they are used for scientific discovery (knowledge generation) as well as for addressing practical problems. Practical applications are of course abundant in the engineering area, such as simulating the flow over an aircraft as part of the design process. In the context of microscale atmospheric science, some examples follow.

- Studies that are aimed at better understanding turbulence can lead to improved parameterizations of the effects of the turbulence on fluxes, for example in the nocturnal stable boundary layer, or within a tree canopy during the day.
- The wind loading on tall buildings is studied by design engineers, to enable safer construction.
- Wind turbines must be located to maximize the available power as well as to minimize the turbulence load on the generator. Decisions about the general area for locating wind farms are often made using analyses that are based on mesoscale models. But, optimizing the locations of the individual turbines in complex terrain requires the use of CFD models.
- Wake turbulence produced by specific types of aircraft on takeoff is studied to define the requirements for safe distances that must be maintained between aircraft in a takeoff sequence.
- The transport in the urban boundary layer of hazardous gases or aerosols, released for example from a transportation or industrial accident, can be studied.

Many, many other examples exist.

## 15.6 Algorithmic approximations to CFD models

Because of the small grid increments, CFD models are very computationally demanding to run. Thus, when solutions are required quickly to meet operational needs of some type, algorithmic approximations to CFD-model solutions are used. For example, if forecasts are needed of street-level winds between buildings in an urban area, say based on the input of rooftop winds from an operational mesoscale model, LES or RANS CFD models may be too computationally demanding to provide building-scale solutions on usable time scales. An approach that has been used to address this problem is to employ LES-model solutions and wind-tunnel studies to define the patterns of the airflow around a variety of obstacle shapes, for different wind directions, stabilities, and vertical shears of the horizontal wind. The resulting catalogue of flow patterns can be used to develop algorithms that define the building-aware wind flow under variable large-scale conditions. An example of such a rule-based system is the Quick Urban & Industrial Complex (QUIC) model (Pardyjak *et al.* 2004), which has a wind-flow component (QUIC-URB) and a QUIC-PLUME code that tracks plumes of air pollutants among buildings.

## SUGGESTED GENERAL REFERENCES FOR FURTHER READING

Mason, P. J., and A. R. Brown (1999). On subgrid models and filter operations in large eddy simulations. *J. Atmos. Sci.*, **56**, 2101–2114.

Moin, P., and K. Mahesh (1998). Direct numerical simulation: A tool in turbulence research. *Annu. Rev. Fluid. Mech.*, **30**, 539–578.

Sagaut, P. (2006). *Large Eddy Simulation for Incompressible Flows*. Berlin, Germany: Springer-Verlag.

Stevens, B., and D. H. Lenschow (2001). Observations, experiments, and large eddy simulation. *Bull. Amer. Meteor. Soc.*, **82**, 283–294.

Wyngaard, J. C. (2004). Toward numerical modeling in the "terra incognita". *J. Atmos. Sci.*, **61**, 1816–1826.

## PROBLEMS AND EXERCISES

1. Describe additional practical applications of LES models.
2. Speculate about the challenges associated with using LES models to simulate the stable boundary layer, versus the neutral or unstable boundary layer. Confirm your ideas with a literature search.
3. What are the similarities between the turbulence-modeling scale issues described in Wyngaard (2004), and the fact that 1–10 km grid increments are considered to be too small for parameterizing convection but too large to resolve it in a model?
4. When CFD models run fast enough in the future to be used operationally, discuss whether there will be a role them in predicting urban weather – that is, the specific weather conditions within the street canyons.

# Climate modeling and downscaling

The term climate modeling, as used here, includes (1) forecasts of climate with global AOGCMs that simulate the physical system's response to radiative-forcing scenarios that assume a specific trajectory for anthropogenic and natural gas and aerosol emissions, (2) initial-value simulations on seasonal to annual time scales, (3) the production of model-based analyses of the present climate, and (4) model experiments that evaluate the response of the climate system to anthropogenic changes in the landscape, say associated with continued urbanization or the expansion of agriculture. Thus, the term climate modeling refers to the use of a model to define the state of Earth's physical system on time scales of seasons to centuries. As we will see, the specifics of the modeling process depend on the time scale. Typically not included are monthly forecasts (e.g., Vitart 2004), which bridge the gap between medium-range forecasting and seasonal forecasting. If the AOGCM forecasts or the global-reanalysis data sets are used as input to a regional (mesoscale) model or a statistical procedure for correlating the large- and small-scale climate of a region, the process is called climate downscaling.

The material about the modeling of weather that has been presented so far in this book also has direct application to the problem of climate modeling. The climate is, after all, just the aggregate behavior of many thousands of individual weather events. So, errors in the model's numerical algorithms, shortcomings in physical-process parameterizations, and incorrectly represented land–ocean–atmosphere interactions may affect climate predictions just as severely as weather predictions. In fact, some model errors that are acceptable for forecasts of a day to a couple of weeks may severely impact model integrations that extend over decades and centuries. Included would be slow rates of mass gain or loss, or errors in the representation of radiation such that there are nonphysical drifts in the temperature. Alternatively, there are serious errors that can develop in weather predictions, such as phase errors in waves, that may have less consequence for climate prediction.

This chapter will begin with a review of global climate modeling, including how the models differ from those used in weather prediction, how their skill is verified using simulations of current or past climates, the differences in approach between seasonal and longer radiatively forced predictions, a summary of the models being employed, and the use of ensemble methods. The section after that summarizes how global models are used to create reanalyses of the current climate. This is followed by a section on climate downscaling, where downscaling is motivated by the often-stated fact that "all climate is local". That is, the human response to climate change takes place at the local level, and depends on local economic, agricultural, and societal factors. Thus, this need for fine-scale information requires the use of LAMs or statistical methods that employ input from AOGCM forecasts or global reanalyses. The final section describes the use of models to estimate the effects on climate of anthropogenic landscape changes.

# 16.1 Global climate prediction

Numerical methods and physical-process parameterizations used for modeling the global atmosphere, with applications to both weather and climate prediction, were described in earlier chapters. However, global climate models must also represent many additional physical processes in the hydrosphere (including ocean circulations), the cryosphere (land and sea ice), the lithosphere (land surface), and the biosphere, and how they interact with each other. For most weather forecasts of up to two weeks, the sea-surface temperature; the health, spatial extent, and types of vegetation; the extent of the permafrost, glaciers, and sea ice; the chemical composition of the atmosphere; etc., can be specified and assumed to be invariant during the model integration. However, this is not the case with climate simulations with durations of years to decades to centuries. To the greatest extent possible, the physical subsystems need to fully interact in the coupled model because they are part of the complex of, sometimes nonlinearly interacting, processes that affect climate. This is why such models are sometimes referred to as climate-system models rather than simply climate models.

## 16.1.1 Experimental designs for global climate-change studies

Before models are employed for forecasting future climates, their ability to reasonably replicate the current or past climates must be confirmed. Naturally, this success is not a guarantee of an accurate climate forecast because some model representations of physical processes are tuned for the current climate, and may not be as accurate in a different climate regime. Nevertheless, for future-climate studies using AOGCMs, and LAMs or statistical methods to downscale from AOGCM simulations, the responsible experimental approach is to first apply the modeling system for present or past climates to quantify the model's performance. This verification process will be described in Section 16.1.3.

The annual-mean weather varies from year to year and decade to decade, partly because of internal variability in the climate system resulting from natural long-time-scale physical processes (related to deep-ocean circulations, the land surface, and ice). For example, the heavy black line in Fig. 16.3, later in the chapter, shows the evolution of the global-average observed surface temperature during the twentieth century. Superimposed on the long-term upward trend in temperature are many scales of variation, from a few years to multiple decades. Research has shown that the long-term trend is probably of anthropogenic origin, while the decadal and shorter oscillations represent internal variability. If the goal is to quantify the effects of anthropogenic forcing – e.g., the Intergovernmental Panel on Climate Change (IPCC) effort – simulated changes associated with internal variability should be filtered in some way in order to avoid misinterpreting natural variability as anthropogenic effects. This is especially important when downscaling from AOGCM simulations, where

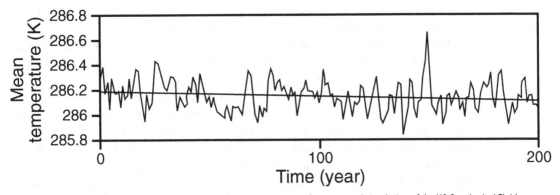

Fig. 16.1 Time series of the annual-mean surface air temperature from a control simulation of the US Geophysical Fluid Dynamics Laboratory AOGCM. From AchutaRao *et al.* (2004).

only short segments of time (referred to as slices) are simulated for the present and future climates. That is, the simulated difference between the current and future downscaled climate will depend on the phase of the internal variations at the times of the slices.

There are a few approaches for modeling climate change, where the method of choice depends on the objective. For estimation of the effects on climate of radiative forcing from optically active gases and aerosols of anthropogenic origin, it is typical to first generate a present-climate AOGCM control, or reference, simulation for a period of centuries, using a constant, present radiative forcing. These simulations sometimes require thousands of years of spin up in order to allow the deep ocean circulations sufficient time to develop. Figure 16.1 illustrates the global-mean surface air temperature from such a control simula- tion, where the multi-year (internal) variability in this current-climate regime results from slow ocean–atmosphere interactions. Then, a future-climate projection is made, starting at an arbitrary time in the control simulation, using a particular future-emissions scenario for aerosols and optically active gases. See Nakicenovic (2000) for a description of the differ- ent scenarios used in the IPCC modeling experiments. Because longer-term trends associ- ated with the climate change will be superimposed on the internal variations, the simulated change in a variable will depend on the phase and amplitude of the internal anomalies at the start time of the simulation. Choosing a different start time will result in a different pattern to the internal variability. Thus, running an ensemble of simulations with different start times, and averaging, will remove some of the effects of the internal variation. The same model can be used for each simulation in the ensemble, but a similar filtering can be achieved by using a variety of different models, such as the suite used in the IPCC simula- tions (IPCC 2007). As evidence of this smoothing, note that the individual simulations (light gray lines) depicted later in Fig. 16.3 have much greater temporal variance than does the ensemble-average (dark gray line).

Coupled AOGCMs can also be used to forecast the change in climate – both internal and anthropogenic – on time scales of seasons to years. For example, the drought in the Sahel during the last few decades of the twentieth century was likely, at least partially, caused by natural internal variability in the climate system. Thus, forecasts must accurately define the prevailing phases and amplitudes of many internal processes such as ENSO, the Pacific decadal oscillation, the North Atlantic oscillation, and changes in the meridional overturning circulation in the Atlantic Ocean basin. To represent the internal oscillations, initial conditions must be employed that define the state of the entire physical system. Accurately defining the state of the deep-ocean waters is clearly a great challenge. Section 16.1.4 summarizes the process by which these initial-value climate predictions are produced.

As with weather prediction, multi-model ensembles can be used to improve the predictability of climate. In particular, the unpredictable aspects in the model solution can cancel when aggregating the simulations, so that the ensemble mean is superior to the individual members. However, unlike ensemble weather prediction, where the members can be chosen so that they have roughly equal skill, that is not possible when members are based on disparate models from various organizations worldwide. The use of such multi-model ensembles is central to the IPCC assessments of climate change (IPCC 2007). Ensemble methods used in climate simulation are discussed in Section 16.1.6.

Another category of experiments has been used to define the strength of the internal feedback processes that amplify or dampen the system's response to the radiative forcing. Metrics for this sensitivity include a quantity called the *equilibrium climate sensitivity*, which is defined as the equilibrium surface temperature change that results from a doubling of the carbon dioxide concentration in a model atmosphere, and it is expressed in degrees Celsius. An alternative measure of the strength of the feedbacks is the *transient climate response*, which is the surface air temperature change that results from a carbon dioxide concentration increase of 1% $yr^{-1}$, until the doubling point is reached. After reaching this point, the system is given sufficient time to come to equilibrium. Differences among models in terms of their future-climate predictions are partially a result of the feedback strengths measured here. See Box 10.2 in Meehl *et al.* (2007) and Section 8.6 in Randall *et al.* (2007) for further discussions of the concept of climate sensitivity.

For estimation of the effects on global (and regional) climate of future anthropogenic landscape changes, models can be run for long periods of time with and without the change. Such studies are important because of the potential climate impacts of future large-scale deforestation, conversion of grassland to agricultural crops, expansion of cities, expansion and contraction of the irrigation of agricultural crops, and diversion of water from lakes causing them to shrink in size or disappear altogether. Depending on the scale of the climate response to be evaluated, a global model can be used for the study, or a LAM can be employed to resolve mesoscale processes.

## 16.1.2  Special model requirements

Climate models differ in a variety of ways from the traditional weather-forecast models described elsewhere in this book. The following sections review some of these differences.

### Land-surface and ice modeling

Surface-process components of weather-prediction models were discussed earlier in Chapter 5. However, there are a number of land-ice processes that operate on longer, climate time scales that must be considered here. Ice (cryospheric) processes, on both land

and sea, are also included in this discussion. The reader should see Sections 8.2.3 and 8.2.4 of Randall *et al.* (2007), and the references in the land and sea-ice columns of Table 16.1 in this chapter, for additional information. The terminology used in this table deserves elaboration. For sea-ice dynamics, the word "leads" refers to the representation in the models of narrow areas of open-water in cracks within the ice. In these areas, the heat and water-vapor fluxes are extremely large. The word "rheology" refers to whether the models represent the slow "flow" of ice sheets. None of the models represents the dynamics of ice-sheet melting (e.g., Greenland and Antarctica), which is why there is great uncertainty about sea-level rise. In the column that pertains to land processes, "canopy" refers to the explicit treatment of vegetation effects, "routing" refers to whether rainwater or snow/ice-melt water are routed into stream channels on the land surface, "layers" refers to the use of a multi-layer soil model, and "bucket" refers to a simple method for treating soil hydrology. Two community-developed land-surface-process models that are commonly used for climate applications are the Community Land Model (Oleson *et al.* 2008) and the Common Land Model (Dai *et al.* 2003).

One of the major important advances in newer-generation climate models is the inclusion of terrestrial-biosphere models that treat some terrestrial carbon sources and sinks. The processes that are represented involve both soil carbon cycling and vegetation. For example, dynamic-vegetation models simulate the response of the vegetation to changes in carbon dioxide concentrations and to climate variables (e.g., precipitation, temperature) that affect vegetation health. In addition, there is higher-resolution modeling of the overland flow of water, the inclusion of plant root dynamics, the use of multi-layer snowpack models, and the prediction of the motion and thickness of sea ice. However, as with efforts to improve other aspects of climate models, it is unclear how well the new representations of these land-surface and cryospheric processes will perform in greatly different climate regimes.

It is important to be reminded of the need to employ surface-process models that can adequately represent land–biosphere–cryosphere–atmosphere feedbacks. For example, the simulated soil moisture influences dynamic-vegetation models, and the state of the vegetation determines its quantitative influence in the carbon cycle. Pielke *et al.* (1999a) illustrate the importance of land–atmosphere interactions by calculating the time after which the initial soil-moisture conditions became unimportant in seasonal weather prediction with an Atmospheric General Circulation Model (AGCM). They concluded that the model's memory of the initial soil moisture lasted 200–300 days. A general discussion is provided, along with a good list of references, of the importance of properly modeling the landscape (e.g., landcover type, leaf-area index, soil moisture) changes associated with drought and climate change. There are also plentiful examples provided of the long-distance impacts on climate of anthropogenic landscape changes, which need to be accounted for as well in climate simulations.

## Ocean-circulation modeling

The ocean and atmosphere interact through fluxes of heat, water vapor, and momentum. For weather-prediction purposes, it is generally sufficient to represent the ocean through specified surface temperature and salinity (which affects saturation vapor pressure) patterns, and the use of a wind-speed dependent roughness length. That is, except as a result of extreme wind speeds, such as in hurricanes, the feedback between the ocean and the atmosphere during the period of a forecast is sufficiently small that most weather-prediction models are not coupled to ocean models. However, on time scales of longer than a few weeks the ocean properties can evolve considerably, and an active ocean-circulation model should be employed. The ocean models run simultaneously with the atmospheric models, out of necessity because of the two-way interaction, but they typically have different horizontal resolutions than do the atmospheric models. See Section 8.2.2 of Randall *et al.* 2007, and the ocean-modeling references in Table 16.1, for additional information about the ocean component of global-climate models.

## Physical-process parameterizations

The challenges of parameterizing physical processes in climate models do not differ greatly from those associated with global weather-prediction models. Exceptions include the fact that small errors in the representation of processes may be acceptable for forecasts of weeks, but over much longer time periods the cumulative error can cause unacceptable drifts in simulated climate. This problem can be addressed to some degree by "tuning" model parameters (in parameterizations) so as to optimize simulation results, such as in the atmosphere–ocean flux corrections described below. But, this process is not intuitively appealing because tuning the model to the current climate does not ensure an equally positive effect for future climates. Furthermore, adjusting a particular parameter does not necessarily give you a better model solution for the right reason. A partial remedy to this problem is to use higher-resolution global models that explicitly resolve processes such as convection. The use of higher-resolution regional models for downscaling may allow better explicit local representation of processes, but that has no benefit for the simulation of the global climate by the parent AOGCM. Lastly, because global climate models typically have coarser horizontal resolution than do global weather-prediction models, parameterizations may be more suitable for one application than the other because their performance is sometimes scale dependent.

## Conservation properties of dynamical cores

The general issues associated with the conservation of properties such as mass and energy by models were discussed in Chapter 3. This need for conservation is clearly more critical for long climate-time-scale simulations than for weather-prediction time scales of days to weeks. That is, small rates of error accumulation may not be damaging for short model integrations, but may be for long ones. For example, Boville (2000) states that, for climate models, energy must be conserved to tenths of a watt per square meter. Williamson (2007) points out that one of the energy conservation problems that must be addressed is the accumulation of energy in small scales through aliasing, discussed in Section 3.4.5. See Thuburn (2008) for a good general discussion of conservation issues for the dynamical cores of climate and NWP models.

## Initial conditions

For IPCC-scenario climate-change simulations, or those involving projected landscape changes, it is sufficient to initialize the model with any realization of the weather associated with current climate conditions. However, the aforementioned seasonal to decadal forecasts that hope to represent the phase and amplitude of internal climate variations are initial-value problems. Thus, the states of the atmosphere, ocean, biosphere, cryosphere, and lithosphere must be defined by initial values.

For experiments that might involve more drastic perturbations to the climate or external forcing, it is useful to keep in mind the distinction between *transitive* and *intransitive* climate systems (Lorenz 1968). A transitive climate system is one in which there is only one permitted set of long-term climate statistics – that is, given a particular set of external forcing parameters for the atmosphere, such as the orography, the solar input, Earth's rotation rate, etc., there is only one stable long-term climate. In contrast, an intransitive system has more than one possible stable climate, with the prevailing climate determined by the present state of the system. A special type of system is an *almost-intransitive* one, in which the climate remains within a regime for a finite time, with the system then migrating into another equally acceptable regime without any change in the external forcing. In other words, climate regimes have sufficient "inertia", in a dynamic sense, to be self-perpetuating for a period of time. This is consistent with the observed situation where distinct periods of prolonged regional drought can transition abruptly into periods with normal or abundant precipitation. Thus, for intransitive and almost-intransitive climate systems, the initial conditions that define the present state of the system can determine the resulting climate regime.

## Flux corrections

Small errors in the simulated fluxes of heat, water vapor, and momentum at the air–sea interface can cause climate-model solutions to drift to an unrealistic climate state. To address this problem of some models, artificial corrections have been added to the flux terms in the equations. This practice is of obvious concern because it is nonphysical, and the corrections cannot be targeted for those physical situations where the errors may dominate. In the first two Climate Model Intercomparison Projects, which took place in the 1990s, over half of the models were flux corrected (Reichler and Kim 2008), whereas in the third and latest comparison, less than one-quarter of the models were flux corrected (for example, see the flux-adjustments column in Table 16.1 later in the chapter).

### 16.1.3  Verification of global climate-change models for past or current climates

As noted above, the only way of gaining confidence in the ability of a global climate model to simulate future climate is to evaluate its ability to replicate the conditions of past climates or the current climate. The advantage of using recent climates for this purpose is that meteorological observations are more plentiful. However, the opportunity to fully test the ability of models to simulate climate change is limited because recent climate variation has been small compared with the potential future changes that must be predicted. Thus, there have also been efforts to test models by running them for paleoclimate periods, during which climates have varied widely. The obvious drawback is that there are large uncertainties in the external forcing as well as in the prevailing climate variables themselves, where various proxies need to be used to estimate the latter. In addition, simulating climate change that takes place over very long, paleoclimate time periods is computationally impractical with traditional full-physics models.

It is arguable that this process should involve verification of the model's ability to simulate individual weather events, as well as the long-term statistics of the events – the climate. That is, both the specific characteristics of the simulated weather events (e.g., tracks, intensity, and frequency of midlatitude storms; characteristics of easterly waves in the tropics; properties of low-level jets near coastlines and mountains) and the climate statistics of the aggregated weather events should be compared with the actual properties of the weather and climate as defined by archived observations. Without evaluating the model's rendering of the events that make up the climate, there is the risk that the statistics could be correct for the wrong reason, leading to errors when applying the model for future climates.

Because climate models are extremely complex, components are often developed and tested individually. For example, the properties of numerical methods can be isolated and evaluated much more effectively without the use of the physical-process parameterizations. And the physical-process parameterizations can be studied through the use of case studies, possibly with special field-program data for verification. Only when a climate model has been tested as thoroughly as possible at the component level, should its performance be evaluated in the context of approximating the entire climate system.

Climate-model verification has employed a number of metrics for comparing the model solution with observations, including global means of variables, composite global indices based on a number of variables, spatial patterns of variables, the temporal variability of regional climates over times scales as large as decades (internal climate-system variability), the ability of the model to replicate specific well-documented features of the current climate (e.g., ENSO), and the ability of the model to simulate regional extremes of variables on various time scales. A good general reference for this subject is Section 8.3 of Randall *et al.* (2007).

## Verification of global-average climate statistics

One of the challenges in climate-model verification is simply deciding upon what variables best represent climate and can serve as metrics of overall errors in its simulation. This choice is not easy because of the many variables associated with the state of the atmosphere, hydrosphere, cryosphere, lithosphere, and biosphere. Some studies simply use global-mean surface air temperature (e.g., Min and Hense 2006). Others use composite error indices that are based on a broad range of climate variables (Murphy *et al.* 2004, Reichler and Kim 2008). Others use a few traditional error statistics, where Boer and Lambert (2001) and Taylor (2001) summarize them in graphical form. Another complication in the model verification process results from the fact that the observations that define the current climate are not an independent measure of the model accuracy because they have already been used to tune the model physical-process parameterizations. Nevertheless, we have no choice but to use them for the verification. This issue may be somewhat mitigated by the use of higher-resolution climate models that require less parameterization of processes.

An example of a thorough verification of models for current climates is described in Reichler and Kim (2008). In this study, numerous models (see Table 16.1) were compared, using a performance index that was based on many variables, for simulations from three different Climate Model Intercomparison Projects (CMIP): CMIP1 (Meehl *et al.* 2000) organized in the mid-1990s; CMIP2 (Covey *et al.* 2003, Meehl *et al.* 2005); and CMIP3 (PCMDI 2007) based on the IPCC Fourth Assessment Report (AR4, Meehl *et al.* 2007, Randall *et al.* 2007) simulations that were produced by the most-current climate models. For the calculation of the multivariate performance index, observations and global gridded analyses were used to compute annual-mean climatologies for the period 1979–1999. From this data set can be calculated errors in the modeled mean states of many different climate variables. To determine a model performance index, a normalized error variance, $e^2$, is calculated by squaring the grid-point differences between simulated and observed climate, normalizing for each grid point with the observed interannual variance, and averaging globally. This can be written as

$$e_{vm}^2 = \sum_n w_n (\bar{s}_{vmn} - \bar{o}_{vn})^2 / \sigma_{vn}^2 ,$$

where $\bar{s}_{vmn}$ is the simulated annual climatological mean for variable $v$, model $m$, and grid point $n$, $\bar{o}_{vn}$ is the corresponding observed climatology; $w_n$ are weights required for area and mass averaging; and $\sigma_{vn}^2$ is the interannual variance based on the observations. One challenge when combining errors for variables with different dimensions is to weight them properly. In this method, $e^2$ is scaled according to the average error for a reference ensemble of models. Specifically, a performance index ($I$) is calculated as follows:

$$I_{vm}^2 = e_{vm}^2 / \overline{e_{vm}^2}^m ,$$

where the overbar represents an average of the climates from all the models for that variable. The final step in calculating the performance index involves averaging over all the variables:

$$I_m^2 = \overline{I_{vm}^2}^v .$$

Figure 16.2 shows the value of the performance index, $I_m^2$, (solid vertical lines) for each of the models, for each of the three generations of CMIP exercises. The average performance index for each generation of models is shown by the dashed vertical line. The value of the index associated with the NCEP-NCAR reanalysis (Kalnay *et al.* 1996), which is a model-based analysis of observations, is 0.4. And, the black circle indicates the performance of the multi-model ensemble mean. The figure depicts a large variation in the ability

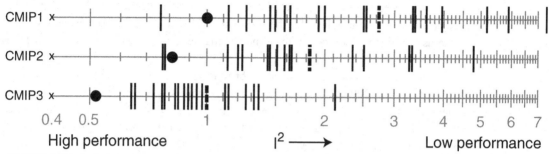

**Fig. 16.2** Performance index, $I^2$, for individual models (vertical black lines) and model generations (CMIP1, CMIP2, and CMIP3). The dashed vertical lines show the average performance index for each model generation, the ×'s at the left edge of the scales indicate the value of the performance index when the NCEP-NCAR reanalysis is used to define the climate, and the black solid circles show the performance index for the multi-model ensemble mean for each generation. Better performing models have lower index values, to the left. Adapted from Reichler and Kim (2008).

of the models of a given generation to replicate the current climate. Also, there is a steady improvement in this ability from one generation to the next, such that the realism of the climate of the best models from CMIP3 approaches that of the atmospheric reanalysis. These generational improvements are at least partly a consequence of improvements in physical-process parameterizations and the greater horizontal and vertical resolution that has been permitted by increases in available computing capacity.

Another example of the relative performance of the individual models used in the Third IPCC Assessment Report (*c.* 2000), and those used in the current IPCC AR4, is provided in Randall *et al.* (2007) (Section 8.3.5). These statistics are reported individually for different variables, such as precipitation, sea-level pressure, and surface air temperature, rather than for a single performance index. Conclusions are that (1) on average, flux-adjusted models have smaller errors than those without flux adjustments, for both the third and fourth assessments, but the smallest errors are from models without flux adjustments; and (2) the mean error from the recent suite of models is smaller than that from the earlier suite, in spite of the fact that all but two of the newer models do not use flux adjustments.

An illustration of historical-temperature change produced by CMIP3 simulations is shown in Fig. 16.3. Depicted is the observed global-average near-surface temperature trace from 1900 to the early twenty-first century (black line), the simulations from the individual models used in the ensemble (light gray lines), and the multi-model ensemble mean (heavy gray line). The model simulations shown in Fig. 16.3a employed both natural and anthropogenic forcings, and those in Fig. 16.3b used only natural forcings. The ensemble-mean model-simulated temperature in Fig. 16.3a closely approximates the observed trend.

The relative skill of different models of an ensemble in replicating the observed historical climate can be used to infer which models will perform best for future climates. For example, Shukla *et al.* (2006) correlated the skill at simulating twentieth-century surface temperature with simulated future-climate temperatures. The models that had the smallest error for the twentieth-century climate produced relatively larger temperature increases for the twenty-first century. Meehl *et al.* (2007) describe the use of observation-based metrics

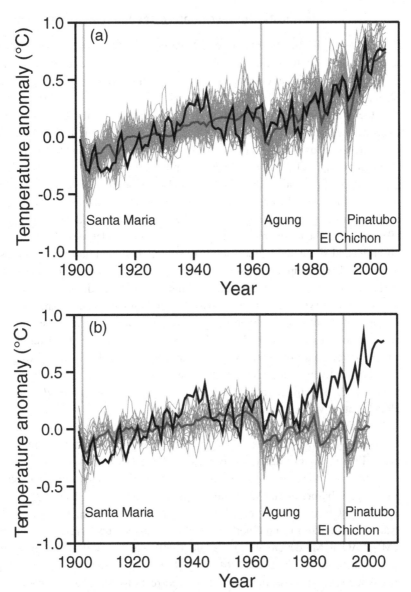

Fig. 16.3 Global-mean, near-surface temperature from observations (heavy black line) and from some of the individual climate models listed in Table 16.1 (light gray lines), based on the use in the models of (a) both natural and anthropogenic forcings and (b) natural forcings only. The temperature is represented as an anomaly relative to the observed 1900–1950 mean. The multi-model ensemble means are shown as relatively smooth heavy dark-gray lines. Vertical lines indicate the timing of major volcanic events. Adapted from Hegerl *et al.* (2007), which should be referenced for additional detail.

to weight the reliability of contributing models when making projections with ensembles, but a challenge is defining a metric, or a set of metrics, that is a reasonable indicator of overall model performance.

## Verification of specific processes and regional features

The above verification involved the calculation of global-average statistics, which facilitated the use of multivariate composite error indices. In contrast, this section provides examples of how global climate models have been evaluated (1) in the context of specific physical processes, such as the ENSO cycle, (2) in terms of the spatial pattern of error over the globe, and (3) for limited geographic areas. The inability of some climate models to faithfully simulate specific observed features of the current atmosphere has historically been the cause for a lack of confidence in the models for use in climate prediction.

As an example of the models' replication of spatial patterns, Fig. 16.4 shows the spatial distribution of annual-mean precipitation based on observations and CMIP3 climate-model simulations. The observation-based climatology is from the NOAA Climate Prediction Center (CPC) Merged Analysis of Precipitation (Xie and Arkin 1997) for the period 1980 to 1999. The model climate is based on a multi-model mean for the same period. The observed and simulated patterns are very similar, but quantitative differences are apparent regionally. For example, the subtropical precipitation deficits to the west of the continents in the Americas are less widespread and intense in the model solution, and the simulated precipitation maxima off the east coasts in midlatitudes are also smaller and weaker. An example of a more-quantitative interpretation of model skill at the regional level, also for precipitation, is provided in Fig. 16.5. It shows the annual cycle of regional-average precipitation for southwestern USA, simulated by the models participating in the CMIP2 comparison. Even though the overall cycle is captured, there is much variation among the models in the specific monthly precipitation values. AchutaRao *et al.* (2004) show similar plots for other variables and geographic areas.

There are many examples of studies that have focussed on the ability of climate models to simulate specific processes in the current climate. For example, precipitation is a primary climate variable, and in the tropics the diurnal cycle dominates its occurrence. However, many models have difficulty simulating the early-evening maximum, instead producing it before noon (Yang and Slingo 2001, Dai 2006). The ENSO cycle has been a process that has received special attention in terms of climate-model verification. For example, AchutaRao and Sperber (2006) compared the ENSO-simulation skill of the CMIP2 AOGCMs developed in the late 1990s with the skill of the more-recent models used as the basis for the IPCC AR4. The AR4 models were better in many respects, but the fact that fewer of those models use flux corrections may have prevented an otherwise greater improvement. Also, Randall *et al.* (2007) review progress with respect to the ability of climate models to simulate the Madden–Julian oscillation, the quasi-biennial oscillation, ENSO, and intraseasonal to interannual variability in monsoons. Also summarized is the ability of the models to reproduce observed extreme events in the current climate, including extreme temperatures, extreme precipitation events, and tropical cyclones.

The extreme values of model dependent variables, in new climate regimes, are often of equal or greater interest than are mean values. For example, greater temperature extremes associated with heat waves would take a larger toll in human lives, wind extremes would affect wind-power generation as well as the engineering design of tall buildings, heavier rains would produce more floods and flash floods that affect public safety, more-prolonged drought

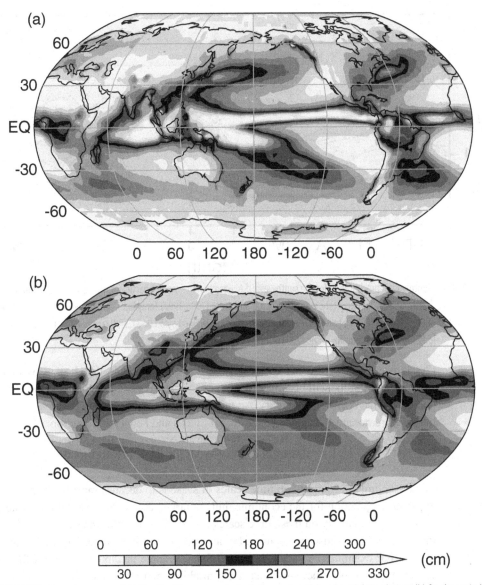

**Fig. 16.4** Annual-mean precipitation (cm) from an analysis of observations (a) and model simulations (b) for the period 1980–1999. The simulation is based on a CMIP3 multi-model mean and the analysis is a merger of gauge observations, satellite estimates, and model output. Adapted from Randall *et al.* (2007).

periods would have water-resources and agricultural implications, and stronger hurricanes would increase the loss of lives and infrastructure in coastal communities. Thus, there is a great emphasis in global and regional climate-change studies on verifying the model simulations of extremes for current climates and forecasting them for future climates. For example, a European project entitled Modeling the Impact of Climate Extremes (MICE) involved the use of both climate models and impact models (Hanson *et al.* 2007). Also, Meehl and Tebaldi

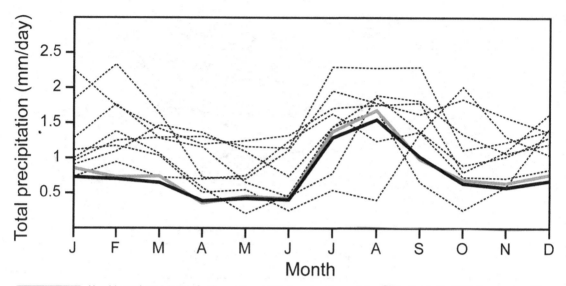

**Fig. 16.5**    Monthly total precipitation for the southwest USA, based on the models participating in the CMIP-2 comparison (fine dashed lines), the precipitation analyses of Xie and Arkin (1996,1997) (gray line), and the NOAA CPC Merged Analysis of Precipitation (CMAP) (wide black line). The region was defined by 30.0–37.5° N latitude and 105–115° W longitude. From AchutaRao *et al.* (2004).

(2004) use an ensemble of AOGCM predictions to conclude that heat waves in North America and Europe will become more intense, more frequent, and longer lasting in the second half of the twenty-first century. Additional examples of the use of ensembles for prediction of extremes are Alexander and Arblaster (2009) and Fowler and Ekström (2009).

### 16.1.4 Seasonal to multi-year initial-value predictions

Regional climate variability on time scales of seasons to decades must be forecast in order to permit preparation for the economic, humanitarian, and environmental consequences of the change. Variability on these time scales can result from anthropogenic, greenhouse-gas forcing and from landscape changes, but also because of internal variability in the climate system. For example, Fig. 16.6 shows the Sahel-average precipitation record for a recent 83-year period, where there is evidence of variability on times scales of a few years to multiple decades. See Barnston and Livezey (1987) for a summary of the low-frequency atmospheric-circulation patterns. To forecast the internal variation, which results from the atmosphere's response to sea-surface temperatures, soil moisture, and snow and sea ice, the full physical system must be initialized. Kanamitsu *et al.* (2002b) suggests that there are at least four major requirements for a successful dynamical seasonal prediction system:

- accurate models of the atmosphere, ocean, land, and sea ice that are coupled in a physically consistent manner;
- initial conditions for the atmosphere, ocean, land, and sea ice;
- a methodology for ensemble prediction; and
- a strategy for correcting the systematic error.

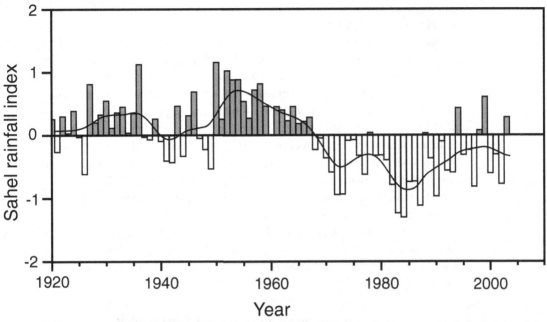

Time series of Sahel (10° N–20° N, 18° W–20° E) regional rainfall during the rainy season (April–October) from 1920 through 2003, illustrating large changes on time scales of years to decades. Negative values of the index correspond to deficits relative to the period mean, and positive values are excesses. It is these changes, which include internal climate variability, that interseasonal to decadal predictions are designed to capture. The black curve shows decadal time-scale variations. From Trenberth *et al.* (2007).

Thus, the modeling system, in addition to encompassing the atmosphere, needs to also represent all the slower components of the variability in the system – the ocean, the land, and the ice. Defining the initial conditions for the forecasts is especially challenging because these slowly varying systems are not well-measured in three dimensions. Thus, data-assimilation systems must be relied upon, where the models themselves provide information that supplements the observations. And, as with longer-range simulations of greenhouse-gas emission scenarios, ensemble methods are important here. However, in addition to the multi-model ensemble methods used for simulation of those scenarios, the fact that this is an initial value problem with some poorly observed variables means that the initial-condition uncertainty must be sampled as well. Lastly, the model bias must be removed by subtracting the model climatology from the seasonal prediction to produce an anomaly field, and then adding the resulting anomaly to the observed climatology. This is equivalent to correcting the prediction using the difference between the model and observed climatology, and significantly increases the forecast skill of the model. A disadvantage of having to correct for this systematic error is that the error is a function of forecast lead times as well as month/season, and is best calculated through performing a large number of predictions with previous cases (reforecasts). Unfortunately, every time changes are made to the model, the systematic error must be recomputed. This problem is similar to the one described in Chapter 13 on the post processing of model output, where the correction of weather forecasts using MOS precludes making frequent improvements and code corrections in the forecast model.

A number of special projects have been aimed at improving seasonal predictions. For example, the DEMETER project employed models from seven institutions in Europe. Using hindcasts, seasonal predictability was assessed for single-model ensembles, and for combining the single-model ensembles to produce multi-model superensembles. Palmer *et al.* (2004) and Hagedorn *et al.* (2005) summarize the project, and Doblas-Reyes *et al.* (2005) discuss the calibration of the ensemble and the combination of the members. A significant amount of the research effort related to prediction on these time scales has been focussed, understandably, on development and verification of methods for the prediction of the ENSO cycle (e.g., Gualdi *et al.* 2005, Keenlyside *et al.* 2005).

There are also seasonal climate-prediction methods that are based on a combination of dynamical and statistical models. For example, O'lenic *et al.* (2008) describe such a system used by the NCEP Climate Prediction Center for producing predictions with lead times of 1 month to 12 months. Also, there are research efforts that are aimed at extending the seasonal and multi-year initial-value prediction methods for use on time scales of decades (e.g., Smith *et al.* 2007, Keenlyside *et al.* 2008).

## 16.1.5  Summary of existing global models

### Global models used for forced anthropogenic climate-change prediction

Table 16.1 lists the AOGCMs, from a variety of different research centers worldwide, that were involved in the IPCC AR4 (Randall *et al.* 2007). Most of the model versions originated in the late 1990s or the first five years of the twenty-first century. For the atmospheric models, the horizontal grid increments varied from about one degree (T106) to five degrees latitude–longitude, and the number of vertical levels varied from 12 to 56. The coupled ocean models typically had similar or better horizontal resolution than that of the corresponding atmospheric models, and the number of vertical ocean levels ranged from 16 to 47. Most of the models represent sea-ice rheology and leads in the ice. Only a few use flux adjustments. Lastly, regarding land-surface processes, most employ multi-layer soil models, route surface water in channels, and have some representation of a vegetation canopy. The main point of showing this table is to emphasize the tremendous effort that is being dedicated to the modeling of radiatively forced climate change.

### Global models used for seasonal to multi-year initial-value predictions

Interseasonal prediction systems sometimes employ atmospheric models that are similar to the AGCMs that are used for medium-range forecasting, but a separately run ocean model provides the SSTs. For longer time-scale predictions, fully coupled AOGCMs are used. Interseasonal forecasts are produced operationally by a number of national centers, on a regular forecast cycle. Ensemble methods are virtually always employed. Table 16.2 lists some of the models that are being used operationally on seasonal to annual time scales.

**Table 16.1** Features of atmosphere–ocean general-circulation models used in the IPCC Fourth Assessment

| Model ID, Vintage | Sponsor(s), Country | Atmosphere Top, Resolution, References | Ocean Resolution, z Coord., Top BC, References | Sea ice Dynamics, Leads, References | Coupling flux adjustments References | Land, Soil, Plants Routing, References |
|---|---|---|---|---|---|---|
| 1. BCC-CM1 2005 | Beijing Climate Center, China | top = 25 hPa T63 (1.9° × 1.9°) L16 CSMD 2005; Xu et al. 2005 | 1.9° × 1.9° L30 depth, free surface Jin et al. 1999 | no rheology or leads Xu et al. 2005 | heat, momentum Yu and Zhang 2000; CSMD 2005 | layers, canopy, routing CSMD 2005 |
| 2. BCCR-BCM2.0 2005 | Bjerknes Centre for Climate Research, Norway | top = 10 hPa T63 (1.9° × 1.9°) L31 Déqué et al. 1994 | 0.5°–1.5° × 1.5° L35 density, free surface Bleck et al. 1992 | rheology, leads Hibler 1979; Harder 1996 | no adjustments Furevik et al. 2003 | layers, canopy, routing Mahfouf et al. 1995; Douville et al. 1995; Oki and Sud 1998 |
| 3. CCSM3 2005 | National Center for Atmospheric Research, USA | top = 2.2 hPa T85 (1.4° × 1.4°) L26. Collins et al. 2004 | 0.3°–1° × 1° L40 depth, free surface Smith and Gent 2002 | rheology, leads Briegleb et al. 2004 | no adjustments Collins et al. 2006a | layers, canopy, routing Oleson et al. 2004; Branstetter 2001 |
| 4. CGCM3.1 (T47) 2005 | Canadian Centre for Climate Modelling and Analysis | top = 1 hPa T47 (2.8° × 2.8°) L31 McFarlane et al. 1992 | 1.9° × 1.9° L29 depth, rigid lid Pacanowski et al. 1993 | rheology, leads Hibler 1979; Flato and Hibler 1992 | heat, fresh water | layers, canopy, routing Verseghy et al. 1993 |
| 5. CGCM3.1 (T63) 2005 | | top = 1 hPa T63 (1.9° × 1.9°) L31 McFarlane et al. 1992 | 0.9° × 1.4° L29 depth, rigid lid Flato and Boer 2001; Kim et al. 2002 | rheology, leads Hibler 1979; Flato and Hibler 1992 | heat, fresh water | layers, canopy, routing Verseghy et al. 1993 |
| 6. CNRM-CM3 2004 | Météo-France/ Centre National de Recherches Météorologiques, France | top = 0.05 hPa T63 (1.9° × 1.9°) L45 Déqué et al. 1994 | 0.5°–2° × 2° L31 depth, rigid lid Madec et al. 1998 | rheology, leads Hunke and Dukowicz 1997; Salas-Mélia 2002 | no adjustments Terray et al. 1998 | layers, canopy, routing Mahfouf et al. 1995; Douville et al. 1995; Oki and Sud 1998 |

**Table 16.1** (continued)

| Model ID, Vintage | Sponsor(s), Country | Atmosphere Top, Resolution, References | Ocean Resolution, z Coord., Top BC, References | Sea ice Dynamics, Leads, References | Coupling flux adjustments References | Land, Soil, Plants Routing, References |
|---|---|---|---|---|---|---|
| 7. CSIRO-MK3.0 2001 | Commonwealth Scientific and Industrial Research Organisation, Australia | top = 4.5 hPa T63 (1.9° × 1.9°) L18 Gordon et al. 2002 | 0.8° × 1.9° L31 depth, rigid lid Gordon et al. 2002 | rheology, leads O'Farrell 1998 | no adjustments Gordon et al. 2002 | layers, canopy Gordon et al. 2002 |
| 8. CHAMS/ MPI-OM 2005 | Max Planck Inst. for Meteorology, Germany | top = 10 hPa T63 (1.9° × 1.9°) L31 Roeckner et al. 2003 | 1.5° × 1.5° L40 depth, free surface Marsland et al. 2003 | rheology, leads Hibler 1979; Semtner 1976 | no adjustments Jungclaus et al. 2006 | bucket, canopy, routing Hagemann 2002; Hagemann and Dümenil-Gates 2001 |
| 9. ECHO-G 1999 | Meteorological Institute of the University of Bonn, Meteorological Research Institute of the Korea Meteorological Administration, and Model and Data Group, Germany/ Korea | top = 10 hPa T30 (3.9° × 3.9°) L19 Roeckner et al. 1996 | 0.5°– 2.8° × 2.8° L20 depth, free surface Wolff et al. 1997 | rheology, leads Wolff et al. 1997 | heat, fresh water Min et al. 2005 | bucket, canopy, routing Roeckner et al. 1996; Dümenil and Todini 1992 |
| 10. FGOALS-g1.0 2004 | National Key Laboratory of Numerical Modeling for Atmospheric Science and Geophysical Fluid Dynamics/Institute of Atmospheric Physics, China | top = 2.2 hPa T42 (2.8° × 2.8°)L26 Wang et al. 2004 | 1.0° × 1.0° L16 eta, free surface Jin et al. 1999; Liu et al. 2004 | rheology, leads Briegleb et al. 2004 | no adjustments Yu et al. 2002, 2004 | layers, canopy, routing Bonan et al. 2002 |

| Model | Sponsor(s), Country | Atmosphere | Ocean | Sea ice | Coupling | Land soil, plants, routing |
|---|---|---|---|---|---|---|
| 11. GFDL-CM2.0 2005 | U.S. Department of Commerce/National Oceanic and Atmospheric Administration/Geophysical Fluid Dynamics Laboratory, USA | top = 3 hPa 2.0° × 2.5° L24 GFDL GAMDT 2004 | 0.3° – 1.0° × 1.0° depth, free surface Gnanadesikan et al. 2006 | rheology, leads Winton 2000; Delworth et al. 2006 | no adjustments Delworth et al. 2006 | bucket, canopy, routing Milly and Shmakin 2002; GFDL GAMDT 2004 |
| 12. GFDL-CM2.1 2005 | Geophysical Fluid Dynamics Laboratory, USA | top = 3 hPa 2.0° × 2.5° L24 GFDL GAMDT 2004 with semi-Lagrangian transports | 0.3° – 1.0° × 1.0° depth, free surface Gnanadesikan et al. 2006 | rheology, leads Winton 2000; Delworth et al. 2006 | no adjustments Delworth et al. 2006 | bucket, canopy, routing Milly and Shmakin 2002; GFDL GAMDT 2004 |
| 13. GISS-AOM 2004 | National Aeronautics and Space Administration (NASA)/Goddard Institute for Space Studies (GISS), USA | top = 10 hPa 3° × 4° L12 Russell et al. 1995 | 3° × 4° L16 mass/area, free surface Russell et al. 1995 | rheology, leads Flato and Hibler 1992 | no adjustments | layers, canopy, routing Abramopoulos et al. 1988; Miller et al. 1994 |
| 14. GISS-EH 2004 | Institute for Space Studies (GISS), USA | top = 0.1 hPa 4° × 5° L20 Schmidt et al. 2006 | 2° × 2° L16 density, free surface Bleck 2002 | rheology, leads Liu et al. 2003; Schmidt et al. 2004 | no adjustments Schmidt et al. 2006 | layers, canopy, routing Friend and Kiang 2005 |
| 15. GISS-ER 2004 | NASA/GISS, USA | top = 0.1 hPa 4° × 5° L20 Schmidt et al. 2006 | 4° × 5° L13 mass/area, free surface Russell et al. 1995 | rheology, leads Liu et al. 2003; Schmidt et al. 2004 | no adjustments Schmidt et al. 2006 | layers, canopy, routing Friend and Kiang 2005 |
| 16. INM-CM3.0 2004 | Institute for Numerical Mathematics, Russia | top = 10 hPa 4° × 5° L21 Galin et al. 2003 | 2° × 2.5° L33 sigma, rigid lid Diansky et al. 2002 | no rheology or leads Diansky et al. 2002 | regional fresh water Diansky and Volodin 2002; Volodin and Diansky 2004 | layers, canopy, no routing Volodin and Lykosoff 1998 |
| 17. IPSL-CM4 2005 | Institute Pierre Simon Laplace, France | top = 4 hPa 2.5° × 3.75° L19 Hourdin et al. 2006 | 2° × 2° L31 depth, free surface Madec et al. 1998 | rheology, leads Fichefet and Morales-Maqueda 1997; Goosse and Fichefet 1999 | no adjustments Marti et al. 2005 | layers, canopy, routing Krinner et al. 2005 |

### Table 16.1 (continued)

| Model ID, Vintage | Sponsor(s), Country | Atmosphere Top, Resolution, References | Ocean Resolution, z Coord., Top BC, References | Sea ice Dynamics, Leads, References | Coupling flux adjustments References | Land, Soil, Plants Routing, References |
|---|---|---|---|---|---|---|
| 18. MIROC3.2 (hires) 2004 | Center for Climate System Research (University of Tokyo), National Institute for Environmental | top = 40 km T106 (1.1° × 1.1°) L56 K-1 Developers 2004 | 0.2° × 0.3° L47 sigma/depth, free surface K-1 Developers 2004 | rheology, leads K-1 Developers 2004 | no adjustments K-1 Developers 2004 | layers canopy, routing K-1 Developers 2004; Oki and Sud 1998 |
| 19. MIROC3.2 (medres) 2004 | Studies, and Frontier Research Center for Global Change, Japan | top = 30 km T42 (2.8° × 2.8°) L20 K-1 Developers 2004 | 0.5°–1.4° × 1.4° L43 sigma/depth, free surface K-1 Developers 2004 | rheology, leads K-1 Developers 2004 | no adjustments K-1 Developers 2004 | layers, canopy, routing K-1 Developers 2004; Oki and Sud 1998 |
| 20. MRI-CGCM 2.3.2 2003 | Meteorological Research Institute, Japan | top = 0.4 hPa T42 (2.8° × 2.8°) L30 Shibata et al. 1999 | 0.5°–2.0° × 2.5° L23 depth, rigid lid Yukimoto et al. 2001 | free drift, leads Mellor and Kantha 1989 | heat, fresh water, momentum (12° S–12°N) Yukimoto et al. 2001 | layers, canopy, routing Sellers et al. 1986; Sato et al. 1989 |
| 21. PCM 1998 | National Center for Atmospheric Research, USA | top = 2.2 hPa T42 (2.8° × 2.8°) L26 Kiehl et al. 1998 | 0.5°–0.7° × 1.1° L40 depth, free surface Maltrud et al. 1998 | rheology, leads Hunke and Dukowicz 1997, 2003; Zhang et al 1999 | no adjustments Washington et al. 2000 | layers, canopy, no routing Bonan 1998 |
| 22. UKMO-HadCM3 1997 | Hadley Centre for Climate Prediction and Research/Met Office, UK | top = 5 hPa 2.5° × 3.75° L19 Pope et al. 2000 | 1.25°×1.25° L20 depth, rigid lid Gordon et al. 2000 | free drift, leads Cattle and Crossley 1995 | no adjustments Gordon et al. 2000 | layers, canopy, routing Cox et al. 1999 |
| 23. UKMO-HadGEM1 2004 | | top = 39.2 km 1.3° × 1.9° L38 Martin et al. 2004 | 1.0° × 1.0° L40 depth, free surface Roberts 2004 | rheology, leads Hunke and Dukowicz 1997; Semtner 1976; Lipscomb 2001 | no adjustments Johns et al. 2006 | layers, canopy, routing Oki and Sud 1998 |

*Source:* Adapted from Randall *et al.* (2007).

The first column shows the IPCC identification, with the calendar year of the first publication describing the model results. The second column lists the sponsoring institution and country. The third shows the pressure at the top boundary of the model, the horizontal resolution in terms of the spectral truncation and/or the latitude–longitude grid increment, and the number of vertical levels. The next column lists the horizontal grid increment of the ocean model, the number of computational levels, and the upper boundary condition. The fifth column shows information about sea-ice dynamics, and the sixth indicates whether there are flux adjustments to the interaction of the atmosphere and ocean. The last

**Table 16.2** Example global models used for operational seasonal to annual initial-value predictions. Note that forecast systems that combine dynamical and statistical models are not listed here.

| Models (where model developed) | Organization producing forecast | Duration of predictions | References |
| --- | --- | --- | --- |
| CFS | NCEP | 9 months | Saha *et al.* 2006 |
| ECHAM (MPI) | IRI | 6+ months | Barnston *et al.* 2003 |
| CCSM (NCAR) | | | |
| MRF (NCEP) | | | |
| NSIPP (NASA) | | | |
| COLA | | | |
| ECPC (Scripps) | | | |
| System-III | ECMWF | 3 months | Anderson *et al.* 2003 |
| | | 1 year | George and Sutton 2006 |
| GloSea | UK Met Office | 3 months | Gordon *et al.* 2000 |
| | | | Graham *et al.* 2005 |

Longer-range (decadal) initial-value predictions are performed on a one-time basis, rather than on a regular cycle. Even though the model products are used for development of adaptation strategies, the modeling is more reasonably defined as a research project rather than being operational.

### 16.1.6 Ensemble climate simulation

The use of ensemble methods in weather prediction is discussed in Chapter 7. The application of similar techniques has also been well established for seasonal, interannual, decadal, and centennial forecasting with AOGCMs. Good summaries are provided in Section 10.5.4 in Meehl *et al.* (2007). In contrast to ensemble weather prediction, predictions of forced anthropogenic climate change and inter-seasonal forecasting, both with AOGCMs, require the sampling of additional sources of uncertainty. For initial-value simulations, the uncertain initial state of the ocean and other components of the system must be sampled. For IPCC-type simulations, future emissions of aerosols and greenhouse gases are unknown, and this uncertainty can be sampled by assuming different scenarios. Also, ensembles can take two forms. In one, the same model can be used for multiple experiments with different choices for poorly constrained internal parameters and for the overall process parameterizations themselves. In the other, multi-model ensembles can be generated through the use of a range of AOGCMs developed at different modeling centers, such as used in the CMIP experiments described above or in the IPCC ensemble.

In experiments that define the climate response to different forcing scenarios, the change between the present state of a variable and the state in some future year is a result of both the start and end times. Because the aim of such experiments is to assess the impact of the forcing on the climate, it is typical to create an ensemble using different initial times within the current climate, and use the average of the projections in order to minimize the influence of the internal variations.

To the degree that the errors from different AOGCMs used in an ensemble are independent of each other, the ensemble mean can be expected to outperform the individual ensemble members. Palmer *et al.* (2004), Hagedorn *et al.* (2005), and Krishnamurti *et al.* (2006b) demonstrate that this is the case for seasonal prediction. For longer-range radiatively forced scenarios, Lambert and Boer (2001), Taylor *et al.* (2004), and Reichler and Kim (2008) show the superiority of the ensemble mean over the use of individual members (e.g., Fig. 16.2). Indeed, the conclusions about global warming reported by the IPCC were based on the multi-model ensemble mean of the CMIP3 models.

As with ensemble weather prediction, ensemble climate predictions can be calibrated (Doblas-Reyes *et al.* 2005). Because of the disparate nature of the multi-national suite of climate models, it is not possible to make the assumption that all the models are equally skillful at predicting each variable at each geographic location. Thus, the optimal solution would not be based on an equally weighted combination of the model solutions. It was noted earlier that models that have verified better against the historical climate record may be given more weight in terms of their projections of future climate (e.g., Krishnamurti *et al.* 1999, Shukla *et al.* 2000, 2006, Goddard *et al.* 2001, Rajagopalan *et al.* 2002, Robertson *et al.* 2004, Yun *et al.* 2005). Clemen (1989) reviews different methods for optimally combining multi-model ensemble climate predictions.

One of the many examples of the benefits of ensemble climate prediction is described in Fedderson and Andersen (2005), who compared the skill of 2-month (seasonal) statistically downscaled multi-model ensemble predictions with the skill of the downscaled individual-model predictions. This comparison was made for 40 years of retrospective seasonal-forecast downscalings over Europe, northwest North America, the contiguous USA, Australia, and Scandinavia. The forecasts employed were part of the DEMETER study, and were performed by Météo-France, ECMWF, and the UK Meteorological Office (Palmer *et al.* 2004). A linear-regression-based downscaling algorithm was constructed using the ensemble mean from the model forecasts (the MOS approach mentioned in Section 16.3.1) and observations from the noted geographic areas. Using a cross-validation approach (Michaelsen 1987), the statistical relationship used in each forecast year was constructed without data from that year. The regression equation obtained from using the ensemble mean was also applied to the downscaling of the individual ensemble members. Table 16.3 shows the verification of the 2-month forecasts in terms of the anomaly correlation, for the individual models as well as the multi-model ensemble, for selected seasons and geographic areas. The predictive skill varies geographically, with season, and with the model. No single model is consistently better than another. The ensemble skill is generally comparable to that of the best model, where the positive scores indicate modest predictive skill beyond that of climatology. The skill varies from year to year, as seen in Fig. 16.7, which shows the time series of the anomaly correlation for the 2-m temperature for Europe in the JAS season. In years where the ensemble prediction shows no skill (negative or zero anomaly correlation), two of the member forecasts have typically failed. In most years, the anomaly correlation of the ensemble prediction is positive.

**Table 16.3** Comparison of the anomaly correlation for individual models and for the ensemble mean, for selected seasons and geographic regions.

| Model | Precipitation | | 2-m Temperature | |
|---|---|---|---|---|
| | Europe JFM | Scandinavia JFM | Europe JAS | Scandinavia AMJ |
| Météo-France | 0.07 | 0.11 | 0.16 | 0.33 |
| ECMWF | 0.30 | 0.09 | 0.35 | 0.14 |
| UK Met Office | 0.03 | 0.28 | 0.25 | 0.15 |
| Ensemble | 0.22 | 0.27 | 0.35 | 0.28 |

*Source:* From Feddersen and Andersen (2005).

Another European ensemble climate-prediction effort is ENSEMBLES (Hewitt 2005). This project involves the development, verification, and application of several AOGCMs and regional models to produce ensemble forecasts on time scales of seasons to decades. In addition to generating the probabilistic forecasts for Europe, the project is linking the outputs with the needs of various sectors such as agriculture, health, energy, water, and food security.

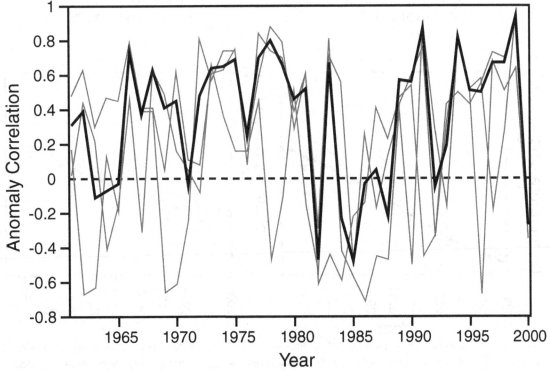

**Fig. 16.7** Anomaly correlations for 2-month downscaled predictions of 2-m temperature for JAS in Europe, averaged by year, for three models (thin lines) and for the ensemble mean (thick line). From Feddersen and Andersen (2005).

## 16.1.7  Example global climate-change predictions

Predictions with AOGCMs have been conducted on seasonal, annual, decadal, and centennial time scales. Even though our focus in this text is not on the results, but rather on the
methods, a couple of examples will be provided of climate change that has been simulated
by radiative-forcing experiments. Figure 16.8 illustrates the change in global-average surface air temperature and precipitation predicted for the twenty-first century by 21
AOGCMs used in the IPCC AR4 experiments. There is a significant spread among the
model predictions, especially for precipitation, but all of the simulations have the same
general trend. Of more direct use for addressing practical questions about adaptation to
climate change are regional interpretations of the global-model output. Such regional analyses of the results of IPCC-ensemble simulations have been generated for a wide range of
applications related to water resources, air quality, etc. For example, Fig. 16.9 shows the
ensemble-mean predicted precipitation change in the Middle East between 2005 and 2050,
based on 18 global climate models that participated in the IPCC AR4. These results indicate a trend toward a drier future to the north and east of the Mediterranean, along the
track of extratropical cyclones. By the end of the century (not shown), the precipitation
decrease exceeds 100 mm, and the temperature increases by 4 °C. Note that these types of
regional analyses are not based on downscalings, but are simply windows within which the
global-model output is displayed and analyzed. For other examples of regional analyses of
AOGCM climate projections, see Gibelin and Déqué (2003), Déqué *et al.* (2005), Cook
and Vizy (2006), d'Orgeval *et al.* (2006), García-Morales and Dubus (2007), and Hanson
*et al.* (2007).

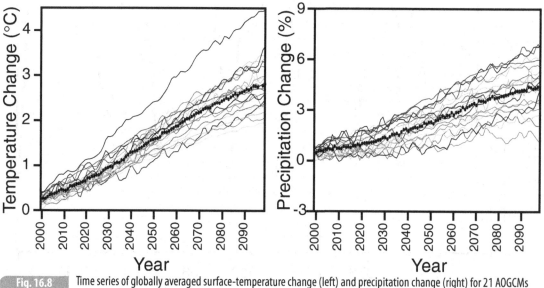

**Fig. 16.8**  Time series of globally averaged surface-temperature change (left) and precipitation change (right) for 21 AOGCMs
employed in the IPCC AR4, for the A1B emissions scenario. The values plotted are annual means, relative to the 1980 to
1999 average from the corresponding twentieth century simulations. The multi-model ensemble means are plotted as
black dots and the individual simulations are shown by the fine lines. From Meehl *et al.* (2007).

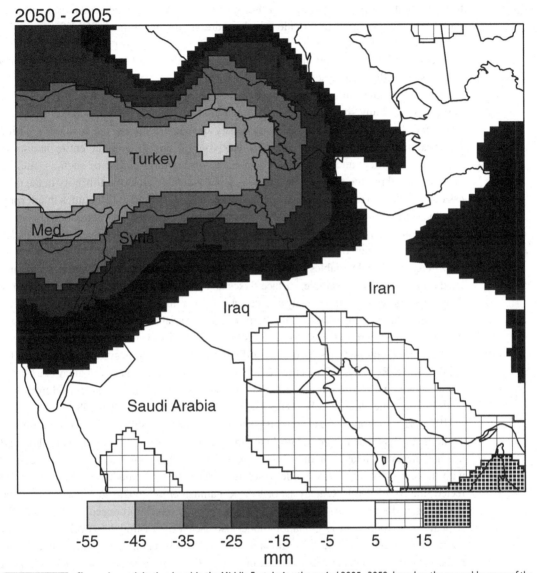

2050 - 2005

Change in precipitation (mm) in the Middle East during the period 2005–2050, based on the ensemble mean of the predictions of 18 global models that participated in the IPCC AR4. Adapted from Evans (2008).

## 16.2  Reanalyses of the current global climate

Global data-assimilation systems typically employ a model to periodically ingest observations, with the result being a gridded set of model dependent variables that are consistent with both model dynamics and the information represented in the observations. The resulting analyses are sometimes called Model-Assimilated Data Sets (MADS). Such data-assimilation systems are used to define long-term analyses of atmospheric fields pertaining to the current and recent climates, for use by the research and climate-monitoring communities, and they are used to provide initial conditions for operational GCM forecasts. The former process of reconstructing historical conditions involves using a contemporary, frozen version of the assimilation system, to generate multi-decadal analyses. These gridded data sets are called reanalyses.

Even though the same assimilation modeling software is used for the entire reanalysis period, to avoid any shifts in analyzed climate that would result from model changes, the types and amounts of ingested data obviously do change during the reanalysis period. Observation locations change, periodic data voids occur throughout entire nations or regions for political and economic reasons, and observation platforms such as satellites change throughout the period. Thus, there is an unavoidable lack of uniformity in the process that can lead to changes in the accuracy of the reanalysis and to difficulties in interpreting the climate. When major new observation platforms are introduced, parallel assimilations are run for short test periods, with and without the new data, to help isolate the impact of the new data on the analyzed climate.

Different analyzed variables have different relative dependencies on the model and assimilated observations. For example, surface fluxes and precipitation are often not assimilated, so their values provided in the reanalysis data set can be entirely products of the model. In contrast, winds, mass-field variables, and thermodynamic variables are assimilated, so the resulting analysis will represent a mix of the constraints of the observations and model dynamics.

The assimilation methods used in the different global reanalysis systems are similar. The NCEP-NCAR Reanalysis Project (NNRP) reanalysis (referred to as R-1, Kalnay *et al.* 1996) was generated with a 6-h intermittent-assimilation method, for the 40-year period from 1957 to 1996, using the NCEP Global Data-Assimilation System (Kanamitsu 1989, Kanamitsu *et al.* 1991). Here, an objective analysis is performed every 6 h, using the previous 6-h forecast as the first guess. The horizontal grid increment of the model and the analysis is about 210 km, and there are 28 layers in the vertical. An updated reanalysis was subsequently produced, with the same data input and model resolution, but with some model and data errors corrected (the NCEP-DOE Reanalysis, or R-2; Kanamitsu *et al.* 2002a). A more-recent global reanalysis is the ERA-40, also a second-generation product, produced by the ECMWF (Uppala *et al.* 2005). The assimilation model also used a 6-h assimilation cycle, the horizontal grid increment was about 125 km, and there were 60 layers in the vertical. Other global reanalyses are the Japanese 25-year Reanalysis Project (JRA-25, Onogi *et al.* 2007) and NASA's Modern Era Retrospective-analysis for Research and Applications (MERRA, Bosilovich *et al.* 2006). The MERRA analysis is defined on a 2/3 degree longitude by 1/2 degree latitude grid, with 72 vertical layers. The JRA-25 analysis has a horizontal grid size of about 120 km and 40 vertical layers.

## 16.3  Climate downscaling

The term *future-climate downscaling* refers to techniques that use AOGCM predictions of future climates as input to methods that produce finer-scale climate information. This process is necessitated by the fact that the AOGCMs that are used for predicting future climate typically have grid spacings of hundreds of kilometers, and thus there is a mismatch between the output data of those models and the needs of, for example, hydrological models that require watershed-scale information. In the process of determining the accuracy of the AOGCMs and the downscaling methods, simulations of current or past climates may be performed, but the ultimate objective in future-climate downscaling studies is to produce high-horizontal-resolution information about future climates. Downscaling methods can be applied to initial-value-based interseasonal or longer forecasts, or to radiatively forced IPCC-type simulations.

There is also a need for gridded, high-horizontal-resolution information about the current climate. For example, wind-energy prospecting and defining source–receptor relationships in air-quality studies benefit from the availability of high-resolution mesoscale data that define the characteristics of the current climate. To satisfy this need, *current-climate downscaling* uses global gridded data sets that are produced by the global-model-based data-assimilation systems described in the last section.

For both present and future climates, there are two basic approaches for accomplishing this downscaling. One is to use statistical-empirical relationships that define the high-resolution subgrid-scale variability based on resolved, grid-scale, values from the global data set. The other uses a LAM whose LBCs are forced by the global data set, or a stretched-grid AGCM. The former process is referred to as *statistical downscaling*, while the latter is called *dynamical downscaling*. The advantages and disadvantages of the two approaches are summarized in Table 16.4.

Topics that are related to the subject of climate downscaling appear elsewhere in this book. For example, Chapter 13 about the post processing of model output discusses statistical downscaling as applied to weather prediction. And, the subject of forcing the LBCs of LAMs with global analyses or global forecast-model output is described in Chapter 3 on numerical methods.

Downscaling, whether it is statistically or dynamically based, is a local diagnostic process that represents a post processing of the global data set. That is, the modulations of the large-scale climate by local forcing, such as orography, are diagnosed by statistical or dynamical methods, but they typically cannot feed back to the global scales to improve the rendering of the climates of other regions. The exception is when variable-resolution global climate models focus higher horizontal resolution over certain geographic areas (discussed in Laprise 2008). Even in this latter situation, there are concerns that teleconnections between two distant geographic areas cannot be treated properly in climate simulations if only one region is modeled with high resolution. For example, if we are modeling the future climate of the Amazon Basin, we would be tempted to represent only that area with higher resolution. However, there is recent evidence (Koren *et al.* 2006) that the dust originating in the relatively small Bodélé Depression in the Sahara Desert provides a large fraction of the nutrients for Amazon vegetation. Thus, inadequate resolution of the orographically generated high winds in this area of Africa could lead to an erroneous devegetation of the Amazon in a future-climate simulation that was able to represent the effects of soil nutrients on the vegetation. Unfortunately, we often don't have sufficient understanding of the current climate

**Table 16.4** Summary of advantages and disadvantages of statistical and dynamical downscaling for current and future climates

|  | Statistical downscaling | Dynamical downscaling |
|---|---|---|
| **Advantages** | • Computationally efficient, and cheap<br>• Can be used to derive variables not available from RCMs (e.g., river discharge)<br>• Easily transferable between different regions<br>• Based on standard and accepted statistical procedures<br>• Able to directly incorporate observations<br>• Can provide climate variables at a point, based on large-scale input | • Response is based on physically consistent process<br>• Gridded output is available for physical-process analysis<br>• Can better capture extreme events and variance |
| **Disadvantages** | • Requires long and reliable observed historical data record for calibration<br>• Success dependent upon choice of predictors<br>• Nonstationarity may exist in the predictor–predictand relationship<br>• Climate-system feedbacks not included<br>• Variance is underestimated, may poorly represent extreme events<br>• GCM biases can cause error, unless the procedure corrects for them<br>• Domain size, region, and season affect skill | • Computationally intensive<br>• Sensitive to location of lateral boundaries<br>• Feedback to large scale generally not considered<br>• Biases in large-scale conditions will cause errors<br>• Domain size, region, and season affect skill |

sensitivities, such as the one just mentioned, let alone teleconnections that will be potentially important in future climates.

Section 16.1.3 describes the need for global climate models to provide information on the extremes in the PDFs of the dependent variables, for use in analyzing both current and future climates. Because atmospheric extremes tend to be associated with small-scale features, such as convective events, strong fronts, terrain-induced downslope winds, etc., it is understandable that coarse-resolution global-model-based analyses or predictions will tend to underestimate extremes, resulting in overly smooth features in analyses of the current climate or predictions of future climates. Thus, there has been much research activity associated with capturing the extremes with the downscaling process. For example, the STAtistical and Regional dynamical Downscaling of EXtremes (STARDEX) project was designed to compare statistical and dynamical downscaling methods in terms of their ability to estimate the extremes, for European future climates. For more information about STARDEX and the downscaling of extremes, see Fowler *et al.* (2007) and references therein. Figure 16.10 shows how downscaling with a Regional Climate Model (RCM) can better define extremes, by comparing the maximum one-day rainfall at a location in

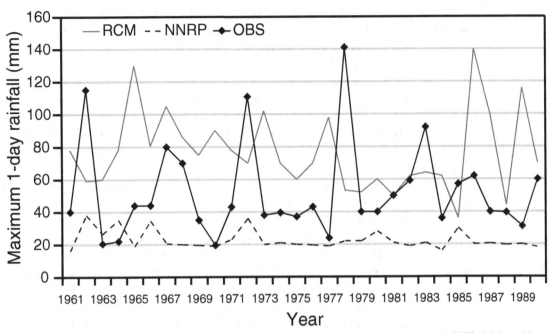

Maximum one-day rainfall for Larissa, Greece, for each year of a 30-year period, based on the NNRP global-model-based reanalysis (NNRP), a downscaling simulation with the HadRM3 RCM (RCM), and rain-gauge observations (OBS). From Hanson *et al.* (2007).

Europe in each year of a 30-year period, based on observations, the NNRP global analysis (Kalnay *et al.* 1996), and a simulation by the HadRM3 RCM. The NNRP reanalysis is model based, with a grid increment of about 210 km. The RCM has a grid increment of about 40 km, with LBCs provided by an AOGCM having about one-fourth the resolution. Clearly, the RCM produces rainfall extremes whose average magnitude is more consistent with the observations than that of the GCM-based NNRP reanalysis.

The following subsections will describe climate downscaling by classifying methods in different ways. First, the statistical and dynamical methods will be described and contrasted in Sections 16.3.1 and 16.3.2, respectively. Then, the use of both of these methods for downscaling future climates and current climates will be described in Sections 16.3.3 and 16.3.4, respectively. Lastly, a summary will be provided of how the downscaling methods have been applied to address different practical climate-dependent problems.

## 16.3.1 Statistical climate downscaling methods

### Spatial statistical downscaling

Spatial statistical downscaling involves using linear or nonlinear statistical/empirical relationships to estimate small-scale local processes (the predictands, such as precipitation rate, temperature, or river discharge) based on features of the large-scale weather or climate (the predictors) that are represented in (1) global reanalyses, (2) seasonal or annual predictions with AOGCMs, or (3) longer climate-forcing simulations with AOGCMs. Whether reanalyses or AOGCM output are used to define the predictors (for current-climate or future-climate downscaling), the general objective is to infer information about fine-scale processes without the use of a high-resolution model. A two-step process is involved. First, the statistical relationships are developed between the local climate varia-bles of interest and the large-scale predictors, using local and large-scale data for the cur-rent climate. If observations of the local variables are not available for a geographic area, a LAM can be run for a range of large-scale regimes and a catalogue developed of local responses. In the second step, the statistical relationships are used to define the local sys-tem response to the large-scale features, for example in a future-climate simulated by an AOGCM. This statistical-downscaling process can be imagined as a parameterization of the local system response to the large-scale forcing, or as the establishment of analogs between the large- and small-scale features.

The choice of the predictor obviously depends on the predictand, given that a physical relationship must prevail. Two requirements exist for this choice: (1) it must be possible to diagnose the predictand accurately from the predictor and (2) the predictor must be well-predicted by the dynamic model or well-defined by the reanalysis. A third requirement, which may be more problematic, is that the predictor–predictand relationship must remain stationary during climate change. For diagnosis of precipitation on small scales, examples of predictors from the large-scale dynamic model are precipitation, sea-level pressure, rel-ative humidity, geopotential height, wind direction, vorticity, and divergence (Wilby and Wigley 2000). The geographic area over which the predictor is defined should be suffi-ciently large to encompass the relevant large-scale processes. For example, Feddersen (2003) had to include a large fraction of the North Atlantic in order to represent the North Atlantic Oscillation (NAO) for the downscaling of seasonal precipitation predictions for Scandinavia.

Statistical relationships used in downscaling are typically calculated from observed or analyzed predictors and predictands. This is analogous to the perfect prognosis (perfect prog) approach to the statistical post processing of model weather forecasts, described in Chapter 13, where observations are used to define the algo-rithms that translate model output into information (e.g., variables) that cannot be obtained directly from the model. However, the large-scale dynamic models have biases, and thus the predictors produced by their forecasts will introduce errors in the downscaled variables. An alternative for defining predictor–predictand relationships for downscaling interseasonal predictions is to employ the approach used to generate MOS (Wilks 2006), also described in Chapter 13. Here, the predictands used to obtain the post-processing relationships are obtained from model forecasts, so the post processing of the model output corrects for the model's systematic errors. In the context of statistically downscaling climate forecasts, if the statistical downscaling relationship is based on model fields, the systematic errors in the model will auto-matically be accounted for. Unfortunately, as with the use of any MOS-based meth-ods, a long series of historical climate reforecasts, or hindcasts, is needed in order to build the statistical relationships. And, these relationships must be regenerated any time the model is upgraded because the systematic errors will change. Examples of the use of this method for defining the predictands in the statistical downscaling of seasonal predictions are found in Feddersen *et al.* (1999), Feddersen (2003), and Feddersen and Andersen (2005).

An obvious issue related to statistical climate downscaling is that it is limited by the assumption of temporal stationarity in the empirical relations. That is, an algorithm that is trained with current-climate data may not apply for future climates. Unfortunately, the nonstationarity in empirical climate relations is well documented (Ramage 1983, Slonosky *et al.* 2001, Charles *et al.* 2004). In contrast, Hewitson and Crane (2006) found that the nonstationarity can be relatively small. Of course the tuning to the current climate of parameters in physical-process parameterizations is of concern as well.

There are a number of different mathematical approaches for statistically relating the predictors on the large scale and the local variables, the predictands. Linear approaches are based upon regression models, while nonlinear approaches are based on weather-typing schemes and weather generators. See Zorita and von Storch (1999), Hanssen-Bauer *et al.* (2005), and Fowler *et al.* (2007) for further discussion of the different methods.

## Regression models

These methods directly quantify the relationship between the predictand and the predictor variables, and are classified as linear methods. A simple example described in Zorita and von Storch (1999) is based on the well-known correlation between the surface air temperature in Scandinavia and the NAO. In this case, a linear regression equation can be established between the anomaly in the NAO index (the sea-level pressure difference between the Azores and Iceland) and the anomaly in the temperature at a Scandinavian station. Thus, given changes in the NAO index in a future climate, the change in the Scandinavian temperature can be obtained from the regression equation. The technical complexity of the regression methods can be considerably greater than in this example, where the following mathematical constructs can be applied. See Zorita and von Storch (1999) and other references for discussion of the assumption, which is implicit in these methods, that the local variables are normally distributed.

- Simple/multiple regression (Hanssen-Bauer *et al.* 2003, Hay and Clark 2003, Johansson and Chen 2003, Matulla *et al.* 2003, Huth 2004, Hessami *et al.* 2008, Huth *et al.* 2008, Tolika *et al.* 2008),
- singular-value decomposition (Huth 1999, 2002; von Storch and Zwiers 1999; Widmann *et al.* 2003; Paul *et al.* 2008),
- canonical-correlation analysis (Wigley *et al.* 1990; von Storch *et al.* 1993; Busuioc *et al.* 2001, 2006, 2008; Chen and Chen 2003; Huth 2004; Xoplaky *et al.* 2004),
- empirical-orthogonal functions (Zorita and von Storch 1999, Benestad 2001, Wilby 2001), and
- principal-component analysis (Cubasch *et al.* 1996, Kidson and Thompson 1998, Hanssen-Bauer *et al.* 2003).

## Weather-typing schemes

Some of these approaches classify the large-scale weather in a region into a specific number of different, dominant weather regimes, patterns, or classes, based on a variable such as sea-level pressure or geopotential height. Then, the local weather (the predictand, say precipitation) is defined for each category based on the historical observation record. In contrast to the previous regression methods, where a grid-box value of a large-scale variable may be the predictor, here the pattern is the predictor. The downscaled climate, current or future, is defined by the frequency-of-occurrence of the predictor regimes. This method assumes that the same weather regimes will exist for future climates, and that the change will be represented by a new occurrence frequency for each of the regimes. The weather typing can be done through self-organizing maps, cluster analysis, or empirical-orthogonal functions. Alternatively, so-called analog or weather-classification schemes are based on a long record of observations of the large-scale weather and the local weather. The output from a GCM is compared to the large-scale observations for the period of record, the historical case where the observations best fit the GCM output is defined, and the simultaneously observed local weather is then associated with the case. Some examples of these methods follow:

- weather-classification/analogs (Zorita and von Storch 1999, Palutikof *et al.* 2002, Díez *et al.* 2005, Timbal and Jones 2008), and
- self-organizing maps, cluster analysis, neural networks (Heimann 2001, Trigo and Palutikov 2001, Cavazos *et al.* 2002, Hewitson and Crane 2002, Gutiérrez *et al.* 2005, Moriondo and Bindi 2006, Huth *et al.* 2008, Tolika *et al.* 2008).

The downscaling relationships can be developed for large-scale data that pertain to different time scales. For example, monthly or seasonal anomalies on the global scales can be downscaled to produce corresponding anomalies on small scales. Or large-scale daily data can be downscaled. Buishand *et al.* (2004) discuss temporal aggregation levels for statistical downscaling of precipitation. For summaries of different statistical-downscaling methods, and their limitations and comparison, see Wilby *et al.* (1998), Haylock *et al.* (2006), and Busuioc *et al.* (2008).

## Temporal statistical downscaling

One of the motivations for the spatial downscaling of climate forecasts is to produce information on the scales that are needed for decision making. Similarly, it is important to provide data with sufficient temporal structure for these decisions. Time series of monthly or seasonal averages of future temperature or precipitation anomalies are useful for broad assessment of water-resources and agricultural impacts, but daily information is really needed for crop-yield models, and even finer time scales are needed, for example, for assessing the risk of flash floods. There are at least two issues. One is that statistical spatial downscaling may only provide monthly or seasonal anomalies. The other is that time series of variables that are associated with large AOGCM grid boxes are smoother than those that apply to single points, simply because of the averaging that is implied over the large area. To address this need, synthetic high-resolution time series can be generated with what are called *stochastic weather generators*. For example, regarding precipitation, these generators can simulate the temporal distribution of wet and dry spells, the typical number of days with and without precipitation, and the intensity of precipitation. The generators can be tuned to apply to particular current and recent weather types. Information about the application of stochastic weather generators for climate-change studies can be found in Katz (1996), Semenov and Barrow (1997), Goddard *et al.* (2001), Huth *et al.* (2001), Palutikof *et al.* (2002), Busuioc and von Storch (2003), Katz *et al.* (2003), Wilby *et al.* (2003), Elshamy *et al.* (2006), Wilks (2006), and Kilsby *et al.* (2007).

## 16.3.2  Dynamical climate downscaling methods

Dynamical downscaling, or the dynamic-model-based generation of high-resolution climate conditions for a particular region and period of time, can be accomplished using a few different approaches (CCSP 2008). Each of the following methods can be used for current-climate or future-climate downscaling.

- Limited-area models (RCMs) are located over a geographical region of interest, and long simulations are produced by defining LBCs with output from an AOGCM or with a global analysis (Jones *et al.* 1995, 1997; Ji and Vernekar 1997; McGregor 1997; Gochis *et al.* 2002, 2003; Frei *et al.* 2003; Hay and Clark 2003; Roads *et al.* 2003a; Boo *et al.* 2004; Liang *et al.* 2004; Castro *et al.* 2005; Díez *et al.* 2005; Kang *et al.* 2005; Misra 2005; Paeth *et al.* 2005; Sotillo *et al.* 2005; Sun *et al.* 2005; Afiesimama *et al.* 2006; Antic *et al.* 2006; De Sales and Xue 2006; Druyan *et al.* 2006; Feser 2006; Liang *et al.* 2006; Moriondo and Bindi 2006; Woth *et al.* 2006; Christensen *et al.* 2007; Xue *et al.* 2007; Jiang *et al.* 2008; Lo *et al.* 2008; Rockel *et al.* 2008; Salathé *et al.* 2008).
- Global stretched-grid AGCMs (discussed in Chapter 3) use enhanced horizontal resolution over a geographic region of interest, and are run for climate time scales (Déqué and Piedelievre 1995, Lorant and Royer 2001, Gibelin and Déqué 2003, Déqué *et al.* 2005, Fox-Rabinovitz *et al.* 2006, Boé *et al.* 2007).
- Uniformly high-resolution AGCMs produce simulations of climate (Branković and Gregory 2001, May and Roeckner 2001, Duffy *et al.* 2003, Coppola and Giorgi 2005, Yoshimura and Kanamitsu 2008).
- Very-high-resolution orographic forcing is used in coarse-grid AOGCMs (Ghan *et al.* 2006, Ghan and Shippert 2006).

Instead of the typical grid increments of hundreds of kilometers for AOGCMs, models used for dynamical downscaling have grid increments of tens of kilometers or less. As with the use of high-resolution LAMs for weather prediction, the benefits of the resolution for climate prediction are attributable to (1) the better representation of fine-scale local forcing such as from orographic or other landscape variability, (2) the ability to explicitly represent processes rather than parameterize them, (3) the nonlinear interactions permitted among a more-complete spectrum of waves, and (4) greater compatibility between the model's vertical and horizontal resolutions.

When the above approaches are used for future-climate downscaling, the downscaling models generally do not have ocean, ice, and vegetation dynamics, and thus the surface properties must be specified from (1) previously run parent, coarser-resolution AOGCM simulations or (2) global analyses for which ocean and surface conditions are specified. In common for all the approaches is the way that the downscaling models are run for time slices of a few decades, for example from 1961 to 1990 for a baseline climate and from 2070 to 2100 for a changed climate.

Even though global, multi-model ensemble climate-prediction methods have proven to be very valuable, opportunities for performing multi-model ensemble downscalings are limited by the fact that AOGCM output for many models is not archived with sufficient frequency to allow the use of the data to drive RCM LBCs. Another disadvantage of dynamic downscaling approaches is that the higher-resolution features represented in the atmosphere cannot interact with the ocean dynamics, and a departure of the large-scale atmospheric features in the downscaling model from those in the AOGCM means that the atmosphere–ocean interaction is negatively affected.

The following sections review the above-listed different approaches for dynamic down-scaling.

## Limited-area models nested within AOGCMs or global analyses

An advantage of downscaling with RCMs is that the models are often very similar to mesoscale weather-prediction models that have already been developed and are well tested. The resolutions of the LAMs are simply adjusted so that they can be integrated for decades to centuries, with the computational resources available. Also, the use of RCMs for future-climate downscaling is appealing because the global-model output used for LBCs is easily accessed from the IPCC runs of past, present, and future climates that are archived at the Program for Climate Model Diagnosis and Intercomparison (PCMDI) at the Lawrence Livermore National Laboratory in the USA. And, for current-climate downscaling, high-quality global reanalyses are publicly available and easily accessible (Kalnay *et al.* 1996, Kanamitsu *et al.* 2002a, Upalla *et al.* 2005). Christensen *et al.* (2007) summarize a series of papers that describe results from the project entitled Prediction of Regional scenarios and Uncertainties for Defining EuropeaN Climate change risks and Effects (PRUDENCE), a European study of dynamic downscaling that evaluated simulations of current and future climates using various RCMs. Also discussed were (1) the modeling of the specialized impacts of regional climate change on water resources, agriculture, ecosystems, energy, and transportation; (2) the modeling of extreme weather events; and (3) the policy implications associated with the availability of high-resolution climate predictions. Several other coordinated projects have applied RCMs for simulating regional climate change in various parts of the world: PIRCS for the USA (Takle *et al.* 1999), RMIP for Asia (Fu *et al.* 2005), ARCMIP for the Arctic (Curry and Lynch 2002, Rinke *et al.* 2005), and over the Pacific Ocean (Stowasser *et al.* 2007).

The impact of lateral boundaries on the solution of a LAM must be viewed somewhat differently in the context of climate downscaling. In Chapter 3 it was stated that, ideally, lateral boundaries should be sufficiently distant from the area of meteorological interest on a grid such that the negative effects of the boundaries do not contaminate the solution during the period of a forecast. However, for climate downscaling that involves running the LAM for years, this is not an option. In fact, there has been plentiful documentation of the strong sensitivity of downscaled climates to the exact position of lateral boundaries, the domain size of the RCM, and the quality of the data used for LBCs (Dickinson *et al.* 1989, Jones *et al.* 1995, Laprise *et al.* 2000, Denis *et al.* 2002, Rojas and Seth 2003, Seth and Rojas 2003, Dimitrijevic and Laprise 2005, Vannitsem and Chomé 2005, Diaconescu *et al.* 2007). In Section 10.4, a method was demonstrated for assessing the impact of the LBCs in RCM simulations. Denis *et al.* (2002) describe what are called Big-Brother–Little-Brother experiments, wherein a large domain RCM (Big-Brother) is used to establish a reference climate over an area, and short waves are filtered so that the RCM climate has scales similar to those of a GCM. This filtered reference climate is then used to provide LBCs for the same RCM (same resolution) that is run for a smaller domain (Little-Brother). The differences between the climate statistics from the Big- and Little-Brother integrations can then be attributed to LBC effects. Dimitrijevic and Laprise (2005), Antic *et al.* (2006), and Køltzow *et al.* (2008) use a similar method to evaluate LBC effects in downscaling with RCMs. Other applications of this method are referenced in Laprise *et al.* (2008) and Laprise (2008).

Because the RCM is intended to simulate the atmospheric response on small scales to the forcing by the global weather patterns, it is desirable if the regional model's large-scale solution does not depart significantly from that in the global data set. Unfortunately, if the models communicate only through the lateral boundaries of the regional model, the large-scale solution of the regional model can drift away from the imposed large scale (e.g., Jones *et al.* 1995). In order to reduce this problem for both future-climate and current-climate downscaling, a method is used called *spectral nudging* (Waldron *et al.* 1996; von Storch *et al.* 2000; Miguez-Macho *et al.* 2004; Castro *et al.* 2005; Kanamaru and Kanamitsu 2007a, 2008; Yoshimura and Kanamitsu 2008; Alexandru *et al.* 2009). Most applications of this method are based on the data-assimilation approach called Newtonian relaxation, or nudging, described in Chapter 6, wherein the model solution is nudged toward observations or gridded data using artificial terms in the prognostic equations. With spectral nudging, however, only the large-scale part of the regional-model solution is nudged toward the global data set, leaving the small scales unaffected.

For future- or present-climate downscaling, if sufficient computational resources exist the regional model can be run for many decades, with LBCs from AOGCMs or global reanalyses. However, to make the efforts more tractable, the downscaling may be done for selected time slices. For example, if a current-climate downscaling is required for a region, but only for, say, one season, the model can be run for time slices consisting of only the particular three months out of each year in the 40-year record of the NNRP reanalysis. For future-climate downscalings using output from AOGCM IPCC-scenario runs, the time-slicing can be more challenging because the existence of internal climate

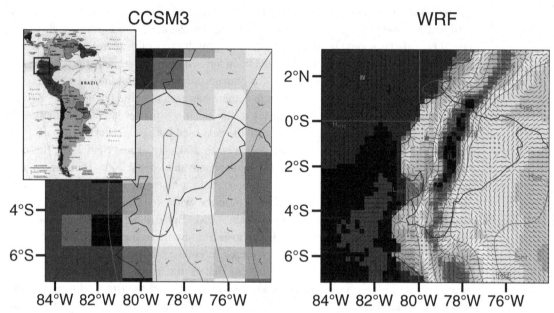

**Fig. 16.11** Illustration of the difference in the horizontal resolution of simulations by the NCAR CCSM3 model with a grid increment of 250 km (left), and the WRF RCM with a grid increment of 10 km (right). The lateral-boundary conditions of WRF were forced by the CCSM simulation. Illustrated are the mean temperature and wind simulations from each model for the same period and geographic area. Darker grays represent lower temperature. The box in the map inset at the upper left shows the model study area over Peru. Provided by Andrew Monaghan, NCAR.

variability in the AOGCM output can cause the implied climate change in the region to depend on the time of the slices. For example, if an internal-variability cycle is at an extreme during the period of a slice, the downscaled climate during the time slice would not be representative of the more temporally averaged climate. This can be avoided by using long time slices, or by first evaluating the internal cycles in the AOGCM solution and avoiding the extremes. Another significant disadvantage of using short time slices is that the RCM does not have sufficient time to fully develop its own regional climate. Even though fine-scale responses to orography and coastlines will develop, recall that Pielke *et al.* (1999a) concluded that almost one year of simulation time may be required for a model to spin up its own soil moisture (for example, in response to fine-scale precipitation).

To illustrate the resolution benefits of downscaling with an RCM, Fig. 16.11 shows the mean temperature and winds over an area along the west coast of South America, for a period simulated by the NCAR Community Climate System Model 3 (CCSM3) AOGCM with a grid increment of 250 km, and the mean temperature and winds for the same region and period simulated by the WRF model with a grid increment of 10 km. The WRF RCM used the CCSM simulation for LBCs. In the CCSM mean temperatures, there is no evidence of the spine of the smoothed Andes Mountains, but the effect is clearly seen in the WRF solution.

## Global variable-resolution (stretched-grid) AGCMs

These variable-resolution GCMs have enhanced, sometimes uniform, horizontal resolution over the geographic region for which a downscaled solution is needed (e.g., Déqué and Piedelievre 1995; Déqué *et al.* 1998; Fox-Rabinovitz *et al.* 2001, 2002, 2005, 2006; Gibelin and Déqué 2003). The fine grid is an integral part of the AGCM, with the transition in grid increment being gradual between the coarse-resolution and fine-resolution regions. Thus, unlike RCMs, there are no traditional LBC issues to deal with when using the stretched grids, and there is less potential error in the propagation of waves from the low-resolution to the high-resolution regions. Other claimed benefits of stretched-grid GCMs, relative to RCMs that are one-way nested within AOGCMs, include (1) the consistency in the parameterizations and the numerics between coarse and fine grids (thus ensuring greater spatial consistency in the model solution) and (2) the fact that processes represented on the regional grid can feed back to the global scales, which they cannot do with the one-way, parasitic nesting that is used with RCMs. Regarding point (1) above, it can, however, be argued that there are some situations where different parameterizations (e.g., convection) *should* be used in the regions with different horizontal resolution. Fox-Rabinovitz *et al.* (2006) describe a GCM Stretched-Grid Model Intercomparison Project (SGMIP).

These models have been run for times scales of seasons to decades in physical-process studies related to current regional climates. For example, Fox-Rabinovitz *et al.* (2001) use a stretched-grid GCM to study the anomalous regional climates associated with the 1988 US summer drought and the 1993 US summer flood. And Barstad *et al.* (2008) spectrally nudged a stretched-grid GCM to the ERA-40 analysis, where the resulting downscaled data set showed large improvements over the ERA-40 analysis. For example, the precipitation bias was reduced from 50% to 11%. For future-climate downscaling, Gibelin and Déqué (2003) obtain sea-surface temperature forecasts from a coarser-resolution coupled AOGCM, allowing their stretched-grid AGCM to simulate the future climate in the Mediterranean region for an IPCC scenario.

## Uniformly high-horizontal-resolution AGCMs

These AGCMs use relatively high horizontal resolution over the entire sphere in order to represent fine-scale processes, for time slices of AOGCM simulations. A disadvantage is obviously the high computational cost, which necessitates the use of time slices of modest length. Advantages include the lack of LBC problems that can be encountered with parasitically nested RCMs, and the uniformly high resolution means that small-scale processes in one region can interact with small-scale processes in another. Various studies have demonstrated the benefit of employing such higher horizontal resolution for climate simulations. Spectral nudging has been used to maintain the larger scales of the high-resolution simulation consistent with those of the coarser-resolution AOGCM simulations, while allowing fine-scale forcing from orography and other landscape features to develop regional climate features (von Storch *et al.* 2000, Yoshimura and Kanamitsu 2008). Figure 16.12 summarizes the relative strengths and weaknesses of the use of RCMs embedded within AOGCMs or global analyses, global variable-resolution

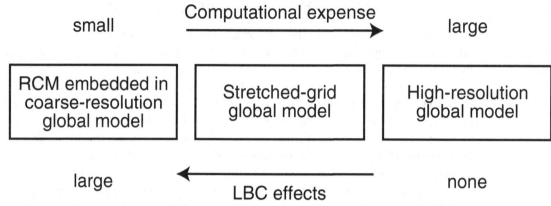

**Fig. 16.12** Schematic comparing the LBC effects and computational costs of three approaches to high-resolution climate modeling. Adapted from original by Jack Katzfey, CSIRO.

(stretched-grid) AGCMs, and uniformly high-resolution AGCMs in terms of computational cost and LBC effects.

### Very-high-resolution orographic forcing in a coarse-grid AOGCM

This method uses a very-high-resolution surface-elevation data set to define the fractional area and mean elevation of a set of elevation classes, for each model grid cell. This information is then used during the simulation to define the vertical displacement of air flowing over the orography, where the effect of the Froude number is accounted for. The heating and moistening rates for each elevation class are area-weighted and applied to the grid-cell-mean conservation equations. Leung and Ghan (1995, 1998) developed and tested this method in a regional climate model, Ghan *et al.* (2002) applied it for a global-model downscaling simulation (with the NCAR CCSM) of western USA, and Ghan *et al.* (2006) tested it for a variety of other geographic regions worldwide.

### 16.3.3 Downscaling future climates

Future climate downscaling, using both dynamical and statistical methods, has been performed for every region of the world. As noted earlier, initial-value-based interseasonal and multi-year predictions can be downscaled, as can simulations of radiatively forced climate change. Many of the examples and citations in the previous two sections pertain to future-climate downscaling, and should be referenced for additional information.

Just as global models that are used for future-climate predictions are verified against observations of past or present climate, a similar process should be undertaken when downscaling with statistical or dynamical methods. That is, observations that define the present or past regional climates of an area should be used to verify the skill of the downscaling method. As noted earlier, for the verification of statistical methods the observations used in training the algorithm should not be used for verification. An example of the verification of a statistical downscaling process, using retrospective simulations of the current climate, is shown in Fig. 16.13. The observations define high average precipitation in northern Spain

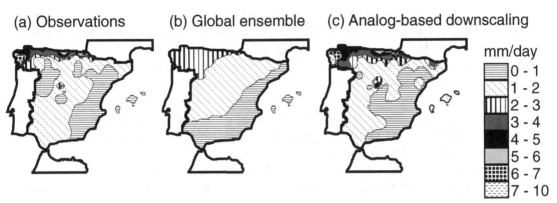

**Fig. 16.13** A comparison, for the current climate, of the regional precipitation for February through April in Spain, based on observations (a), the ensemble-mean of ECMWF and UK Met Office global-model simulations (b), and an analog-based statistical downscaling. Adapted from Palmer *et al.* (2004).

for February through April, with a maximum on the Atlantic coast (a). The analogous climate that is based on the ensemble-mean of ECMWF and UK Met Office global-model simulations (b) has no regional detail. An analog-based statistical downscaling reasonably reproduces the higher precipitation in the north, and the maximum to the west.

The following list provides examples of future-climate downscaling studies, for additional reading. Included are studies to define future regional climates, as well as evaluations of future-climate-downscaling methods using the current climate. One set of papers summarized by Iversen (2008) is related to the project Regional Climate Development under Global Warming (RegClim) that focusses on Northern Europe. See Fowler *et al.* (2007) for additional examples.

- *Asia* – Boo *et al.* (2004), Rupa Kumar *et al.* (2006), Chu *et al.* (2008), Ghosh and Mujumdar (2008), Paul *et al.* (2008), Zhu *et al.* (2008)
- *Europe* – Déqué and Piedelievre (1995), Jones *et al.* (1995, 1997), Zorita and von Storch (1999), Gibelin and Déqué (2003), Haylock *et al.* (2006), Boé *et al.* (2007), Bronstert *et al.* (2007), Ådlandsvik (2008), Beldring *et al.* (2008), Busuioc *et al.* (2008), Debernard and Røed (2008), Haugen and Iversen (2008), Hundecha and Bárdossy (2008), Huth *et al.* (2008), Tolika *et al.* (2008).
- *North America* – Wilby *et al.* (1998), Leung *et al.* (2004), Duffy *et al.* (2006), Liang *et al.* (2006), Gachon and Dibike (2007), Salathé *et al.* (2008)
- *South America* – Druyan *et al.* (2002)
- *Australia* – Timbal and Jones (2008)
- *Africa* – Lynn *et al.* (2005)

## 16.3.4 Downscaling current climates

To provide a historical context to current-climate downscaling, note that atmospheric models have been used for decades for filling space and time gaps among observations. For example, on global scales, long-term reanalyses have been generated with global-model data-assimilation systems. The resulting MADS are used for analyzing trends, studying physical processes, and identifying erroneous observational data. On smaller scales, mesoscale models have been used in short simulations for case study analyses, using LBCs from global MADS, where good model-simulation skill at observation locations has been justification for believing the simulation in the space and time gaps between the observations. In both cases, the strength of using the models to fill the observation gaps is that the fields are dynamically consistent and they are defined on a regular grid. Additionally, the models respond to local forcing that adds information beyond what can be represented by the observations. More recently, the availability of greater computing power has allowed the generation of long-term mesoscale analyses using downscaling methods, where the resulting gridded data sets are used for various applications. For example, mesoscale analyses can be used to define the statistical distribution of wind speeds for wind-energy "prospecting". Boundary-layer climatologies can be used with transport and diffusion or air-quality models to define prevailing patterns in source–receptor relationships. Also, such boundary-layer climatologies can allow an assessment of the statistical risk to populations of the release of hazardous material into the atmosphere from chemical or nuclear-power-generation facilities. Lastly, the automated interpretation of long-period reconstructions of the atmosphere can be used to define physical processes in ways that are much more robust than obtainable from the use of a few case studies.

Mesoscale reanalyses are used to fill both small and large space and time gaps. In regions with plentiful radiosonde data, the spatial voids may be only a few hundred kilometers in size. In other areas of the world, there are large expanses without data, either because there are no observations or because data have not been archived. Of course the oceans of the world represent large voids where models and satellite data must be relied upon for estimates of atmospheric properties and processes.

For future-climate downscaling, discussed in the last section, the RCM's solution is determined by the LBCs and the surface forcing. But, for current-climate downscaling, observations are available to help define the model's solution for the regional climate. The model can be restarted periodically, in an intermittent-assimilation process, or observations can be assimilated continuously through Newtonian relaxation or other methods. For example Hahmann et al. (2010) describe the use of Newtonian relaxation (Stauffer and Seaman 1994) in a WRF-based RCM used for current-climate downscaling. And Nunes and Roads (2007a,b) showed the benefits of precipitation assimilation in regional downscaling. When RCMs are restarted at regular intervals from new initial conditions, for example to assimilate observations or to maintain consistency with the large-scale analysis, the model may never fully spin up its own internal regional climate. Even though the RCM solution will contain small-scale thermally or orographically forced features, which is obviously value added to what is available from the coarse analysis, the restarts can prevent the model from developing its own soil moisture equilibrium and land–atmosphere feedbacks.

In order for a MADS to be used for a particular purpose, the veracity of the analysis needs to be verified in the context of the application. For example, for wind-energy-prospecting purposes, the model must be able to reasonably reproduce the PDF of the wind speed at the height of the generator. Because the generation equipment is vulnerable to speeds above a threshold, it would be especially important that the model capture the upper tail of the speed PDF. Similarly, for air-quality or other transport and diffusion modeling applications, the boundary-layer depth and static stability must be simulated accurately, as must the mean wind in the boundary layer.

Specific applications of a MADS also dictate how the data must be interpreted. That is, translating a reanalysis data set into useful and intuitively understandable climatological information generally goes beyond the calculation of simple, conventional climate statistics. Further processing may be necessary to classify the data into different weather regimes as a function of time of day and season. It may be necessary to define a "typical day" in July over a particular area, in terms of the wind field, or the precipitation distribution, and this product is clearly not the climatological mean. And, for many applications, the complete PDF of a variable is needed; for example, for hydrologic purposes the PDF of rainfall intensity is essential.

An example of the use of an RCM (MM5) for current-climate downscaling is shown in Fig. 16.14. In this case, the ability of the model to replicate the climate of 60-m AGL

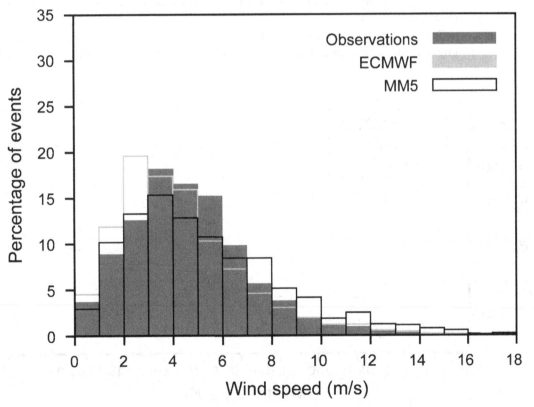

**Fig. 16.14** Frequency distribution of 60-m AGL wind speed at a location on the Eastern Mediterranean coast, based on observations (dark gray), a downscaling simulation by the MM5 RCM (black line), and the ECMWF global model (light gray line) for 10 Januaries from 1998 to 2007. Adapted from Hahmann *et al.* (2010).

winds has implications for wind-power generation. The figure depicts the frequency distribution of observed wind speed at a location on the eastern coast of the Mediterranean, for the month of January during the period 1998–2007. Also shown are the frequency distributions based on the ECMWF analysis, and the MM5 model that was used to downscale from the NNRP analysis. This clearly shows that added model resolution does not always produce improved PDFs. The RCM better defines the low-speed part of the spectrum, but the global reanalysis is better at the higher speeds.

Another example of a current-climate dynamic downscaling by an RCM (WRF model) is shown in Fig. 16.15. Here, the WRF model's LBCs are forced by the North American Regional Reanalysis (NARR, Mesinger *et al.* 2006) in a six-month simulation for a region of the Rocky Mountains in western Colorado that is the source of the water for the Colorado River. The WRF grid increment was 2 km and the NARR's is 32 km. Shown in the figure is the cumulative precipitation (liquid equivalent) for the period of the winter- and spring-season simulation, averaged over 120 snow observation locations (SNOw TELemetry - SNOTEL), based on the NARR and the WRF RCM simulation. The NARR and WRF precipitation values were bilinearly interpolated to the SNOTEL locations. In this case, the RCM clearly added value to the coarser reanalysis. The expectation is that such a downscaling with an RCM could be similarly beneficial for downscaling future climates.

Some dynamic downscalings of the current climate have been performed to provide gridded publicly available data sets for community use, in the same way that the NNRP,

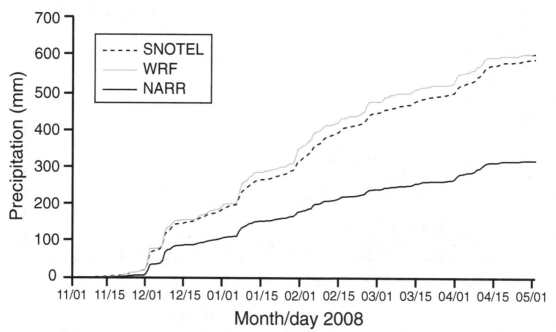

**Fig. 16.15**  Cumulative liquid-equivalent precipitation over the headwaters of the Colorado River in Western Colorado, for six winter and spring months, based on observations (120 SNOTEL sites), the NARR, and the WRF RCM (2-km grid increment). Observed and analyzed values are averages for the 120 locations. Provided by Roy Rasmussen, NCAR.

NCEP-DOE, MERRA, and ERA-40 analyses have been generated on the global scale. For example, the NARR has been created for North America and adjacent ocean areas using NCEP's Eta-model (Janjić 1994) Data-Assimilation System (EDAS). Unlike the global analysis systems, the EDAS that was used to construct the NARR assimilated high-quality precipitation observations. Thus, the land-surface state is more-accurately defined, as is the land–atmosphere interaction. This makes the analysis more valuable for hydrologic studies. The horizontal grid increment of the NARR is 32 km, there are 45 levels in the vertical, and LBCs are provided by the NCEP-DOE (R-2) global reanalysis. The reanalysis spans the 25-year period from 1979 to 2003, but its production has been continued beyond 2003 in near-real time using a Regional Climate Data Assimilation System. Another long-period current-climate downscaling for North America was performed for the period 1950–2002 using the RAMS model (Castro et al. 2007a,b). For the area of California, a 57-year regional reanalysis has been produced, with a 10-km grid increment, for applications in various climate studies (Kanamitsu and Kanamaru 2007, Kanamaru and Kanamitsu 2007b). Publicly available regional reanalyses are also being prepared for other areas, such as Europe and the Arctic.

The following list provides example references to additional efforts that used current-climate downscaling methods to study the regional climate of an area, or the prevailing physical processes.

- *Asia* – Ji and Vernekar (1997), Fox-Rabinovitz *et al.* (2002), Kang *et al.* (2005)
- *Europe* – Heimann (2001), Frei *et al.* (2003), Fil and Dubus (2005), Sotillo *et al.* (2005), Žagar *et al.* (2006), Boé *et al.* (2007)
- *North America* – Stensrud *et al.* (1995), Fox-Rabinovitz *et al.* (2001, 2002, 2005), Gochis *et al.* (2002, 2003), Hay and Clark (2003), Widmann *et al.* (2003), Liang *et al.* (2004), Duffy *et al.* (2006), Xue *et al.* (2007)
- *South America* – Fox-Rabinovitz *et al.* (2002), Roads *et al.* (2003b), Rojas and Seth (2003), Seth and Rojas (2003), Misra (2005), Sun *et al.* (2005), Rauscher *et al.* (2007)
- *Australia* – Fox-Rabinovitz *et al.* (2002), Mehrotra *et al.* (2004)
- *Africa* – Fox-Rabinovitz *et al.* (2002), Song *et al.* (2004), Paeth *et al.* (2005), Afiesimama *et al.* (2006), Druyan *et al.* (2006, 2007), Anyah and Semazzi (2007)

## 16.3.5 Examples of practical problems addressed by climate downscaling

It was noted earlier that climate impacts must typically be forecast, understood, and dealt with at the local and regional levels. This need has motivated many of the current-climate and future-climate downscaling activities. Some examples of the practical problems that have been addressed by the hundreds of downscaling studies are provided in Table 16.5. This table does not include the many other downscaling studies that have been used for (1) the evaluation and comparison of downscaling methods, (2) long-term mesoscale physical-process studies, and (3) the verification of downscaling schemes for the present climate, so that they may be used with greater confidence in downscalings of future climates. Fowler and Wilby (2007) introduce a series of papers that describe downscaling techniques employed in hydrological impact studies.

**Table 16.5** Examples of the practical problems that have been addressed by downscaling studies. Listed are (1) the application of the study; (2) whether the downscaling was dynamically (D) or statistically (S) based; (3) if the downscaling was applied to future-climate simulations (not current), was the simulation for forced climate change (GG, greenhouse gas) or based on initial conditions (IC); (4) the relevant geographic area; and (5) references.

| Application | Dynamical or statistical | GG-forced or IC global forecast, or current | Geographic area | References |
|---|---|---|---|---|
| Water resources, drought, flood | S | IC | UK | Wilby *et al.* 2004, 2006 |
| | S and D | GG | UK | Haylock *et al.* 2006, Bell *et al.* 2007 |
| | D | GG | Europe | Hanson *et al.* 2007, Blenkinsop and Fowler 2007 |
| | S and D | GG | Germany | Bronstert *et al.* 2007 |
| | S | GG | Australia | Charles *et al.* 2007, Timbal and Jones 2008 |
| | S and D | IC | Spain | Diez *et al.* 2005 |
| | S | GG | Greece | Tolika *et al.* 2008 |
| | D | current | North America | Brochu and Laprise 2007 |
| | D | GG | North America | Salathé Jr. *et al.* 2007, 2008 |
| | S | GG | East Asia | Paul *et al.* 2008 |
| Air quality | D | GG | Houston, USA | Jiang *et al.* 2008 |
| Wind energy | S and D | GG | Europe | Pryor *et al.* 2005a, b; 2006 |
| | S | current | Europe | Heimann 2001, Landberg *et al.* 2003 |
| Wave, storm surge | D | GG | North Sea | Woth *et al.* 2006, Debernard and Røed 2008 |
| General weather | D | GG | USA | Liang *et al.* 2006, Salathé Jr. *et al.* 2008, Leung *et al.* 2004, Duffy *et al.* 2006 |
| Crop development, agriculture | S and D | GG | Italy | Moriondo and Bindi 2006, Marletto *et al.* 2005 |
| Radioactive waste disposal | D | current | USA | Dickinson *et al.* 1989 |
| Forestry | D | GG | Europe | Hanson *et al.* 2007 |
| Energy use | D | GG | Europe | Hanson *et al.* 2007 |
| Tourism | D | GG | Mediterranean | Hanson *et al.* 2007 |
| Insurance | D | GG | Europe | Hanson *et al.* 2007 |
| Temperature extremes | D | current | Canada | Gachon and Dibike 2007 |
| | S | IC | Spain | Frías *et al.* 2005 |
| Hurricanes | D | GG | Atlantic | Knutson *et al.* 2008 |

# 16.4 Modeling the climate impacts of anthropogenic landscape changes

Atmospheric models are good tools for evaluating the impact of historical or future landscape changes on climate. Specifically, a simulation is performed with the landscape that prevailed before the change, a second simulation is performed based on the landscape that prevailed after the change, and the differences in the associated weather and climate are documented. Sometimes, short simulations focus on weather impacts, from which can be inferred climate impacts. Or, such pairs of simulations can be conducted for longer periods to allow more-direct calculation of climate statistics. If the experiments are performed for historical periods, the motivation is generally to develop a better understanding of land–atmosphere interactions that could have led to regional-climate change in the recent past. Alternatively, estimates of possible future changes in the landscape can be used in "what if" experiments – that is, what will be the impact on the regional climate if specific anthropogenic changes occur, such as continued urbanization, conversion of an area of grassland to farming, changing irrigation practices, or desertification. These studies are generally conducted using GCM, or coupled GCM-RCM, simulations of the present large-scale climate, but sometimes climate-forcing effects are included along with prescribed, expected anthropogenic landscape changes.

Even though there clearly are local-climate effects of landscape change, it is estimated that humans have modified one-third to one-half of the land surface area of Earth, and the aggregate effect of this could influence global-scale circulations as well. For example, Pielke (2005) states that landscape effects may be just as important in altering the weather as changes in climate patterns associated with greenhouse gases, and he further points out that anthropogenic landscape changes are not adequately accounted for in IPCC simulations.

## 16.4.1 Modeling the effects on local and regional climate of specific anthropogenic landscape changes

The following sections describe a few of the different types of anthropogenic landscape change for which the climate impacts have been modeled.

### Urbanization and suburbanization

Even though urban regions cover only about 0.2% of Earth's land surface, their impacts on local and regional climate can be substantial. Thus, urban landscapes must be accurately mapped (Jin and Shepherd 2005), and their physical properties must be included in the land-surface components of RCMs. The urban heat island is a well-known impact of urbanization on local climate, where the thermal effect can be 5–10 °C. An example of a study that looked at both the anthropogenic landscape changes associated with future urbanization as well as greenhouse forcing is Jiang *et al.* (2008). For the city of Houston

(USA), the coupled CCSM and WRF-Chem models were run for a current time slice and for a future time slice (2051–2053) for the A1B emissions scenario. The WRF RCM's land surface for the future-climate simulation was defined based on the landscape changes expected from the growth of the city. The WRF-Chem model was used because the objective was to predict the impacts on surface ozone concentrations. Other examples are Pielke *et al.* (1999b) and Marshall *et al.* (2004) who use multi-month RCM simulations to show the significant impact on regional climate of the development of the Florida Peninsula during the last century.

## Deforestation and the expansion of agriculture

The area involved in this type of landscape change is immense, and numerous modeling studies have evaluated its impacts on regional climate. For example, Strack *et al.* (2008) used an RCM and a landcover data set for the eastern USA to estimate the regional climate for the years 1650, 1850, 1920, and 1992, which span the deforestation and expansion of agriculture that has occurred since European settlement began. Since 1650, the model simulations showed an increase in maximum and minimum daily temperatures by 0.3–0.4°C, with most of the change occurring before 1920. Adegoke *et al.* (2006) found similar important climate effects of landscape change in the US High Plains. For example, cloud development occurred almost two hours earlier over agricultural land than over forested areas. Even though it might be imagined that desert climates would remain unaffected by anthropogenic impacts, Beltrán-Przekurat *et al.* (2008) demonstrated that the conversion of grassland to shrubland in the Chihuahuan Desert, through overgrazing in the last 150 years, had significant impacts on the regional climate. And, in a study of tropical landscape modification, Lawton *et al.* (2001) showed the impact of deforestation in Costa Rica on ecosystems in adjacent mountains. And for Africa, Semazzi and Song (2001) evaluated the potential effect on climate of the deforestation of the tropical rain forests.

## Agricultural irrigation

As agriculture has expanded into semi-arid lands, large areas have become irrigated so that their soil moisture far exceeds natural values. Models have shown a significant impact on regional climate. Segal *et al.* (1989) employed a LAM and observations to study the atmospheric effects of irrigation in eastern Colorado (USA), and demonstrated a significant impact. Yeh *et al.* (1984) used a simple global model to show that large-scale irrigation has an effect on regional climate, and especially precipitation. Chang and Wetzel (1991) employed a LAM to show that spatial variations in soil moisture and vegetation affect the evolution of the prestorm convective environment in the eastern Great Plains of North America. Beljaars *et al.* (1996) documented that model precipitation forecasts of extreme rainfall events in July 1993 in the Midwest USA were strongly related to soil-moisture anomalies about one day upstream. Paegle *et al.* (1996) related model-simulated rainfall for the same period to local evaporation, where the link was through effects on the low-level jet. Chen and Avissar (1994) demonstrated that landscape discontinuities (such as those associated with boundaries between irrigated and nonirrigated land) enhance shallow convective precipitation. Chase *et al.* (1999) and Chen *et al.* (2001) show how conversion of semi-arid grasslands to dry farmland and irrigated farmland in northeastern Colorado affected the local atmospheric conditions as well as the rainfall in the mountains to the west.

## 16.4.2 Modeling the effects on global climate of anthropogenic landscape change

Because of the large land-surface area that has been modified by humans, it is reasonable that the aggregate effect could be sufficiently large to influence global-scale weather patterns and climate. Feddema *et al.* (2005) describe experiments with an AOGCM that were designed to evaluate this large-scale impact. By adding the effects of global land cover changes in simulations of IPCC scenarios A2 and B1, they showed the influence of landscape-change impacts on many aspects of regional climate, as well as effects on the global circulation. For example, in simulations to 2100 for the A2 scenario, agricultural expansion over the Amazon influenced the Hadley circulation, monsoon circulations, and the location of the ITCZ, which in turn affected extratropical climates. Also, in simulations with an AGCM, Chase *et al.* (2000) showed that tropical landscape changes altered the high-latitude Northern Hemisphere winter climate, including pushing the westerly jet farther north. Pielke (2002b) claims that anthropogenic landscape change has been overlooked in the IPCC assessments.

### SUGGESTED GENERAL REFERENCES FOR FURTHER READING

#### AOGCM simulations

AchutaRao, K., C. Covey, C. Doutriaux, *et al.* (2004). *An Appraisal of Coupled Climate Model Simulations*. Lawrence Livermore National Laboratory report UCRL-TR-202550, 16 August 2004.

Giorgi, F. (2005). Climate change prediction. *Climate Change*, **73**, 239–265, doi: 10.1007/s10584-005-6857-4.

Hurrell, J., G. A. Meehl, D. Bader, *et al.* (2009). A unified modeling approach to climate system prediction. *Bull. Amer. Meteor. Soc.*, **90**, 1819–1832.

IPCC, 2007: Climate Change 2007: *The Physical Science Basis. Contribution of Working Group I to the Fourth Assessment Report of the Intergovernmental Panel on Climate Change,* S. Solomon, D. Qin, M. Manning, *et al.* (eds.). Cambridge, UK: Cambridge University Press.

McGuffie, K. and A. Henderson-Sellers (2001). Forty years of numerical climate modeling. *Int. J. Climatol.*, **21**, 1067–1109.

#### Seasonal to interannual simulations

Goddard, L., S. J. Mason, S. E. Zebiak, *et al.* (2001). Current approaches to seasonal-to-interannual climate predictions. *Int. J. Climatol.*, **21**, 1111–1152.

Shukla, J., J. Anderson, D. Baumhefner, *et al.* (2000). Dynamical seasonal prediction. *Bull. Amer. Meteor. Soc.*, **81**, 2593–2606.

## Climate downscaling

Feser, F. (2006). Enhanced detectability of added value in limited-area model results separated into different spatial scales. *Mon. Wea. Rev.*, **134**, 2180–2190.

Fowler, H. J., S. Blenkinsop, and C. Tebaldi (2007). Linking climate change modelling to impacts studies: Recent advances in downscaling techniques for hydrological modelling. *Int. J. Climatol.*, **27**, 1547–1568.

Hanssen-Bauer, I., E. J. Førland, J. E. Haugen, and O. E. Tveito (2003). Temperature and precipitation scenarios for Norway: comparison of results from dynamical and empirical downscaling. *Climate Res.*, **25**, 15–27.

Hewitson, B. C., and R. G. Crane (1996). Climate downscaling: Techniques and application. *Climate Res.*, **7**, 85–95.

Laprise, R. (2008). Regional climate modelling. *J. Comput. Phys.*, **227**, 3641–3666.

Laprise, R., R. de Elía, D. Caya, *et al.* (2008). Challenging some tenets of regional climate modeling. *Meteorol. Atmos. Phys.*, **100**, 3–22.

Wilby, R. L., and T. M. L. Wigley (1997). Downscaling general circulation model output: a review of methods and limitations. *Prog. Phys. Geog.*, **21**, 530–548.

## Modeling the impacts of anthropogenic landscape changes on climate

Feddema, J. J., K. W. Oleson, G. B. Bonan, *et al.* (2005). The importance of land-cover change in simulating future climates. *Science*, **310**, 1674–1678.

Pielke, R., Sr. (2005). Land use and climate change. *Science*, **310**, 1625–1626.

Pielke, R. A., Sr., J. Adegoke, A. Beltrán-Przekurat, *et al.* (2007). An overview of regional land-use and land-cover impacts on rainfall. *Tellus*, **59B**, 587–601.

Pielke, R. A., G. Marland, R. A. Betts, *et al.* (2002). The influence of land-use change and landscape dynamics on the climate system: relevance to climate-change policy beyond the radiative effect of greenhouse gases. *Phil. Trans. R. Soc. Lond.*, **360**, 1705–1719.

## PROBLEMS AND EXERCISES

1. Learn how to access the IPCC-model output at a website provided by your instructor, and use the data to briefly illustrate climate change for a particular geographic area.
2. In interpreting output from current- or future-climate downscalings, it is common to want to define a "typical day" rather than simply looking at mean values or PDFs of variables. Explain possible approaches for doing this.
3. Summarize the types of land–atmosphere–biosphere feedbacks that should be represented in models that are used for studies of the impacts on regional climate of anthropogenic landscape changes.
4. Explain specific examples of the component-level testing of climate models.
5. Summarize the types and time scales of the different cycles that contribute to the internal variability of the climate system.
6. Based on your knowledge of physical processes, suggest some reasonable choices for predictors, for different predictand variables.

7. Describe the types of applications of climate models for which PDFs of model variables would be especially important.

8. Internal variability in the climate system, on time scales of years to decades, is relevant to the problem of modeling radiatively forced climate change. Is there variability on shorter time scales that must be understood and accounted for when interpreting seasonal climate predictions?

9. How can Rossby phase-speed errors associated with a particular dynamical core have consequences for climate prediction?

Many NWP courses involve the coding by students of one- or two-dimensional shallow-fluid models, and the use of these models in experiments to evaluate the influence of different numerical methods on model solutions (described in Chapter 3). This allows students to become familiar with the structural components of models, to gain experience in debugging model code, and to conduct experiments to confirm concepts discussed in the text.

This appendix suggests an overall framework for coding the shallow-fluid equations that are described in Section 2.3.3, as well as some experiments that can be part of a laboratory component of an NWP course. Because the specific programming language used will determine the details of the model code, only a high-level outline will be provided here. The best approach is to start with the development of a one-dimensional model. Figure A.1 shows a schematic of the procedure for solving such a system, using an advection equation as a simple example. The abscissa is the space dimension and the ordinate is time. A predictive equation would of course be required for $u$, unless a constant mean speed is employed.

Components, or subroutines, of the model could be organized as follows.

- *Set parameters* – Define physical constants and quantities that establish the structure of the model. These would include the gravitational constant ($g$), the Coriolis parameter ($f$), the grid increment ($\Delta x$), the time step ($\Delta t$), the length of the simulation (timemax), the dimension of the computational array in terms of the number of grid points ($idim$), the output frequency, etc. Additional quantities may need to be defined here, depending on the specific form of the equations being used.
- *Initialization* – Define the initial value of the model dependent variables; $u(1 \rightarrow idim)$, $v(1 \rightarrow idim)$, and $h(1 \rightarrow idim)$.
- *Tendency calculation* – For example,

$$UTEND_i^\tau = u_i^\tau \frac{u_{i+1}^\tau - u_{i-1}^\tau}{2\Delta x} + fv_i^\tau - g\frac{h_{i+1}^\tau - h_{i-1}^\tau}{2\Delta x}, \text{ for grid points } i = 2 \rightarrow idim - 1.$$

- *Extrapolation* – For example,

$$u_i^{\tau+1} = u_i^{\tau-1} + 2\Delta t\, UTEND_i^\tau, \text{ for grid points } i = 2 \rightarrow idim - 1. \tag{A.1}$$

- *Define lateral-boundary conditions (LBCs)* – constant, periodic, etc., for $i$ equal 1 and $idim$.

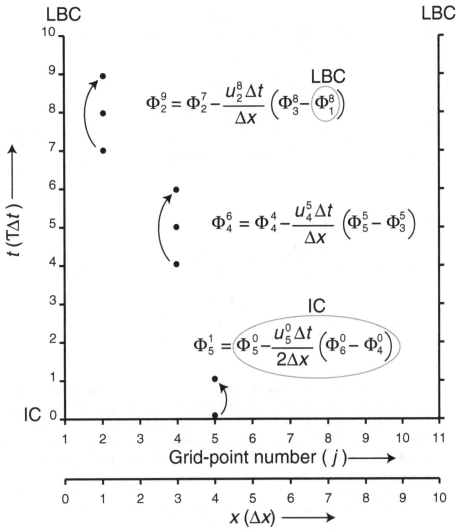

**Fig. A.1** Schematic showing the method of solving a one-dimensional shallow-fluid model that is based on an advection equation for a variable $\Phi$ . The subscript is the grid-point number (abscissa) and the superscript is the time-step number (ordinate).

- *Output plotting* – graphical display of dependent variables.
- *Digital save* – archive for analysis, restart. The restart file contains all the information needed to seamlessly start the model in the middle of a simulation in the event of a hardware- or software-related failure. Without this file, the simulation would need to be restarted from the beginning, wasting computing resources. For simple experiments, this capability is typically unnecessary.

Figure A.2 shows a standard sequence in which these operations take place in a model integration.

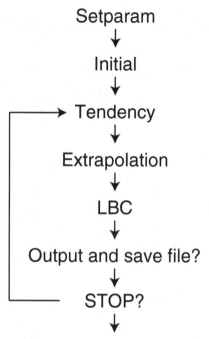

A standard sequence (flow chart) for executing components of a simple model.

The extrapolation step can be problematic to program without the use of temporary arrays. That is, in Eq. A.1 the grid-point number ($i$) can be used as the index of the variable array, but the time-step number ($\tau$) should not also be employed as an array index. This is because the value of a variable does not need to be saved for every time step, and to do so would require the use of a prohibitive amount of storage. Thus temporary arrays can be used in the following way. For each time step, solve

$$ua_i = ub_i + 2\Delta t UTEND_i^\tau, \text{ for grid points } i = 2 \rightarrow idim - 1,$$

where $ua$ represents an "advance" value of $u$, and $ub$ is a "back" value. After $ua_i$ is calculated for each interior grid point, the following exchange is made:

$$u_i \rightarrow ub_i \text{ and}$$

$$ua_i \rightarrow u_i.$$

This process is repeated every time step.

Some configuration suggestions follow, for initial experiments.

- The total number of grid points ($idim$) may be 100.
- Let $\Delta x = 10$ km.
- The LBCs should be periodic (cyclic). See Fig. 3.49 for an illustration of how information is exchanged. The boundary values on each edge of the grid are defined based on

the penultimate points on the opposite end. What leaves the grid on one edge comes back in on the other, where essentially the computational domain is wrapped around on itself.
- Choose a midlatitude value for the Coriolis parameter.
- Define the time step to be 80% of the value calculated using the CFL criterion.

Lastly, the following list is of experiments that can be performed with the shallow-fluid model. Note that the model-check-out process extends throughout virtually all of the experiments.

- *Linear advection tests* – The initial experiment consists of the advection of a perturbation in the height field. Use three-point time and space differencing, with no explicit diffusion. Other terms in the equations should not be included, and a constant $u$ should be employed. That is, there is no predictive equation for $u$. Different perturbation shapes can be employed, such as a Gaussian function, a square wave, and a triangular wave. The shapes with first-order discontinuities have more short-wavelength energy, and thus numerical dispersion should be more evident.
- *CFL violation test* – Choose a time step that violates the linear stability criterion, perhaps using a Courant number of 1.1, and print the model solution each time step to observe the instability.
- *Effect of Courant number on the solution* – Using the equation for the linear-advection of $h$, vary the time step over a range of stable values to evaluate how the use of different Courant numbers affects the model solution. Test Courant numbers ranging from 0.1 to 0.99.
- *Diffusion-term tests* – Add an explicit second-order diffusion term to the equation for the linear advection of $h$, and show its effect on the model solution for different stable diffusion coefficients. Repeat with higher-order forms of the diffusion term.
- *Gravity-wave tests* – Use the complete forms of all three predictive equations, and simulate a gravity wave. Define the initial wind components to be zero, and the initial height field as a smooth perturbation (maximum) in the middle of the grid, superimposed on the mean value. The model solution will show the mass being transported in both directions by gravity waves. Repeat the experiment with different mean depths for the fluid.
- *Horizontal-resolution tests* – For the same initial conditions (perturbation wavelength and amplitude), evaluate the effect of horizontal resolution on the model solution. Begin with a grid increment that resolves the wave very well, and progressively increase the grid increment in subsequent experiments.
- *Geostrophic adjustment experiments* – Establish in the initial conditions a geostrophic balance between a perturbation in $h$ (e.g., Gaussian) and the $v$ component of the wind. Run the model to determine if the balance is correct. Then run the model for 48 h with the $h$ perturbation but with an initial $v$ of zero, and observe the adjustment. Perform the analogous experiment with the initial $v$ that is in geostrophic balance with the original $h$ perturbation, but do not include the $h$ perturbation. Again run the simulation for 48 h and observe the adjustment. Perform the experiments for both synoptic-scale and mesoscale perturbations, and observe the differences in the adjustment process.
- *Advanced experiment* – Program a spectral version of the shallow-fluid equations, and compare the solution with the equivalent grid-point model.

Table A.1 summarizes these experiments.

**Table A.1** Suggested experiments with the one-dimensional shallow-fluid model

| Experiment | Initial conditions | Equation set | Comments |
|---|---|---|---|
| **1. Linear advection tests** | | | |
| 1a Wave shape | Gaussian wave | Linear advection of $h$, second-order time and space differencing | Courant number = 0.8 |
| 1b Wave shape | Square wave | same | same |
| 1c Wave shape | Triangular wave | same | same |
| 1d CFL violation test | Gaussian wave | same | Courant number = 1.1 |
| 1e Courant number effect | Gaussian wave | same | Courant number from 0.1 to 0.99 |
| 1f Horizontal resolution tests | Gaussian wave | same | Vary $\Delta x$ so that $L = 4\,\Delta x$ to $L = 20\,\Delta x$ |
| 1g Higher-order differencing | Gaussian wave | Linear advection, higher-order space differencing, and multi-step methods | |
| **2. Diffusion tests** | | | |
| 2a Stable diffusion | Gaussian wave | same + 2nd-order diffusion | Use $K$ that is stable and damps the same fraction of the $2\Delta x$ wave |
| 2b Stable diffusion | Gaussian wave | same + 4th-order diffusion | |
| 2c Stable diffusion | Gaussian wave | same + 6th-order diffusion | |
| 2d Unstable diffusion | Gaussian wave | same + 6th-order diffusion | Use slightly unstable $\Delta t$ |
| **3. Gravity wave tests** | | | |
| 3a Standard depth | Gaussian $h$, $u = v = 0$, $H = 8$ km | Complete, with diffusion | |
| 3b Reduced depth | same, $H = 2$ km | same | |
| **4. Geostrophic-adjustment experiments** | | | |
| Balanced ICs | Gaussian $h$, $u = 0$, $v = v_g$ | Complete, with diffusion | |
| Unbalanced ICs | Gaussian $h$, $u = v = 0$ | Complete, with diffusion | Perform for both synoptic-scale and mesoscale perturbations |
| Unbalanced ICs | $h = u = 0$, $v = v_g$ for Gaussian $h$ | Complete, with diffusion | same |

# References

Abdalla, S., and L. Cavaleri (2002). Effect of wind variability and variable air density on wave modeling. *J. Geophys. Res.*, **107**(C7), 3080, doi:10.1029/2000JC000639.

Aberson, S. D. (2003). Targeted observations to improve operational tropical cyclone track forecast guidance. *Mon. Wea. Rev.*, **131**, 1613–1628.

Abramopoulos, F., C. Rosenzweig, and B. Choudhury (1988). Improved ground hydrology calculations for global climate models (GCMs): Soil water movement and evapotranspiration. *J. Climate*, **1**, 921–941.

AchutaRao, K., and K. R. Sperber (2006). ENSO simulation in coupled ocean-atmosphere models: Are the current models better? *Climate Dyn.*, **27**, 1–15.

AchutaRao, K., C. Covey, C. Doutriaux, *et al.* (2004). *An Appraisal of Coupled Climate Model Simulations.* Lawrence Livermore National Laboratory report UCRL-TR-202550, 16 August 2004.

Adcroft, A., C. Hill, and J. Marshall (1997). Representation of topography by shaved cells in a height coordinate ocean model. *Mon. Wea. Rev.*, **125,** 2293–2315.

Adcroft, A., J.-M. Campin, C. Hill, and J. Marshall (2004). Implementation of an atmosphere-ocean general circulation model on the expanded spherical cube. *Mon. Wea. Rev.*, **132**, 2845–2863.

Adegoke, J. O., R. A. Pielke Sr, and A. M. Carleton (2006). Observational and modeling studies of the impacts of agriculture-related land use change on planetary boundary layer processes in the central U.S. *Agric. Forest Meteor.*, **142**, 201–215.

Ådlandsvik, B. (2008). Marine downscaling of a future climate scenario for the North Sea. *Tellus*, **60A**, 451–458.

Afiesimama, E. A., J. S. Pal, B. J. Abiodun, W. J. Gutowski Jr, and A. Adedoyin (2006). Simulation of West African monsoon using the RegCM3. Part I: Model validation and interannual variability. *Theor. Appl. Climatol.*, **86**, 23–37.

Albrecht, B. A. (1989). Aerosols, cloud microphysics, and fractional cloudiness. *Science*, **245**, 1227–1230.

Albrecht, B. A., A. K. Betts, W. H. Schubert, and S. K. Cox (1979). A model for the thermodynamic structure of the trade-wind boundary layer: I. Theoretical formulation and sensitivity tests. *J. Atmos. Sci.*, **36**, 73–89.

Alexander, L. V., and J. M. Arblaster (2009). Assessing trends in observed and modelled climate extremes over Australia in relation to future projections. *Int. J. Climatol.*, **29**, 417–435.

Alexandru, A., R. de Elia, R. Laprise, L. Separovic, and S. Biner (2009). Sensitivity study of regional climate model simulations to large-scale nudging parameters. *Mon. Wea. Rev.*, **137**, 1666–1686.

Alpert, P., S. O. Krichak, T. N. Krishnamurti, U. Stein, and M. Tsidulko (1996). The relative roles of lateral boundaries, initial conditions, and topography in mesoscale simulations of lee cyclogenesis. *J. Appl. Meteor.*, **35**, 1091–1099.

Alpert, P., M. Tsidulko, and D. Itzigsohn (1999). A shallow, short-lived meso-β cyclone over the Gulf of Antalya, eastern Mediterranean. *Tellus*, **51A**, 249–262.

Anderson, D., T. Stockdale, M. Balmaseda, *et al.* (2003). *Comparison of the ECMWF Seasonal Forecast Systems 1 and 2, Including the Relative Performance for the 1997/8 El Niño.* Technical Memo. 404, European Centre for Medium-Range Weather Forecasts, Reading, UK.

Anderson, J. L. (1996). A method for producing and evaluating probabilistic forecasts from ensemble model integrations. *J. Climate*, **9**, 1518–1530.

Anderson, J. L. (2003). A local least squares framework for ensemble filtering. *Mon. Wea. Rev.*, **131**, 634–642.

Anthes, R. A. (1970). Numerical experiments with a two-dimensional horizontal variable grid. *Mon. Wea. Rev.*, **98**, 810–822.

Anthes, R. A. (1974). Data assimilation and initialization of hurricane prediction models. *J. Atmos. Sci.*, **31**, 702–719.

Anthes, R. A. (1977). A cumulus parameterization scheme utilizing a one-dimensional cloud model. *Mon. Wea. Rev.*, **105**, 270–286.

Anthes, R. A., and J. W. Trout (1971). Three dimensional particle trajectories in a model hurricane. *Weatherwise*, **24**, 174–178.

Anthes, R. A., and T. T. Warner (1978). Development of hydrodynamic models suitable for air pollution and other mesometeorological studies. *Mon. Wea. Rev.*, **106**, 1045–1078.

Anthes, R. A., Y.-H. Kuo, and J. R. Gyakum (1983). Numerical simulations of a case of explosive marine cyclogenesis. *Mon. Wea. Rev.*, **111**, 1174–1188.

Anthes, R. A., Y. H. Kuo, D. P. Baumhefner, R. M. Errico, and T. W. Bettge (1985). Predictability of mesoscale motions. *Adv. in Geophys.*, Vol. 28, Academic Press, 159–202.

Anthes, R. A., E.-Y. Hsie, and Y.-H. Kuo (1987). *Description of the Penn State/NCAR Mesoscale Model Version 4 (MM4).* NCAR Tech. Note NCAR/TN-282+STR.

Anthes, R. A., P. A. Bernhardt, Y. Chen, *et al.* (2008). The COSMIC/FORMOSAT-3 mission: Early results. *Bull. Amer. Meteor. Soc.*, **89**, 313–333.

Antic, S., R. Laprise, B. Denis, and R. de Elía (2006). Testing the downscaling ability of a one-way nested regional climate model in regions of complex topography. *Climate Dyn.*, **26**, 305–325.

Anyah, R. O., and F. H. M. Semazzi (2007). Variability of East African rainfall based on multiyear Regcm3 simulations. *Int. J. Climatol.*, **27**, 357–371.

Arakawa, A. (1993). Closure assumptions in the cumulus parameterization problem. In *The Representation of Cumulus Convection in Numerical Models of the Atmosphere,* eds. K. A. Emanuel and D. J. Raymond. Meteorological Monograph, No. 46, Boston, USA: American Meteorological Society, pp. 1–15.

Arakawa, A., and V. R. Lamb (1977). Computational design of the basic dynamical processes of the UCLA general circulation model. *Methods Comput. Phys.*, **17**, 173–265.

Arakawa, A., and W. H. Schubert (1974). Interaction of a cumulus cloud ensemble with the large-scale environment. Part I. *J. Atmos. Sci.*, **31**, 674–701.

Asselin, R. (1972). Integration of a semi-implicit model with time dependent boundary conditions. *Atmosphere*, **10**, 44–55.

Atlas, R., R. N. Hoffman, S. M. Leidner, *et al.* (2001). The effects of marine winds from scatterometer data on weather analysis and forecasting. *Bull. Amer. Meteor. Soc.*, **82**, 1965–1990.

Bacon, D. P., N. N. Ahmad, Z. Boybeyi, *et al.* (2000). A dynamically adaptive weather and dispersion model: The Operational Multiscale Environmental Model with Grid Adaptivity (OMEGA). *Mon. Wea. Rev.*, **128**, 2044–2076.

Baer, F., and J. J. Tribbia (1977). On complete filtering of gravity modes through nonlinear initialization. *Mon. Wea. Rev.*, **105**, 1536–1539.

Bagnold, R. A. (1954). *The Physics of Blown Sand and Desert Dunes*. London, UK: Methuen.

Ballish, B., X. Cao, E. Kalnay, and M. Kanamitsu (1992). Incremental nonlinear normal-mode initialization. *Mon. Wea. Rev.*, **120**, 1723–1734.

Bao, J.-W., and R. M. Errico (1997). An adjoint examination of a nudging method for data assimilation. *Mon. Wea. Rev.*, **125**, 1355–1373.

Bao, J.-W., J. M. Wilczak, J.-K. Choi, and L. A. Kantha (2000). Numerical simulation of air–sea interaction under high wind conditions using a coupled model: A study of hurricane development. *Mon. Wea. Rev.*, **128**, 2190–2210.

Barnes, S. (1964). A technique for maximizing details in numerical map analysis. *J. Appl. Meteor.*, **3**, 396–409.

Barnes, S. (1978). Oklahoma thunderstorms on 29–30 April 1970, Part I: Morphology of a tornadic storm. *Mon. Wea. Rev.*, **108**, 673–684.

Barnes, S. (1994a). Applications of the Barnes objective analysis scheme. Part I: Effects of undersampling, wave position, and station randomness. *J. Atmos. Oceanic Technol.*, **11**, 1433–1448.

Barnes, S. (1994b). Applications of the Barnes objective analysis scheme. Part II: Improving derivative estimates. *J. Atmos. Oceanic Technol.*, **11**, 1449–1458.

Barnett, T. P., and R. Preisendorfer (1987). Origins and levels of monthly and seasonal forecast skill for United States surface air temperatures determined by canonical correlation analysis. *Mon. Wea. Rev.*, **115**, 1825–1850.

Barnston, A. G., and R. E. Livezey (1987). Classification, seasonality and persistence of low-frequency atmospheric circulation patterns. *Mon. Wea. Rev.*, **115**, 1083–1126.

Barnston, A. G., S. J. Mason, L. Goddard, D. G. Dewitt, and S. E. Zebiak (2003). Multi-model ensembling in seasonal climate forecasting at IRI. *Bull. Amer. Meteor. Soc.*, **84**, 1783–1796.

Barnum, B. H., N. S. Winstead, J. Wesely, *et al.* (2004). Forecasting dust storms using the CARMA-dust model and MM5 weather data. *Environ. Model. Software*, **19**, 129–140.

Barstad, I., A. Sorteberg, F. Flatøy, and M. Déqué (2009). Precipitation, temperature and wind in Norway: dynamical downscaling of ERA40. *Climate Dyn.*, **33**, 769–776, doi: 10.1007/s00382-008-0476-5.

Bartello, P., and H. L. Mitchell (1992). A continuous three-dimensional model of short-range error covariances. *Tellus*, **44A**, 217–235.

Baumgardner, J. R., and P. O. Frederickson (1985). Icosahedral discretization of the two-sphere. *SIAM J. Numer. Anal.*, **22**, 1107–1115.

Baumhefner, D. P., and D. J. Perkey (1982). Evaluation of lateral boundary errors in a limited-domain model. *Tellus*, **34**, 409–428.

Beare, R. J., M. K. MacVean, A. A. M. Holtslag, *et al.* (2006). An intercomparison of large-eddy simulations of the stable boundary layer. *Bound.-Layer Meteor.*, **118**, 247–272.

Beck, C., J. Grieser, and B. Rudolf (2005). A new monthly precipitation climatology for the global land areas for the period 1951 to 2000. *Klimastatusbericht*, Deutscher Wetterdienst (DWD), 181–190.

Belair, S., D.-L. Zhang, and J. Mailhot (1994). Numerical prediction of the 10–11 June 1985 squall line with the Canadian Regional Finite-Element model. *Wea. Forecasting*, **9**, 157–172.

Béland, M., and C. Beaudoin (1985). A global spectral model with a finite-element formulation for the vertical discretization: Adiabatic formulation. *Mon. Wea. Rev.*, **113**, 1910–1919.

Béland, M., J. Côté, and A. Staniforth (1983). The accuracy of a finite-element vertical discretization scheme for primitive equation models: Comparison with a finite-difference scheme. *Mon. Wea. Rev.*, **111**, 2298–2318.

Beldring, S., T. Engen-Skaugen, E. J. Førland, and L. A. Roald (2008). Climate change impacts on hydrologic processes in Norway based on two methods for transferring regional climate model results to meteorological station sites. *Tellus*, **60A**, 439–450.

Beljaars, A. C. M., P. Viterbo, and M. J. Miller (1996). The anomalous rainfall over the United States during July 1993: Sensitivity to land surface parameterization and soil moisture anomalies. *Mon. Wea. Rev.*, **124**, 362–383.

Bell, V. A., A. L. Kay, R. G. Jones, and R. J. Moore (2007). Use of a grid-based hydrological model and regional climate model outputs to assess changing flood risk. *Int. J. Climatol.*, **27**, 1657–1671.

Beltrán-Przekurat, A., R. A. Pielke Sr, D. P. C. Peters, K. A. Snyder, and A. Rango (2008). Modeling the effects of historical vegetation change on near-surface atmosphere in the northern Chihuahuan Desert. *J. Arid Environ.*, **72**, 1897–1910.

Benestad, R. E. (2001). A comparison between two empirical downscaling strategies. *Int. J. Climatol.*, **21**, 1645–1668.

Benjamin, S. G. (1983). Some effects of surface heating and topography on the regional severe-storm environment. Ph.D. Thesis, Department of Meteorology, The Pennsylvania State University, PA, USA.

Benjamin, S. G. (1989). An isentropic meso-alpha scale analysis system and its sensitivity to aircraft and surface observations. *Mon. Wea. Rev.*, **117**, 1586–1603.

Benjamin, S. G., and N. L. Seaman (1985). A simple scheme for objective analyses in curved flow. *Mon. Wea. Rev.*, **113**, 1184–1198.

Benjamin, S. G., D. Dévényi, S. S. Weygandt, *et al.* (2004a). An hourly assimilation–forecast cycle: The RUC. *Mon. Wea. Rev.*, **132**, 495–518.

Benjamin, S. G., G. A. Grell, J. M. Brown, and T. G. Smirnova (2004b). Mesoscale weather prediction with the RUC hybrid isentropic-terrain-following coordinate model. *Mon. Wea. Rev.*, **132**, 473–494.

Benoit, R., J. Côté, and J. Mailhot (1989). Inclusion of a TKE boundary layer parameterization in the Canadian regional finite element model. *Mon. Wea. Rev.*, **117**, 1726–1750.

Berger, M., and J. Oliger (1984). Adaptive mesh refinement for hyperbolic partial differential equations. *J. Comput. Phys.*, **53**, 484–512.

Bergot, T. (1999). Adaptive observations during FASTEX: A systematic survey of upstream flights. *Quart. J. Roy. Meteor. Soc.*, **125**, 3271–3298.

Bergot, T. (2001). Influence of the assimilation scheme on the efficiency of adaptive observations. *Quart. J. Roy. Meteor. Soc.*, **127**, 635–660.

Bergot, T., and A. Doerenbecher (2002). A study on the optimization of the deployment of targeted observations using adjoint-based methods. *Quart. J. Roy. Meteor. Soc.*, **128**, 1689–1712.

Bergthorsson, P., and B. Doos (1955). Numerical weather map analysis. *Tellus*, **7**, 329–340.

Berliner, L. M., Z.-Q. Lu, and C. Snyder (1999). Statistical design for adaptive weather observations. *J. Atmos. Sci.*, **56**, 2536–2552.

Berner, J., G. J. Shutts, M. Leutbecher, and T. N. Palmer (2009). A spectral stochastic kinetic energy backscatter scheme and its impact on flow-dependent predictability in the ECMWF ensemble prediction system. *J. Atmos. Sci.*, **66**, 603–626.

Bernstein, A. B. (1966). Examination of certain terms appearing in Reynolds' equations under unsteady conditions and their implications for micrometeorology. *Quart. J. Roy. Meteor. Soc.*, **92**, 533–544.

Bernstein, B. C., F. McDonough, M. K. Politovich, *et al.* (2005). Current icing potential: Algorithm description and comparison with aircraft observations. *J. Appl. Meteor.*, **44**, 969–986.

Betts, A. K. (1986). A new convective adjustment scheme. Part I: Observational and theoretical basis. *Quart. J. Roy. Meteor. Soc.*, **112**, 677–692.

Betts, A. K., and M. J. Miller (1986). A new convective adjustment scheme. Part II: Single column tests using GATE wave, BOMEX, ATEX, and Arctic air-mass data sets. *Quart. J. Roy. Meteor. Soc.*, **112**, 693–709.

Betts, A. K., and M. J. Miller (1993). The Betts-Miller scheme. In *The Representation of Cumulus Convection in Numerical Models of the Atmosphere*, eds. K. A. Emanuel and D. J. Raymond. Meteorological Monograph, No. 46, Boston, USA: American Meteorological Society, pp. 107–121.

Bidlot, J. R., D. J. Holmes, P. A. Wittmann, R. Lalbeharry, and H. S. Chen (2002). Intercomparison of the performance of operational ocean wave forecasting systems with buoy data. *Wea. Forecasting*, **17**, 287–310.

Bishop, C. H., and Z. Toth (1999). Ensemble transformation and adaptive observations. *J. Atmos. Sci.*, **56**, 1748–1765.

Bishop, C. H., B. J. Etherton, and S. J. Majumdar (2001). Adaptive sampling with the ensemble transform Kalman filter. Part I: Theoretical aspects. *Mon. Wea. Rev.*, **129**, 420–436.

Black, T. L. (1994). The new NMC mesoscale Eta model: Description and forecast examples. *Wea. Forecasting*, **9**, 265–278.

Black, T. L., D. G. Deaven, and G. J. DiMego (1993). *The Step-mountain Eta Coordinate Model: 80 km "Early" Version and Objective Verifications*. Technical Procedures

Bulletin 412. NOAA/NWS. [Available from National Weather Service, Office of Meteorology, 1325 East–West Highway, Silver Spring, MD 20910.]

Blackadar, A. K. (1978). Modeling pollutant transfer during daytime convection. *Preprints, Fourth Symposium on Atmospheric Turbulence, Diffusion, and Air Quality*, Reno, American Meteorological Society, pp. 443–447.

Bleck, R. (2002). An oceanic general circulation model framed in hybrid isopycnic-Cartesian coordinates. *Ocean Modelling*, **4**, 55–88.

Bleck, R., and M. A. Shapiro (1976). Simulation and numerical weather prediction framed in isentropic coordinates. In *Weather Forecasting and Weather Forecasts: Models, Systems, and Users*. Vol. 1, NCAR Colloquium, National Center for Atmospheric Research, 154–168.

Bleck, R., C. Rooth, D. Hu, and L. T. Smith (1992). Salinity-driven thermocline transients in a wind- and thermohaline-forced isopycnic coordinate model of the North Atlantic. *J. Phys. Oceanogr.*, **22**, 1486–1505.

Blenkinsop, S., and H. J. Fowler (2007). Changes in European drought characteristics projected by the PRUDENCE regional climate models. *Int. J. Climatol.*, **27**, 1595–1610.

Bloom, S. C., L. L. Takacs, A. M. da Silva, and D. Ledvina (1996). Data assimilation using incremental analysis updates. *Mon. Wea. Rev.*, **124**, 1256–1271.

Bluestein, H. (1992a). *Synoptic-dynamic Meteorology in Midlatitudes. Vol. 1: Principles of Dynamics and Kinematics*. New York, USA: Oxford University Press.

Bluestein, H. (1992b). *Synoptic-dynamic Meteorology in Midlatitudes. Vol. 2: Observations and Theory of Weather Systems*. New York, USA: Oxford University Press.

Boé, J, L. Terra, F. Habets, and E. Martin (2007). Statistical and dynamical downscaling of the Seine basin climate for hydro-meteorological studies. *Int. J. Climatol.*, **27**, 1643–1655.

Boer, G. J., and S. J. Lambert (2001). Second order space-time climate difference statistics. *Climate Dyn.*, **17**, 213–218.

Bonan, G. B. (1998). The land surface climatology of the NCAR land surface model (LSM 1.0) coupled to the NCAR Community Climate Model (CCM3). *J. Climate*, **11**, 1307–1326.

Bonan, G. B., K. W. Oleson, M. Vertenstein, and S. Levis (2002). The land surface climatology of the Community Land Model coupled to the NCAR Community Climate Model. *J. Climate*, **15**, 3123–3149.

Bonavita, M., L. Torrisi, and F. Marcucci (2008). The ensemble Kalman filter in an operational regional NWP system: Preliminary results with real observations. *Quart. J. Roy. Meteor. Soc.*, **134**, 1733–1744.

Boo, K.-O., W.-T. Kwon, J.-H. Oh, and H.-J. Baek (2004). Response of global warming on regional climate change over Korea: An experiment with the MM5 model. *Geophys. Res. Lett.*, **31**, L21206, doi:10.1029/2004GL021171.

Bosart, L. F. (1975). SUNYA experimental results in forecasting daily temperature and precipitation. *Mon. Wea. Rev.*, **103**, 1013–1020.

Bosilovich, M. G., S. D. Schubert, M. Rienecker, *et al.* (2006). NASA's Modern Era Retrospective-analysis for Research and Applications. *U.S. CLIVAR Variations*, **4**, 5–8.

Bourassa, M. A., D. M. Legler, J. J. O'Brien, and S. R. Smith (2003). SeaWinds validation with research vessels. *J. Geophys. Res.*, **108**, 3019, doi:10.1029/2001JC001028.

Bourke, W. (1974). A multi-level spectral model. I. Formulation and hemispheric integrations. *Mon. Wea. Rev.*, **102**, 687–701.

Boussinesq, J. (1903). *Théorie Analytique de la Chaleur, Vol. II*. Paris, France: Gauthier-Villars.

Bouttier, F. (1994). A dynamical estimation of forecast error covariances in an assimilation system. *Mon. Wea. Rev.*, **122**, 2376–2390.

Boville, B. A. (2000). Toward a complete model of the climate system. In *Numerical Modeling of the Global Atmosphere in the Climate System,* eds. P. Mote and A. O'Neill. Dordrecht, the Netherlands: Kluwer Academic Publishers, pp. 419–442.

Boyd, J. P. (2005). Limited-area Fourier spectral models and data analysis schemes: Windows, Fourier extension, Davies relaxation, and all that. *Mon. Wea. Rev.*, **133**, 2030–2042.

Braham, R. R., Jr, and P. Squires (1974). Cloud physics: 1974. *Bull. Amer. Meteor. Soc.*, **55**, 543–586.

Branković, Č., and D. Gregory (2001). Impact of horizontal resolution on seasonal integrations. *Climate Dyn.*, **18**, 123–143.

Branstetter, M. L. (2001). Development of a parallel river transport algorithm and application to climate studies. Ph.D. Dissertation, University of Texas, Austin, USA.

Bratseth, A. M. (1986). Statistical interpolation by means of successive corrections. *Tellus*, **38A**, 439–447.

Bremnes, J. B. (2004). Probabilistic forecasts of precipitation in terms of quantiles using NWP model output. *Mon. Wea. Rev.*, **132**, 338–347.

Bretherton, C. S., C. Smith, and J. M. Wallace (1992). An intercomparison of methods for finding coupled patterns in climate data. *J. Climate*, **5**, 541–560.

Bretherton, C. S., J. R. McCaa, and H. Grenier (2004). A new parameterization for shallow cumulus convection and its application to marine subtropical cloud-topped boundary layers. Part I: Description and 1D results. *Mon. Wea. Rev.*, **132**, 864–882.

Briegleb, B. P., C. M. Bitz, E. C. Hunke, *et al.* (2004). *Scientific Description of the Sea Ice Component in the Community Climate System Model, Version Three*. Technical Note TN-463STR, NTIS #PB2004-106574, National Center for Atmospheric Research, Boulder, CO.

Brochu, R., and R. Laprise (2007). Surface water and energy budgets over the Mississippi and Columbia River Basins as simulated by two generations of the Canadian regional climate model. *Atmos.-Ocean*, **45**, 19–35.

Brock, F. V., K. C. Crawford, R. L. Elliott, *et al.* (1995). The Oklahoma Mesonet: a technical overview. *J. Atmos. Oceanic Technol.*, **12**, 5–19.

Bronstert, A., V. Kolokotronis, D. Schwandt, and H. Straub (2007). Comparison and evaluation of regional climate scenarios for hydrological impact analysis: General scheme and application example. *Int. J. Climatol.*, **27**, 1579–1594.

Brown, J. M. (1979). Mesoscale unsaturated downdrafts driven by rainfall evaporation: A numerical study. *J. Atmos. Sci.*, **36**, 313–338.

Browning, G., H.-O. Kreiss, and J. Oliger (1973). Mesh refinement. *Math. Comp.*, **27**, 29–39.

Browning, G. L., J. J. Hack, and P. N. Swarztrauber (1989). A comparison of three numerical methods for solving differential equations on the sphere. *Mon. Wea. Rev.*, **117**, 1058–1075.

Brunet, N., R. Verret, and N. Yacowar (1988). An objective comparison of model output statistics and perfect prog systems in producing numerical weather element forecasts. *Wea. Forecasting*, **3**, 273–283.

Bryan, G. H., J. C. Wyngaard, and J. M. Fritsch (2003). Resolution requirements for the simulation of deep moist convection. *Mon. Wea. Rev.*, **131**, 2394–2416.

Buishand, T. A, M. V. Shabalova, and T. Brandsma (2004). On the choice of the temporal aggregation level for statistical downscaling of precipitation. *J. Climate*, **17**, 1816–1827.

Buizza, R. (1997). Potential forecast skill of ensemble prediction and spread and skill distributions of the ECMWF Ensemble Prediction System. *Mon. Wea. Rev.*, **125**, 99–119.

Buizza, R. (2001). Chaos and weather prediction – A review of recent advances in numerical weather prediction: Ensemble forecasting and adaptive observation targeting. *Il Nuovo Cimento*, **24C**, 273–301.

Buizza, R. (2008). Comparison of a 51-member low-resolution ($T_L 399L62$) ensemble with a 6-member high-resolution ($T_L 799L91$) lagged-forecast ensemble. *Mon. Wea. Rev.*, **136**, 3343–3362.

Buizza, R., and A. Montani (1999). Targeting observations using singular vectors. *J. Atmos. Sci.*, **56**, 2965–2985.

Buizza, R., M. Miller, and T. N. Palmer (1999). Stochastic representation of model uncertainties in the ECMWF Ensemble Prediction System. *Quart. J. Roy. Meteor. Soc.*, **125**, 2887–2908.

Burgers, G., P. J. van Leeuwen, and G. Evensen (1998). Analysis scheme in the ensemble Kalman filter. *Mon. Wea. Rev.*, **126**, 1719–1724.

Burridge, D. M., J. Steppeler, and J. Strufing (1986). *Finite Element Schemes for the Vertical Discretization of the ECMWF Forecast Model Using Linear Elements*. ECMWF Technical Report 54.

Busuioc, A., and H. von Storch (2003). Conditional stochastic model for generating daily precipitation time series. *Climate Res.*, **24**, 181–195.

Busuioc, A., D. Chen, and C. Hellström (2001). Performance of statistical downscaling models in GCM validation and regional climate change estimates: Application for Swedish precipitation. *Int. J. Climatol.*, **21**, 557–578.

Busuioc, A., F. Giorgi, X. Bi, and M. Ionita (2006). Comparison of regional climate model and statistical downscaling simulations of different winter precipitation change scenarios over Romania. *Theor. Appl. Climatol.*, **86**, 101–123.

Busuioc, A., R. Tomozeiu, and C. Cacciamani (2008). Statistical downscaling model based on canonical correlation analysis for winter extreme precipitation events in the Emilia-Romagna region. *Int. J. Climatol.*, **28**, 449–464.

Byrne, M. A., A. G. Laing, and C. Connor (2007). Predicting tephra dispersion with a mesoscale atmospheric model and a particle fall model: Application to Cerro Negro volcano. *J. Appl. Meteor. Climatol.*, **46**, 121–135.

Byun, D., and K. L. Schere (2006). Review of the governing equations, computational algorithms, and other components of the Models-3 Community Multiscale Air Quality modeling system. *Appl. Mech. Rev.*, **59**, 51–77.

Cantelaube, P., and J.-M. Terres (2005). Seasonal weather forecasts for crop yield modeling in Europe. *Tellus*, **57A**, 476–487.

Carlson, T. N., and F. E. Boland (1978). Analysis of urban-rural canopy using a surface heat flux/temperature model. *J. Appl. Meteor.*, **17**, 998–1013.

Carmichael, G. R., A. Sandu, T. Chai, *et al.* (2008). Predicting air quality: Improvements through advanced methods to integrate models and measurements. *J. Comput. Phys.*, **227**, 3540–3571.

Carroll, E. B., and T. D. Hewson (2005). NWP grid editing at the Met Office. *Wea. Forecasting*, **20**, 1021–1033.

Case, J. L., J. Manobianco, A. V. Dianic, *et al.* (2002). Verification of high-resolution RAMS forecasts over east-central Florida during the 1999 and 2000 summer months. *Wea. Forecasting*, **17**, 1133–1151.

Cassano, J. J., P. Uotila, and A. Lynch (2006). Changes in synoptic weather patterns in the polar regions in the twentieth and twenty-first centuries, Part 1: Arctic. *Int. J. Climatol.*, **26**, 1027–1049.

Castro, C. L., R. A. Pielke Sr, and G. Leoncini (2005). Dynamical downscaling: An assessment of value retained and added using the Regional Atmospheric Modeling System (RAMS). *J. Geophys. Res.*, **110**, D05108, doi:10.1029/2004JD004721.

Castro, C. L., R. A. Pielke Sr, and J. O. Adegoke (2007a). Investigation of the summer climate of the contiguous United States and Mexico using the regional atmospheric modeling system (RAMS). Part I: Model climatology (1950-2002). *J. Climate*, **20**, 3844–3865.

Castro, C. L., R. A. Pielke Sr, J. O. Adegoke, S. D. Schubert, and P. J. Pegion (2007b). Investigation of the summer climate of the contiguous United States and Mexico using the regional atmospheric modeling system (RAMS). Part II: Model climate variability. *J. Climate*, **20**, 3866–3887.

Catry, B., J.-F. Geleyn, F. Bouyssel, *et al.* (2008). A new sub-grid scale lift formulation in a mountain drag parameterisation scheme. *Meteorol. Zeitschrift*, **17**, 193–208.

Cattle, H., and J. Crossley (1995). Modelling Arctic climate change. *Phil. Trans. R. Soc. Lond. Ser. A*, **352**, 201–213.

Cavazos, T., A. C. Comrie, and D. M. Liverman (2002). Intraseasonal variability associated with wet monsoons in southeast Arizona. *J. Climate*, **15**, 2477–2490.

Caya, D., and R. Laprise (1999). A semi-implicit, semi-Lagrangian regional climate model: The Canadian RCM. *Mon. Wea. Rev.*, **127**, 341–362.

CCSP (2008). *Climate Models: An Assessment of Strengths and Limitations*. A Report by the U.S. Climate Change Science Program and the Subcommittee on Global Change Research. Authors D. C. Bader, C. Covey, W. J. Gutowski Jr, *et al.* Department of Energy, Office of Biological and Environmental Research, Washington, DC, USA.

Challinor, A. J., J. M. Slingo, T. R. Wheeler, and F. J. Doblas-Reyes (2005). Probabilistic simulations of crop yield over western India using the DEMETER seasonal hindcast ensembles. *Tellus*, **57A**, 498–512.

Chang, J. -T., and P. J. Wetzel (1991). Effects of spatial variations of soil moisture and vegetation on the evolution of a prestorm environment: A numerical case study. *Mon. Wea. Rev.*, **119**, 1368–1390.

Charles, S. P., B. C. Bates, I. N. Smith, and J. P. Hughes (2004). Statistical downscaling of daily precipitation from observed and modeled atmospheric fields. *Hydrological Processes*, **18**, 1373–1394.

Charles, S. P., M. A. Bari, A. Kitsios, and B. C. Bates (2007). Effect of GCM bias on downscaled precipitation and runoff projections for the Serpentine catchment, Western Australia. *Int. J. Climatol.*, **27**, 1673–1690.

Chase, T. N., R. A. Pielke Sr, T. G. F. Kittell, J. S. Baron, and T. J. Stohlgren (1999). Potential impacts on Colorado Rocky Mountain weather due to land use changes on the adjacent Great Plains. *J. Geophys. Res.*, **104**, 16673–16690.

Chase, T. N., R. A. Pielke Sr, T. G. F. Kittel, R. R. Nemani, and S. W. Running (2000). Simulated impacts of historical land-cover changes on global climate in northern winter. *Climate Dyn.*, **16**, 93–105.

Chaves, R. R., R. S. Ross, and T. N. Krishnamurti (2005). Weather and seasonal climate prediction for South America using a multi-model superensemble. *Int. J. Climatol.*, **25**, 1881–1914.

Chen, D. L., and Y. M. Chen (2003). Association between winter temperature in China and upper air circulation over East Asia revealed by canonical correlation analysis. *Global Planet. Change*, **37**, 315–325.

Chen, F., and R. Avissar (1994). The impact of shallow convective moist processes on mesoscale heat fluxes. *J. Appl. Meteor.*, **33**, 1382–1401.

Chen, F., and J. Dudhia (2001). Coupling an advanced land surface-hydrology model with the Penn State - NCAR MM5 modeling system. Part I: Model implementation and sensitivity. *Mon. Wea. Rev.*, **129**, 569–585.

Chen, F., K. Mitchell, J. Schaake, *et al.* (1996). Modeling of land surface evaporation by four schemes and comparison with FIFE observations. *J. Geophys. Res.*, **101**, 7251–7266.

Chen, F., T. T. Warner, and K. Manning (2001). Sensitivity of orographic moist convection to landscape variability: A study of the Buffalo Creek, Colorado, flash flood case of 1996. *J. Atmos. Sci.*, **58**, 3204–3223.

Chen, F., K. W. Manning, M. A. LeMone, *et al.* (2007). Description and evaluation of the characteristics of the NCAR High-Resolution Land Data Assimilation System during IHOP-02. *J. Appl. Meteor. Climatol.*, **46**, 694–713.

Chen, T. H., A. Henderson-Sellers, P. C. D. Milly, *et al.* (1997). Cabauw experimental results from the project for intercomparison of land-surface parameterization schemes. *J. Climate*, **10**, 1194–1215.

Cheong, H.-B. (2000). Application of double Fourier series to the shallow-water equations on a sphere. *J. Comput. Phys.*, **165**, 261–287.

Cheong, H.-B. (2006). A dynamical core with double Fourier series: Comparison with the spherical harmonics method. *Mon. Wea. Rev.*, **134**, 1299–1315.

Chin, H.-N. S., M. Leach, G. A. Sugiyama, *et al.* (2005). Evaluation of an urban canopy parameterization in a mesoscale model using VTMX and URBAN 2000 data. *Mon. Wea. Rev.*, **133**, 2043–2068.

Christensen, J. H., T. R. Carter, M. Rummukainen, and G. Amanatidis (2007). Evaluating the performance and utility of regional climate models: the PRUDENCE project. *Climatic Change*, **81**, 1–6.

Chu, J.-L., H. Kang, C.-Y. Tam, C.-K. Park, and C.-T. Chen (2008). Seasonal forecast for local precipitation over northern Taiwan using statistical downscaling. *J. Geophys. Res.*, **113**, D12118, doi:10.1029/2007JD009424.

Chuang, H. Y., and P. J. Sousounis (2000). A technique for generating idealized initial and boundary conditions for the PSU–NCAR model MM5. *Mon. Wea. Rev.*, **128**, 2875–2884.

Clark, M. P., and L. E. Hay (2004). Use of medium-range numerical weather prediction model output to produce forecasts of streamflow. *J. Hydrometeor.*, **5**, 15–32.

Clark, T. L., and R. D. Farley (1984). Severe downslope windstorm calculations in two and three spatial dimensions using anelastic interactive grid nesting: A possible mechanism for gustiness. *J. Atmos. Sci.*, **41**, 329–350.

Clark, T. L., and W. D. Hall (1991). Multi-domain simulations of the time dependent Navier Stokes equations: Benchmark error analysis of some nesting procedures. *J. Comp. Phys.*, **92**, 456–481.

Clark, T. L., and W. D. Hall (1996). The design of smooth, conservative vertical grids for interactive grid nesting with stretching. *J. Appl. Meteor.*, **35**, 1040–1046.

Clark, A. J., W. A. Gallus Jr, and T.-C. Chen (2008a). Contributions of mixed physics versus perturbed initial/lateral boundary conditions to ensemble-based precipitation forecast skill. *Mon. Wea. Rev.*, **136**, 2140–2156.

Clark, P. A., S. A. Harcourt, B. MacPherson, *et al.* (2008b). Prediction of visibility and aerosol within the operational Met Office Unified Model. I: Model formulation and variational assimilation. *Quart. J. Roy. Meteor. Soc.*, **134**, 1801–1816.

Clemen, R. T. (1989). Combining forecasts: a review and annotated bibliography. *Int. J. Forecasting*, **5**, 559–583.

Cocke, S. (1998). Case study of Erin using the FSU nested regional spectral model. *Mon. Wea. Rev.*, **126**, 1337–1346.

Cocke, S., T. E. LaRow, and D. W. Shin (2007). Seasonal rainfall predictions over the southeast United States using the Florida State University nested regional spectral model. *J. Geophys. Res.*, **112**, D04106, doi:10:1029/2006JD007535.

Coen, J. L. (2005). Simulation of the Big Elk fire using coupled atmosphere-fire modeling. *Int. J. Wildland Fire*, **14**, 49–59.

Coirier, W. J., D. M. Fricker, M. Furmaczyk, and S. Kim (2005). A computational fluid dynamics approach for urban area transport and dispersion modeling. *Environ. Fluid Mech.*, **15**, 443–479.

Colle, B. A., and S. E. Yuter (2007). The impact of coastal boundaries and small hills on the precipitation distribution across southern Connecticut and Long Island, New York. *Mon. Wea. Rev.*, **135**, 933–954.

Collins, W. D., P. J. Rasch, B. A. Boville, *et al.* (2004). *Description of the Community Atmosphere Model (CAM 3.0)*. Technical Note TN-464+STR, National Center for Atmospheric Research, Boulder, CO, USA.

Collins, W. D., C. M. Bitz, M. L. Blackmon, *et al.* (2006a). The Community Climate System Model: CCSM3. *J. Climate*, **19**, 2122–2143.

Collins, W. D., P. J. Rasch, B. A. Boville, *et al.* (2006b). The formulation and atmospheric simulation of the Community Atmospheric Model Version 3 (CAM3). *J. Climate*, **19**, 2144–2161.

Cook, K. H., and E. K. Vizy (2006). Coupled model simulations of the West African monsoon system: Twentieth and twenty-first century simulations. *J. Climate*, **19**, 3681–3703.

Cope, M. E., G. D. Hess, S. Lee, *et al.* (2004). The Australian air quality forecasting system. Part I: Project description and early outcomes. *J. Appl. Meteor.*, **43**, 649–662.

Coppola, E., and F. Giorgi (2005). Climate change in tropical regions from high-resolution time-slice AGCM experiments. *Quart. J. Roy. Meteor. Soc.*, **131**, 3123–3145.

Côté, J., M. Roch, A. Staniforth, and L. Fillion (1993). A variable-resolution semi-Lagrangian finite-element global model of the shallow-water equations. *Mon. Wea. Rev.*, **121**, 231–243.

Côté, J., S. Gravel, A. Méthot, *et al.* (1998a). The operational CMC-MRB global environmental multiscale (GEM) model. Part I: Design considerations and formulation. *Mon. Wea. Rev.*, **126**, 1373–1395.

Côté, J., J.-G. Desmarais, S. Gravel, *et al.* (1998b). The operational CMC-MRB global environmental multiscale (GEM) model. Part II: Results. *Mon. Wea. Rev.*, **126**, 1397–1418.

Cotton, W. R., and R. A. Anthes (1989). *Storm and Cloud Dynamics*. London, UK: Academic Press.

Courant, R., K. Friedrichs, and H. Lewy (1928). Über die partiellen Differenzengleichungen der mathematischen Physik. *Mathematische Annalen*, **100**, 32–74.

Covey, C., K. M. AchutaRao, U. Cubasch, *et al.* (2003). An overview of results from the Coupled Model Intercomparison Project (CMIP). *Global Planet. Change*, **37**, 103–133.

Cox, P. M., R. A. Betts, C. B. Bunton, *et al.* (1999). The impact of new land surface physics on the GCM simulation of climate and climate sensitivity. *Climate Dyn.*, **15**, 183–203.

Cressman, G. P. (1959). An operational objective analysis system. *Mon. Wea. Rev.*, **87**, 367–374.

Crook, N. A. (2001). Understanding Hector: The dynamics of island thunderstorms. *Mon. Wea. Rev.*, **129**, 1550–1563.

CSMD (Climate System Modeling Division) (2005). An introduction to the first operational climate model at the National Climate Center. *Advances in Climate System Modeling*, 1, National Climate Center, China Meteorological Administration (in English and Chinese).

Cubasch, U., H. von Storch, J. Waszkewitz, and E. Zorita (1996). Estimates of climate change in southern Europe using different downscaling techniques. *Climate Res.*, **7**, 129–149.

Cullen, M. J. P. (1979). The finite-element method. In *Numerical Methods Used in Atmospheric Models, Volume II*, ed. A. Kasahara, Global Atmospheric Research Programme, GARP Publication Series No. 17. Geneva: World Meteorological Organization, pp. 300–337.

Curry, J. A., and A. H. Lynch (2002). Comparing Arctic regional climate models. *EOS, Trans. Amer. Geophys. Union*, **83**, 87.

Dai, A. (2006). Precipitation characteristics in eighteen coupled climate models. *J. Climate*, **19**, 4605–4630.

Dai, Y., X. Zeng, R. E. Dickinson, *et al.* (2003). The Common Land Model. *Bull. Amer. Meteor. Soc.*, **84**, 1013–1023.

Dalcher, A., E. Kalnay, and R. N. Hoffman (1988). Medium range lagged average forecasts. *Mon. Wea. Rev.*, **116**, 402–416.

Daley, R. (1991). *Atmospheric Data Analysis*. Cambridge, UK: Cambridge University Press.

Danard, M. (1985). On the use of satellite estimates of precipitation in initial analyses for numerical weather prediction. *Atmos.-Ocean*, **23**, 23–42.

Darby, L. S., R. M. Banta, and R. A. Pielke Sr (2002). Comparison between mesoscale model terrain sensitivity studies and Doppler lidar measurements of the sea breeze at Monterey Bay. *Mon. Wea. Rev.*, **130**, 2813–2838.

Darmenova, K., and I. Sokolik (2007). Assessing uncertainties in dust emission in the Aral Sea region caused by meteorological fields predicted with a mesoscale model. *Global Planet. Change*, **56**, 297–310.

Davies, H. C. (1976). A lateral boundary formulation for multilevel prediction models. *Quart. J. Roy. Meteor. Soc.*, **102**, 405–418.

Davies, H. C. (1983). Limitations of some common lateral boundary schemes used in regional NWP models. *Mon. Wea. Rev.*, **111**, 1002–1012.

Davies, H. C., and R. E. Turner (1977). Updating prediction models by dynamical relaxation: An examination of the technique. *Quart. J. Roy. Meteor. Soc.*, **103**, 225–245.

Davis, C., T. Warner, E. Astling, and J. Bowers (1999). Development and application of an operational, relocatable, mesogamma-scale weather analysis and forecasting system. *Tellus*, **51A**, 710–727.

Davis, C., B. Brown, and R. Bullock (2006a). Object-based verification of precipitation forecasts, Part I: Methodology and application to mesoscale rain areas. *Mon. Wea. Rev.*, **134**, 1772–1784.

Davis, C., B. Brown, and R. Bullock (2006b). Object-based verification of precipitation forecasts, Part II: Application to convective rain systems. *Mon. Wea. Rev.*, **134**, 1785–1795.

Debernard, J. B., and L. P. Røed (2008). Future wind, wave and storm surge climate in the Northern Sea: a revisit. *Tellus*, **60A**, 427–438.

Delle Monache, L., X. Deng, Y. Zhou, and R. Stull (2006a). Ozone ensemble forecasts: 1. A new ensemble design. *J. Geophys. Res.*, **111**, D05307, doi: 10.1029/2005JD006310.

Delle Monache, L., T. Nipen, X. Deng, and Y. Zhou (2006b). Ozone ensemble forecasts: 2. A Kalman filter predictor bias correction. *J. Geophys. Res.*, **111**, D05308, doi: 10.1029/2005JD006311.

Delle Monache, L., J. P. Hacker, Y. Zhou, X. Deng, and R. B. Stull (2006c). Probabilistic aspects of meteorological and ozone regional ensemble forecasts. *J. Geophys. Res.*, **111**, D24307, doi:10.1029/2005JD006917.

Delle Monache, L., J. Wilczak, S. McKeen, *et al.* (2008). A Kalman-filter bias correction method applied to deterministic, ensemble averaged and probabilistic forecasts of surface ozone. *Tellus*, **60B**, 238–249.

Delworth, T. L., A. J. Broccoli, A. Rosati, *et al.* (2006). GFDL's CM2 global coupled climate models – Part I: Formulation and simulation characteristics. *J. Climate*, **19**, 643–674.

Deng, A., N. L. Seaman, and J. S. Kain (2003). A shallow convection parameterization for mesoscale models. Part I: Submodel description and preliminary applications. *J. Atmos. Sci.*, **60**, 34–56.

Denis, B., R. Laprise, D. Caya, and J. Côté (2002). Downscaling ability of one-way nested regional climate models: the Big-Brother experiment. *Climate Dyn.*, **18**, 627–646.

Déqué, M., and J. P. Piedelievre (1995). High resolution climate simulations over Europe. *Climate Dyn.*, **11**, 321–339.

Déqué, M., C. Dreveton, A. Braun, and D. Cariolle (1994). The ARPEGE/IFS atmosphere model: A contribution to the French community climate modeling. *Climate Dyn.*, **10**, 249–266.

Déqué, M., P. Marquet, and R. Jones (1998). Simulation of climate change over Europe using a global variable resolution general circulation model. *Climate Dyn.*, **14**, 173–189.

Déqué, M., R. G. Jones, M. Wild, *et al.* (2005). Global high resolution versus limited area model climate change projections over Europe: quantifying confidence level from PRUDENCE results. *Climate Dyn.*, **25**, 653–670.

De Sales, F., and Y. Xue (2006). Investigation of seasonal prediction of the South American regional climate using the nested modeling system. *J. Geophys. Res.*, **111**, D20107, doi:10.1029/2005JD006989.

Dévényi, D., and T. Schlatter (1994). Statistical properties of three-hour prediction "errors" derived from the Mesoscale Analysis and Prediction System. *Mon. Wea. Rev.*, **122**, 1263–1280.

Diaconescu, E. P., R. Laprise, and L. Sushama (2007). The impact of lateral boundary data errors on the simulated climate of a nested regional climate model. *Climate Dyn.*, **28**, 333–350.

Diansky, N. A., and E. M. Volodin (2002). Simulation of the present-day climate with a coupled atmosphere-ocean general circulation model. *Izv. Atmos. Ocean. Phys.*, **38**, 732–747 (English translation).

Diansky, N. A., A. V. Bagno, and V. B. Zalesny (2002). Sigma model of global ocean circulation and its sensitivity to variations in wind stress. *Izv. Atmos. Ocean. Phys.*, **38**, 477–494 (English translation).

Dickinson, R. E., R. M. Errico, F. Giorgi, and G. T. Bates (1989). A regional climate model for the western United States. *Climatic Change*, **15**, 383–422.

Dietachmayer, G. S., and K. K. Droegemeier (1992). Application of continuous dynamic grid adaption techniques to meteorological modeling. I. Basic formulation and accuracy. *Mon. Wea. Rev.*, **120**, 1675–1706.

Díez, E., C. Primo, J. A. García-Moya, J. M. Gutiérrez, and B. Orfila (2005). Statistical and dynamical downscaling of precipitation over Spain from DEMETER seasonal forecasts. *Tellus*, **57A**, 409–423.

Di Giuseppe, F., M. Elementi, D. Cesari, and T. Paccagnella (2009). The potential of variational retrieval of temperature and humidity profiles from Meteosat Second Generation observations. *Quart. J. Roy. Meteor. Soc.*, **135**, 225–237.

Dimitrijevic, M., and R. Laprise (2005). Validation of the nesting technique in a regional climate model and sensitivity tests to the resolution of the lateral boundary conditions during summer. *Climate Dyn.*, **25**, 555–580.

Doblas-Reyes, F. J., R. Hagedorn, and T. N. Palmer (2005). The rationale behind the success of multi-model ensembles in seasonal forecasting: II. Calibration and combination. *Tellus*, **57A**, 234–252.

Doms, G., and U. Schättler (1999). *The Nonhydrostatic Limited-area Model LM (Lokal Modell) of DWD. Part 1. Scientific Documentation*. Deutcher Wetterdienst (DWD), Offenbach, Germany.

d'Orgeval, T., J. Polcher, and L. Li (2006). Uncertainties in modeling future hydrologic change over West Africa. *Climate Dyn.*, **26**, 93–108.

Doswell, C. A., III (2004). Weather forecasting by humans: Heuristics and decision making. *Wea. Forecasting*, **19**, 1115–1126.

Douville, H., J.-F. Royer, and J.-F. Mahfouf (1995). A new snow parameterization for the Meteo-France climate model. *Climate Dyn.*, **12**, 21–35.

Driese, K. L., and W. A. Reiners (1997). Aerodynamic roughness parameters for semi-arid natural shrub communities of Wyoming, USA. *Agric. Forest Meteor.*, **88**, 1–14.

Druyan, L. M., M. Fulakeza, and P. Lonergan (2002). Dynamic downscaling of seasonal climate predictions over Brazil. *J. Climate*, **15**, 3411–3426.

Druyan, L. M., M. Fulakeza, and P. Lonergan (2006). Mesoscale analyses of West African summer climate: focus on wave disturbances. *Climate Dyn.*, **27**, 459–481.

Druyan, L. M., M. Fulakeza, and P. Lonergan (2007). Spatial variability of regional model simulated June-September mean precipitation over West Africa. *Geophys. Res. Lett.*, **34**, L18709, doi:10.1029/2007GL031270.

Dudhia, J. (1989). Numerical study of convection observed during the winter monsoon experiment using a mesoscale two-dimensional model. *J. Atmos. Sci.*, **46**, 3077–3107.

Dudhia, J. (1993). A nonhydrostatic version of the Penn State - NCAR mesoscale model: Validation tests and simulation of an Atlantic cyclone and cold front. *Mon. Wea. Rev.*, **121**, 1493–1513.

Dudhia, J., and J. F. Bresch (2002). A global version of the PSU-NCAR mesoscale model. *Mon. Wea. Rev.*, **130**, 2980–3007.

Duffy, P. B., B. Govindasamy, J. P. Iorio, *et al.* (2003). High-resolution simulations of global climate, part 1: Present climate. *Climate Dyn.*, **21**, 371–390.

Duffy, P. B., R. W. Arritt, J. Coquard, *et al.* (2006). Simulations of present and future climates in the western United States with four nested regional climate models. *J. Climate*, **19**, 873–895.

Dümenil, L., and E. Todini (1992). A rainfall-runoff scheme for use in the Hamburg climate model. In *Advances in Theoretical Hydrology: A Tribute to James Dooge*, ed. J. P. O'Kane, European Geophysical Society Series on Hydrological Sciences, Vol. 1, Amsterdam, the Netherlands: Elsevier Press, pp. 129–157.

Durran, D. R. (1989). Improving the anelastic approximation. *J. Atmos. Sci.*, **46**, 1453–1461.

Durran, D. R. (1991). The third-order Adams-Bashforth method: An attractive alternative to leapfrog time differencing. *Mon. Wea. Rev.*, **119**, 702–720.

Durran, D. R. (1999). *Numerical Methods for Wave Equations in Geophysical Fluid Dynamics*. New York, USA: Springer.

Durran, D. R., and J. B. Klemp (1983). A compressible model for the simulation of moist mountain waves. *Mon. Wea. Rev.*, **111**, 2341–2361.

Dutton, J. A. (1976). *The Ceaseless Wind*. New York, USA: McGraw-Hill.

Easter, R. C., S. J. Ghan, Y. Zhang, *et al.* (2004). MIRAGE: Model description and evaluation of aerosols and trace gases. *J. Geophys. Res.*, **109**, D20210, doi:10.1029/2004JD004571.

Ebert, E., and J. L. McBride (2000). Verification of precipitation in weather systems: Determination of systematic errors. *J. Hydrology*, **239**, 179–202.

Eckel, F. A., and C. F. Mass (2005). Aspects of effective mesoscale, short-range ensemble forecasting. *Wea. Forecasting*, **20**, 328–350.

Eckel, F. A., and M. K. Walters (1998). Calibrated probabilistic quantitative precipitation forecasts based on the MRF ensemble. *Wea. Forecasting*, **13**, 1132–1147.

Efron, B., and R. J. Tibshirani (1993). *An Introduction to the Bootstrap*. Dordrecht, the Netherlands: Chapman and Hall.

Ehrendorfer, M. (1997). Predicting the uncertainty of numerical weather forecasts: a review. *Meteorol. Zeitschrift*, **6**, 147–183.

Ehrendorfer, M., and J. J. Tribbia (1997). Optimal prediction of forecast error covariances through singular vectors. *J. Atmos. Sci.*, **54**, 286–313.

Ek, M., and R. H. Cuenca (1994). Variation in soil parameters: Implications for modeling surface fluxes and atmospheric boundary-layer development. *Bound.-Layer Meteor.*, **70**, 369–383.

Eliasen, E., B. Machenhauer, and E. Rasmussen (1970). *On a Numerical Method for Integration of the Hydrodynamic Equations with a Spectral Representation of the Horizontal Fields*. Report No. 2, Institute for Teoretisk Meteorologi, University of Copenhagen.

Elshamy, M. E., H. S. Wheater, N. Gedney, and C. Huntingford (2006). Evaluation of the rainfall component of a weather generator for climate impact studies. *J. Hydrology (Amsterdam)*, **326**, 1–24.

Emanuel, K. A., and R. Langland (1998). FASTEX adaptive observations workshop. *Bull. Amer. Meteor. Soc.*, **79**, 1915–1919.

Errico, R. M. (1997). What is an adjoint model? *Bull. Amer. Metor. Soc.*, **78**, 2577–2591.

Errico, R. M., and D. Baumhefner (1987). Predictability experiments using a high-resolution limited-area model. *Mon. Wea. Rev.*, **115**, 488–504.

Errico, R. M., and T. Vukicevic (1992). Sensitivity analysis using an adjoint of the PSU–NCAR mesoscale model. *Mon. Wea. Rev.*, **120**, 1644–1660.

Errico, R. M., T. Vukicevic, and K. Raeder (1993). Comparison of initial and lateral boundary condition sensitivity for a limited-area model. *Tellus*, **45A**, 539–557.

Evans, J. P. (2008). 21st century climate change in the Middle East. *Climatic Change*, **92**, 417–432, doi:10.1007/s10584-008-9438-5.

Evensen, G. (2003). The ensemble Kalman filter: theoretical formulation and practical implementation. *Ocean Dyn.*, **53**, 343–367.

Evensen, G. (2007). *Data Assimilation: The Ensemble Kalman Filter*. Berlin, Germany: Springer.

Fanelli, P. F., and P. R. Bannon (2005). Nonlinear atmospheric adjustment to thermal forcing. *J. Atmos. Sci.*, **62**, 4253–4272.

Fast, J. D. (1995). Mesoscale modeling and four-dimensional data assimilation in areas of highly complex terrain. *J. Appl. Meteor.*, **34**, 2762–2782.

Feddema, J. J., K. W. Oleson, G. B. Bonan, *et al.* (2005). The importance of land-cover change in simulating future climates. *Science*, **310**, 1674–1678.

Feddersen, H. (2003). Predictability of seasonal precipitation in the Nordic region. *Tellus*, **55A**, 385–400.

Feddersen, H., and U. Andersen (2005). A method for statistical downscaling of seasonal ensemble predictions. *Tellus*, **57A**, 398–408.

Feddersen, H., A. Navarra, and M. N. Ward (1999). Reduction of model systematic error by statistical correction for dynamical seasonal predictions. *J. Climate*, **14**, 1974–1989.

Feser, F. (2006). Enhanced detectability of added value in limited-area model results separated into different spatial scales. *Mon. Wea. Rev.*, **134**, 2180–2190.

Fichefet, T., and M. A. Morales-Maqueda (1997). Sensitivity of a global sea ice model to the treatment of ice thermodynamics and dynamics. *J. Geophys. Res.*, **102**, 12609–12646.

Fil, C., and L. Dubus (2005). Winter climate regimes over the North Atlantic and European region in ERA40 reanalysis and DEMETER seasonal hindcasts. *Tellus*, **57A**, 290–307.

Fiorino, M. F., and T. T. Warner (1981). Incorporating surface winds and rainfall rates into the initialization of a mesoscale hurricane model. *Mon. Wea. Rev.*, **109**, 1914–1929.

Fiorucci, P., F. Gaetani, A. Lanorte, and R. Lasaponara (2007). Dynamic fire danger mapping from satellite imagery and meteorological forecast data. *Earth Interactions*, **11**, 1–17.

Flato, G. M., and G. J. Boer (2001). Warming asymmetry in climate change simulations. *Geophys. Res. Lett.*, **28**, 195–198.

Flato, G. M., and W. D. Hibler (1992). Modeling pack ice as a cavitating fluid. *J. Phys. Oceanogr.*, **22**, 626–651.

Fleagle, R., and J. Businger (1963). *An Introduction to Atmospheric Physics*. New York, USA: Academic Press.

Fleming, R. J. (1971a). On stochastic-dynamic prediction. Part I, The energetics of uncertainty and the question of closure. *Mon. Wea. Rev.*, **99**, 851–872.

Fleming, R. J. (1971b). On stochastic-dynamic prediction. Part II, Predictability and utility. *Mon. Wea. Rev.*, **99**, 927–938.

Fletcher, N. H. (1962). *The Physics of Rain Clouds*. Cambridge, UK: Cambridge University Press.

Fowler, H. J., and M. Ekström (2009). Multi-model ensemble estimates of climate change impacts on UK seasonal precipitation extremes. *Int. J. Climatol.*, **29**, 385–416.

Fowler, H. J., and R. L. Wilby (2007). Beyond the downscaling comparison study (Editorial). *Int. J. Climatol.*, **27**, 1543–1545.

Fowler, H. J., S. Blenkinsop, and C. Tebaldi (2007). Linking climate change modelling to impacts studies: Recent advances in downscaling techniques for hydrological modelling. *Int. J. Climatol.*, **27**, 1547–1568.

Fox-Rabinovitz, M. S. (1996). Diabatic dynamic initialization with an iterative time integration scheme as a filter. *Mon. Wea. Rev.*, **124**, 1544–1557.

Fox-Rabinovitz, M. S., and B. D. Gross (1993). Diabatic dynamic initialization. *Mon. Wea. Rev.*, **121**, 549–564.

Fox-Rabinovitz, M. S., G. L. Stenchikov, M. J. Suarez, and L. L. Takacs (1997). A finite-difference GCM dynamical core with a variable-resolution stretched grid. *Mon. Wea. Rev.*, **125**, 2943–2968.

Fox-Rabinovitz, M. S., G. L. Stenchikov, M. J. Suarez, L. L. Takacs, and R. C. Govindaraju (2000). A uniform- and variable-resolution stretched-grid GCM dynamical core with realistic orography. *Mon. Wea. Rev.*, **128**, 1883–1898.

Fox-Rabinovitz, M. S., L. L. Takacs, R. C. Govindaraju, and M. J. Suarez (2001). A variable-resolution stretched-grid general circulation model: Regional climate simulations. *Mon. Wea. Rev.*, **129**, 453–469.

Fox-Rabinovitz, M. S., L. L. Takacs, and R. C. Govindaraju (2002). A variable-resolution stretched-grid general circulation model and data-assimilation system with multiple areas of interest: Studying the anomalous regional climate events of 1998. *J. Geophys. Res.*, **107** (D24), 4768, doi:10.1029/2002JD002177.

Fox-Rabinovitz, M. S., E. H. Berbery, L. L. Takacs, and R. C. Govindaraju (2005). A multiyear ensemble simulation of the U.S. climate with a stretched-grid GCM. *Mon. Wea. Rev.*, **133**, 2505–2525.

Fox-Rabinovitz, M. S., J. Côté, B. Dugas, M. Déqué, and J. L. McGregor (2006). Variable-resolution general circulation models: Stretched-grid model intercomparison project (SGMIP). *J. Geophys. Res.*, **111**, D16104, doi:10.1029/2005JD006520.

Frank, W. M. (1983). Review: The cumulus parameterization problem. *Mon. Wea. Rev.*, **111**, 1859–1871.

Frank, W. M., and C. Cohen (1987). Simulation of tropical convective systems. *J. Atmos. Sci.*, **44**, 3787–3799.

Frank, W. M., and E. A. Ritchie (1999). Effects of environmental flow upon tropical cyclone structure. *Mon. Wea. Rev.*, **127**, 2044–2061.

Frehlich, R., and R. Sharman (2008). The use of structure functions and spectra from numerical model output to determine effective model resolution. *Mon. Wea. Rev.*, **136**, 1537–1553.

Frei, C., J. H. Christensen, M. Déqué, *et al.* (2003). Daily precipitation statistics in regional climate models: Evaluation and intercomparison for the European Alps. *J. Geophys. Res.*, **108**(D3), 4124, doi:10.1029/2002JD002287.

Frías, M. D., J. Fernández, J. Sáenz, and C. Rodríguez-Puebla (2005). Operational predictability of monthly average maximum temperature over the Iberian Peninsula using DEMETER simulations and downscaling. *Tellus*, **57A**, 448–463.

Friend, A. D., and N. Y. Kiang (2005). Land surface model development for the GISS GCM: Effects of improved canopy physiology on simulated climate. *J. Climate*, **18**, 2883–2902.

Fritsch, J. M., and C. F. Chappell (1980). Numerical prediction of convectively driven mesoscale pressure systems. Part I: Convective parameterization. *J. Atmos. Sci.*, **37**, 1722–1733.

Fritsch, J. M., J. Hilliker, J. Ross, and R. L. Vislocky (2000). Model consensus. *Wea. Forecasting*, **15**, 571–582.

Fu, C., S. Wang, Z. Xiong, *et al.* (2005). Regional climate model intercomparison project for Asia. *Bull. Amer. Meteor. Soc.*, **86**, 257–266.

Fujioka, F. M. (2002). A new method for the analysis of fire spread modeling errors. *Int. J. Wildland Fire*, **11**, 193–203.

Fujita, T., D. J. Stensrud, and D. C. Dowell (2007). Surface data assimilation using an ensemble Kalman filter approach with initial condition and model physics uncertainties. *Mon. Wea. Rev.*, **135**, 1846–1868.

Fuller, D. O., A. Troyo, and J. C. Beier (2009). El Niño Southern Oscillation and vegetation dynamics as predictors of dengue fever cases in Costa Rica. *Environ. Res. Lett.*, **4**, 014011, doi: 10.1088/1748-9326/4/1/014011.

Fulton, R. A., J. P. Breidenbach, D.-J. Seo, and D. A. Miller (1998). The WSR-88D rainfall algorithm. *Wea. Forecasting*, **13**, 377–395.

Furevik, T., M. Bentsen, H. Drange, *et al.* (2003). Description and evaluation of the Bergen climate model: ARPEGE coupled with MICOM. *Climate Dyn.*, **21**, 27–51.

Gachon, P., and Y. Dibike (2007). Temperature change signals in northern Canada: convergence of statistical downscaling results using two driving GCMs. *Int. J. Climatol.*, **27**, 1623–1641.

Gal-Chen, T., and R. C. J. Somerville (1975). On the use of a coordinate transformation for the solution of the Navier-Stokes equations. *J. Comput. Phys.*, **17**, 209–228.

Galin, V. Ya., E. M. Volodin, and S. P. Smyshliaev (2003). Atmospheric general circulation model of INM RAS with ozone dynamics. *Russ. Meteorol. Hydrol.*, **5**, 13–22.

Gall, R. L., R. T. Williams, and T. L. Clark (1988). Gravity waves generated during frontogenesis. *J. Atmos. Sci.*, **45**, 2204–2219.

García-Morales, M. B., and L. Dubus (2007). Forecasting precipitation for hydroelectric power management: how to exploit GCM's seasonal ensemble forecasts. *Int. J. Climatol.*, **27**, 1691–1705.

García-Pintado, J., G. G. Barberá, M. Erena, and V. M. Castillo (2009). Rainfall estimation by rain gauge-radar combination: A concurrent multiplicative-additive approach. *Water Resour. Res.*, **45**, W01415, doi:10.1029/2008WR007011.

George, S. E., and R. T. Sutton (2006). Predictability and skill of boreal winter forecasts made with the ECMWF Seasonal Forecast System II. *Quart. J. Roy. Meteor. Soc.*, **132**, 2031–2053.

Georgelin, M., P. Bougeault, T. Black, *et al.* (2000). The second COMPARE exercise: A model intercomparison using a case of a typical mesoscale orographic flow, the PYREX IOP3. *Quart. J. Roy. Meteor. Soc.*, **126**, 991–1029.

GFDL GAMDT (The GFDL Global Atmospheric Model Development Team) (2004). The new GFDL global atmosphere and land model AM2-LM2: Evaluation with prescribed SST simulations. *J. Climate*, **17**, 4641–4673.

Ghan, S. J., and T. Shippert (2006). Physically based global downscaling: Climate change projections for a full century. *J. Climate*, **19**, 1589–1604.

Ghan, S. J., X. Bian, A. G. Hunt, and A. Coleman (2002). The thermodynamic influence of subgrid orography in a global climate model. *Climate Dyn.*, **20**, 31–44, doi:10.1007/s00382-002-0257-5.

Ghan, S. J., T. Shippert, and J. Fox (2006). Physically based global downscaling: Regional evaluation. *J. Climate*, **19**, 429–445.

Ghil, M., and P. Malanotte-Rizzoli (1991). Data assimilation in meteorology and oceanography. *Adv. Geophys.*, **33**, 141–266.

Ghosh, S., and P. P. Mujundar (2008). Statistical downscaling of GCM simulations to streamflow using relevance vector machine. *Adv. Water Resour.*, **31**, 132–146.

Gibelin, A. L., and M. Déqué (2003). Anthropogenic climate change over the Mediterranean region simulated by a global variable resolution model. *Climate Dyn.*, **20**, 327–339.

Gilleland, E., D. Ahijevych, B. G. Brown, B. Casati, and E. E. Ebert (2009). Intercomparison of spatial forecast verification metrics. *Wea. Forecasting*, **24**, 1416–1430.

Gilmour, I., L. A. Smith, and R. Buizza (2001). Linear regime duration: Is 24 hours a long time in synoptic weather forecasting? *J. Atmos. Sci.*, **58**, 3525–3539.

Glahn, B. (2008). Comments – Reforecasts: An important data set for improving weather predictions. *Bull. Amer. Meteor. Soc.*, **89**, 1373–1378.

Glahn, H., and R. Lowry (1972). The use of model output statistics (MOS) in objective weather forecasting. *J. Appl. Meteor.*, **11**, 1203–1211.

Gnanadesikan, A., K. W. Dixon, S. M. Griffies, *et al.* (2006). GFDL's CM2 global coupled climate models – Part 2: The baseline ocean simulation. *J. Climate*, **19**, 675–697.

Gochis, D. J., W. J. Shuttleworth, and Z.-L. Yang (2002). Sensitivity of the modeled North American monsoon regional climate to convective parameterization. *Mon. Wea. Rev.*, **130**, 1282–1298.

Gochis, D. J., W. J. Shuttleworth, and Z.-L. Yang (2003). Hydrological response of the modeled North American monsoon to convective parameterization. *Mon. Wea. Rev.*, **130**, 235–250.

Goddard, L., S. J. Mason, S. E. Zebiak, *et al.* (2001). Current approaches to seasonal-to-interannual climate predictions. *Int. J. Climatol.*, **21**, 1111–1152.

Goerss, J. S. (2009). Impact of satellite observations on the tropical cyclone track forecasts of the Navy Operational Global Atmospheric Prediction System. *Mon. Wea. Rev.*, **137**, 41–50.

Goosse, H., and T. Fichefet (1999). Importance of ice-ocean interactions for the global ocean circulation: A model study. *J. Geophys. Res.*, **104**, 23337–23355.

Gordon, C., C. Cooper, C. A. Senior, *et al.* (2000). The simulation of SST, sea ice extents and ocean heat transports in a version of the Hadley Centre coupled model without flux adjustments. *Climate Dyn.*, **16**, 147–168.

Gordon, H. B., L. D. Rotstayn, J. L. McGregor, *et al.* (2002). *The CSIRO Mk3 Climate System Model*. CSIRO Atmospheric Research Technical Paper No. 60, Commonwealth Scientific and Industrial Research Organisation, Aspendale, Victoria, Australia.

Graham, R. J., M. Gordon, P. J. Mclean, *et al.* (2005). A performance comparison of coupled and uncoupled versions of the Met Office seasonal prediction general circulation model. *Tellus*, **57A**, 320–339.

Gray, D., J. Daniels, S. Nieman, S. Lord, and G. Dimego (1996). NESDIS and NWS Assessment of GOES 8/9 Operational satellite motion vectors. *Proceedings 3rd International Winds Workshop*, Ascona, Switzerland. EUMETSAT Pub. EUM P18, pp.175–183.

Gregory, D., and P. R. Rowntree (1990). A mass flux convection scheme with representation of cloud ensemble characteristics and stability-dependent closure. *Mon. Wea. Rev.*, **118**, 1483–1506.

Grell, G. A. (1993). Prognostic evaluation of assumptions used by cumulus parameterizations. *Mon. Wea. Rev.*, **121**, 764–787.

Grell, G. A., and D. Dévényi (2002). A generalized approach to parameterizing convection combining ensemble and data assimilation techniques. *Geophys. Res. Lett.*, **29**, 1693, doi:10.1029/2002GL015311.

Grell, G. A., J. Dudhia, and D. R. Stauffer (1994). *A Description of the Fifth Generation Penn State/NCAR Mesoscale Model* (MM5). NCAR Technical Note NCAR/TN-398+STR. [Available from NCAR, P.O. Box 3000, Boulder, CO 80307- 3000.]

Grell, G. A., S. E. Peckham, R. Smitz, *et al.* (2005). Fully coupled "online" chemistry within the WRF model. *Atmos. Environ.*, **39**, 6957–6975.

Grimit, E. P., and C. F. Mass (2007). Measuring the ensemble spread-error relationship with a probabilistic approach: Stochastic ensemble results. *Mon. Wea. Rev.*, **135**, 203–221.

Grimmer, M., and D. B. Shaw (1967). Energy-preserving integrations of the primitive equations on the sphere. *Quart. J. Roy. Meteor. Soc.*, **93**, 337–349.

Gualdi, S., A. Alessandri, and A. Navarra (2005). Impact of atmospheric horizontal resolution on El Niño Southern Oscillation forecasts. *Tellus*, **57A**, 357–374.

Gustafsson, N. (1990). Sensitivity of limited area model data assimilation to lateral boundary condition fields. *Tellus,* **42A**, 109–115.

Gutiérrez, J. M., R. Cano, A. S. Cofiño, and C. Sordo (2005). Analysis and downscaling multi-model seasonal forecasts in Peru using self-organizing maps. *Tellus*, **57A**, 435–447.

Gutman, G., and A. Ignatov (1998). The derivation of green vegetation fraction from NOAA/AVHRR data for use in numerical weather prediction models. *Int. J. Remote Sens.*, **19**, 1533–1543.

Gyakum, J. R. (1986). Experiments in temperature and precipitation forecasting. *Wea. Forecasting*, **1**, 77–88.

Ha, S.-Y., Y.-H. Kuo, Y.-R. Guo, and G.-H. Lim (2003). Variational assimilation of slant-path wet delay measurements from a hypothetical ground-based GPS network. Part I: Comparison with precipitable water assimilation. *Mon. Wea. Rev.*, **131**, 2635–2655.

Hacker, J. (2010). Spatial and temporal scales of boundary-layer wind predictability in response to small-amplitude land-surface uncertainty. *J. Atmos. Sci.*, **67**, 217–233.

Hacker, J. P., and D. L. Rife (2007). A practical approach to sequential estimation of systematic error on near-surface mesoscale grids. *Wea. Forecasting*, **22**, 1257–1273.

Hagedorn, R., F. J. Doblas-Reyes, and T. N. Palmer (2005). The rationale behind the success of multi-model ensembles in seasonal forecasting – I. Basic concept. *Tellus*, **57A**, 219–233.

Hagemann, S. (2002). *An Improved Land Surface Parameter Data set for Global and Regional Climate Models*. Max Planck Institute for Meteorology Report 162, MPI for Meteorology, Hamburg, Germany.

Hagemann, S., and L. Dümenil-Gates (2001). Validation of the hydrological cycle of ECMWF and NCEP reanalyses using the MPI hydrological discharge model. *J. Geophys. Res.*, **106**, 1503–1510.

Hahmann, A. N., D. Rostkier-Edelstein, T. T. Warner, *et al.* (2010). A dynamical downscaling system for the generation of mesoscale climatographies. *J. Appl. Meteor. Climatol.*, in press.

Haidvogel, D. B., and A. Beckmann (1999). *Numerical Ocean Modeling.* London, UK: Imperial College Press.

Hall, M. C. G., and D. G. Cacuci (1983). Physical interpretation of adjoint functions for sensitivity analysis of atmospheric models. *J. Atmos. Sci.*, **40**, 2537–2546.

Haltiner, G. J., and R. T. Williams (1980). *Numerical Prediction and Dynamic Meteorology.* New York, USA: John Wiley and Sons.

Hamill, T. M. (2001). Interpretation of rank histograms for verifying ensemble forecasts. *Mon. Wea. Rev.*, **129**, 550–560.

Hamill, T. M. (2006). Ensemble-based atmospheric data assimilation. In *Predictability of Weather and Climate*, eds. T. Palmer and R. Hagedorn. Cambridge, UK: Cambridge University Press, pp. 123–156.

Hamill, T. M., and S. J. Colucci (1997). Verification of Eta-RSM short-range ensemble forecasts. *Mon. Wea. Rev.*, **125**, 1312–1327.

Hamill, T. M., and S. J. Colucci (1998). Evaluation of Eta-RSM ensemble probabilistic precipitation forecasts. *Mon. Wea. Rev.*, **126**, 711–724.

Hamill, T. M., and C. Snyder (2000). A hybrid ensemble Kalman filter / 3d variational analysis scheme. *Mon. Wea. Rev.*, **128**, 2905–2919.

Hamill, T. M., and J. S. Whitaker (2006). Probabilistic quantitative precipitation forecasts based on reforecast analogs: Theory and application. *Mon. Wea. Rev.*, **134**, 3209–3229.

Hamill, T. M., J. S. Whitaker, and C. Snyder (2001). Distance-dependent filtering of background-error covariance estimates in an ensemble Kalman filter. *Mon. Wea. Rev.*, **129**, 2776–2790.

Hamill, T. M., J. S. Whitaker, and X. Wei (2004). Ensemble reforecasting: Improving medium-range forecast skill using retrospective forecasts. *Mon. Wea. Rev.*, **132**, 1434–1447.

Hamill, T. M., J. S. Whitaker, and S. L. Mullen (2006). Reforecasts: An important data set for improving weather predictions. *Bull. Amer. Meteor. Soc.*, **87**, 33–46.

Han, J., and H.-L. Pan (2006). Sensitivity of hurricane intensity to convective momentum transport parameterization. *Mon. Wea. Rev.*, **134**, 664–674.

Han, J., and J. Roads (2004). U.S. climate sensitivity simulated with the NCEP regional spectral model. *Climatic Change*, **62**, 115–154.

Hannachi, A., I. T. Jolliffe, and D. B. Stephenson (2007). Empirical orthogonal functions and related techniques in atmospheric science: A review. *Int. J. Climatol.*, **27**, 1119–1152.

Hanson, C. E., J. P. Palutikof, M. T. J. Livermore, *et al.* (2007). Modelling the impact of climate extremes: an overview of the MICE project. *Climate Change*, **81**, 163–177.

Hanssen-Bauer, I., E. J. Førland, J. E. Haugen, and O. E. Tveito (2003). Temperature and precipitation scenarios for Norway: comparison of results from dynamical and empirical downscaling. *Climate Res.*, **25**, 15–27.

Hanssen-Bauer, I., C. Achberger, R. E. Benestad, D. Chen, and E. J. Førland (2005). Statistical downscaling of climate scenarios over Scandinavia. *Climate Res.*, **29**, 255–268.

Harder, M. (1996). *Dynamik, Rauhigkeit und Alter des Meereises in der Arktis.* Ph.D. Dissertation, Alfred-Wegener-Institut für Polar und Meeresforschung, Bremerhaven, Germany.

Harper, K. C. (2008). *Weather by the Numbers: The Genesis of Modern Meteorology.* Cambridge, MA, USA: MIT Press.

Harrison, E. J., and R. L. Elsberry (1972). A method for incorporating nested grids in the solution of systems of geophysical equations. *J. Atmos. Sci.*, **29**, 1235–1245.

Hart, K. A., W. J. Steenburgh, D. J. Onton, and A. J. Siffert (2004). An evaluation of mesoscale-model-based Model Output Statistics (MOS) during the 2002 Olympic and Paralympic Winter Games. *Wea. Forecasting*, **19**, 200–218.

Hartmann, D. L. (1988). On the comparison of finite-element to finite-difference methods for the representation of vertical structure in model atmospheres. *Mon. Wea. Rev.*, **116**, 269–273.

Haugen, J. E., and T. Iversen (2008). Response in extremes of daily precipitation and wind from a downscaled multi-model ensemble of anthropogenic global climate change scenarios. *Tellus*, **60A**, 411–426.

Hay, L. E., and M. P. Clark (2003). Use of statistically and dynamically downscaled atmospheric model output for hydrologic simulations in three mountainous basins in the western United States. *J. Hydrol.*, **282**, 56–75.

Haylock, M. R., G. C. Cawley, C. Harpham, R. L. Wilby, and C. M. Goodess (2006). Downscaling heavy precipitation over the United Kingdom: A comparison of dynamical and statistical methods and their future scenarios. *Int. J. Climatol.*, **26**, 1397–1415.

Haywood, J., M. Bush, S. Abel, *et al.* (2008). Prediction of visibility and aerosol within the operational Met Office Unified Model. II: Validation of model performance using operational data. *Quart. J. Roy. Meteor. Soc.*, **134**, 1817–1832.

Heckley, W. A., G. Kelly, and M. Tiedtke (1990). On the use of satellite-derived heating rates for data-assimilation within the tropics. *Mon. Wea. Rev.*, **118**, 1743–1757.

Heemink, A. W., M. Verlaan, and A. J. Segers (2001). Variance-reduced ensemble Kalman filtering. *Mon. Wea. Rev.*, **129**, 1718–1728.

Heffter, J. L., and B. J. B. Stunder (1993). Volcanic Ash Forecast Transport and Dispersion (VAFTAD) Model. *Wea. Forecasting*, **8**, 533–541.

Hegerl, G. C., F. W. Zwiers, P. Braconnot, *et al.* (2007). Understanding and Attributing Climate Change. In *Climate Change 2007: The Physical Science Basis. Contribution of Working Group I to the Fourth Assessment Report of the Intergovernmental Panel on Climate Change*, eds. S. Solomon, D. Qin, M. Manning, *et al.* Cambridge, UK and New York, USA: Cambridge University Press.

Heimann, D. (2001). A model-based wind climatology of the eastern Adriatic coast. *Meteorol. Zeitschrift*, **10**, 5–16.

Heintzenberg, J., and R. J. Charlson (eds.) (2007). *Clouds in the Perturbed Climate System: Their Relationship to Energy Balance, Atmospheric Dynamics, and Precipitation.* Cambridge, USA: MIT Press.

Henderson, S. T. (1977). *Daylight and its Spectrum.* Bristol, UK: Adam Hilger.

Henderson-Sellers, A., A. J. Pitman, P. K. Love, P. Irannejad, and T. H. Chen (1995). The Project for Intercomparison of Land-surface Parameterization Schemes (PILPS): Phases 2 and 3. *Bull. Amer. Meteor. Soc.*, **76**, 489–503.

Herceg, D., A. H. Sobel, L. Sun, and S. E. Zebiak (2006). The Big Brother Experiment and seasonal predictability in the NCEP regional spectral model. *Climate Dyn.*, **27**, 69–82.

Hessami, M., P. Gachon, T. B. M. J. Ouarda, and A. St-Hilaire (2008). Automated regression-based statistical downscaling tool. *Environ. Modeling and Software*, **23**, 813–834.

Hewitson, B. C., and R. G. Crane (2002). Self-organizing maps: application to synoptic climatology. *Climate Res.*, **22**, 13–26.

Hewitson, B. C., and R. G. Crane (2006). Consensus between GCM climate change projections with empirical downscaling: precipitation downscaling over South Africa. *Int. J. Climatol.*, **26**, 1315–1337.

Hewitt, C. D. (2005). The ENSEMBLES Project: Providing ensemble-based predictions of climate changes and their impacts. *EGGS Newsletter*, **13**, 22–25.

Hibler, W. D. (1979). A dynamic thermodynamic sea ice model. *J. Phys. Oceanogr.*, **9**, 817–846.

Hodur, R. M. (1997). The Naval Research Laboratory's Coupled Ocean/Atmosphere Mesoscale Prediction System (COAMPS). *Mon. Wea. Rev.*, **125**, 1414–1430.

Hoffman, R. N., and E. Kalnay (1983). Lagged average forecasting, an alternative to Monte Carlo forecasting. *Tellus*, **35A**, 100–118.

Hoffman, R. N., and S. M. Leidner (2005). An introduction to the near-real-time Quik-SCAT data. *Wea. Forecasting*, **20**, 476–493.

Hoffman, R. N., C. Grassotti, R. G. Isaacs, J.-F. Louis, and T. Nehrkorn (1990). Assessment of the impact of simulated satellite lidar wind and retrieved 183 GHz water vapor observations on a global data assimilation system. *Mon. Wea. Rev.*, **118**, 2513–2542.

Hollingsworth, A., and P. Lönnberg (1986). The statistical structure of short-range forecast errors as determined from radiosonde data, Part I: The wind field. *Tellus*, **38A**, 111–136.

Hollingsworth, A., D. B. Shaw, P. Lönnberg, *et al.* (1986). Monitoring of observation and analysis quality by a data-assimilation system. *Mon. Wea. Rev.*, **114**, 861–879.

Holt, T., and J. Pullen (2007). Urban canopy modeling of the New York City metropolitan area: A comparison and validation of single- and multilayer parameterizations. *Mon. Wea. Rev.*, **135**, 1906–1930.

Holt, T., J. Pullen, and C. H. Bishop (2009). Urban and ocean ensembles for improved meteorological and dispersion modelling of the coastal zone. *Tellus*, **61A**, 232–249.

Holton, J. R. (2004). *An Introduction to Dynamic Meteorology.* Oxford, UK: Elsevier, Academic Press.

Holtslag, B. (2006). GEWEX Atmospheric Boundary-Layer Study (GABLS) on stable boundary layers. *Bound.-Layer Meteor.*, **118**, 243–246.

Homar, V., R. Romero, D. J. Stensrud, C. Ramis, and S. Alonso (2003). Numerical diagnosis of a small, quasi-tropical cyclone over the western Mediterranean: Dynamical vs. boundary factors. *Quart. J. Roy. Meteor. Soc.*, **129**, 1469–1490.

Horel, J., M. Splitt, L. Dunn, *et al.* (2002). MesoWest: Cooperative mesonets in the western United States. *Bull. Amer. Meteor. Soc.*, **83**, 211–226.

Hortal, M., and A. J. Simmons (1991). Use of reduced Gaussian grids in spectral models. *Mon. Wea. Rev.*, **119**, 1057–1074.

Horvath, K., L. Fita, R. Romero, and B. Ivancan-Picek (2006). A numerical study of the first phase of a deep Mediterranean cyclone: Cyclogenesis in the lee of the Atlas Mountains. *Meteorol. Zeitschrift*, **15**, 133–146.

Hoshen, M. B., and A. P. Morse (2004). A weather-driven model of malaria transmission. *Malaria J.*, **3**, 32, doi:10.1186/1475-2875-3-32.

Hourdin, F., I. Musat, S. Bony, *et al.* (2006). The LMDZ4 general circulation model: Climate performance and sensitivity to parameterized physics with emphasis on tropical convection. *Climate Dyn.*, **27**, 787–813.

Houtekamer, P. L., and H. L. Mitchell (1998). Data assimilation using an ensemble Kalman filter technique. *Mon. Wea. Rev.*, **126**, 796–811.

Houtekamer, P. L., and H. L. Mitchell (1999). Reply to comment on "Data assimilation using an ensemble Kalman filter technique". *Mon. Wea. Rev.*, **127**, 1378–1379.

Houtekamer, P. L., and H. L. Mitchell (2001). A sequential ensemble Kalman filter for atmospheric data assimilation. *Mon. Wea. Rev.*, **129**, 123–137.

Houtekamer, P. L., H. L. Mitchell, G. Pellerin, *et al.* (2005). Atmospheric data assimilation with the ensemble Kalman filter: results with real observations. *Mon. Wea. Rev.*, **133**, 604–620.

Houtekamer, P. L., H. L. Mitchell, and X. Deng (2009). Model error representation in an operational ensemble Kalman filter. *Mon. Wea. Rev.*, **137**, 2126–2143.

Houze, R. A., Jr (1993). *Cloud Dynamics*. London, UK: Academic Press.

Huffman, G. J., R. F. Adler, D. T. Bolvin, *et al.* (2007). The TRMM multi-satellite precipitation analysis: Quasi-global, multi-year, combined-sensor precipitation estimates at fine scale. *J. Hydrometeor.*, **8**, 38–55.

Hundecha, Y., and A. Bárdossy (2008). Statistical downscaling of extremes of daily precipitation and temperature and construction of their future scenarios. *Int. J. Climatol.*, **28**, 589–610.

Hunke, E. C., and J. K. Dukowicz (1997). An elastic-viscous-plastic model for sea ice dynamics. *J. Phys. Oceanogr.*, **27**, 1849–1867.

Hunke, E. C., and J. K. Dukowicz (2003). *The Sea Ice Momentum Equation in the Free Drift Regime*. Technical Report LA-UR-03-2219, Los Alamos National Laboratory, Los Alamos, USA.

Huth, R. (1999). Statistical downscaling in central Europe: evaluation of methods and potential predictors. *Climate Res.*, **13**, 91–101.

Huth, R. (2002). Statistical downscaling of daily temperature in Central Europe. *J. Climate*, **15**, 1731–1741.

Huth, R. (2004). Sensitivity of local daily temperature change estimates to the selection of downscaling models and predictors. *J. Climate*, **17**, 640–652.

Huth, R., J. Kysely, and M. Dubrovsky (2001). Time structure of observed GCM-simulated downscaled and stochastically generated daily temperature series. *J. Climate*, **14**, 4047–4061.

Huth, R., S. Kliegrová, and L. Metelka (2008). Non-linearity in statistical downscaling: Does it bring an improvement for daily temperature in Europe? *Int. J. Climatol.*, **28**, 465–477.

Ide, K., P. Courtier, M. Ghil, and A. C. Lorenc (1997). Unified notation for data assimilation: Operational, sequential, and variational. *J. Meteor. Soc. Japan*, **75**, 181–189.

Inclán, M. G., R. Forkel, R. Dlugi, and R. B. Stull (1996). Application of transilient turbulent theory to study the interactions between the atmospheric boundary layer and forest canopies. *Bound.-Layer Meteor.*, **79**, 315–344.

IPCC (2007). *Climate Change 2007: The Physical Science Basis. Contribution of Working Group I to the Fourth Assessment Report of the Intergovernmental Panel on Climate Change*, eds. S. Solomon, D. Qin, M. Manning, *et al.* Cambridge, UK and New York, USA: Cambridge University Press.

Isard, S. A., J. M. Russo, and E. D. DeWolf (2006). The establishment of a National Pest Information Platform for extension and education. *Plant Health Progress*, doi.1094/PHP-2006-0915-01-RV.

Israeli, M., and S. A. Orzag (1981). Approximation of radiation boundary conditions. *J. Comp. Phys.*, **41**, 115–135.

Iversen, T. (2008). Preface to special issue on RegClim. *Tellus*, **60A**, 395–397.

Jacks, E., J. B. Bower, V. J. Dagostaro, *et al.* (1990). New NGM-based MOS guidance for maximum/minimum temperature, probability of precipitation, cloud amount, and surface wind. *Wea. Forecasting*, **5**, 128–138.

Jacobson, M. Z. (1999). *Fundamentals of Atmospheric Modeling*. Cambridge, UK: Cambridge University Press.

Janjić, Z. I. (1994). The step-mountain Eta coordinate model: Further developments of the convection, viscous sublayer, and turbulence closure schemes. *Mon. Wea. Rev.*, **122**, 927–945.

Janjić, Z. I. (2000). Comments on "Development and evaluation of a convection scheme for use in climate models". *J. Atmos. Sci.*, **57**, 3686.

Jankov, I., W. A. Galus Jr, M. Segal, B. Shaw, and S. E. Koch (2005). The impact of different WRF model physical parameterizations and their interactions on warm season MCS rainfall. *Wea. Forecasting*, **20**, 1048–1060.

Jankov, I., W. A. Galus Jr, M. Segal, and S. E. Koch (2007). Influence of initial conditions on the WRF-ARW model QPF response to physical parameterization changes. *Wea. Forecasting*, **22**, 501–519.

Janssen, P. A. E. M. (2008). Progress in ocean wave forecasting. *J. Comp. Phys.*, **227**, 3572–3594.

Jastrow, R., and M. Halem (1970). Simulation studies related to GARP. *Bull. Amer. Meteor. Soc.*, **51**, 490–513.

Ji, Y., and A. D. Vernekar (1997). Simulation of the Asian summer monsoons of 1987 and 1988 with a regional model nested in a global GCM. *J. Climate*, **10**, 1965–1979.

Jiang, X., C. Wiedinmyer, F. Chen, Z.-L. Yang, and J. C.-F. Lo (2008). Predicted impacts of climate and land-use change on surface ozone in the Houston, Texas area. *J. Geophys. Res.*, **113**, D20312, doi:10.1029/2008JD009820.

Jin, M., and J. M. Shepherd (2005). Inclusion of urban landscape in a climate model: How can satellite data help? *Bull. Amer. Meteor. Soc.*, **86**, 681–689.

Jin, X. Z., X. H. Zhang, and T. J. Zhou (1999). Fundamental framework and experiments of the third generation of the IAP/LASG World Ocean General Circulation Model. *Adv. Atmos. Sci.*, **16**, 197–215.

Johansson, B., and D. Chen (2003). The influence of wind and topography on precipitation distribution in Sweden: statistical analysis and modelling. *Int. J. Climatol.*, **23**, 1523–1535.

Johns, T. C., C. F. Durman, H. T. Banks, *et al.* (2006). The new Hadley Centre climate model HadGEM1: Evaluation of coupled simulations. *J. Climate*, **19**, 1327–1353.

Jolliffe, I. T. (2002). *Principal Component Analysis*. Berlin, Germany: Springer.

Jolliffe, I. T., and D. B. Stephenson (2003). *Forecast Verification: A Practitioner's Guide in Atmospheric Science*. Chichester, UK: Wiley and Sons Ltd.

Joly, A., K. A. Browning, P. Bessemoulin, *et al.* (1999). Overview of the field phase of the Fronts and Atlantic Storm-Track EXperiment (FASTEX) project. *Quart. J. Roy. Meteor. Soc.*, **125**, 3131–3163.

Jones, C. D., and B. Macpherson (1997). A latent heat nudging scheme for the assimilation of precipitation data into an operational mesoscale model. *Meteor. Applic.*, **4**, 269–277.

Jones, M. S., B. A. Colle, and J. S. Tongue (2007). Evaluation of a mesoscale short-range ensemble forecast system over the northeast United States. *Wea. Forecasting*, **22**, 36–55.

Jones, R. G., J. M. Murphy, and M. Noguer (1995). Simulation of climate change over Europe using a nested regional-climate model. I: Assessment of control climate, including sensitivity to location of lateral boundaries. *Quart. J. Roy. Meteor. Soc.*, **121**, 1413–1449.

Jones, R. G., J. M. Murphy, M. Noguer, and A. B. Keen (1997). Simulation of climate change over Europe using a nested regional-climate model. II: Comparison of driving and regional model responses to a doubling of carbon dioxide. *Quart. J. Roy. Meteor. Soc.*, **123**, 265–292.

Juang, H.-M. H. (2000). The NCEP mesoscale spectral model: A revised version of the nonhydrostatic regional spectral model. *Mon. Wea. Rev.*, **128**, 2329–2362.

Juang, H.-M. H., and S.-Y. Hong (2001). Sensitivity of the NCEP regional spectral model to domain size and nesting strategy. *Mon. Wea. Rev.*, **129**, 2904–2922.

Juang, H.-M. H., and M. Kanamitsu (1994). The NMC nested regional spectral model. *Mon. Wea. Rev.*, **122**, 3–26.

Juang, H.-M. H., S.-Y. Hong, and M. Kanamitsu (1997). The NMC nested regional spectral model. An update. *Bull. Amer. Meteor. Soc.*, **78**, 2125–2143.

Jung, T., T. N. Palmer, and G. J. Shutts (2005). Influence of a stochastic parameterization on the frequency of occurrence of North Pacific weather regimes in the ECMWF model. *Geophys. Res. Lett.*, **32**, L23811, doi:10.1029/2005GL024248.

Jungclaus, J. H., N. Keenlyside, M. Botzet, *et al.* (2006). Ocean circulation and tropical variability in the AOGCM ECHAM5/MPI-OM. *J. Climate*, **19**, 3952–3972.

K-1 Developers (2004). *K-1 Coupled Model (MIROC) Description*. K-1 Technical Report 1. eds. H. Hasumi, and S. Emori. Center for Climate System Research, University of Tokyo, Tokyo, Japan.

Kain, J. S. (2004). The Kain-Fritsch convective parameterization: An update. *J. Appl. Meteor.*, **43**, 170–181.

Kain, J. S., and J. M. Fritsch (1992). The role of the convective "trigger function" in numerical forecasts of mesoscale convective systems. *Meteor. Atmos. Phys.*, **49**, 93–106.

Kain, J. S., and J. M. Fritsch (1993). Convective parameterization for mesoscale models: The Kain-Fritsch scheme. In *The Representation of Cumulus Convection in Numerical Models*. Meteorological Monograph, No. 46, American Meteorological Society, Boston, USA, pp. 165–170.

Kalnay de Rivas, E. (1972). On the use of nonuniform grids in finite difference equations. *J. Comput. Phys.*, **10**, 202–210.

Kalnay, E. (2003). *Atmospheric Modeling, Data Assimilation and Predictability*. Cambridge, UK: Cambridge University Press.

Kalnay, E., M. Kanamitsu, R. Kistler, *et al.* (1996). The NCEP/NCAR 40-year reanalysis project. *Bull. Amer. Meteor. Soc.*, **77**, 437–472.

Kalnay, E., D. L. T. Anderson, A. F. Bennett, *et al.* (1997). Data assimilation in the ocean and atmosphere: What should be next. *J. Meteor. Soc. Japan*, **75**, 489–496.

Kanamaru, H., and M. Kanamitsu (2007a). Scale-selective bias correction in a downscaling of global analysis using a regional model. *Mon. Wea. Rev.*, **135**, 334–350.

Kanamaru, H., and M. Kanamitsu (2007b). Fifty-seven-year California Reanalysis downscaling at 10 km (CaRD10). Part II: Comparison with North American Regional Reanalysis. *J. Climate*, **20**, 5572–5592.

Kanamaru, H., and M. Kanamitsu (2008). Dynamical downscaling of global analysis and simulation over the Northern Hemisphere. *Mon. Wea. Rev.*, **136**, 2796–2803.

Kanamitsu, M. (1989). Description of the NMC global data assimilation and forecast system. *Wea. Forecasting*, **4**, 335–342.

Kanamitsu, M., and H. Kanamaru (2007). Fifty-seven-year California Reanalysis downscaling at 10 km (CaRD10). Part I: System detail and validation with observations. *J. Climate*, **20**, 5553–5571.

Kanamitsu, M., J. C. Alpert, K. A. Campana, *et al.* (1991). Recent changes implemented into the global forecast system at NMC. *Wea. Forecasting*, **6**, 425–435.

Kanamitsu, M., W. Ebisuzaki, J. Woollen, *et al.* (2002a). NCEP-DOE AMIP-II reanalysis (R-2). *Bull. Amer. Meteor. Soc.*, **83**, 1631–1643.

Kanamitsu, M., A. Kumar, H.-M. H. Juang, *et al.* (2002b). NCEP dynamical seasonal forecast system 2000. *Bull. Amer. Meteor. Soc.*, **83**, 1019–1037.

Kang, H.-S., D.-H. Cha, and D.-K. Lee (2005). Evaluation of the mesoscale model/land surface model (MM5/LSM) coupled model for East Asian summer monsoon simulations. *J. Geophys. Res.*, **110**, D10105, doi:10.1029/2004JD005266.

Kar, S. K., and R. P. Turco (1995). Formulation of a lateral sponge layer for limited-area shallow water models and an extension for the vertically stratified case. *Mon. Wea. Rev.*, **123**, 1542–1559.

Karyampudi, V. M., and H. F. Pierce (2002). Synoptic-scale influence of the Saharan air layer on tropical cyclogenesis over the Eastern Atlantic. *Mon. Wea. Rev.*, **130**, 3100–3128.

Kasahara, A. (1974). Various vertical coordinate systems used for numerical weather prediction. *Mon. Wea. Rev.*, **102**, 509–522.

Kasahara, A., J.-I. Tsutsui, and H. Hirakuchi (1996). Inversion methods of three cumulus parameterizations for diabatic initialization of a tropical cyclone model. *Mon. Wea. Rev.*, **124**, 2304–2321.

Katz, R. W. (1996). Use of conditional stochastic models to generate climate change scenarios. *Climatic Change*, **32**, 237–255.

Katz, R. W., M. B. Parlange, and C. Tebaldi (2003). Stochastic modelling of the effects of large-scale circulation on daily weather in the southeastern US. *Climatic Change*, **60**, 189–216.

Kaufman, Y. J., and T. Nakajima (1993). Effect of Amazon smoke on cloud microphysics and albedo: Analysis from satellite imagery. *J. Appl. Meteor.*, **32**, 729–744.

Kawai, Y., and A. Wada (2007). Diurnal sea surface temperature variation and its impact on the atmosphere and ocean: A review. *J. Oceanog.*, **63**, 721–744.

Keenlyside, N., M. Latif, M. Botzet, J. Jungclaus, and U. Schulzweida (2005). A coupled method for initializing El Niño Southern Oscillation forecasts using sea surface temperature. *Tellus*, **57A**, 340–356.

Keenlyside, N. S., M. Latif, J. Jungclaus, L. Kornblueh, and E. Roeckner (2008). Advancing decadal-scale climate prediction in the North Atlantic sector. *Nature*, **453**, 84–88, doi:10.1038/nature06921.

Keppenne, C. L. (2000). Data assimilation into a primitive equation model with a parallel ensemble Kalman filter. *Mon. Wea. Rev.*, **128**, 1971–1981.

Keppenne, C. L., and M. M. Rienecker (2002). Initial testing of a massively parallel ensemble Kalman filter with the Poseidon isopycnal ocean general circulation model. *Mon. Wea. Rev.*, **130**, 2951–2965.

Kesel, P. G., and F. J. Winninghoff (1972). The Fleet Numerical Weather Central operational primitive equation model. *Mon. Wea. Rev.*, **100**, 360–373.

Kessler, E., III (1969). *On the Distribution and Continuity of Water Substance in Atmospheric Circulations*. Meteorological Monograph No. 32, American Meteorological Society, Boston, USA.

Kidson, J. W., and C. S. Thompson (1998). A comparison of statistical and model-based downscaling techniques for estimating local-climate variations. *J. Climate*, **11**, 735–753.

Kiehl, J. T., and K. E. Trenberth (1997). Earth's annual global mean energy budget. *Bull. Amer. Meteor. Soc.*, **78**, 197–208.

Kiehl, J. T., J. J. Hack, G. B. Bonan, *et al.* (1998). The National Center for Atmospheric Research Community Climate Model: CCM3. *J. Climate*, **11**, 1131–1149.

Kilsby, C. G., P. D. Jones, A. Burton, *et al.* (2007). A daily weather generator for use in climate change studies. *Environ. Model. Software*, **22**, 1705–1719.

Kim, S.-J., G. M. Flato, G. J. Boer, and N. A. McFarlane (2002). A coupled climate model simulation of the Last Glacial Maximum, Part 1: Transient multi-decadal response. *Climate Dyn.*, **19**, 515–537.

Kinnmark, I. (1985). *The Shallow Water Wave Equations: Formulation, Analysis and Application*. Lecture Notes in Engineering, Vol. 15. Berlin, Germany: Springer-Verlag.

Klein, T., and G. Heinemann (2001). On the forcing mechanisms of mesocyclones in the eastern Weddell Sea region, Antarctica: process studies using a mesoscale numerical model. *Meteorol. Zeitschrift*, **10**, 113–122.

Klein, W. H., B. M. Lewis, and I. Enger (1959). Objective prediction of five-day mean temperatures during winter. *J. Atmos. Sci.*, **16**, 672–682.

Klemp, J. B., and D. R. Durran (1983). An upper boundary condition permitting internal gravity-wave radiation in numerical mesoscale models. *Mon. Wea. Rev.*, **111**, 430–444.

Klemp, J. B., and D. K. Lilly (1978). Numerical simulation of hydrostatic mountain waves. *J. Atmos. Sci.*, **35**, 78–107.

Klemp, J. B., and R. Wilhelmson (1978). The simulation of three-dimensional convective-storm dynamics. *J. Atmos. Sci.*, **35**, 1070–1096.

Klemp, J. B., W. C. Skamarock, and J. Dudhia (2007). Conservative split-explicit time integration methods for the compressible nonhydrostatic equations. *Mon. Wea. Rev.*, **135**, 2897–2913.

Knievel, J. C., G. H. Bryan, and J. P. Hacker (2007). Explicit numerical diffusion in the WRF model. *Mon. Wea. Rev.*, **135**, 3808–3824.

Knutson, T. R., J. J. Sirutis, S. T. Garner, G. A. Vecchi, and I. M. Held (2008). Simulated reduction in Atlantic hurricane frequency under twenty-first-century warming conditions. *Nature Geoscience*, **1**, 359–364.

Kohonen, T. (2000). *Self-Organizing Maps*. New York, USA: Springer.

Køltzow, M., T. Iversen, and J. E. Haugen (2008). Extended Big-Brother experiments: the role of lateral boundary data quality and size of integration domain in regional climate modelling. *Tellus*, **60A**, 398–410.

Kondo, H., Y. Genchi, Y. Kikegawa, *et al.* (2005). Development of a multi-layer urban canopy model for the analysis of energy consumption in a big city: Structure of the urban canopy model and its basic performance. *Bound.-Layer Meteor.*, **116**, 395–421.

Koren, I., Y. J. Kaufman, R. Washington, *et al.* (2006). The Bodélé depression: a single spot in the Sahara that provides most of the mineral dust for the Amazon forest. *Environ. Res. Lett.*, **1**, doi:10.1088/1748-9326/1/1/014005.

Koren, V., J. Schaake, K. Mitchell, *et al.* (1999). A parameterization of snowpack and frozen ground intended for NCEP weather and climate models. *J. Geophys. Res.*, **104**, 19569–19585.

Koshyk, J. N., and K. Hamilton (2001). The horizontal kinetic energy spectrum and spectral budget simulated by a high-resolution troposphere-stratosphere-mesosphere GCM. *J. Atmos. Sci.*, **58**, 329–348.

Koster, R. D., and M. J. Suarez (1996). *Energy and Water Balance Calculations in the Mosaic LSM*. NASA Tech. Memo. 104606, Vol. 9.

Kreitzberg, C. W., and D. Perkey (1976). Release of potential instability. Part I: A sequential plume model within a hydrostatic primitive equation model. *J. Atmos. Sci.*, **33**, 456–475.

Krinner, G., N. Viovy, N. de Noblet-Ducoudré, *et al.* (2005). A dynamic global vegetation model for studies of the coupled atmosphere-biosphere system. *Global Biogeochem. Cycles*, **19**, GB1015, doi:10.1029/2003GB002199.

Krishnamurti, T. N., J. Xue, H. S. Bedi, K. Ingles, and D. Oosterhof (1991). Physical initialization for numerical weather prediction in the tropics. *Tellus*, **43AB**, 53–81.

Krishnamurti, T. N., G. Rohaly, and H. S. Bedi (1994). On the improvement of precipitation forecast skill from physical initialization. *Tellus*, **46A**, 598–614.

Krishnamurti, T. N., C. M. Kishtawal, T. E. LaRow, *et al.* (1999). Improved weather and seasonal climate forecasts from multimodel superensemble. *Science*, **285**, 1548–1550.

Krishnamurti, T. N., H. S. Bedi, V. M Hardiker, and L. Ramaswamy (2006a). *An Introduction to Global Spectral Modeling*. New York, USA: Springer.

Krishnamurti, T. N., A. Chakraborty, R. Krishnamurti, W. K. Dewar, and C. A. Clayson (2006b). Seasonal prediction of sea surface temperature anomalies using a suite of 13 coupled atmosphere-ocean models. *J. Climate*, **19**, 6069–6088.

Krishnamurti, T. N., S. Pattnaik, and D. V. Bhaskar Rao (2007). Mesoscale moisture initialization for monsoon and hurricane forecasts. *Mon. Wea. Rev.*, **135**, 2716–2736.

Kuhn, K., D. Campbell-Lendrum, A. Haines, and J. Cox (2005). *Using Climate to Predict Infectious Disease Epidemics*. Geneva: World Health Organization.

Kuo, H.-L. (1965). On the formation and intensification of tropical cyclones through latent heat release by cumulus convection. *J. Atmos. Sci.*, **22**, 40–63.

Kuo, H.-L. (1974). Further studies of the parameterization of the influence of cumulus convection on large-scale flow. *J. Atmos. Sci.*, **31**, 1232–1240.

Kuo, Y.-H., and R. A. Anthes (1984). Accuracy of diagnostic heat and moisture budgets using SESAME-79 field data as revealed by observing system simulation experiments. *Mon. Wea. Rev.*, **112**, 1465–1481.

Kuo, Y.-H., and Y.-R. Guo (1989). Dynamic initialization using observations from a hypothetical network of profilers. *Mon. Wea. Rev.*, **117**, 1975–1998.

Kuo, Y.-H., M. Skumanich, P. L. Haagenson, and J. S. Chang (1985). The accuracy of trajectory models as revealed by the observing system simulation experiments. *Mon. Wea. Rev.*, **113**, 1852–1867.

Kuo, Y.-H., E. G. Donall, and M. A. Shapiro (1987). Feasibility of short-range numerical weather prediction using observations from a network of profilers. *Mon. Wea. Rev.*, **115**, 2402–2427.

Kuo, Y.-H., X. Zou, and W. Huang (1998). The impact of global positioning system data on the prediction of an extratropical cyclone: An observing system simulation experiment. *Dyn. Atmos. Oceans*, **27**, 439–470.

Kusaka, H., H. Kondo, Y. Kikegawa, and F. Kimura (2001). A simple single-layer urban canopy model for atmospheric models: Comparisons with multi-layer and slab models. *Bound.-Layer Meteor.*, **101**, 329–358.

Kutzbach, J. (1967). Empirical eigenvectors of sea-level pressure, surface temperature, and precipitation complexes over North America. *J. Appl. Meteor.*, **6**, 791–802.

Lacarra, J. F., and O. Talagrand (1988). Short-range evolution of small perturbations in a barotropic model. *Tellus*, **40A**, 81–95.

Lackmann, G. M., D. Keyser, and L. F. Bosart (1999). Energetics of an intensifying jet streak during the Experiment on Rapidly Intensifying Cyclones over the Atlantic (ERICA). *Mon. Wea. Rev.*, **127**, 2777–2795.

Laîné, A., M. Kageyama, D. Salas-Mélia, *et al.* (2009). An energetic study of wintertime Northern Hemisphere storm tracks under $4 \times CO_2$ conditions in two ocean-atmosphere coupled models. *J. Climate*, **22**, 819–839.

Lakhtakia, M. N., and T. T. Warner (1987). A real-data numerical study of the development of the precipitation along the edge of an elevated mixed layer. *Mon. Wea. Rev.*, **115**, 156–168.

Lambert, S. J., and G. J. Boer (2001). CMIP1 evaluation and intercomparison of coupled climate models. *Climate Dyn.*, **17**, 83–116.

Landberg, L., L. Myllerup, O. Rathmann, *et al.* (2003). Wind-resource estimation: An overview. *Wind Energy*, **6**, 261–271.

Langland, R. H. (2005). Observation impact during the North Atlantic TReC-2003. *Mon. Wea. Rev.*, **133**, 2297–2309.

Langland, R. H., Z. Toth, R. Gelaro, *et al.* (1999). The North Pacific Experiment (NOR-PEX-98). Targeted observations for improved North American weather forecasts. *Bull. Amer. Meteor. Soc.*, **80**, 1363–1384.

Lanzinger, A., and R. Steinacker (1990). A fine mesh analysis scheme designed for mountainous terrain. *Meteor. Atmos. Phys.*, **43**, 213–219.

Lapeyre, G., and I. M. Held (2004). The role of moisture in the dynamics and energetics of turbulent baroclinic eddies. *J. Atmos. Sci.,* **61,** 1693 1710.

Laprise, R. (1992). The resolution of global spectral models. *Bull. Amer. Meteor. Soc.*, **73**, 1453–1454.

Laprise, R. (2008). Regional climate modelling. *J. Comput. Phys.*, **227**, 3641–3666.

Laprise, R., M. R. Varma, B. Denis, D. Caya, and I. Zawadzki (2000). Predictability of a nested limited-area model. *Mon. Wea. Rev.*, **128**, 4149–4154.

Laprise, R., R. de Elía, D. Caya, *et al.* (2008). Challenging some tenets of regional climate modeling. *Meteor. Atmos. Phys.*, **100**, 3–22.

Laursen, L., and E. Eliasen (1989). On the effects of the damping mechanisms in an atmospheric general circulation model. *Tellus*, **41A**, 385–400.

Lavoie, R. L. (1972). A mesoscale numerical model of lake-effect storms. *J. Atmos. Sci.*, **29**, 1025–1040.

Lawton, R. O., U. S. Nair, R. A. Pielke Sr, and R. M. Welch (2001). Climatic impacts of tropical lowland deforestation on nearby montane cloud forests. *Science*, **294**, 584–587.

Lax, P. D., and B. Wendroff (1960). Systems of conservation laws. *Comm. Pure Appl. Math.*, **13**, 217–237.

Lean, H. W., and P. A. Clark (2003). The effects of changing resolution on mesoscale modelling of line convection and slantwise circulations in FASTEX IOP 16. *Quart. J. Roy. Meteor. Soc.*, **129**, 2255–2278.

Leavesley, G. H., R. W. Lichty, B. M. Troutman, and L. G. Saindon (1983). *Precipitation-runoff Modeling System: User's Manual.* U.S. Geological Survey Water Investment Rep. 83–4238.

LeDimet, F.-X., and O. Talagrand (1986). Variational algorithms for analysis and initialization of meteorological observations: theoretical aspects. *Tellus*, **38A**, 97–110.

Legg, T. P., and K. R. Mylne (2004). Early warnings of severe weather from ensemble forecast information. *Wea. Forecasting*, **19**, 891–906; Corrigendum, **22**, 216–219.

Legg, T. P., K. R. Mylne, and C. Woodcock (2002). Use of medium-range ensembles at the Met Office I: PREVIN – a system for the production of probabilistic forecast information from the ECMWF EPS. *Meteor. Applic.*, **9**, 255–271.

Le Marshall, J. F., L. M. Leslie, and C. Spinoso (1997). The generation and assimilation of cloud-drift winds in numerical weather prediction. *J. Meteor. Soc. Japan*, **75**, Special Issue 1B, 383–393.

Leslie, L. M., J. F. LeMarshall, R. P. Morrison, *et al.* (1998). Improved hurricane track forecasting from the continuous assimilation of high quality satellite wind data. *Mon. Wea. Rev.*, **126**, 1248–1257.

Leuenberger, D., and A. Rossa (2007). Revisiting the latent heat nudging scheme for the rainfall assimilation of a simulated convective storm. *Meteor. Atmos. Phys.*, **98**, 195–215.

Leung, L. R., and S. J. Ghan (1995). A subgrid parameterization of orographic precipitation. *Theor. Appl. Climatol.*, **52**, 95–118.

Leung, L. R., and S. J. Ghan (1998). Parameterizing subgrid orographic precipitation and surface cover in climate models. *Mon. Wea. Rev.*, **126**, 3271–3291.

Leung, L. R., Y. Qian, X. Bian, *et al.* (2004). Mid-century ensemble regional climate change scenarios for the western United States. *Climate Change*, **62**, 75–113.

Leutbecher, M., and T. N. Palmer (2008). Ensemble forecasting. *J. Comput. Phys.*, **227**, 3515–3539.

Levin, Z., and W. R. Cotton (eds.) (2009). *Air Pollution Impact on Precipitation*. Berlin, Germany: Springer.

Li, P. W., and E. S. T. Lai (2004). Short-range quantitative precipitation forecasting in Hong Kong. *J. Hydrol.*, **288**, 189–209.

Liang, X.-Z., L. Li, K. E. Kunkel, M. Ting, and J. X. L. Wang (2004). Regional climate model simulation of U.S. precipitation during 1982-2002. Part I: Annual cycle. *J. Climate*, **17**, 3510–3529.

Liang, X.-Z., J. Pan, J. Zhu, *et al.* (2006). Regional climate model downscaling of the U.S. summer climate and future change. *J. Geophys. Res.*, **111**, D10108, doi:10.1029/2005JD006685.

Liljegren, J. C., S. Tschopp, K. Rogers, *et al.* (2009). Quality control of meteorological data for the Chemical Stockpile Emergency Preparedness Program. *J. Atmos. and Oceanic Technol.*, **26**, 1510–1526.

Lin, J. W.-B., and J. D. Neelin (2000). Influence of a stochastic moist convective parameterization on tropical climate variability. *Geophys. Res. Lett.*, **27**, 3691–3694.

Lin, Y., and K. Mitchell (2005). The NCEP Stage II/IV hourly precipitation analyses: Development and applications. Preprints, *19th Conference on Hydrology*, San Diego, CA, American Meteorological Society, Paper 1.2.

Lin, Y.-L., R. D. Farley, and H. D. Orville (1983). Bulk parameterization of the snow field in a cloud model. *J. Climate Appl. Meteor.*, **22**, 1065–1092.

Lindzen, R. S., and M. Fox-Rabinovitz (1989). Consistent vertical and horizontal resolution. *Mon. Wea. Rev.*, **117**, 2575–2583.

Liou, K.-N. (1980). *An Introduction to Atmospheric Radiation*. London, UK: Academic Press.

Lipps, F., and R. Hemler (1982). A scale analysis of deep moist convection and some related numerical calculations. *J. Atmos. Sci.*, **29**, 2192–2210.

Lipscomb, W. H. (2001). Remapping the thickness distribution in sea ice models. *J. Geophys. Res.*, **106**, 13989–14000.

List, R. J. (1966). *Smithsonian Meteorological Tables*. Washington, DC, USA: Smithsonian Institution Press.

Liu, H., and M. Xue (2008). Prediction of convective initiation and storm evolution on 12 June 2002 during IHOP 2002. Part I: Control simulation and sensitivity experiments. *Mon. Wea. Rev.*, **136**, 2261–2282.

Liu, H., X. Zhang, W. Li, Y. Yu, and R. Yu (2004). An eddy-permitting oceanic general circulation model and its preliminary evaluations. *Adv. Atmos. Sci.*, **21**, 675–690.

Liu, J., G. A. Schmidt, D. Martinson, *et al.* (2003). Sensitivity of sea ice to physical parameterizations in the GISS global climate model. *J. Geophys. Res.*, **108**, 3053, doi:10.1029/2001JC001167.

Liu, M., D. L. Westphal, A. L. Walker, *et al.* (2007). COAMPS real-time dust storm forecasting during Operation Iraqi Freedom. *Wea. Forecasting*, **22**, 192–206.

Liu, Y., F. Chen, T. Warner, and J. Basara (2006). Verification of a mesoscale data-assimilation and forecasting system for the Oklahoma City area during the Joint Urban 2003 Field Project. *J. Appl. Meteor. Climatol.*, **45**, 912–929.

Liu, Y., T. T. Warner, J. F. Bowers, *et al.* (2008a). The operational mesogamma-scale analysis and forecast system of the U.S. Army Test and Evaluation Command. Part 1: Overview of the modeling system, the forecast products, and how the products are used. *J. Appl. Meteor. Climatol.*, **47**, 1077–1092.

Liu, Y., T. T. Warner, E. G. Astling, *et al.* (2008b). The operational mesogamma-scale analysis and forecast system of the U.S. Army Test and Evaluation Command. Part 2: Verification of model analyses and forecasts. *J. Appl. Meteor. Climatol.*, **47**, 1093–1104.

Liu, Z.-Q., and F. Rabier (2003). The potential of high-density observations for numerical weather prediction: A study with simulated observations. *Quart. J. Roy. Meteor. Soc.*, **129**, 3013–3035.

Lo, J. C.-F., Z.-L. Yang, and R. A. Pielke Sr (2008). Assessment of three dynamical climate downscaling methods using the Weather Research and Forecasting (WRF) model. *J. Geophys. Res.*, **113**, D09112, doi:10.1029/2007JD009216.

Lönnberg, P., and A. Hollingsworth (1986). The statistical structure of short-range forecast errors as determined from radiosonde data, Part II: The covariance of height and wind errors. *Tellus*, **38A**, 137–161.

Lorant, V., and J. F. Royer (2001). Sensitivity of equatorial convection to horizontal resolution in aqua-planet simulations with variable resolution. *Mon. Wea. Rev.*, **129**, 2730–2745.

Lorenc, A. (1986). Analysis methods for numerical weather prediction. *Quart. J. Roy. Meteor. Soc.*, **112**, 1177–1194.

Lorenc, A. C. (2003). The potential of the ensemble Kalman filter for NWP: A comparison with 4D-Var. *Quart. J. Roy. Meteor. Soc.*, **129**, 3183–3203.

Lorenc, A. C., R. S. Bell, and B. Macpherson (1991). The Meteorological Office analysis correction data assimilation scheme. *Quart. J. Roy. Meteor. Soc.*, **117**, 59–89.

Lorenz, E. N. (1963a). Deterministic nonperiodic flow. *J. Atmos. Sci.*, **20**, 130–141.

Lorenz, E. N. (1963b). The predictability of hydrodynamic flow. *Trans. NY Acad. Sci.*, *Series II*, **25**, 409–432.

Lorenz, E. N. (1968). Climate determinism. In *Causes of Climatic Change*, ed. J. M. Mitchell. Meteorological Monograph No. 8, Boston, USA: American Meteorological Society, 1–3.

Lott, F., and M. J. Miller (1997). A new subgrid-scale orographic drag parameterization: Its formulation and testing. *Quart. J. Roy. Meteor. Soc.*, **123**, 101–127.

Loveland, T. R., J. W. Merchant, J. F. Brown, *et al.* (1995). Seasonal land-cover regions of the United States. *Ann. Assoc. Amer. Geogr.*, **85**, 339–355.

Lu, C., H. Yuan, B. E. Schwartz, and S. G. Benjamin (2007). Short-range numerical weather prediction using time-lagged ensembles. *Wea. Forecasting*, **22**, 580–595.

Lynch, P. (2007). The origins of computer weather prediction and climate modeling. *J. Comp. Phys.*, **227**, 3431–3444.

Lynn, B. H., L. Druyan, C. Hogrefe, *et al.* (2005). Sensitivity of present and future surface temperature to precipitation characteristics. *Climate Res.*, **28**, 53–65.

Machenhauer, B. (1977). On the dynamics of gravity oscillations in a shallow water model, with applications to normal mode initialization. *Contrib. Atmos. Phys.*, **50**, 253–271.

Machenhauer, B. (1979). The spectral method. In *Numerical Methods Used in Atmospheric Models, Volume II,* ed. A. Kasahara. Global Atmospheric Research Programme, GARP Publication Series No. 17. Geneva: World Meteorological Organization. pp 121–275.

Machenhauer, B., E. Kaas, and P. H. Lauritzen (2008). Finite volume methods in meteorology. In *Computational Methods for the Atmosphere and Oceans*, eds. R. Tenan and J. Tribia. Amsterdam, the Netherlands: Elsevier.

Madec, G., P. Delecluse, M. Imbard, and C. Lévy (1998). *OPA Version 8.1 Ocean General Circulation Model Reference Manual*. Notes du Pôle de Modélisation No. 11, Institut Pierre-Simon Laplace, Paris, France.

Magarey, R. D, G. A. Fowler, D. M. Borchert, *et al.* (2007). NAPPFAST: An internet system for the weather-based mapping of plant pathogens. *Plant Disease*, **91**, 336–345.

Mahfouf, J.-F., A. O. Manzi, J. Noilhan, H. Giordani, and M. DéQué (1995). The land surface scheme ISBA within the Meteo-France climate model ARPEGE. Part 1: Implementation and preliminary results. *J. Climate*, **8**, 2039–2057.

Majewski, D., D. Liermann, P. Prohl, *et al.* (2002). The operational global icosahedral-hexagonal gridpoint model GME: Description and high-resolution tests. *Mon. Wea. Rev.*, **130**, 319–338.

Majumdar, S. J., C. H. Bishop, R. Buizza, and R. Gelaro (2002a). A comparison of ensemble transform Kalman filter targeting guidance with ECMWF and NRL total energy singular vector guidance. *Quart. J. Roy. Meteor. Soc.*, **128**, 2527–2549.

Majumdar, S. J., C. H. Bishop, B. J. Etherton, and Z. Toth (2002b). Adaptive sampling with the ensemble transform Kalman filter. Part II: Field program implementation. *Mon. Wea. Rev.*, **130**, 1356–1369.

Majumdar, S. J., S. D. Aberson, C. H. Bishop, *et al.* (2006). A comparison of adaptive observing guidance for Atlantic tropical cyclones. *Mon. Wea. Rev.*, **134**, 2354–2372.

Maltrud, M. E., R. D. Smith, A. J. Semtner, and R. C. Malone (1998). Global eddy-resolving ocean simulations driven by 1985–1995 atmospheric winds. *J. Geophys. Res.*, **103**, 30825–30853.

Mann, G. E., R. B. Wagenmaker, and P. J. Sousounis (2002). The influence of multiple lake interactions upon lake-effect storms. *Mon. Wea. Rev.*, **130**, 1510–1530.

Mao, Q., R. T. McNider, S. F. Mueller, and H. H. Juang (1999). An optimal model output calibration algorithm suitable for objective temperature forecasting. *Wea. Forecasting*, **14**, 190–202.

Mapes, B. E. (1997). Equilibrium vs. activation control of large-scale variations of tropical deep convection. In *The Physics and Parameterization of Moist Atmospheric Convection*, ed. R. K. Smith. Dordrecht, the Netherlands: Kluwer Academic Publishers. pp. 321–358.

Marchuk, G. I. (1965). *A New Approach to the Numerical Solution of Differential Equations of Atmospheric Processes*. World Meteorological Organization Technical Note No. 66, Geneva, pp. 286–294.

Marchuk, G. I. (1974). *Numerical Methods in Weather Prediction.* New York, USA: Academic Press.

Marletto, V., F. Zinoni, L. Criscuolo, *et al.* (2005). Evaluation of downscaled DEMETER multi-model ensemble seasonal hindcasts in a northern Italy location by means of a model of wheat growth and soil water balance. *Tellus*, **57A**, 488–497.

Marshall, C. H., R. A. Peilke Sr, L. T. Steyaert, and D. A. Willard (2004). The impact of anthropogenic land-cover change on the Florida Peninsula sea breezes and warm season sensible weather. *Mon. Wea. Rev.*, **132**, 28–52.

Marsigli, C., F. Boccanera, A. Montani, and T. Paccagnella (2005). The COSMO-LEPS mesoscale ensemble system: validation of the methodology and verification. *Nonlinear Processes Geophys.*, **12**, 527–536.

Marsland, S. J., H. Haak, J. H. Jungclaus, M. Latif, and F. Röske (2003). The Max-Planck-Institute global ocean/sea ice model with orthogonal curvilinear coordinates. *Ocean Modelling*, **5**, 91–127.

Marti, O., P. Braconnot, J. Bellier, *et al.* (2005). *The New IPSL Climate System Model: IPSL-CM4*. Note du Pôle de Modélisation No. 26, Institut Pierre Simon Laplace des Sciences de l'Environnement Global, Paris, France.

Martilli, A. (2007). Current research and future challenges in urban mesoscale modelling. *Int. J. Climatol.*, **27**, 1909–1918.

Martilli, A., A. Clappier, and M. W. Rotach (2002). An urban surface exchange parameterization for mesoscale models. *Bound.-Layer Meteor.*, **104**, 261–304.

Martin, G. M., C. Dearden, C. Greeves, *et al.* (2004). *Evaluation of the Atmospheric Performance of HadGAM/GEM1*. Hadley Centre Technical Note No. 54, Hadley Centre for Climate Prediction and Research/Met Office, Exeter, UK.

Marzban, C., and S. Sandgathe (2006). Cluster analysis for verification of precipitation fields. *Wea. Forecasting*, **21**, 824–838.

Marzban, C., and S. Sandgathe (2008). Cluster analysis for object-oriented verification of fields: A variation. *Mon. Wea. Rev.*, **136**, 1013–1025.

Mass, C. F., D. Ovens, K. Westrick, and B. A. Colle (2002). Does increasing horizontal resolution produce more skillful forecasts? *Bull. Amer. Meteor. Soc.*, **83**, 407–430.

Masson, V. (2000). A physically based scheme for the urban energy budget in atmospheric models. *Bound.-Layer Meteor.*, **94**, 357–397.

Matsuno, T. (1966). Numerical integrations of the primitive equations by a simulated backward difference method. *J. Meteor. Soc. Japan,* **44**, 76–84.

Matulla, C., H. Scheifinger, A. Menzel, and E. Koch (2003). Exploring two methods for statistical downscaling of Central European phenological time series. *Int. J. Biometeor.,* **48**, 56–64.

May, W., and E. Roeckner (2001). A time-slice experiment with the ECHAM4 AGCM at high resolution: The impact of horizontal resolution on annual mean climate change. *Climate Dyn.,* **17**, 407–420.

McCollor, D., and R. Stull (2008a). Hydrometeorological short-range ensemble forecasts in complex terrain. Part I: Meteorological evaluation. *Wea. Forecasting,* **23**, 533–556.

McCollor, D., and R. Stull (2008b). Hydrometeorological short-range ensemble forecasts in complex terrain. Part II: Economic evaluation. *Wea. Forecasting,* **23**, 557–574.

McCollor, D., and R. Stull (2008c). Hydrometeorological accuracy enhancement via post-processing of numerical weather forecasts in complex terrain. *Wea. Forecasting,* **23**, 131–144.

McFarlane, N. A., G. J. Boer, J.-P. Blanchet, and M. Lazare (1992). The Canadian Climate Centre second-generation general circulation model and its equilibrium climate. *J. Climate,* **5**, 1013–1044.

McGregor, J. L. (1996). Semi-Lagrangian advection on conformal-cubic grids. *Mon. Wea. Rev.,* **124**, 1311–1322.

McGregor, J. L. (1997). Regional climate modelling. *Meteorol. Atmos. Phys.,* **63**, 105–117.

McGregor, J. L., and M. R. Dix (2001). The CSIRO conformal-cubic atmospheric GCM. *Proc. IUTAM Symposium on Advances in Mathematical Modeling of Atmospheric and Ocean Dynamics,* Limerick, Ireland. Dordrecht, the Netherlands: Kluwer Academic Publishers, pp. 197–202.

McWilliams, J. C. (2006). *Fundamentals of Geophysical Fluid Dynamics.* Cambridge, UK: Cambridge University Press.

Meehl, G. A., and C. Tebaldi (2004). More intense, more frequent, and longer lasting heat waves in the 21st century. *Science,* **305**, 994–997.

Meehl, G. A., G. J. Boer, C. Covey, M. Latif, and R. J. Stouffer (2000). The Coupled Model Intercomparison Project (CMIP). *Bull. Amer. Meteor. Soc.,* **81**, 313–318.

Meehl, G. A., C. Covey, B. McAvaney, M. Latif, and R. J. Stouffer (2005). Overview of the coupled model intercomparison project. *Bull. Amer. Meteor. Soc.,* **86**, 89–93.

Meehl, G. A., T. F. Stocker, W. D. Collins, *et al.* (2007). Global Climate Projections. In *Climate Change 2007: The Physical Science Basis. Contribution of Working Group I to the Fourth Assessment Report of the Intergovernmental Panel on Climate Change,* eds. S. Solomon, D. Qin, M. Manning, *et al.,* Cambridge, UK and New York, USA: Cambridge University Press.

Mehrotra, R., A. Sharma, and I. Cordery (2004). Comparison of two approaches for downscaling synoptic atmospheric patterns to multisite precipitation occurrence. *J. Geophys. Res.,* **109**, D14107, doi:10.1029/2004JD004823.

Mellor, G. L., and L. Kantha (1989). An ice-ocean coupled model. *J. Geophys. Res.,* **94**, 10937–10954.

Menendez, C. G., V. Serafini, and H. Le Treut (1999). The storm tracks and the energy cycle of the Southern Hemisphere: sensitivity to sea-ice boundary conditions. *Ann. Geophys. Atmos. Hydros. Space Sci.*, **17**, 1478–1492.

Meng, Z., and F. Zhang (2008). Tests of an ensemble Kalman filter for mesoscale and regional-scale data assimilation. Part IV: Comparison with 3DVAR in a month-long experiment. *Mon. Wea. Rev.*, **136**, 3671–3682.

Mera, R. J., D. Niyogi, G. S. Buol, G. G. Wilkerson, and F. H. M. Semazzi (2006). Potential individual versus simultaneous climate change effects on soybean (C3) and maize (C4) crops: An agrotechnology model based study. *Global Planet. Change*, **54**, 163–182.

Mesinger, F., Z. I. Janjić, S. Nicković, D. Gavrilov, and D. G. Deaven (1988). The step-mountain coordinate: model description and performance for cases of Alpine lee cyclogenesis and for a case of an Appalachian redevelopment. *Mon. Wea. Rev.*, **116**, 1491–1520.

Mesinger, F., G. DiMego, E. Kalnay, *et al.* (2006). North American Regional Reanalysis. *Bull. Amer. Meteor. Soc.*, **87**, 343–360.

Michaelsen, J. (1987). Cross-validation in statistical climate forecast models. *J. Climate Appl. Meteor.*, **26**, 1589–1600.

Miguez-Macho, G., G. L. Stenchikov, and A. Robock (2004). Spectral nudging to eliminate the effects of domain position and geometry in regional climate simulations. *J. Geophys. Res.*, **109**, D13104, doi:10.1029/2003JD004495.

Milan, M., V. Venema, D. Schüttemeyer, and C. Simmer (2008). Assimilation of radar and satellite data in mesoscale models: A physical initialization scheme. *Meteorol. Zeitschrift*, **17**, 887–902.

Miller, D. A., and R. A. White (1998). A conterminous United States multilayer soil characteristics data set for regional climate and hydrology modeling. *Earth Interactions*, **2**, 1–26.

Miller, D. H. (1981). *Energy at the Surface of the Earth: An Introduction to the Energetics of Ecosystems*. New York, USA: Academic Press.

Miller, J. R., G. L. Russell, and G. Caliri (1994). Continental-scale river flow in climate models. *J. Climate*, **7**, 914–928.

Miller, N. L., K. Hayhoe, J. Jin, and M. Auffhammer (2008). Climate, extreme heat, and electricity demand in California. *J. Appl. Meteor. Climatol.*, **47**, 1834–1844.

Miller, P. A., and S. G. Benjamin (1992). A system for the hourly assimilation of surface observations in mountainous and flat terrain. *Mon. Wea. Rev.*, **120**, 2342–2359.

Miller, R. N. (2007). *Numerical Modeling of Ocean Circulation*. Cambridge, UK: Cambridge University Press.

Milly, P. C. D., and A. B. Shmakin (2002). Global modeling of land water and energy balances, Part I: The Land Dynamics (LaD) model. *J. Hydrometeor.*, **3**, 283–299.

Min, S.-K., and A. Hense (2006). A Bayesian approach to climate model evaluation and multi-model averaging with an application to global mean surface temperatures from IPCC AR4 coupled climate models. *Geophys. Res. Lett.*, **33**, L08708, doi:10.1029/2006GL025779.

Min, S.-K., S. Legutke, A. Hense, and W.-T. Kwon (2005). Climatology and internal variability in a 1000-year control simulation with the coupled climate model

ECHO-G–I. Near-surface temperature, precipitation and mean sea level pressure. *Tellus*, **57A**, 605–621.

Misra, V. (2005). Simulation of the intraseasonal variance of the South American summer monsoon. *Mon. Wea. Rev.*, **133**, 663–676.

Mitchell, H. L., and P. L. Houtekamer (2000). An adaptive ensemble Kalman filter. *Mon. Wea. Rev.*, **128**, 416–433.

Mitchell, H. L., P. L. Houtekamer, and G. Pellerin (2002). Ensemble size, balance, and model-error representation in an ensemble Kalman filter. *Mon. Wea. Rev.*, **130**, 2791–2808.

Mitchell, J. F. B. (1989). The "greenhouse" effect and climate change. *Rev. Geophys.*, **27**, 115–139.

Mittermaier, M. P. (2007). Improving short-range high-resolution model precipitation forecast skill using time lagged ensembles. *Quart. J. Roy. Meteor. Soc.*, **133**, 1487–1500.

Miyakoda, K., and A. Rosati (1977). One way nested grid models: The interface conditions and the numerical accuracy. *Mon. Wea. Rev.*, **105**, 1092–1107.

Mohanty, U. C., R. K. Paliwal, A. Tyagi, and A. John (1990). Evaluation of a limited area model for short range prediction over Indian region: Sensitivity studies. *Mausam,* **41**, 251–256.

Molinari, J., and T. Corsetti (1985). Incorporation of cloud-scale and mesoscale downdrafts into a cumulus parameterization: Results of one- and three-dimensional integrations. *Mon. Wea. Rev.*, **113**, 485–501.

Molteni, F., R. Buizza, T. N. Palmer, and T. Petroliagis (1996). The new ECMWF Ensemble Prediction System: methodology and validation. *Quart. J. Roy. Meteor. Soc.*, **122**, 73–119.

Moninger, W. R., R. D. Mamrosh, and P. M Pauley (2003). Automated meteorological reports from commercial aircraft. *Bull. Amer. Meteor. Soc.*, **84**, 203–216.

Monobianco, J., S. Koch, V. M. Karyampudi, and A. J. Negri (1994). The impact of assimilating satellite-derived precipitation rates on numerical simulations of the ERICA IOP 4 cyclone. *Mon. Wea. Rev.*, **122**, 341–365.

Monobianco, J., J. G. Dreher, R. J. Evans, J. L. Case, and M. L. Adams (2008). The impact of simulated super pressure balloon data on regional weather analyses and forecasts. *Meteor. Atmos. Phys.*, **101**, 21–41.

Montani, A., A. J. Thorpe, R. Buizza, and P. Undén (1999). Forecast skill of the ECMWF model using targeted observations during FASTEX. *Quart. J. Roy. Meteor. Soc.*, **125**, 3219–3240.

Monteith, J. L., and M. H. Unsworth (1990). *Principles of Environmental Physics*. London, UK: Edward Arnold.

Moore, R. W., and M. T. Montgomery (2004). Reexamining the dynamics of short-scale, diabatic Rossby waves and their role in midlatitude moist cyclogenesis. *J. Atmos. Sci.*, **61**, 754–768.

Moore, R. W., and M. T. Montgomery (2005). Analysis of an idealized, three-dimensional diabatic Rossby vortex: A coherent structure of the moist baroclinic atmosphere. *J. Atmos. Sci.*, **62**, 2703–2725.

Moriondo, M., and M. Bindi (2006). Comparisons of temperatures simulated by GCMs, RCMs and statistical downscaling: potential application in studies of future crop development. *Climate Res.*, **30**, 149–160.

Morse, A. P., F. J. Doblas-Reyes, M. B. Hoshen, R. Hagedorn, and T. N. Palmer (2005). A forecast quality assessment of an end-to-end probabilistic multi-model seasonal forecast system using a malaria model. *Tellus*, **57A**, 464–475.

Mu, M., and G. J. Zhang (2006). Energetics of Madden-Julian oscillations in the National Center for Atmospheric Research Community Atmosphere Model version 3 (NCAR CAM3). *J. Geophys. Res.*, **111**, D24112, doi:10.1029/2005JD007003.

Mullen, S. L., and D. P. Baumhefner (1989). The impact of initial condition uncertainty on numerical simulations of large-scale explosive cyclogenesis. *Mon. Wea. Rev.*, **117**, 2800–2821.

Mullen, S. L., and R. Buizza (2002). The impact of horizontal resolution and ensemble size on probabilistic forecasts of precipitation by the ECMWF Ensemble Prediction System. *Wea. Forecasting*, **17**, 173–191.

Murphy, A. H. (1988). Skill scores based on the mean square error and their relationships to the correlation coefficient. *Mon. Wea. Rev.*, **116**, 2417–2424.

Murphy, A. H., B. G. Brown, and Y.-S. Chen (1989). Diagnostic verification of temperature forecasts. *Wea. Forecasting*, **4**, 485–501.

Murphy, J. M., D. M. H. Sexton, D. N. Barnett, *et al.* (2004). Quantification of modelling uncertainties in a large ensemble of climate simulations. *Nature*, **430**, 768–772.

Nachamkin, J. E., S. Chen, and J. Schmidt (2005). Evaluation of heavy precipitation forecasts using composite-based methods: A distributions-oriented approach. *Mon. Wea. Rev.*, **133**, 2163–2177.

Nadiga, B. T., L. G. Margolin, and P. K. Smolarkiewicz (1996). Different approximations of shallow fluid flow over an obstacle. *Phys. Fluids*, **8**, 2066–2077.

Nair, U. S., H. R. Mark, and R. A. Pielke Sr (1997). Numerical simulation of the 9–10 June 1972 Black Hills storm using CSU RAMS. *Mon. Wea. Rev.*, **125**, 1753–1766.

Nakicenovic, N. (ed.) (2000). *Special Report on Emissions Scenarios*. Cambridge, UK: Cambridge University Press.

Nehrkorn, T., R. N. Hoffman, J.-F. Louis, R. G. Isaacs, and J.-L. Moncet (1993). Analysis and forecast improvements from simulated satellite water vapor profiles and rainfall using a global data assimilation system. *Mon. Wea. Rev.*, **121**, 2727–2739.

Nickovic, S. (1994). On the use of hexagonal grids for simulation of atmospheric processes. *Beitr. Phys. Atmos.*, **67**, 103–107.

Nickovic, S., G. Kallos, A. Papadopoulos, and O. Kakaliagou (2001). A model for prediction of desert dust cycle in the atmosphere. *J. Geophys. Res.*, **106**, 18113–18129.

Nieman, S. J., W. P. Menzel, C. M. Hayden, D. Gray, S. Wanzong, C. Velden, and J. Daniels (1997). Fully automated cloud-drift winds in NESDIS operations. *Bull. Amer. Meteor. Soc.*, **78**, 1121–1133.

Ninomiya, K., and K. Kurihara (1987). Forecast experiment of a long-lived mesoscale convective system in Baiu frontalzone. *J. Meteor. Soc. Japan*, **65**, 885–899.

Nitta, T., and J. B. Hovermale (1969). A technique of objective analysis and initialization for the primitive forecast equations. *Mon. Wea. Rev.*, **97**, 652–658.

Novak, M. D. (1986). Theoretical values of daily atmospheric and soil thermal admittances. *Bound. Layer Meteor.*, **34**, 17–34.

NRC (2001). *Under the Weather: Climate, Ecosystems, and Infectious Disease*. National Research Council, Washington, DC, USA: National Academy Press.

Nunes, A. M. B., and S. Cocke (2004). Implementing a physical initialization procedure in a regional spectral model: impact on the short-range rainfall forecasting over South America. *Tellus*, **56A**, 125–140.

Nunes, A. M. B., and J. O. Roads (2007a). Dynamical influences of precipitation assimilation on regional downscaling. *Geophys. Res. Lett.*, **34**, L16817, doi:10.1029/2007GL030247.

Nunes, A. M. B., and J. O. Roads (2007b). Influence of precipitation assimilation on a regional climate model's surface water and energy budgets. *J. Hydrometeor.*, **8**, 642–664.

Nuss, W. A., and R. A. Anthes (1987). A numerical investigation of low-level processes in rapid cyclogenesis. *Mon. Wea. Rev.*, **115**, 2728–2743.

Nutter, P. A. (2003). Effects of nesting frequency and lateral boundary perturbations on the dispersion of limited-area ensemble forecasts. Ph.D. Dissertation, University of Oklahoma, Norman, USA.

O'Farrell, S. P. (1998). Investigation of the dynamic sea ice component of a coupled atmosphere sea-ice general circulation model. *J. Geophys. Res.*, **103**, 15751–15782.

Ogura, Y., and N. Phillips (1962). Scale analysis for deep and shallow convection in the atmosphere. *J. Atmos. Sci.*, **19**, 173–179.

Okamoto, K., and J. C. Derber (2006). Assimilation of SSM/I radiances in the NCEP Global Data Assimilation System. *Mon. Wea. Rev.*, **134**, 2612–2631.

Oke, T. R. (1987). *Boundary Layer Climates*. London, UK: Methuen.

Oki, T., and Y. C. Sud (1998). Design of total runoff integrating pathways (TRIP): A global river channel network. *Earth Interactions*, **2**, 1–37.

O'lenic, E. A., D. A. Unger, M. S. Halpert, and K. S. Pelman (2008). Developments in operational long-range climate prediction at CPC. *Wea. Forecasting*, **23**, 496–515.

Oleson, K. W., G. Bonan, S. Levis, *et al.* (2004). *Technical Description of the Community Land Model (CLM)*. NCAR Technical Note NCAR/TN-461+STR, National Center for Atmospheric Research, Boulder, CO, USA.

Oleson, K. W., G.-Y. Niu, Z.-L. Yang, *et al.* (2008). Improvements to the Community Land Model and their impact on the hydrologic cycle. *J. Geophys. Res.*, **113**, G01021, doi:10.1029/2007JG000563.

Oliger, J., and A. Sundstrom (1978). Theoretical and practical aspects of some initial boundary value problems in fluid dynamics. *S.I.A.M. J. Appl. Math.*, **35**, 419–446.

Onogi, K., J. Tsutsui, H. Koide, *et al.* (2007). The JRA-25 Reanalysis. *J. Meteor. Soc. Japan*, **85**, 369–432.

Ooyama, K. V. (1982). Conceptual evolution of the theory and modeling of the tropical cyclone. *J. Meteor. Soc. Japan*, **60**, 369–379.

Orlanski, I., and J. Katzfey (1991). The life cycle of a cyclone wave in the Southern Hemisphere. Part I: Eddy energy budget. *J. Atmos. Sci.*, **48**, 1972–1998.

Orlanski, I., and J. Katzfey (1995). Stages in the energetics of baroclinic systems. *Tellus*, **47A**, 605–628.

Orzag, S. A. (1970). Transform method for the calculation of vector-coupled sums: Application to the spectral form of the vorticity equation. *J. Atmos. Sci.*, **27**, 890–895.

Otte, T. L., N. L. Seaman, and D. R. Stauffer (2001). A heuristic study on the importance of anisotropic error distributions in data assimilation. *Mon. Wea. Rev.*, **129**, 766–783.

Pacanowski, R. C., K. Dixon, and A. Rosati (1993). *The GFDL Modular Ocean Model Users Guide, Version 1.0*. GFDL Ocean Group Technical Report No. 2, Geophysical Fluid Dynamics Laboratory, Princeton, NJ, USA.

Paegle, J., J. Nogues-Paegle, and K. C. Mo (1996). Dependence of simulated precipitation on surface evaporation during the 1993 United States summer floods. *Mon. Wea. Rev.*, **124**, 345–361.

Paeth, H., K. Born, R. Podzun, and D. Jacob (2005). Regional dynamical downscaling over West Africa: Model evaluation and comparison of wet and dry years. *Meteorol. Zeitschrift*, **14**, 349–367.

Pagowski, M., and G. A. Grell (2006). Ensemble-based ozone forecasts: Skill and economic value. *J. Geophys. Res.*, **111**, D23S30, doi: 10.1029/2006JD007124.

Palmer, T. N. (1993). Extended-range atmospheric prediction and the Lorenz model. *Bull. Amer. Meteor. Soc.*, **74**, 49–65.

Palmer, T. N. (2001). A nonlinear dynamical perspective on model error: A proposal for non-local stochastic-dynamic parameterization in weather and climate prediction models. *Quart. J. Roy. Meteor. Soc.*, **127**, 279–304.

Palmer, T. N. (2002). The economic value of ensemble forecasts as a tool for risk assessment: From days to decades. *Quart. J. Roy. Meteor. Soc.*, **128**, 747–774.

Palmer, T. N., R. Gelaro, J. Barkmeijer, and R. Buizza (1998). Singular vectors, metrics, and adaptive observations. *J. Atmos. Sci.*, **55**, 633–653.

Palmer, T. N., A. Alessandri, U. Andersen, *et al.* (2004). Development of a European multimodel ensemble system for seasonal to interannual prediction (DEMETER). *Bull. Amer. Meteor. Soc.*, **85**, 853–872.

Palmer, T. N., G. J. Shutts, R. Hagedorn, *et al.* (2005). Representing model uncertainty in weather and climate prediction. *Annu. Rev. Earth Planet. Sci.*, **33**, 163–193.

Palutikof, J. P., C. M. Goodess, S. J. Watkins, and T. Holt (2002). Generating rainfall and temperature scenarios at multiple sites: examples from the Mediterranean. *J. Climate*, **15**, 3529–3548.

Pardyjak, E. R., M. J. Brown, and N. L. Bagal (2004). Improved velocity deficit parameterizations for a fast response urban wind model. Preprints, *Symposium on Planning, Nowcasting, and Forecasting in the Urban Zone*, Seattle, WA, American Meteorological Society.

Parrish, D. F., and J. D. Derber (1992). The National Meteorological Center spectral statistical interpolation analysis system. *Mon. Wea. Rev.*, **120**, 1747–1763.

Paul, S., C. M. Liu, J. M. Chen, and S. H. Lin (2008). Development of a statistical downscaling model for projecting monthly rainfall over East Asia from a general circulation model output. *J. Geophys. Res.*, **113**, D15117, doi:10.1029/2007JD009472.

PCMDI (2007). IPCC Model Output. [Available online at http://www-pcmdi.llnl.gov/ipcc/about_ipcc.php]

Pecnick, M. J., and D. Keyser (1989). The effect of spatial resolution on the simulation of upper-tropospheric frontogenesis using a sigma-coordinate primitive-equation model. *Meteor. Atmos. Phys.*, **40**, 137–149.

Pedlosky, J. (1987). *Geophysical Fluid Dynamics*. Berlin, Germany: Springer.

Perkey, D. J., and C. W. Kreitzberg (1976). A time-dependent lateral boundary scheme for limited area primitive equation models. *Mon. Wea. Rev.*, **104**, 744–755.

Perkey, D. J., and R. A. Maddox (1985). A numerical investigation of a mesoscale convective system. *Mon. Wea. Rev.*, **113**, 553–566.

Persson, A. (2005). Early operational numerical weather prediction outside the USA: an historical introduction. Part 1: Internationalism and engineering NWP in Sweden, 1952–69. *Meteorol. Appl.*, **12**, 135–159.

Persson, P. O. G., and T. T. Warner (1991). Model generation of spurious gravity waves due to the inconsistency of the vertical and horizontal resolution. *Mon. Wea. Rev.*, **119**, 917–935.

Persson, P. O. G., and T. T. Warner (1995). The nonlinear evolution of idealized, unforced, conditional symmetric instability: A numerical study. *J. Atmos. Sci.*, **52**, 3449–3474.

Petersen, E. L., N. G. Mortensen, L. Landberg, J. Højstrup, and H. P. Frank (1998a). Wind power meteorology. Part I: Climate and turbulence. *Wind Energy*, **1**, 2–22.

Petersen, E. L., N. G. Mortensen, L. Landberg, J. Højstrup, and H. P. Frank (1998b). Wind power meteorology. Part II: Siting and models. *Wind Energy*, **1**, 55–72.

Phillips, N. A. (1957a). A map-projection system suitable for large-scale numerical weather prediction. *J. Meteor. Soc. Japan*, **35**, 262–267.

Phillips, N. A. (1957b). A coordinate system having some special advantages for numerical forecasting. *J. Meteor.*, **14**, 184–185.

Phillips, N. A. (1962). Numerical integration of the hydrostatic system of equations with a modified version of the Eliassen finite-difference grid. *Proc. International Symposium on Numerical Weather Prediction*, Tokyo, Meteorological Society of Japan, 109–120.

Phillips, N., and J. Shukla (1973). On the strategy of combining coarse and fine grid meshes in numerical weather prediction. *J. Appl. Meteor.*, **12**, 763–770.

Pielke, R. A. (1991). A recommended specific definition of "resolution". *Bull. Amer. Meteor. Soc.*, **72**, 1914.

Pielke, R. A. (2001). Influence of the spatial distribution of vegetation and soils on the prediction of cumulus convective rainfall. *Rev. Geophys.*, **39**, 151–177.

Pielke, R. A. (2002a). *Mesoscale Meteorological Modeling*. London, UK: Academic Press.

Pielke, R. A., Sr (2002b). Overlooked issues in the U.S. National climate and IPCC assessments. *Climate Change*, **52**, 1–11.

Pielke, R., Sr (2005). Land use and climate change. *Science*, **310**, 1625–1626.

Pielke, R. A., Sr, G. E. Liston, J. L. Eastman, and L. Lu (1999a). Seasonal weather prediction as an initial value problem. *J. Geophys. Res.*, **104**, 19463–19479.

Pielke, R. A., Sr, R. L. Walko, L. T. Steyaert, *et al.* (1999b). The influence of anthropogenic landscape changes on weather in South Florida. *Mon. Wea. Rev.*, **127**, 1663–1673.

Plant, R. S., and G. C. Craig (2008). A stochastic parameterization for deep convection based on equilibrium statistics. *J. Atmos. Sci.*, **65**, 87–105.

Pope, V. D., M. L. Gallani, P. R. Rowntree, and R. A. Stratton (2000). The impact of new physical parametrizations in the Hadley Centre climate model: HadAM3. *Climate Dyn.*, **16**, 123–146.

Powers, J. G., A. J. Monaghan, A. M. Cayette, *et al.* (2003). Real-time mesoscale modeling over Antarctica. *Bull. Amer. Meteor. Soc.*, **84**, 1533–1545.

Preisendorfer, R. W. (1988). In *Principal Component Analysis in Meteorology and Oceanography,* ed. C. Mobley. Amsterdam, the Netherlands: Elsevier Press.

Priestley, C. H. B. (1959). *Turbulent Transfer in the Lower Atmosphere.* Chicago, USA: University of Chicago Press.

Pruppacher, H. R., and J. D. Klett (2000). *Microphysics of Clouds and Precipitation.* Dordrecht, the Netherlands: Kluwer Academic Publishers.

Pryor, S. C., R. J. Barthelmie, and E. Kjellström (2005a). Potential climate change impact on wind energy resources in northern Europe: analyses using a regional climate model. *Climate Dyn.*, **25**, 815–835.

Pryor, S. C., J. T. Schoof, and R. J. Barthelmie (2005b). Climate change impacts on wind speeds and wind energy density in northern Europe: empirical downscaling of multiple AOGCMs. *Climate Res.*, **29**, 183–198.

Pryor, S. C., J. T. Schoof, and R. J. Barthelmie (2006). Winds of change?: Projections of near-surface winds under climate change scenarios. *Geophys. Res. Lett.*, **33**, L11702, doi:10.1029/2006GL026000.

Pu, Z. X., S. Lord, and E. Kalnay (1998). Forecast sensitivity with dropwindsonde data and targeted observations. *Tellus,* **50A**, 391–410.

Pudykiewicz, J. (1988). Numerical simulation of the transport of radioactive cloud from the Chernobyl nuclear accident. *Tellus,* **40B,** 241–259.

Pullen, J., J. D. Doyle, and R. P. Signell (2006). Two-way air-sea coupling: A study of the Adriatic. *Mon. Wea. Rev.*, **134**, 1465–1483.

Puri, K. (1987). Some experiments on the use of tropical diabatic heating information for initial state specification. *Mon. Wea. Rev.*, **115**, 1394–1406.

Purser, R. J. (2007). Accuracy considerations of time-splitting methods for models using two-time-level schemes. *Mon. Wea. Rev.*, **135**, 1158–1164.

Purser, R. J., and M. Rančić (1997). Conformal octagon: An attractive framework for global models offering quasi-uniform regional enhancement of resolution. *Meteor. Atmos. Phys.*, **62**, 33–48.

Purser, R. J., and M. Rančić (1998). Smooth quasi-homogeneous gridding of the sphere. *Quart. J. Roy. Meteor. Soc.*, **124**, 637–647.

Rabier, F., A. McNally, E. Andersson, *et al.* (1998). The ECMWF implementation of three-dimensional variational assimilation (3D-Var). II: Structure function. *Quart. J. Roy. Meteor. Soc.*, **124**, 1809–1830.

Raftery, A. E., T. Gneiting, F. Balabdaoui, and M. Polakowski (2005). Using Bayesian model averaging to calibrate forecast ensembles. *Mon. Wea. Rev.*, **133**, 1155–1174.

Rajagopalan, B., U. Lall, and S. E. Zebiak (2002). Categorical climate forecasts through regularization and optimal combination of multiple GCM ensembles. *Mon. Wea. Rev.*, **130**, 1792–1811.

Ramage, C. S. (1983). Teleconnections and the siege of time. *J. Climatol.*, **3**, 223–231.

Ramos da Silva, R., G. Bohrer, D. Werth, M. J. Otte, and R. Avissar (2006). Sensitivity of ice storms in the southeastern United States to Atlantic SST: Insights from a case study of the December 2002 storm. *Mon. Wea. Rev.*, **134**, 1454–1464.

Rančić, M., R. J. Purser, and F. Mesinger (1996). A global shallow-water model using an expanded spherical cube: Gnomonic versus conformal coordinates. *Quart. J. Roy. Meteor. Soc.*, **122**, 959–982.

Randall, D. A., T. D. Ringler, R. P. Heikes, P. Jones, and J. Baumgardner (2002). Climate modeling with spherical geodesic grids. *Computing Sci. Eng.*, **4**, 32–41.

Randall, D. A., R. A. Wood, S. Bony, *et al.* (2007). Climate models and their evaluation. In *Climate Change 2007: The Physical Science Basis. Contribution of Working Group I to the Fourth Assessment Report of the Intergovernmental Panel on Climate Change*, eds. S. Solomon, D. Qin, M. Manning, *et al.* Cambridge, UK and New York, USA: Cambridge University Press.

Rauscher, S. A., A. Seth, B. Liebmann, J.-H. Qian, and S. J. Camargo (2007). Regional climate model-simulated timing and character of seasonal rains in South America. *Mon. Wea. Rev.*, **135**, 2642–2657.

Raymond, T. M., and S. N. Pandis (2002). Cloud activation of single-component organic aerosol particles. *J. Geophys. Res.*, **107**, 4787, doi:10.1029/2002JD002159.

Raymond, W. H. (1988). High-order low-pass implicit tangent filters for use in finite area calculations. *Mon. Wea. Rev.,* **116,** 2132–2141.

Raymond, W. H., and A. Garder (1976). Selective damping in a Galerkin method for solving wave problems with variable grids. *Mon. Wea. Rev.,* **104,** 1583–1590.

Raymond, W. H., and A. Garder (1988). A spatial filter for use in finite area calculations. *Mon. Wea. Rev.*, **116**, 209–222.

Raymond, W. H., W. S. Olson, and G. Callan (1995). Diabatic forcing and initialization with assimilation of cloud water and rainwater in a forecast model. *Mon. Wea. Rev.*, **123**, 366–382.

Reale, O., J. Terry, M. Masutani, *et al.* (2007). Preliminary evaluation of the European Centre for Medium-Range Weather Forecasts' (ECMWF) Nature Run over the tropical Atlantic and African monsoon region. *Geophys. Res. Lett.*, **34**, L22810, doi:10.1029/2007GL031640.

Reichler, T., and J. Kim (2008). How well do coupled models simulate today's climate? *Bull. Amer. Meteor. Soc.*, **89**, 303–311.

Reisner, J., R. M. Rasmussen, and R. T. Bruintjes (1998). Explicit forecasting of super-cooled liquid water in winter storms using the MM5 mesoscale model. *Quart. J. Roy. Meteor. Soc.*, **124**, 1071–1107.

Reynolds. O. (1895). On the dynamical theory of incompressible fluids and the determination of the criterion. *Phil. Trans. R. Soc., A*, **186**, 123–164.

Reynolds, R. W., N. A. Rayner, T. M. Smith, D. C. Stokes, and W. Wang (2002). An improved in situ and satellite SST analysis for climate. *J. Climate*, **11**, 3320–3323.

Richardson, D. S. (2000). Skill and relative economic value of the ECMWF ensemble prediction system. *Quart. J. Roy. Meteor. Soc.*, **126**, 649–667.

Richardson, D. S. (2001). Measures of skill and value of ensemble prediction systems, their relationship and the effect of ensemble size. *Quart. J. Roy. Meteor. Soc.*, **127**, 2473–2489.

Richardson, L. F. (1922). *Weather Prediction by Numerical Process*. Cambridge, UK: Cambridge University Press.

Riemer, M., S. C. Jones, and C. A. Davis (2008). The impact of extratropical transition on the downstream flow: An idealized modeling study with a straight jet. *Quart. J. Roy. Meteor. Soc.*, **134**, 69–91.

Rife, D. L., and C. A. Davis (2005). Verification of temporal variations in mesoscale numerical wind forecasts. *Mon. Wea. Rev.*, **133**, 3368–3381.

Rife, D. L., T. T. Warner, F. Chen, and E. A. Astling (2002). Mechanisms for diurnal boundary-layer circulations in the Great Basin Desert. *Mon. Wea. Rev.*, **130**, 921–938.

Rife, D. L., C. A. Davis, Y. Liu, and T. T. Warner (2004). Predictability of low-level winds by mesoscale meteorological models. *Mon. Wea. Rev.*, **132**, 2553–2569.

Ringler, T. D., and D. A. Randall (2002). A potential enstrophy and energy conserving numerical scheme for solution of the shallow-water equations on a geodesic grid. *Mon. Wea. Rev.*, **130**, 1397–1410.

Ringler, T. D., R. P. Heikes, and D. A. Randall (2000). Modeling the atmospheric general circulation using a spherical geodesic grid: A new class of dynamical cores. *Mon. Wea. Rev.*, **128**, 2471–2490.

Rinke, A., K. Dethloff, J. J. Cassano, *et al.* (2005). Evaluation of an ensemble of Arctic regional climate models: spatiotemporal fields during the SHEBA year. *Climate Dyn.*, **26**, 459–472.

Roads, J. (2000). The second annual regional spectral model workshop. *Bull. Amer. Meteor. Soc.*, **81**, 2979–2981.

Roads, J. (2004). Experimental weekly to seasonal U.S. forecasts with the regional spectral model. *Bull. Amer. Meteor. Soc.*, **85**, 1887–1902.

Roads, J., S.-C. Chen, and M. Kanamitsu (2003a). U.S. regional climate simulations and seasonal forecasts. *J. Geophys. Res.*, **108**(D16), 8606, doi:10.1029JD002232.

Roads, J., S. Chen, S. Cocke, *et al.* (2003b). International Research Institute/Applied Research Centers (IRI/ARCs) regional model intercomparison over South America. *J. Geophys. Res.*, **108**(D14), doi:10.1029/2002JD003201.

Robert, A. (1979). The semi-implicit method. In *Numerical Methods Used in Atmospheric Models, Vol. II*, ed. A. Kasahara. Global Atmospheric Research Programme, GARP Publication Series No. 17. Geneva: World Meteorological Organization, 417–436.

Robert, A. (1981). A stable numerical integration scheme for the primitive meteorological equations. *Atmos.-Ocean*, **19**, 35–46.

Robert, A. (1982). A semi-Lagrangian and semi-implicit numerical integration scheme for the primitive meteorological equations. *J. Meteor. Soc. Japan*, **60**, 319–324.

Roberts, M. J. (2004). *The Ocean Component of HadGEM1*. GMR Report Annex IV.D.3, Exeter, UK: Met Office.

Robertson, A. W., U. Lall, S. E. Zebiak, and L. Goddard (2004). Improved combination of multiple atmospheric GCM ensembles for seasonal prediction. *Mon. Wea. Rev.*, **132**, 2732–2744.

Rockel, B., C. L. Castro, R. A. Pielke Sr, H. von Storch, and G. Leoncini (2008). Dynamical downscaling: Assessment of model system dependent retained and added variability for two different regional climate models. *J. Geophys. Res.*, **113**, D21107, doi:10.1029/2007JD009461.

Rodell, M., P. R. Houser, U. Jambor, *et al.* (2004). The global land data assimilation system. *Bull. Amer. Meteor. Soc.*, **85**, 381–394.

Roebber, P. J., D. M. Schultz, B. A. Colle, and D. J. Stensrud (2004). Toward improved prediction: High-resolution and ensemble modeling systems in operations. *Wea. Forecasting*, **19**, 936–949.

Roeckner, E., K. Arpe, L. Bengtsson, *et al.* (1996). *The Atmospheric General Circulation Model ECHAM4: Model Description and Simulation of Present-Day Climate.* MPI Report No. 218, Max-Planck-Institut für Meteorologie, Hamburg, Germany.

Roeckner, E., G. Bäuml, L. Bonaventura, *et al.* (2003). *The Atmospheric General Circulation Model ECHAM5. Part I: Model Description.* MPI Report No. 349, Max-Planck-Institute fur Meteorologie, Hamburg, Germany.

Rogers, R. R. (1976). *A Short Course in Cloud Physics*, 2nd edn. Oxford, UK: Pergamon Press.

Rogers, R. R., and M. K. Yau (1989). *A Short Course in Cloud Physics,* 3rd edn. Oxford, UK: Butterworth-Heinemann.

Rojas, M., and A. Seth (2003). Simulation and sensitivity in a nested modeling system for South America. Part II: GCM boundary forcing. *J. Climate*, **16**, 2454–2471.

Romero, R., C. Ramis, S. Alonso, C. A. Doswell III, and D. J. Stensrud (1998). Mesoscale model simulations of three heavy precipitation events in the western Mediterranean region. *Mon. Wea. Rev.*, **126**, 1859–1881.

Romero, R., C. A. Doswell III, and C. Ramis (2000). Mesoscale numerical study of two cases of long-lived quasi-stationary convective systems over eastern Spain. *Mon. Wea. Rev.*, **128**, 3731–3751.

Ronchi, C., R. Iaconno, and P. S. Paolucci (1996). The 'cubed sphere': A new method for the solution of partial differential equations in spherical geometry. *J. Comput. Phys.* **124**, 93–114.

Rosenfeld, D., U. Lohmann, G. B. Raga, *et al.* (2008). Flood or drought: How do aerosols affect precipitation? *Science*, **321**, 1309–1313.

Ross, R. S., and T. N. Krishnamurti (2005). Reduction of forecast error for global numerical weather prediction by the Florida State University (FSU) superensemble. *Meteor. Atmos. Phys.*, **88**, 215–235.

Roulin, E. (2007). Skill and relative economic value of medium-range hydrological ensemble predictions. *Hydrol. Earth Syst. Sci.*, **11**, 725–737.

Roulston, M. S. (2005). A comparison of predictors of the error of weather forecasts. *Nonlinear Proc. Geophys.*, **12**, 1021–1032.

Rupa Kumar, K., A. K. Sahai, K. Krishna Kumar, *et al.* (2006). High-resolution climate change scenarios for India for the 21st century. *Current Science*, **90**, 334–345.

Russell, A., and R. Dennis (2000). NARSTO critical review of photochemical models and modeling. *Atmos. Environ.*, **34**, 2283–2324.

Russell, G. L., J. R. Miller, and D. Rind (1995). A coupled atmosphere-ocean model for transient climate change studies. *Atmos.-Ocean*, **33**, 683–730.

Sadourny, R. (1972). Conservative finite-difference approximations of the primitive equations on quasi-uniform spherical grids. *Mon. Wea. Rev.*, **100**, 136–144.

Sadourny, R., A. Arakawa, and Y. Mintz (1968). Integration of the nondivergent barotropic vorticity equation with an icosahedral-hexagonal grid for the sphere. *Mon. Wea. Rev.*, **96**, 351–356.

Saha, S., S. Nadiga, C. Thiaw, *et al.* (2006). The NCEP climate forecast system. *J. Climate*, **19**, 3483–3517.

Salas-Mélia, D. (2002). A global coupled sea ice-ocean model. *Ocean Modelling*, **4**, 137–172.

Salathé, E. P., Jr, P. W. Mote, and M. W. Wiley (2007). Review of scenario selection and downscaling methods for the assessment of climate change impacts on hydrology in the United States Pacific Northwest. *Int. J. Climatol.*, **27**, 1611–1621.

Salathé, E. P., Jr, R. Steed, C. F. Mass, and P. H. Zahn (2008). A high-resolution climate model for the U.S. Pacific Northwest: Mesoscale feedbacks and local responses to climate change. *J. Climate*, **21**, 5708–5726.

Salmon, E. M., and T. T. Warner (1986). Short-term numerical precipitation forecasts initialized using a diagnosed divergent wind component. *Mon. Wea. Rev.*, **114**, 2122–2132.

San José, R., J. L. Pérez, and R. M. González (2006). Air quality real-time operational forecasting system for Europe: an application of the MM5-CMAQ-EMIMO modelling system. *Air Pollution XIV*, **1**, 75–84.

Sanders, F. (1963). On subjective probability forecasting. *J. Appl. Meteor.*, **2**, 191–201.

Sanders, F. (1973). Skill in forecasting daily temperature and precipitation: some experimental results. *Bull. Amer. Meteor. Soc.*, **54**, 1171–1179.

Sato, N., P. J. Sellers, D. A. Randall, *et al.* (1989). Effects of implementing the simple biosphere model in a general circulation model. *J. Atmos. Sci.*, **46**, 2757–2782.

Satoh, M., T. Matsuno, H. Tomita, H. Miura, T. Nasuno, and S. Iga (2008). Nonhydrostatic icosahedral atmospheric model (NICAM) for global cloud-resolving simulations. *J. Comput. Phys.*, **227**, 3486–3514.

Saucier, W. J. (1955). *Principles of Meteorological Analysis*. New York, USA: Dover Publications.

Savijärvi, H. (1990). Fast radiation parameterization schemes for mesoscale and short-range forecast models. *J. Appl. Meteor.*, **29**, 437–447.

Schlatter, T. W. (1975). Some experiments with a multivariate statistical objective analysis scheme. *Mon. Wea. Rev.*, **103**, 246–257.

Schmidt, G. A., C. M. Bitz, U. Mikolajewicz, and L. B. Tremblay (2004). Ice-ocean boundary conditions for coupled models. *Ocean Modelling*, **7**, 59–74.

Schmidt, G. A., R. Ruedy, J. E. Hansen, *et al.* (2006). Present day atmospheric simulations using GISS ModelE: Comparison to in-situ, satellite and reanalysis data. *J. Climate*, **19**, 153–192.

Seaman, N. L., D. R. Stauffer, and A. M. Lario-Gibbs (1995). A multiscale four-dimensional data assimilation system applied in the San Joaquin valley during SARMAP. Part I: Modeling design and basic performance characteristics. *J. Appl. Meteor.*, **34**, 1739–1761.

Seefeldt, M. W., and J. J. Cassano (2008). An analysis of low-level jets in the greater Ross Ice Shelf region based on numerical simulations. *Mon. Wea. Rev.*, **136**, 4188–4205.

Seemann, S. W., J. Li, W. P. Menzel, and L. E. Gumley (2003). Operational retrieval of atmospheric temperature, moisture, and ozone from MODIS infrared radiances. *J. Appl. Meteor.*, **42**, 1072–1091.

Segal, M., W. E. Schreiber, G. Kallos, *et al.* (1989). The impact of crop areas in northeast Colorado on midsummer mesoscale thermal circulations. *Mon. Wea. Rev.*, **117**, 809–825.

Sellers, P. J., Y. Mintz, Y. C. Sud, and A. Dalcher (1986). A simple biosphere model (SiB) for use within general circulation models. *J. Atmos. Sci.*, **43**, 505–531.

Semazzi, F. H. M., and Y. Song (2001). A GCM study of climate change induced by deforestation in Africa. *Climate Res.*, **17**, 169–182.

Semenov, M. A., and E. M. Barrow (1997). Use of a stochastic weather generator in the development of climate change scenarios. *Climatic Change*, **35**, 397–414.

Semtner, A. J. (1976). A model for the thermodynamic growth of sea ice in numerical investigations of climate. *J. Phys. Oceanogr.*, **6**, 379–389.

Seth, A., and M. Rojas (2003). Simulation and sensitivity in a nested modeling system for South America. Part I: Reanalyses boundary forcing. *J. Climate,* **16**, 2437–2453.

Shafer, M. A., C. A. Fiebrich, D. S. Arndt, S. E. Fredrickson, and T. W. Hughes (2000). Quality assurance procedures in the Oklahoma mesonetwork. *J. Atmos. Oceanic Technol.*, **17**, 474–494.

Shao, Y., and A. Henderson-Sellers (1996). Modeling soil moisture: A Project for Intercomparison of Land Surface Parameterization Schemes Phase 2(b). *J. Geophys. Res.*, **101**, 7227–7250.

Shapiro, M. A., and J. T. Hastings (1973). Objective cross section analysis by Hermite polynomial interpolation on isentropic surfaces. *J. Appl. Meteor.*, **12**, 753–762.

Shapiro, M. A., and J. J. O'Brien (1970). Boundary conditions for fine-mesh limited-area forecasts. *J. Appl. Meteor.*, **9**, 345–349.

Shapiro, R. (1970). Smoothing, filtering and boundary effects. *Rev. Geophys. Space Phys.*, **8**, 359–387.

Shapiro, R. (1975). Linear filtering. *Math. Comp.*, **29**, 1094–1097.

Sharman, R. D., and M. G. Wurtele (1983). Ship waves and lee waves. *J. Atmos. Sci.*, **40**, 396–427.

Sharman, R., C. Tebaldi, G. Wiener, and J. Wolff (2006). An integrated approach to mid- and upper-level turbulence forecasting. *Wea. Forecasting*, **21**, 268–287.

Sharman, R., Y. Liu, R.-S. Sheu, *et al.* (2008). The operational mesogamma-scale analysis and forecast system of the U.S. Army Test and Evaluation Command. Part 3: Coupling of special applications models with the meteorological model. *J. Appl. Meteor. Climatol.*, **47**, 1105–1122.

Shibata, K., H. Yoshimura, M. Ohizumi, M. Hosaka, and M. Sugi (1999). A simulation of troposphere, stratosphere and mesosphere with an MRI/JMA98 GCM. *Papers Meteor. Geophys.*, **50**, 15–53.

Shin, D. W., and T. N. Krishnamurti (1999). Improving precipitation forecasts over the global tropical belt. *Meteor. Atmos. Phys.*, **70**, 1–14.

Shukla, J., J. Anderson, D. Baumhefner, *et al.* (2000). Dynamical seasonal prediction. *Bull. Amer. Meteor. Soc.*, **81**, 2593–2606.

Shukla, J., T. DelSole, M. Fennessy, J. Kinter, and D. Paolino (2006). Climate model fidelity and projections of climate change. *Geophys. Res. Lett.*, **33**, L07702, doi:10.1029/2005GL025579.

Shuman, F. (1989). History of numerical weather prediction at the National Meteorological Center. *Wea. Forecasting*, **4**, 286–296.

Shutts, G. J. (2005). A kinetic energy backscatter algorithm for use in ensemble prediction systems. *Quart. J. Roy. Meteor. Soc.*, **131**, 3079–3102.

Sills, D. M. L. (2009). On the MSC forecasters forums and the future role of the human forecaster. *Bull. Amer. Meteor. Soc.*, **90**, 619–627.

Singleton, A. T., and C. J. C. Reason (2007). A numerical model study of an intense cutoff low pressure system over South Africa. *Mon. Wea. Rev.*, **135**, 1128–1150.

Skamarock, W. C. (2004). Evaluating mesoscale NWP models using kinetic energy spectra. *Mon. Wea. Rev.*, **132**, 3019–3032.

Skamarock, W. C., and J. B. Klemp (1992). The stability of time-split numerical methods for the hydrostatic and nonhydrostatic elastic equations. *Mon. Wea. Rev.*, **120**, 2109–2127.

Skamarock, W. C., and J. B. Klemp (1993). Adaptive grid refinement for two-dimensional and three-dimensional nonhydrostatic atmospheric flow. *Mon. Wea. Rev.*, **121**, 788–804.

Skamarock, W. C., and J. B. Klemp (2008). A time-split nonhydrostatic atmospheric model for weather research and forecasting applications. *J. Comput. Phys.*, **227**, 3465–3485.

Skamarock, W. C., J. B. Klemp, J. Dudhia, *et al.* (2008). *A Description of the Advanced Research WRF Version 3*. NCAR/TN-475+STR. [Available from NCAR, P.O. Box 3000, Boulder, CO 80307- 3000.]

Slingo, J. M. (1980). A cloud parameterization scheme derived from GATE data for use with a numerical-model. *Quart, J. Roy. Meteor. Soc.*, **106**, 747–770.

Slingo, J. M. (1987). The development and verification of a cloud prediction scheme for the ECMWF model. *Quart. J. Roy. Meteor. Soc.*, **113**, 899–927.

Slonosky, V. C., P. D. Jones, and T. D. Davies (2001). Atmospheric circulation and surface temperature in Europe from the 18th century to 1995. *Int. J. Climatol.*, **21**, 63–75.

Smagorinsky, J. (1983). The beginnings of numerical weather prediction and general circulation modeling: Early recollections. In *Theory of Climate*, ed. B. Saltzman. Advances in Geophysics, Vol. 25, pp. 3–37.

Smith, D. M., S. Cusack, A. W. Colman, *et al.* (2007). Improved surface temperature prediction for the coming decade from a global climate model. *Science*, **317**, doi: 10.1126/science.1139540.

Smith, G. L., P. E. Mlynczak, D. A. Rutan, and T. Wong (2008). Comparison of the diurnal cycle of outgoing longwave radiation from a climate model with results from ERBE. *J. Appl. Meteor. Climatol.*, **47**, 3188–3201.

Smith, R. D., and P. R. Gent (2002). *Reference Manual for the Parallel Ocean Program (POP), Ocean Component of the Community Climate System Model (CCSM2.0 and 3.0).* Technical Report LA-UR-02-2484, Los Alamos National Laboratory, Los Alamos, USA.

Snyder, C., and F. Zhang (2003). Assimilation of simulated Doppler radar observations with an ensemble Kalman filter. *Mon. Wea. Rev.*, **131**, 1663–1677.

Snyder, C., W. C. Skamarock, and R. Rotunno (1993). Frontal dynamics near and following frontal collapse. *J. Atmos. Sci.*, **50**, 3194–3211.

Song, Y., F. H. M. Semazzi, L. Xie, and L. J. Ogallo (2004). A coupled regional climate model for the Lake Victoria basin of East Africa. *Int. J. Climatol.*, **24**, 57–75.

Sotillo, M. G., A. W. Ratsimandresy, J. C. Carretero, *et al.* (2005). A high-resolution 44-year atmospheric hindcast for the Mediterranean Basin: contribution to the regional improvement of global reanalysis. *Climate Dyn.*, **25**, 219–236.

Srivastava, R. K., D. S. McRae, and M. T. Odman (2000). An adaptive grid algorithm for air-quality modeling. *J. Comput. Phys.*, **165**, 437–472.

Staniforth, A. (1984). The application of the finite-element method to meteorological simulations: A review. *Int. J. Numer. Methods Fluids*, **4**, 1–12.

Staniforth, A., and J. Côté (1991). Semi-Lagrangian integration schemes for atmospheric models: A review. *Mon. Wea. Rev.*, **119**, 2206–2223.

Staniforth, A. N., and R. W. Daley (1977). A finite element formulation for the discretization of sigma-coordinate primitive equation models. *Mon. Wea. Rev.*, **105**, 1108–1118.

Staniforth, A. N., and R. W. Daley (1979). A baroclinic finite-element model for regional forecasting with the primitive equations. *Mon. Wea. Rev.*, **107**, 107–121.

Staniforth, A. N., and J. Mailhot (1988). An operational model for regional weather forecasting. *Comput. Math. Applic.*, **16**, 1–22.

Staniforth, A. N., and H. L. Mitchell (1978). A variable-resolution finite-element technique for regional forecasting with the primitive equations. *Mon. Wea. Rev.*, **106**, 439–447.

Stauffer, D. R., and J.-W. Bao (1993). Optimal determination of nudging coefficients using the adjoint equations. *Tellus*, **45A**, 358–369.

Stauffer, D. R., and N. L. Seaman (1990). Use of four-dimensional data assimilation in a limited-area mesoscale model. Part I: Experiments with synoptic-scale data. *Mon. Wea. Rev.*, **118**, 1250–1277.

Stauffer, D. R., and N. L. Seaman (1994). Multiscale four-dimensional data assimilation. *J. Appl. Meteor.*, **33**, 416–434.

Stauffer, D. R., N. L. Seaman, and F. S. Binkowski (1991). Use of four-dimensional data assimilation in a limited-area mesoscale model. Part II: Effects of data assimilation within the planetary boundary layer. *Mon. Wea. Rev.*, **119**, 734–754.

Stauffer, D. R., N. L. Seaman, G. K. Hunter, *et al.* (2000). A field-coherence technique for meteorological field-program design for air quality studies. Part I: Description and interpretation. *J. Appl. Meteor.*, **39**, 297–316.

Steenburgh, W. J., C. R. Neuman, and G. L. West (2009). Discrete frontal propagation over the Sierra-Cascade Mountains and Intermountain West. *Mon. Wea. Rev.*, **137**, 2000–2020.

Stefanova, L., and T. N. Krishnamurti (2002). Interpretation of seasonal climate forecast using Brier skill score, the Florida State University superensemble, and the AMIP-I data set. *J. Climate*, **15**, 537–544.

Stein, U., and P. Alpert (1993). Factor separation in numerical simulations. *J. Atmos. Sci.*, **50**, 2107–2115.

Stensrud, D. J. (2007). *Parameterization Schemes: Keys to Understanding Numerical Weather Prediction Models*. Cambridge, UK: Cambridge University Press.

Stensrud, D. J., and J. M. Fritsch (1994). Mesoscale convective systems in weakly forced large-scale environments. Part III: Numerical simulations and implications for operational forecasting. *Mon. Wea. Rev.*, **122**, 2084–2104.

Stensrud, D. J., and J. A. Skindlov (1996). Gridpoint predictions of high temperature from a mesoscale model. *Wea. Forecasting*, **11**, 103–110.

Stensrud, D. J., and N. Yussouf (2003). Short-range ensemble predictions of 2-m temperature and dewpoint temperature over New England. *Mon. Wea. Rev.*, **131**, 2510–2524.

Stensrud, D. J., and N. Yussouf (2005). Bias-corrected short-range ensemble forecasts of near surface variables. *Meteorol. Appl.*, **12**, 217–230.

Stensrud, D. J., R. L. Gall, S. L. Mullen, and K. W. Howard (1995). Model climatology of the Mexican monsoon. *J. Climate*, **8**, 1775–1794.

Stensrud, D. J., J.-W. Bao, and T. T. Warner (2000). Using initial condition and model physics perturbations in short-range ensemble simulations of mesoscale convective systems. *Mon. Wea. Rev.*, **128**, 2077–2107.

Stephens, G. L. (1984). The parameterization of radiation for numerical weather prediction and climate models. *Mon. Wea. Rev.*, **112**, 826–867.

Stephenson, D. B., C. A. S. Coelho, F. J. Doblas-Reyes, and M. Balmaseda (2005). Forecast assimilation: a unified framework for the combination of multi-model weather and climate predictions. *Tellus*, **57A**, 253–264.

Steppeler, J. (1987). Quadratic Galerkin finite-element schemes for the vertical discretization of sigma-coordinate primitive-equation models. *Mon. Wea. Rev.*, **115**, 1575–1588.

Steppeler, J., H.-W. Bitzer, M. Minotte, and L. Bonaventura (2002). Nonhydrostatic atmospheric modeling using a z-coordinate representation. *Mon. Wea. Rev.*, **130**, 2143–2149.

Stoelinga, M. T., and T. T. Warner (1999). Nonhydrostatic, mesobeta-scale model simulations of cloud ceiling and visibility for an East Coast winter precipitation event. *J. Appl. Meteor.*, **38**, 385–404.

Stoker, J. J., and E. Isaacson (1975). Final Report 1, Courant Institute of Mathematical Sciences, IMM 407, New York University, USA.

Stowasser, M., Y. Wang, and K. Hamilton (2007). Tropical cyclone changes in the Western North Pacific in a global warming scenario. *J. Climate*, **20**, 2378–2396.

Strack, J. E., R. A. Pielke Sr, L. T. Steyaert, and R. G. Knox (2008). Sensitivity of June near-surface temperatures in the eastern United States to historical land cover changes since European settlement. *Water Resour. Res.*, **44**, W11401, doi:10.1029/2007WR006546.

Straka, J. M. (2009). *Cloud and Precipitation Microphysics: Principles and Parameterization*. Cambridge, UK: Cambridge University Press.

Strassberg, D., M. LeMone, T. Warner, and J. Alfieri (2008). Comparison of observed 10-m wind speeds to those based on Monin-Obukhov similarity theory using IHOP 2002 aircraft and surface data. *Mon. Wea. Rev.*, **136**, 964–972.

Straus, D. M., and D. Paolino (2009). Intermediate time error growth and predictability: tropics versus mid-latitudes. *Tellus*, **61A**, 579–586.

Stull, R. B. (1988). *An Introduction to Boundary Layer Meteorology*. Dordrecht, the Netherlands: Kluwer Academic Publishers.

Stull, R. B. (1991). Static stability: An update. *Bull. Amer. Meteor. Soc.*, **72**, 1521–1529.

Stull, R. B. (1993). Review of nonlocal mixing in turbulent atmospheres: Transilient turbulence theory. *Bound.-Layer Meteor.* **62**, 21–96.

Su, L., and O. B. Toon (2009). Numerical simulation of Asian dust storms using a coupled climate-aerosol microphysical model. *J. Geophys. Res.*, **114**, D14202, doi:10.1029/2008JD010956.

Suarez, M. J., and L. L. Takacs (1995). *Documentation of the Aries/GEOS Dynamical Core Version 2*. NASA Technical Memo. 104606, Vol. 5. [Available from NASA, Goddard Space Flight Center, Greenbelt, MD 20771.]

Sun, L., D. F. Moncunill, H. Li, A. D. Moura, and F. de A. de S. Filho (2005). Climate downscaling over Nordeste, Brazil using the NCEP RSM97. *J. Climate*, **18**, 551–567.

Sundqvist, H. (1979). Vertical coordinates and related discretization. In *Numerical Methods Used in Atmospheric Models, Vol. II*, ed. A. Kasahara. Global Atmospheric Research Programme, GARP Publication Series No. 17. Geneva: World Meteorological Organization, pp. 3–50.

Sundqvist, H., E. Berge, and J. E. Kristjansson (1989). Condensation and cloud parameterization studies with a mesoscale numerical weather prediction model. *Mon. Wea. Rev.*, **117**, 1641–1657.

Sutton, C., T. M. Hamill, and T. T. Warner (2006). Will perturbing soil moisture improve warm-season ensemble forecasts? A proof of concept. *Mon. Wea. Rev.*, **134**, 3172–3187.

Swarztrauber, P. N. (1996). Spectral transform methods for solving the shallow-water equations on the sphere. *Mon. Wea. Rev.*, **124**, 730–744.

Sykes, R. I., S. F. Parker, D. S. Henn, and W. S. Lewellen (1993). Numerical simulation of ANATEX tracer data using a turbulent closure model for long-range dispersion. *J. Appl. Meteor.*, **32**, 929–947.

Szunyogh, I., Z. Toth, K. A. Emanuel, *et al.* (1999). Ensemble-based targeting experiments during FASTEX: The effect of dropsonde data from the Lear jet. *Quart. J. Roy. Meteor. Soc.*, **125**, 3189–3218.

Szunyogh, I., Z. Toth, R. E. Morss, *et al.* (2000). The effect of targeted dropsonde observations during the 1999 Winter Storm Reconnaissance Program. *Mon. Wea. Rev.*, **128**, 3520–3537.

Szunyogh, I., Z. Toth, A. Zimin, S. J. Majumdar, and A. Persson (2002). On the propagation of the effect of targeted observations: The 2000 Winter Storm Reconnaissance Program. *Mon. Wea. Rev.*, **130**, 1144–1165.

Takacs, L. L., and M. J. Suarez (1996). *Dynamical Aspects of Climate Simulations Using the GEOS GCM*. NASA Technical Memo. 104606, Vol. 10. [Available from NASA, Goddard Space Flight Center, Greenbelt, MD 20771.]

Takle, E. S., W. J. Gutowski Jr, R. W. Arritt, *et al.* (1999). Project to intercompare regional climate simulations (PIRCS): description and initial results. *J. Geophys. Res.*, **104**, 443–462.

Talagrand, O., R. Vautard, and B. Strauss (1997). Evaluation of probabilistic prediction systems. *Proceedings, ECMWF Workshop on Predictability*. ECMWF, 1–25.

Tanguay, M., A. Simard, and A. Staniforth (1989). Three-dimensional semi-Lagrangian scheme for the Canadian regional finite-element forecast model. *Mon. Wea. Rev.*, **117**, 1861–1871.

Tanrikulu, S., D. R. Stauffer, N. L. Seaman, and A. J. Ranzieri (2000). A field-coherence technique for meteorological field-program design for air quality studies. Part II: Evaluation in the San Joaquin Valley. *J. Appl. Meteor.*, **39**, 317–334.

Tarbell, T. C., T. T. Warner, and R. A. Anthes (1981). The initialization of the divergent component of the horizontal wind in mesoscale numerical weather prediction models and its effect on initial precipitation rates. *Mon. Wea. Rev.*, **109**, 77–95.

Tatsumi, Y. (1986). A spectral limited-area model with time-dependent lateral boundary conditions and its application to a multi-level primitive-equation model. *J. Meteor. Soc. Japan*, **64**, 637–663.

Taylor, J. P., and A. S. Ackerman (1999). A case study of pronounced perturbations to cloud properties and boundary-layer dynamics due to aerosol emissions. *Quart. J. Roy. Meteor. Soc.*, **125**, 2543–2661.

Taylor, J. W., and R. Buizza (2003). Using weather ensemble predictions in electricity demand forecasting. *Intl. J. Forecasting*, **19**, 57–70.

Taylor, K. E. (2001). Summarizing multiple aspects of model performance in a single diagram. *J. Geophys. Res.*, **106** (D7), 7183–7192.

Taylor, K. E., P. J. Gleckler, and C. Doutriaux (2004). Tracking changes in the performance of AMIP models. *Proc. AMIP2 Workshop*, Toulouse, France, Meteo-France, pp. 5–8.

Teixeira., J., and C. A. Reynolds (2008). Stochastic nature of physical parameterizations in ensemble prediction: A stochastic convection approach. *Mon. Wea. Rev.*, **136**, 483–496.

Teng, Q., J. C. Fyfe, and A. H. Monahan (2007). Northern Hemisphere circulation regimes: observed, simulated and predicted. *Climate Dyn.*, **28**, 867–879.

Terray, L., S. Valcke, and A. Piacentini (1998). *OASIS 2.2 Guide and Reference Manual*. Technical Report TR/CMGC/98-05, Centre Européen de Recherche et de Formation Avancée en Calcul Scientifique, Toulouse, France.

Thiebaux, H. J., and M. A. Pedder (1987). *Spatial Objective Analysis*. London, UK: Academic Press.

Thiebaux, H. J., H. L. Mitchell, and D. W. Shantz (1986). Horizontal structure of hemispheric forecast error correlations for geopotential and temperature. *Mon. Wea. Rev.*, **114**, 1048–1066.

Thiebaux, J., E. Rogers, W. Wang, and B. Katz (2003). A new high-resolution blended global sea surface temperature analysis. *Bull. Amer. Meteor. Soc.*, **84**, 645–656.

Thompson, P. D. (1961). *Numerical Weather Analysis and Prediction*. New York, USA: Macmillan.

Thompson, P. D. (1983). History of numerical weather prediction in the United States. *Bull. Amer. Meteor. Soc.*, **64**, 755–769.

Thomson, M. C., and S. J. Connor (2001). The development of malaria early warning systems for Africa. *Trends in Parasitology*, **17**, 9438–9445.

Thomson, M. C., T. N. Palmer, A. P. Morse, M. Cresswell, and S. J. Connor (2000). Forecasting disease risk with seasonal climate predictions. *Lancet*, **355**, 1559–1560.

Thomson M. C., A. M. Molesworth, M. H. Djingarey, *et al.* (2006). Potential of environmental models to predict meningitis epidemics in Africa. *Trop. Med. Intl. Health*, **11**, 1–9.

Thuburn, J. (2008). Some conservation issues for the dynamical cores of NWP and climate models. *J. Comput. Phys.*, **227**, 3715–3730.

Tibaldi, S. (1986). Envelope orography and the maintenance of quasi-stationary waves in the ECMWF model. *Adv. Geophys.*, **29**, 339–374.

Tiedtke, M. (1989). A comprehensive mass flux scheme for cumulus parameterization in large-scale models. *Mon. Wea. Rev.*, **117**, 1779–1800.

Timbal, B., and D. A. Jones (2008). Future projections of winter rainfall in southeast Australia using a statistical downscaling technique. *Climate Change*, **86**, 165–187.

Tippett, M. K., T. DelSole, S. J. Mason, and A. G. Barnston (2008). Regression-based methods for finding coupled patterns. *J. Climate*, **21**, 4384–4398.

Tolika, K., C. Anagnostopoulou, P. Maheras, and M. Vafiadis (2008). Simulation of future changes in extreme rainfall and temperature conditions over the Greek area: A comparison of two statistical downscaling approaches. *Global Planet. Change*, **63**, 132–151.

Tomita, H., and M. Satoh (2004). A new dynamical framework of nonhydrostatic global model using the icosahedral grid. *Fluid Dyn. Res.*, **34**, 357–400.

Tompkins, A. M. (2002). A prognostic parameterization for the subgrid-scale variability of water vapor and clouds in large-scale models and its use to diagnose cloud cover. *J. Atmos. Sci.*, **59**, 1917–1942.

Tompkins, A. M. (2005). *The Parameterization of Cloud Cover*. ECMWF Technical Memorandum.

Tompkins, A. M., and M. Janisková (2004). A cloud scheme for data assimilation: Description and initial tests. *Quart. J. Roy. Meteor. Soc.*, **130**, 2495–2518.

Toon, O. B., R. P. Turco, D. Westphal, R. Malone, and M. S. Liu (1988). A multidimensional model for aerosols: Description of computational analogs. *J. Atmos. Sci.*, **45**, 2124–2143.

Torn, R. D., and G. J. Hakim (2008). Performance characteristics of a pseudo-operational ensemble Kalman filter. *Mon. Wea. Rev.*, **136**, 3947–3963.

Torrence, C., and G. P. Compo (1998). A practical guide to wavelet analysis. *Bull. Amer. Meteor. Soc.*, **79**, 61–78.

Toth, Z., and E. Kalnay (1993). Ensemble forecasting at NMC: The generation of perturbations. *Bull. Amer. Meteor. Soc.*, **74**, 2317–2330.

Toth, Z., and E. Kalnay (1997). Ensemble forecasting at NCEP and the breeding method. *Mon. Wea. Rev.*, **125**, 3297–3319.

Toth, Z., E. Kalnay, S. M. Tracton, R. Wobus, and J. Irwin (1997). A synoptic evaluation of the NCEP ensemble. *Wea. Forecasting*, **12**, 140–153.

Toth, Z., M. Peña, and A. Vintzileos (2007). Bridging the gap between weather and climate forecasting. *Bull. Amer. Meteor. Soc.*, **88**, 1427–1429.

Tracton, M. S. (1990). Predictability and its relationship to scale interaction processes in blocking. *Mon. Wea. Rev.*, **118**, 1666–1695.

Tracton, M. S., K. Mo, W. Chen, *et al.* (1989). Dynamical extended range forecasting (DERF) at the National Meteorological Center. *Mon. Wea. Rev.*, **117**, 1606–1637.

Treadon, R. E. (1996). Physical initialization in the NMC Global Data Assimilation System. *Meteor. Atmos. Phys.*, **60**, 57–86.

Treadon, R. E., and R. A. Petersen (1993). Domain size sensitivity experiments using the NMC Eta model. Preprints, *Proceedings 13th Conference on Weather Analysis and Forecasting*, Vienna, VA, USA, American Meteorological Society, pp. 176–177.

Tremback, C. J. (1990). Numerical simulation of a mesoscale convective complex: Model development and numerical results. Ph.D. Dissertation, Colorado State University, Fort Collins, CO, USA.

Trenberth, K. E., P. D. Jones, P. Ambenje, *et al.* (2007). Observations: Surface and atmospheric climate change. In *Climate Change 2007: The Physical Science Basis. Contribution of Working Group I to the Fourth Assessment Report of the Intergovernmental Panel on Climate Change*, eds. S. Solomon, D. Qin, M. Manning, *et al.* Cambridge, UK and New York, USA: Cambridge University Press.

Trigo, R., and J. Palutikof (2001). Precipitation scenarios over Iberia. A comparison between direct GCM output and different downscaling techniques. *J. Climate*, **14**, 4422–4446.

Turner, R., and T. Hurst (2001). Factors influencing volcanic ash dispersal from the 1995 and 1996 eruptions of Mount Ruapehu, New Zealand. *J. Appl. Meteor.*, **40**, 56–135.

Turpeinen, O. M. (1990). Diabatic initialization of the Canadian Regional Finite-Element (RFE) Model using satellite data. Part II: Sensitivity to humidity enhancement, latent-heating profile and rain rates. *Mon. Wea. Rev.*, **118**, 1396–1407.

Turpeinen, O. M., L. Garand, R. Benoit, and M. Roch (1990). Diabatic initialization of the Canadian Regional Finite-Element (RFE) Model using satellite data. Part I: Methodology and application to a winter storm. *Mon. Wea. Rev.*, **118**, 1381–1395.

Unsworth, M. H., and J. L. Monteith (1975). Long-wave radiation at the ground. I. Angular distribution of incoming radiation. *Quart. J. Roy. Meteor. Soc.*, **101**, 13–24.

Uppala, S. M., P. W. Kållberg, A. J. Simmons, *et al.* (2005). The ERA-40 re-analysis. *Quart. J. Roy. Meteor. Soc.*, **131**, 2961–3012.

Valent, R. A., P. H. Downey, and J. L. Donnelly (1977). *Description and Operational Aspects of the NCAR Segmented Limited Area Model (LAMSEG)*, NCAR Report #0501-77/08. [Available from NCAR, P.O. Box 3000, Boulder, CO 80307-3000.]

van den Dool, H. M., and S. Saha (1990). Frequency dependence in forecast skill. *Mon. Wea. Rev.*, **118**, 128–137.

Van Tuyl, A. H. (1996). Physical initialization with the Arakawa-Schubert scheme in the Navy's operational global forecast model. *Meteor. Atmos. Phys.*, **60**, 47–55.

Vannitsem, S., and F. Chomé (2005). One-way nested regional climate simulations and domain size. *J. Climate*, **18**, 229–233.

Verseghy, D. L., N. A. McFarlane, and M. Lazare (1993). A Canadian land surface scheme for GCMs: II. Vegetation model and coupled runs. *Int. J. Climatol.*, **13**, 347–370.

Vitart, F. (2004). Monthly forecasting at ECMWF. *Mon. Wea. Rev.*, **132**, 2761–2779.

Volodin, E. M., and N. A. Diansky (2004). El-Niño reproduction in a coupled general circulation model of atmosphere and ocean. *Russ. Meteor. Hydrol.*, **12**, 5–14.

Volodin, E. M., and V. N. Lykossov (1998). Parameterization of heat and moisture processes in the soil-vegetation system: 1. Formulation and simulations based on local
observational data. *Izv. Atmos. Ocean Phys.*, **34**, 453–465.

von Storch, H., and F. Zwiers (1999). *Statistical Analysis in Climate Research*. Cambridge,
UK: Cambridge University Press.

von Storch, H., E. Zorita, and U. Cubasch (1993). Downscaling of global climate change
estimates to regional scale: An application to Iberian rainfall in wintertime. *J. Climate*,
**6**, 1161–1171.

von Storch, H., H. Langenberg, and F. Feser (2000). A spectral nudging technique for
dynamical downscaling purposes. *Mon. Wea. Rev.*, **128**, 3664–3673.

Vukicevic, T., and R. M. Errico (1990). The influence of artificial and physical factors
upon predictability estimates using a complex limited-area model. *Mon. Wea. Rev.*, **118**,
1460–1482.

Vukicevic, T., and J. Paegle (1989). The influence of one-way interacting boundary conditions upon predictability of flow in bounded numerical models. *Mon. Wea. Rev.*, **117**,
340–350.

Waldron, K. M., J. Paegle, and J. D. Horel (1996). Sensitivity of a spectrally filtered
and nudged limited area model to outer model options. *Mon. Wea. Rev.*, **124**,
529–547.

Walko, R., and R. Avissar (2008a). The Ocean-Land-Atmosphere Model (OLAM). Part I:
Shallow water tests. *Mon. Wea. Rev.*, **136**, 4033–4044.

Walko, R., and R. Avissar (2008b). The Ocean-Land-Atmosphere Model (OLAM).
Part II: Formulation and tests of the nonhydrostatic dynamic core. *Mon. Wea. Rev.*,
**136**, 4045–4062.

Walko, R., W. R. Cotton, M. P. Meyers, and J. Y. Harrington (1995a). New RAMS
cloud microphysics parameterization. Part I: The single-moment scheme. *Atmos. Res.*,
**38**, 29–62.

Walko, R. L., C. J. Tremback, R. A. Pielke, and W. R. Cotton (1995b). An interactive
nesting algorithm for stretched grids and variable nesting ratios. *J. Appl. Meteor.*, **34**,
994–999.

Wallace, J. M., S. Tibaldi, and A. J. Simmons (1983). Reduction of systematic forecast
errors in the ECMWF model through the introduction of an envelope orography. *Quart.
J. Roy. Meteor. Soc.*, **109**, 683–717.

Wang, B., H. Wan, Z. Ji, *et al.* (2004). Design of a new dynamical core for global atmospheric models based on some efficient numerical methods. *Science in China*, *Ser. A*, **47**,
Suppl., 4–21.

Wang, W., and N. L. Seaman (1997). A comparison study of convective parameterization
schemes in a mesoscale model. *Mon. Wea. Rev.*, **125**, 252–278.

Wang, W., and T. T. Warner (1988). Use of four-dimensional data assimilation by
Newtonian relaxation and latent-heat forcing to improve a mesoscale-model precipitation forecast: A case study. *Mon. Wea. Rev.*, **116**, 2593–2613.

Warner, T. T., and H. M. Hsu (2000). Nested-model simulation of moist convection: The
impact of coarse-grid parameterized convection on fine-grid resolved convection. *Mon.
Wea. Rev.*, **128**, 2211–2231.

Warner, T. T., R. A. Anthes, and A. L. McNab (1978). Numerical simulations with a three-dimensional mesoscale model. *Mon. Wea. Rev.*, **106**, 1079–1099.

Warner, T. T., L. E. Key, and A. M. Lario (1989). Sensitivity of mesoscale model forecast skill to some initial data characteristics: Data density, data position, analysis procedure and measurement error. *Mon. Wea. Rev.*, **117**, 1281–1310.

Warner, T. T., R. A. Petersen, and R. E. Treadon (1997). A tutorial on lateral boundary conditions as a basic and potentially serious limitation to regional numerical weather prediction. *Bull. Amer. Meteor. Soc.*, **78**, 2599–2617.

Warner, T. T., R.-S. Sheu, J. Bowers, *et al.* (2002). Ensemble simulations with coupled atmospheric dynamic and dispersion models: Illustrating uncertainties in dosage simulations. *J. Appl. Meteor.*, **41**, 488–504.

Warner, T. T., B. E. Mapes, and M. Xu (2003). Diurnal patterns of rainfall in northwestern South America. Part II: Model simulations. *Mon. Wea. Rev.*, **131**, 813–829.

Washington, W. M., and A. Kasahara (1970). A January simulation experiment with the two layer version of the NCAR global circulation model. *Mon. Wea. Rev.*, **98**, 559–580.

Washington, W. M., J. W. Weatherly, G. A. Meehl, *et al.* (2000). Parallel Climate Model (PCM) control and transient simulations. *Climate Dyn.*, **16**, 755–774.

Weigel, A. P., M. A. Liniger, and C. Appenzeller (2007). The discrete Brier and ranked probability skill scores. *Mon. Wea. Rev.*, **135**, 118–124.

Weigel, A. P., M. A. Liniger, and C. Appenzeller (2009). Seasonal ensemble forecasts: Are recalibrated single models better than multimodels? *Mon. Wea. Rev.*, **137**, 1460–1479.

Weisman, M. L., W. C. Skamarock, and J. B. Klemp (1997). The resolution dependence of explicitly modeled convective systems. *Mon. Wea. Rev.*, **125**, 527–548.

Welander, P. (1955). Studies on the general development of motion in a two-dimensional, ideal fluid. *Tellus*, **7**, 141–156.

Weller, H., H. G. Weller, and A. Fournier (2009). Voronoi, Delaunay, and block-structured mesh refinement for solution of the shallow-water equations on the sphere. *Mon. Wea. Rev.*, **137**, 4208–4224.

Wergen, W. (1988). The diabatic ECMWF normal mode initialization scheme. *Beitr. Phys. Atmos.*, **61**, 274–302.

Westphal, D. L., C. A. Curtis, M. Liu, and A. L. Walker (2009). Operational aerosol and dust storm forecasting. *IOP Conf. Ser.: Earth Environ. Sci.* **7**, 012007, doi:10.1088/1755-1307/7/1/012007

Whitaker, J. S., G. P. Compo, X. Wei, and T. M. Hamill (2004). Reanalysis without radiosondes using ensemble data assimilation. *Mon. Wea. Rev.*, **132**, 1190–1200.

Wicker, L. J., and W. C. Skamarock (1998). A time-splitting scheme for the elastic equations incorporating second-order Runge-Kutta time differencing. *Mon. Wea. Rev.*, **126**, 1992–1999.

Wicker, L. J., and W. C. Skamarock (2002). Time splitting methods for elastic models using forward time schemes. *Mon. Wea. Rev.*, **130**, 2088–2097.

Widmann, M., C. S. Bretherton, and E. P. Salathé Jr (2003). Statistical precipitation downscaling over the Northwestern United States using numerically simulated precipitation as a predictor. *J. Climate*, **16**, 799–816.

Wigley, T. M. L., P. D. Jones, K. R. Briffa, and G. Smith (1990). Obtaining subgrid-scale information from coarse-resolution general circulation model output. *J. Geophys. Res.*, **95**, 1943–1953.

Wilby, R. L. (2001). Downscaling summer rainfall in the UK from North Atlantic ocean temperatures. *Hydrol. Earth Syst. Sci.*, **5**, 245–257.

Wilby, R. L., and T. M. L. Wigley (2000). Precipitation predictors for downscaling: observed and general circulation model relationships. *Int. J. Climatol.*, **20**, 641–661.

Wilby, R. L., T. M. L. Wigley, D. Conway, *et al.* (1998). Statistical downscaling of general circulation model output: A comparison of methods. *Water Resour. Res.*, **34**, 2995–3008.

Wilby, R. L., O. J. Tomlinson, and C. W. Dawson (2003). Multi-site simulation of precipitation by conditional resampling. *Climate Res.*, **23**, 183–194.

Wilby, R. L., C. S. Wedgbrow, and H. R. Fox (2004). Seasonal predictability of the summer hydrometeorology of the River Thames, UK. *J. Hydrol.*, **295**, 1–16.

Wilby, R. L., P. G. Whitehead, A. J. Wade, *et al.* (2006). Integrated modelling of climate change impacts on water resources and quality in a lowland catchment: River Kennet, UK. *J. Hydrol.*, **330**, 204–220.

Wilks, D. S. (2006). *Statistical Methods in the Atmospheric Sciences*. San Diego, USA: Academic Press.

Williamson, D. L. (1968). Integration of the barotropic vorticity equation on a spherical geodesic grid. *Tellus*, **20**, 642–653.

Williamson, D. L. (2007). The evolution of dynamical cores for global atmospheric models. *J. Meteor. Soc. Japan*, **85B**, 241–269.

Williamson, D., and J. Rosinski (2000). Accuracy of reduced-grid calculations. *Quart. J. Roy. Meteor. Soc.*, **126**, 1619–1640.

Williamson, D. L., J. B. Drake, J. J. Hack, R. Jakob, and P. N. Swarztrauber (1992). A standard test set for numerical approximations to the shallow water equations in spherical geometry. *J. Comput. Phys.*, **102**, 211–224.

Wilson, L. J., and M. Vallée (2002). The Canadian Updatable Model Output Statistics (UMOS) system: Design and development tests. *Wea. Forecasting*, **17**, 206–222.

Wilson, L. J., and M. Vallée (2003). The Canadian Updatable Model Output Statistics (UMOS) system: Validation against Perfect Prog. *Wea. Forecasting*, **18**, 288–302.

Winton, M. (2000). A reformulated three-layer sea ice model. *J. Atmos. Ocean. Technol.*, **17**, 525–531.

Wolff, J.-O., E. Maier-Reimer, and S. Lebutke (1997). *The Hamburg Ocean Primitive Equation Model*. DKRZ Technical Report No. 13, Deutsches KlimaRechenZentrum, Hamburg, Germany.

Woodcock, F., and C. Engel (2005). Operational consensus forecasts. *Wea. Forecasting*, **20**, 101–111.

Woth, K., R. Weisse, and H. von Storch (2006). Climate change and North Sea storm surge extremes: an ensemble study of storm surge extremes expected in a changed climate projected by four different regional climate models. *Ocean Dyn.*, **56**, 3–15.

Wu, X., and X. Li (2008). A review of cloud-resolving model studies of convective processes. *Adv. Atmos. Sci.*, **25**, 202–212.

Wyman, B. L. (1996). A step-mountain coordinate general circulation model: Description and validation of medium-range forecasts. *Mon. Wea. Rev.*, **124**, 102–121.

Wyngaard, J. C. (2004). Toward numerical modeling in the "terra incognita". *J. Atmos. Sci.*, **61**, 1816–1826.

Xie, P., and P. A. Arkin (1996). Analysis of global monthly precipitation using gauge observations, satellite estimates, and numerical model predictions. *J. Climate*, **9**, 840–858.

Xie, P., and P. A. Arkin (1997). Global precipitation: A 17-year monthly analysis based on gauge observation, satellite estimates, and numerical model outputs. *Bull. Amer. Meteor. Soc.*, **78**, 2539–2558.

Xoplaky, E., J. F. Gonzales-Rouco, J. Luterbacher, and M. Wanner (2004). Wet season Mediterranean precipitation variability: Influence of large-scale dynamics and trends. *Climate Dyn.*, **23**, 63–78.

Xu, K.-M., and D. A. Randall (1996). A semiempirical cloudiness parameterization for use in climate models. *J. Atmos. Sci.*, **53**, 3084–3102.

Xu, Q., and L. Wei (2001). Estimation of three-dimensional error covariances. Part II: Analysis of wind innovation vectors. *Mon. Wea. Rev.*, **129**, 2939–2954.

Xu, Q., and L. Wei (2002). Estimation of three-dimensional error covariances. Part III: Height-wind forecast error correlation and related geostrophy. *Mon. Wea. Rev.*, **130**, 1052–1062.

Xu, Q., L. Wei, A. Van Tuyl, and E. H. Barker (2001). Estimation of three-dimensional error covariances. Part I: Analysis of height innovation vectors. *Mon. Wea. Rev.*, **129**, 2126–2135.

Xu, Y., Y. Luo, Z. C. Zhao, *et al.* (2005). Detection of climate change in the 20th century by the NCC T63. *Acta Meteorol. Sin.*, Special Report on Climate Change, **4**, 1–15.

Xue, M. (2000). High-order monotonic numerical diffusion and smoothing. *Mon. Wea. Rev.*, **128**, 2853–2864.

Xue, M., K. K. Droegemeier, and V. Wong (2000). The Advanced Regional Prediction System (ARPS): a multiscale nonhydrostatic atmospheric simulation and prediction tool. Part 1. Model dynamics and verification. *Meteor. Atmos. Phys.*, **75**, 161–193.

Xue, Y., R. Vasic, Z. Janjic, F. Mesinger, and K. E. Mitchell (2007). Assessment of dynamic downscaling of the continental U.S. regional climate using the Eta/SSib regional climate model. *J. Climate*, **20**, 4172–4193.

Yakimiw, E., and A. Robert (1990). Validation experiments for a nested grid-point regional forecast model. *Atmos.-Ocean*, **28**, 466–472.

Yang, G. Y., and J. Slingo (2001). The diurnal cycle in the tropics. *Mon. Wea. Rev.*, **129**, 784–801.

Yates, D. N., T. T. Warner, and G. H. Leavesley (2000). Prediction of a flash flood in complex terrain: Part 2. A comparison of flood discharge simulations using rainfall input from radar, a dynamic model, and an automated algorithmic system. *J. Appl. Meteor.*, **39**, 815–825.

Yeh, K.-S., J. Côté, S. Gravel, *et al.* (2002). The CMC-MRB global environmental multiscale (GEM) model. Part III. *Mon. Wea. Rev.*, **130**, 339–356.

Yeh, T.-C., R. T. Wetherald, and S. Manabe (1984). The effect of soil moisture on the short-term climate and hydrology change: A numerical experiment. *Mon. Wea. Rev.*, **112**, 474–490.

Yoshimura, K., and M. Kanamitsu (2008). Dynamical global downscaling of global reanalyses. *Mon. Wea. Rev.*, **136**, 2983–2998.

Yu, Y., and X. Zhang (2000). Coupled schemes of flux adjustments of the air and sea. In *Investigations on the Model System of the Short-Term Climate Predictions*, eds. Y. Ding, *et al.* Beijing, China: China Meteorological Press, pp. 201–207 (in Chinese).

Yu, Y., R. Yu, X. Zhang, and H. Liu (2002). A flexible global coupled climate model. *Adv. Atmos. Sci.*, **19**, 169–190.

Yu, Y., Z. Zhang, and Y. Guo (2004). Global coupled ocean-atmosphere general circulation models in LASG/IAP. *Adv. Atmos. Sci.*, **21**, 444–455.

Yuan, H., S. Mullen, X. Gao, *et al.* (2005). Verification of probabilistic quantitative precipitation forecasts over the southwest United States during winter 2002/03 by the RSM ensemble system. *Mon. Wea. Rev.*, **133**, 279–294.

Yuan, H., J. A. McGinley, P. J. Schultz, C. J. Anderson, and C. Lu (2008). Short-range precipitation forecasts from time-lagged multimodel ensembles during HMT-West-2006 Campaign. *J. Hydrometeor.*, **9**, 477–491.

Yuan, H., C. Lu, J. A. McGinley, *et al.* (2009). Evaluation of short-range quantitative precipitation forecasts from a time-lagged multimodel ensemble. *Wea. Forecasting*, **24**, 18–38.

Yukimoto, S., A. Noda, A. Kitoh, *et al.* (2001). The new Meteorological Research Institute global ocean-atmosphere coupled GCM (MRI-CGCM2): Model climate and variability. *Papers Meteor. Geophys.*, **51**, 47–88.

Yun, W. T., L. Stefanova, A. K. Mitra, *et al.* (2005). A multi-model superensemble algorithm for seasonal climate prediction using DEMETER forecasts. *Tellus*, **57A**, 280–289.

Žagar, N., M. Žagar, J. Cedilnik, G. Gregorič, and J. Rakovec (2006). Validation of mesoscale low-level winds obtained by dynamical downscaling of ERA40 over complex terrain. *Tellus*, **58A**, 445–455.

Žagar, N., A. Stoffelen, G. Marseille, C. Accadia, and P. Schlüessel (2008). Impact assessment of simulated Doppler wind lidars with a multivariate variational assimilation in the tropics. *Mon. Wea. Rev.*, **136**, 2443–2460.

Zhang, D.-L., and R. A. Anthes (1982). A high-resolution model of the planetary boundary layer: Sensitivity tests and comparison with SESAME-79 data. *J. Appl. Meteor.*, **21**, 1594–1609.

Zhang, D.-L., H.-R. Chang, N. L. Seaman, T. T. Warner, and J. M. Fritsch (1986). A two-way interactive nesting procedure with variable terrain resolution. *Mon. Wea. Rev.*, **114**, 1330–1339.

Zhang, F., M. Zhang, and J. A. Hansen (2009). Coupling ensemble Kalman filter with four-dimensional variational data assimilation. *Adv. Atmos. Sci.*, **26**, 1–8.

Zhang, H., and M. Rančić (2007). A global Eta model on quasi-uniform grids. *Quart J. Roy. Meteor. Soc.*, **133**, 517–528.

Zhang, Y. (2008). Online coupled meteorology and chemistry models: history, current status, and outlook. *Atmos. Chem. Phys. Disc.*, **8**, 1833–1912.

Zhang, Y., W. Maslowski, and A. J. Semtner (1999). Impacts of mesoscale ocean currents on sea ice in high-resolution Arctic ice and ocean simulations. *J. Geophys. Res.*, **104**(C8), 18409–18429.

Zhang, Y., M. K. Dubey, and S. C. Olsen (2009). Comparisons of WRF/Chem simulations in Mexico City with ground-based RAMA measurements during the MILAGRO-2006 period. *Atmos. Chem. Phys. Disc.*, **9**, 1329–1376.

Zheng, X. (2009). An adaptive estimation of forecast error covariance parameters for Kalman filtering data assimilation. *Adv. Atmos. Sci.*, **26**, 154–160.

Zheng, Y., Q. Xu, and D. J. Stensrud (1995). A numerical simulation of the 7 May 1985 mesoscale convective system. *Mon. Wea. Rev.*, **123**, 1781–1799.

Zhu, C., C.-K. Park, W.-S. Lee, and W.-T. Yun (2008). Statistical downscaling for multi-model ensemble prediction of summer monsoon rainfall in the Asia-Pacific region using geopotential height field. *Adv. Atmos. Sci.*, **25**, 867–884.

Zorita, E., and H. von Storch (1999). The analog method as a simple statistical downscaling technique: Comparison with more complicated methods. *J. Climate*, **12**, 2474–2489.

Zou, X., I. M. Navon, and F. X. LeDimet (1992). An optimal nudging data assimilation scheme using parameter estimation. *Quart. J. Roy. Meteor. Soc.*, **118**, 1163–1186.

# Index